Functional and molecular glycobiology

Functional and molecular glycobiology

S.A. Brooks
School of Biological and Molecular Sciences, Oxford Brookes University, Oxford, UK

M.V. Dwek
Department of Surgery, Royal Free and University College Medical School, London, UK

U. Schumacher
Institute for Anatomy, University Hospital, Hamburg-Eppendorf, Hamburg, Germany

First published 2002

A CIP catalogue record for this book is available from the British Library.

ISBN 1 85996 022 7

BIOS Scientific Publishers Ltd
9 Newtec Place, Magdalen Road, Oxford OX4 1RE, UK
Tel. +44 (0)1865 726286. Fax +44 (0)1865 246823
World Wide Web home page: http://www.bios.co.uk/

Distributed exclusively in the United States of America, its dependent territories and Canada, Mexico, Central and South America, and the Caribbean by Springer-Verlag New York Inc., 175 Fifth Avenue, New York, NY 10010-7858, by arrangement with BIOS Scientific Publishers Ltd, 9 Newtec Place, Magdalen Road, Oxford OX4 1RE, UK.

Production Editor: Phil Dines
Typeset by Charon Tec Pvt. Ltd, Chennai, India
Printed by TJ International Ltd., Padstow, UK

Contents

Acknowledgements

The authors would like to thank Dr Pauline Rudd of the Glycobiology Institute, Department of Biochemistry, University of Oxford for her careful and thorough review of the manuscript and for her many helpful suggestions. We are also grateful to Dr Peter Grebenik, School of Biological and Molecular Sciences, Oxford Brookes University for his invaluable help with Chapter 1, 'An introduction to carbohydrate chemisty'. Miriam Dwek gratefully acknowledges funding from Against Breast Cancer, Registered Charity No. 1020967.

Abbreviations

ADP	adenosine diphosphate
AMPA	α-amino-3-hydroxy-methylisoxazole-4-propionic acid
Ara	arabinose
Asn	asparagine
ASGP-1	ascites sialoglycoprotein-1
ATP	adenosine triphosphate
BHK-21	baby hamster kidney cells
cAMP	cyclic adenosine-5′-monophosphate
CDG	congenital disorder of glycosylation or carbohydrate deficient glycoprotein syndrome
CE	capillary electrophoresis
CHO	Chinese hamster ovary (cell line)
CMP	cytidine monophosphate
CNX	calnexin
CPI	carbohydrate-processing inhibitor
CRT	calreticulin
DNA	deoxyribonucleic acid
DNJ	N-nonyl-deoxynojirimycin
ECL	enhanced chemiluminescence
ECM	extracellular matrix
EGF	epidermal growth factor
ELISA	enzyme-linked immunosorbent assay
ELLA	enzyme-linked lectin assay
ER	endoplasmic reticulum
ERAD	ER-associated protein degradation
ESI	electrospray ionization
FAB	fast atom bombardment
FACE	fluorophore-assisted carbohydrate electrophoresis
Fuc	fucose
GABA	gamma aminobutyrate
GalCer	galactosyl ceramide
GalNAc	N-acetylgalactosamine
GDP	guanosine diphosphate
Glc	glucose
GLC	gas liquid chromatography
GlcCer	glucosyl ceramide
GlcNAc	N-acetylglucosamine
GPI	glycosylphosphatidylinositol
GSL	glycosphingolipid
GTP	guanosine-5′-triphosphate
HBV	hepatitis B virus
hCG	human chorionic gonadotrophin
HEMPAS	hereditary erythroblastic multinuclearity with a positive acidified serum lysis test
hGH	human growth hormone
HIV	human immunodeficiency virus
HPAEC	high pH anion-exchange chromatography
HPLC	high-performance liquid chromatography
ICAM-1	intercellular adhesion molecule-1
IFN	interferon
IGF	insulin-like growth factor
IL	interleukin
KDO	3-deoxy-D-manno-octulosonic acid
LacdiNAc	N′N′-diacetyl lactosamine GalNAc(β1→4)GlcNAc
LacNAc	Gal(β1→4)GlcNAc
LAD-II	leukocyte adhesion deficiency II syndrome
Le^{a}	Lewis a
Le^{b}	Lewis b
Le^{y}	Lewis y
Le^{x}	Lewis x
MAG	myelin-associated glycoprotein

MALDI-MS	matrix-assisted laser desorption ionization–mass spectrometry
Man	mannose
MAP kinases	microtubule-associated protein kinases
MBL	mannose-binding lectin
mRNA	messenger ribonucleic acid
MS	mass spectrometry
NB-DNJ	N-butyldeoxynojirimycin
N-CAM	neural cell adhesion molecule
NeuAc	neuraminic acid (sialic acid)
NMR	nuclear magnetic resonance
OGT	O-GlcNAc transferase or UDP-N-acetylglucosamine: polypeptide β-N-acetylglucosaminyltransferase
OST	oligosaccharyltransferase
PAD	pulsed amperometric detector
PAS	periodic acid Schiff
PET	positron emision tomography
PSA	polysialic acid
RAAM	reagent array analysis method
Rha	rhamnose
RIP	ribose-inactivating protein
RNA	ribonucleic acid
RER	rough endoplasmic reticulum
RNase	ribonuclease
SA	sialic acid
SCR	short consensus repeat
SDS-PAGE	sodium dodecyl sulfate–polyacrylamide gel electrophoresis
Se locus	secretor locus
Ser	serine
Siglecs	sialic acid-binding immunoglobulin-like lectins
sLe^x	sialyl Lewis x
SSEA-1	stage-specific antigen-1
T antigen	Thomsen–Friedenreich antigen
Thr	threonine
TFF domain peptides	Trefoil factor family domain peptides
TGF	transforming growth factor
TLC	thin-layer chromatography
TOF	time-of-flight
tPA	tissue plasminogen activator
uPA	urinary type plasminogen activator
UDP	uridine diphosphate
Xyl	xylose

Glossary of terms

Adhesin: bacterial lectin.
Aglycon: a protein that has been stripped of its glycan component.
Asialo-: containing no sialic acid.
Carbohydrate: literally means a 'hydrate of carbon' deriving from their general formula, $C_n(H_2O)_n$ (where n is three or more). A more accurate definition is that carbohydrates are polyhydroxyaldehydes or polyhydroxyketones with three or more carbon atoms, or substances which can be hydrolyzed to form these.
Carbohybrid: a synthetic hybrid molecule comprising an organic core bearing carbohydrate groups, which mimics the three-dimensional structure of a complex oligosaccharide.
CDGs: congenital disorders of glycosylation or carbohydrate-deficient glycoprotein syndromes are clinically heterogeneous autosomal recessive genetic defects in assembly, attachment or processing of N- and/or O-glycans, resulting in multi-systemic diseases characterized by defective glycosylation of glycoproteins.
Disaccharide: two monosaccharides linked together by an O-glycosidic bond.
Endoglycosidase: an enzyme that catalyzes the cleavage of the bond between two monosaccharides situated within an oligosaccharide chain.
Epithelium: cells that form the epidermis, line the respiratory tract and are specialized for secretion.
Exoglycosidase: an enzyme that trims a specific terminal monosaccharide from the outermost part of an oligosaccharide chain.
Glycan: means polysaccharide, but it is loosely used in the glycobiology literature to refer to any carbohydrate structure, for example, the carbohydrate component of a glycoprotein or glycolipid.
Glycoconjugates: molecules that contain carbohydrates linked to another molecular species, for example, glycoproteins, proteoglycans and glycolipids.
Glycoforms: variants of a glycoconjugate in which the number and/or position of monosaccharides in the glycan moieties exhibit heterogeneity.
Glycolipid: a glycoconjugate-containing carbohydrate (one or more monosaccharides or oligosaccharides) bound to a lipid.
Glycopeptide: a compound consisting of carbohydrate (one or more monosaccharides or oligosaccharides) covalently linked to an oligopeptide.
Glycoprotein: a compound containing carbohydrate covalently linked to protein; the carbohydrate may be in the form of mono-, di-, oligo-, or polysaccharide(s) or their derivatives (e.g. sulfo- or phospho-substituted). One, a few or many glycans may be present.
Glycosaminoglycans: heteropolysaccharide chains usually consisting of a hexosamine and glucuronic acid, which are the glycans of proteoglycans.
Glycosidases: a glycosidase is an enzyme that cleaves a bond between two monosaccharides. Glycosidases are responsible for trimming specific saccharides from precursors to form intermediate structures. Includes endoglycosidases which catalyze the cleavage of internal glycosidic bonds and exoglycosidases which catalyze the stepwise linkage-specific removal of sugars from non-reducing termini of carbohydrate polymers.
Glycosphingolipids: glycolipids comprising carbohydrate (one or more monosaccharides or oligosaccharides) bound to the lipid sphinganine.

Glycosylated: term used to indicate that a monosaccharide molecule, usually part of an oligosaccharide, is covalently linked to another molecule, for example, a protein.
Glycosyltransferases: specific, biosynthetic sugar-transferring enzymes.
Glycotope: a carbohydrate epitope.
Glycozymes: enzymes of glycosylation – the glycosyltransferases and glycosidases.
GPI-anchored proteins: a subset of membrane proteins incorporated into the outer leaflet of the lipid bilayer of the plasma membrane by glycosylphosphatidylinositol.
Lactosamine (or N-acetyllactosamine) unit: the Gal(β1→4)GlcNAc-disaccharide, also called the type 2 chain.
Lectins: naturally occurring proteins or glycoproteins of non-immune origin which have the ability to bind non-covalently to carbohydrates with fine specificity.
Lewis system sugars: includes Lewis a (Le^a), b (Le^b), c (Le^c), d (Le^d), x (Le^x) and y (Le^y). Le^x is also called CD15 and stage-specific embryonic antigen-1 (SSEA-1). A family of fucosylated blood group sugars.
Metastasis: the process by which cancer cells spread from their primary site to other parts of the body.
Monosaccharide: carbohydrates that cannot be hydrolyzed are called monosaccharides. They form the basic building blocks of more complex carbohydrates. Sometimes referred to as simple sugars.
***MUC1* gene product/mucin:** a membrane-associated mucin, coded for by the *MUC1* gene, and of interest in that it is over-expressed and abnormally glycosylated in a number of cancers.
Mucins: a large group of glycoproteins characterized by the attachment of multiple O-glycosidically linked oligosaccharide chains to a polypeptide backbone known as the apomucin. Generally, the proportion of carbohydrate far exceeds that of the protein. When secreted, these compounds form a highly viscous material.
N-acetyllactosamine: *see lactosamine.*
Neolactosamine unit: the glycan Gal(β1→3)GlcNAc, the basic building block of the type 1 chain.
N-linked glycoprotein: a glycoprotein in which an oligosaccharide is linked to the protein via a GlcNAc molecule. The GlcNAc is attached through an N-glycosidic-type bond to the nitrogen of the amide group of the side chain of an asparagine (Asn) amino acid on the polypeptide chain. The Asn residue must be in the Asn-Xaa-Ser triplet where Xaa is any amino acid except Pro.
Oligosaccharide: a linear or branching chain of between two and about ten to twenty monosaccharides.
O-linked glycoproteins: glycoproteins in which the first monosaccharide of the oligosaccharide chain, usually GalNAc, is attached through an α-O-glycosidic linkage to the hydroxyl oxygen in the side chain of an amino acid residue, usually serine or threonine, in the polypeptide chain of a protein.
Oncofoetal (carbohydrate) antigens: carbohydrate structures not normally present in adult tissues, being characteristic of foetal cells, but which are sometimes aberrantly expressed by cancer cells.
Polylactosamine: a glycan structure composed of repeating lactosamine units Gal(β1→4) GlcNAc(β1→3)Gal, also known as poly-N-acetyllactosamine.
Poly-N-acetyllactosamine: *see polylactosamine.*
Polysaccharide: a polymer of more than ≈10–20 monosaccharides, but usually composed of many hundreds or thousands of monosaccharide units, usually with a simple, repetitive structure.
Proteoglycans: a subclass of glycoproteins in which the carbohydrate units are heteroglycan polysaccharides usually consisting of a hexosamine and glucuronic acid. The heteropolysaccharide chains of these molecules are also called glycosaminoglycans.

Saccharide: derived from the Greek word '*sakcharon*', meaning 'sugar'; a general term for any carbohydrate structure.
Selectins: a group of mammalian cell-surface carbohydrate receptors (lectins) that mediate interaction with saccharide ligands on other cell surfaces, for example, on circulating white blood cells, selectin interactions mediate their recruitment to epithelial cells as a response to inflammation.
Sialic acids: a family of more than 40 monosaccharides, all derivatives of neuraminic acid, which share a basic 9-carbon carboxylated skeleton.
Sialylated: bearing sialic acid residues.
Thomsen–Friedenreich antigen: also called T or TF antigen, this is Gal(β1$\rightarrow$3)GalNAc-O-α-Ser/Thr.
Tn antigen: GalNAc-O-Ser/Thr, this structure results from the addition of the first GalNAc residue to the polypeptide backbone in O-linked glycoprotein synthesis.
Trimannosyl core: the common $Man_3GlcNAc_2$ pentasaccharide core of all N-linked glycans.
Trisaccharide: three monosaccharides linked together.
Type 1 chain: in O-glycans, the structure Gal(β1$\rightarrow$3)GlcNAc, which is also called the neo-lactosamine unit.
Type 2 chain: in O-linked glycans, the structure Gal(β1$\rightarrow$4)GlcNAc, which is also called the lactosamine unit.

Foreword

The human genome contains approximately 60 000 genes. This relatively small number suggests that the mechanisms that generate much of the complexity and diversity that is characteristic of different cell types may well be found in post-transcriptional events. Post-translational modifications play an indispensable role in many biological systems, and amongst eukaryotes, glycosylation is the most complex. Current estimates indicate that, in humans, genes that regulate N- and O-glycosylation represent 1–2% of the total genome. Moreover, the majority of proteins destined for the cell surface or secretion are glycosylated, and over two thirds of all proteins in the SWISS-PROT database contain glycosylation sites. It is therefore important that biologists and biochemists are aware of the basic principles of glycobiology, whether it is the main focus of their research or not.

This clear introduction assumes very little prior knowledge of the field yet nevertheless discusses many of the exciting discoveries which are being made at the cutting edge of the field in a way that can be easily understood. It is a book which can be read quickly and which has good 'further reading' lists so that those whose imaginations are captured can find their way to more extensive literature.

It is a major undertaking to write such a book, particularly since there is a fine balance to be maintained between being, on the one hand, too superficial or, on the other, too obscure. The authors have managed to steer between these two extremes and are to be congratulated on their achievement.

Glycobiology is an exciting yet demanding field in which to work. It is relevant to almost every field of biology and medicine and deals with the biochemistry of some of the most dramatic and beautiful molecules that have been discovered. I am sure that anyone who reads this book with care will find their thinking enriched and their preconceptions about the roles of sugars challenged.

Dr Pauline Rudd
Oxford 2002

Preface

The enigma of the human genetic code has now been unravelled. With the aid of modern computer science, data mining has started to make some sense out of the flood of information which has poured out of the sequencing machines. As we enter the post-genomic era, glycome analysis is set to become one of the most challenging and exciting research areas.

Glycobiology has a reputation for being technically difficult and intellectually inpenetrable. The underlying sugar chemistry is difficult for the bioscientist and molecular biologist to understand, and is poorly taught – if at all – in many school and university courses. Furthermore, oligosaccharide synthesis is not under direct genetic control and a bewilderingly large number of structures are theoretically possible. We are only just beginning to appreciate their diversity, the fascinating mechanisms by which they are produced and by which their synthesis is regulated, and their myriad and complex functions in every organism from the simplest to the most complex.

This book aims to provide ready access to the area of molecular and functional glycobiology for the advanced beginner. It is aimed at medics, bioscientists and chemists at undergraduate, postgraduate and research level. It assumes little or no previous knowledge of this field and guides the reader from the basic chemistry of carbohydrates, their synthesis, expression and function in organisms as diverse as viruses, bacteria, fungi, parasites, plants, animals and humans, and discusses their relevance in health and disease, in biotechnology and in medicine. The chapters are designed to stand alone, but are cross referenced to other related parts of the volume. Suggestions for further reading ease access into the sometimes overwhelming research literature.

Our hope is that this book will open the door to a complex, fascinating, field that promises to reveal the secrets of some of the most structurally diverse, biologically significant and enigmatic molecules produced by living organisms.

Susan A Brooks
Miriam V Dwek
Udo Schumacher

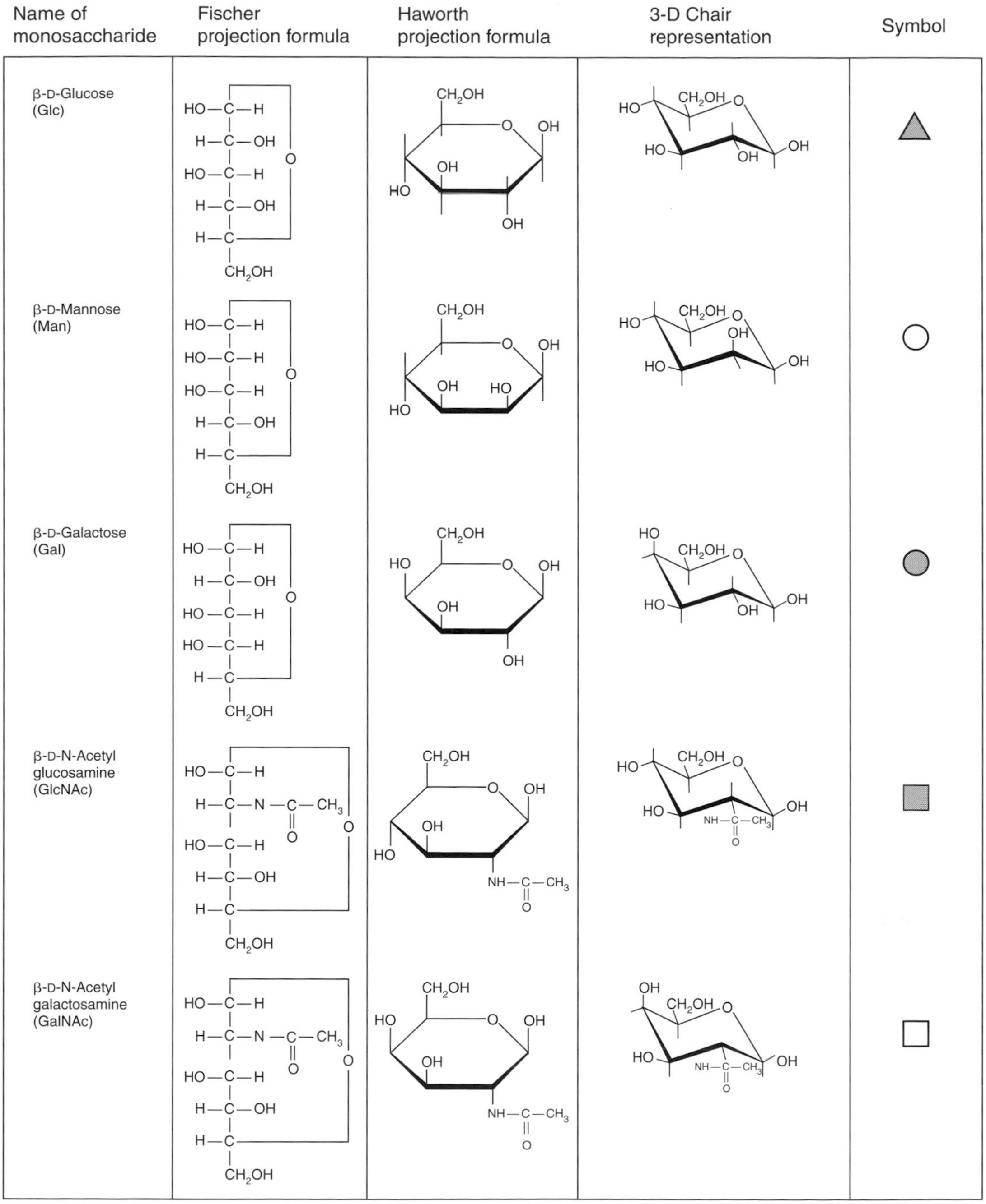

Name of monosaccharide	Fischer projection formula	Haworth projection formula	3-D Chair representation	Symbol
β-D-Glucose (Glc)	HO—C—H H—C—OH HO—C—H H—C—OH H—C CH_2OH			
β-D-Mannose (Man)	HO—C—H HO—C—H HO—C—H H—C—OH H—C CH_2OH			
β-D-Galactose (Gal)	HO—C—H H—C—OH HO—C—H HO—C—H H—C CH_2OH			
β-D-N-Acetyl glucosamine (GlcNAc)	HO—C—H H—C—N—C—CH_3 (C=O) HO—C—H H—C—OH H—C CH_2OH			
β-D-N-Acetyl galactosamine (GalNAc)	HO—C—H H—C—N—C—CH_3 (C=O) HO—C—H H—C—OH H—C CH_2OH			

Plate 1

This illustrates the Fischer projection formula, Haworth projection formula and 3-dimensional 'chair' representation of the most common monosaccharides, with the symbols used to represent them in this volume. This plate is meant for easy reference in comparing the different ways in which monosaccharides are presented in this book, and in the glycobiology literature in general.

Name of monosaccharide	Fischer projection formula	Haworth projection formula	3-D-Chair representation	Symbol
β-D-Xylose (Xyl)	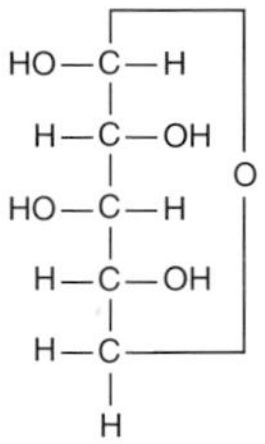 HO—C—H H—C—OH HO—C—H H—C—OH H—C H	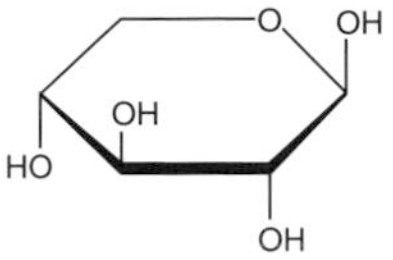	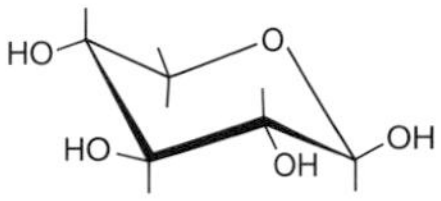	
Neuraminic acid (β-Neu5Ac)	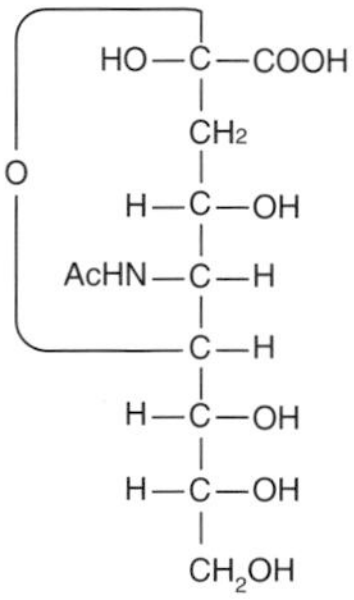 HO—C—COOH CH_2 H—C—OH AcHN—C—H C—H H—C—OH H—C—OH CH_2OH	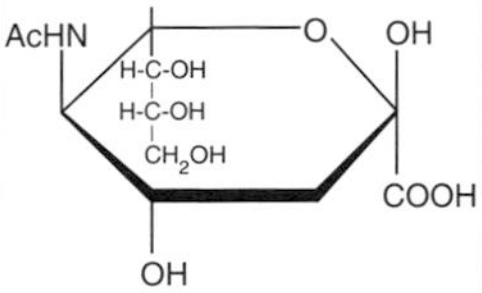	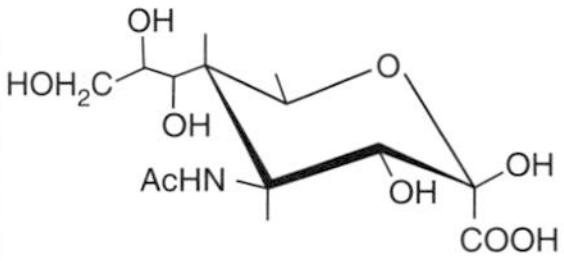	
β-D-Glucuronic acid (GlcA)	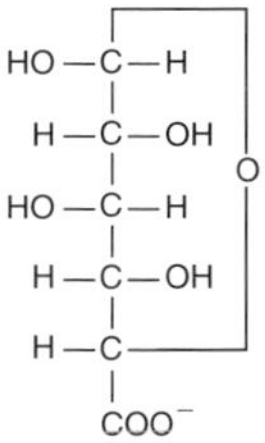 HO—C—H H—C—OH HO—C—H H—C—OH H—C COO^-	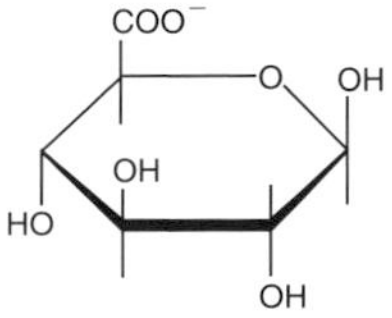	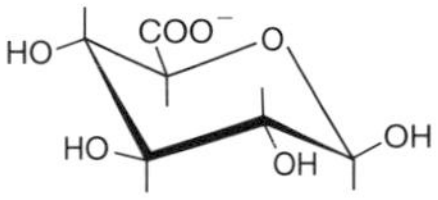	
L-β-Iduronic acid (IdoA)	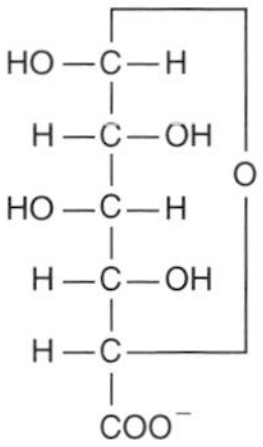 HO—C—H H—C—OH HO—C—H H—C—OH H—C COO^-	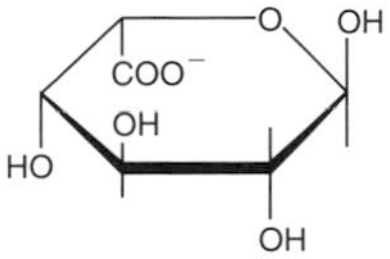	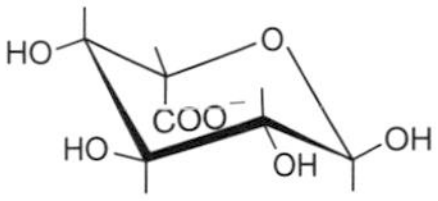	
β-L-Fucose (Fuc)	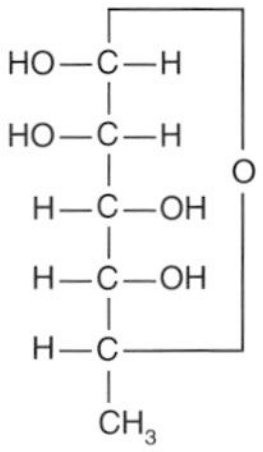 HO—C—H HO—C—H H—C—OH H—C—OH H—C CH_3	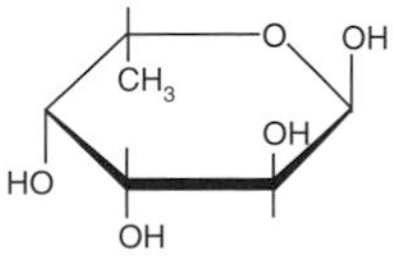	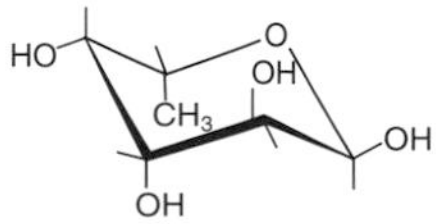	

Plate 1 (Continued)

An introduction to carbohydrate chemistry

1

1.1 Introduction

Carbohydrates are the most abundant organic substances produced by living organisms. They form fuels, metabolic intermediates and energy stores. For many organisms, such as insects and plants, carbohydrate chains, such as chitin and cellulose, form the principle structural components. In all livings organisms, the two monosaccharides ribose and deoxyribose form part of the structure of the nucleic acids DNA and RNA in which sugar rings (deoxyribose and ribose, respectively) form the backbone to which the bases encoding the genetic information are covalently linked. Carbohydrates are also linked to proteins and lipids where they play important structural roles and are involved in many cell communication and signalling events.

In spite of their biological importance, the characterization of carbohydrate structures, and the elucidation of their function, has lagged behind that of other major classes of biological molecules such as proteins and nucleic acids. There are several reasons for this. Carbohydrates are extremely heterogeneous in their size (ranging from a single monosaccharide to polysaccharides of up to 500 000 monosaccharide units), in their composition, and in the different linkages between the monosaccharide subunits, making their chemical and structural analysis challenging. Furthermore, their synthesis is not under direct genetic control, and therefore they are not amenable to the types of molecular biological analysis that have been so powerful in the study of proteins and nucleic acids.

1.2 What are carbohydrates?

The name carbohydrate literally means a 'hydrate of carbon' and derives from their general formula, $(CH_2O)_n$ (where n is 3 or more). Although all carbohydrates could be described by this formula when the term was coined in the 19th century, today many substances which are classed as carbohydrates do not fit this general formula. Some are modified with nitrate, phosphate or sulfate groups and are still called carbohydrates. Other substances which do conform to this formula, such as lactic acid ($C_3H_6O_3$), are not recognized as carbohydrates. ➲ A better defini-tion is that carbohydrates are polyhydroxy (i.e. having many OH groups) aldehydes or polyhydroxyketones with three or more carbon atoms, or substances which can be hydrolysed to form these.

> ➲ Carbohydrates are polyhydroxyaldehydes or polyhydroxyketones.

The basic building blocks of carbohydrates are monosaccharides. 'Saccharide' is derived from the Greek word *'sakcharon'*, meaning 'sugar'. Fundamental to carbohydrate chemistry is the concept that monosaccharides can be linked together to form disaccharides (two monosaccharides linked together), trisaccharides (three monosaccharides linked together) and so on, to form linear or branching chains. A chain of 2–10 monosaccharide residues is termed an oligosaccharide, and a polymer of more than ≈10–20 residues is termed a polysaccharide.

Functional and Molecular Glycobiology, Susan A. Brooks, Miriam V. Dwek and Udo Schumacher

➲ The basic building blocks of carbohydrates are monosaccharides. These can be linked together to form linear or branching chains. Two monosaccharides linked together form a disaccharide. Two to ten are an oligosaccharide. More than ten is a polysaccharide.

The borderline between what is considered an oligosaccharide and what is considered a polysaccharide is not strictly defined, but oligosaccharide is usually used to describe a defined carbohydrate structure, and polysaccharide a polymeric molecule of unspecified length and repetitive composition. ➲ Some polysaccharides are composed of a very large number of monosaccharide units, giving molecules with molecular masses of almost ten million, although their structure is usually simple and repetitive. Polysaccharides are specifically the subject of *Chapter 3*. The oligosaccharides of glycoproteins and glycolipids are usually much smaller, but are structurally heterogeneous in composition forming the basis of their diverse functional roles. These molecules are specifically the subject of *Chapters 4, 5* and *9*.

The range of monosaccharides available, and the different ways in which they can be linked together, α- or β-linkages and linkages between different carbon atoms, all explained later, results in a potentially staggering number of different oligo- and polysaccharide structures. In practice, biosynthetic pathways restrict the number of possible structures so not all occur in nature. It is not known precisely what proportion of possible structures *do* occur, as newly discovered oligo- and polysaccharide structures, and newly discovered glycosyltransferase enzymes responsible for their synthesis, continue to be described. What is certain is that carbohydrates are far more diverse in structure than proteins. In order to understand how this complexity arises, and in order to understand much of the glycobiology literature, it is necessary to appreciate the basic chemistry of carbohydrates; which is the subject of this chapter.

1.3 Monosaccharides are aldoses and ketoses

Carbohydrates that cannot be hydrolysed are called monosaccharides or simple sugars. The backbone of monosaccharides is an unbranched chain of carbon atoms linked together by single bonds. Monosaccharides can occur in two forms: an open chain form and a ring form (*see 1.10*).

In the open chain form, one carbon has a double bond to an oxygen atom forming a carbonyl (C=O) group, and each of the other carbon atoms carries a hydroxyl (OH) group. The carbon atoms carrying the hydroxyl groups are chiral centres (*see 1.4*), which give rise to the many stereoisomers of monosaccharides found in nature. ➲ If the carbonyl group is at the end of the chain (i.e. in an aldehyde group), the monosaccharide is an aldose; if the carbonyl group is at any other position (i.e. in a ketone group), the monosaccharide is a ketose. The two main families of monosaccharides are therefore aldoses and ketoses. From this, carbohydrates usually have trivial names that end in '-ose', for example, glucose, sucrose and cellulose.

➲ A monosaccharide is an aldose if the carbonyl (C=O) group is at the end of the chain and a ketose if it is at any other position.

The simplest monosaccharides are the three carbon trioses, glyceraldehyde (an aldotriose) and dihydroxyacetone (a ketotriose), illustrated in *Figure 1.1*. Monosaccharides with four carbon atoms are tetroses, those with five carbons are pentoses, those with six carbons are hexoses, and those with seven carbon atoms are heptoses. There are aldoses and ketoses with all of these different chain lengths, and some are listed in *Table 1.1*. The aldohexose glucose is the commonest naturally occurring aldose and the ketohexose fructose is the commonest naturally occurring ketose.

1.4 A carbon atom may be asymmetric – the concept of chirality

Central to the understanding of carbohydrate chemistry is the idea that a carbon atom can be asymmetric. This is illustrated in *Figure 1.2*. The two molecules (a) and (b) in *Figure 1.2* are

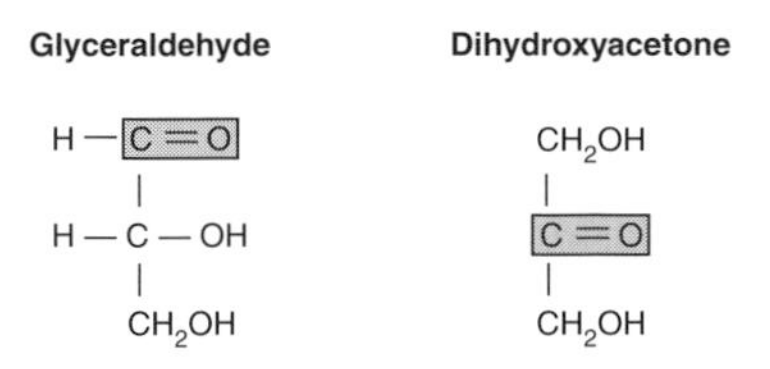

Figure 1.1

The structures of the simplest monosaccharides, glyceraldehyde (an aldotriose) and dihydroxyacetone (a ketotriose). Note that glyceraldehyde contains a carbonyl group (C=O, shaded) at the end of the carbon chain, making it an aldose, and dihydroxyacetone contains a carbonyl group (C=O, shaded) within the carbon chain, making it a ketose.

Table 1.1 Naming monosaccharides on the basis of the number of carbon atoms

Number of carbon atoms	Name	Examples of aldoses	Examples of ketoses
3	Triose	Glyceraldehyde	Dihydroxyacetone
4	Tetrose	Erythrose, threose	Erythrulose
5	Pentose	**Ribose, deoxyribose, arabinose, xylose**	Ribulose, xylulose
6	Hexose	**Glucose, mannose, galactose**	**Fructose**, sorbose

Monosaccharides that occur commonly are given in bold.

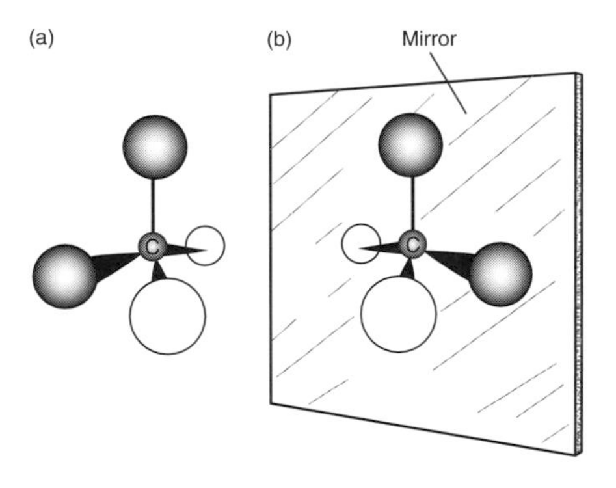

Figure 1.2

The carbon atom can be an asymmetric one. The central C in the middle of the pyramid carries four different groups, so that the compounds (a) and (b) are non-superimposable mirror images of each other (the shaded area represents the mirror).

said to have different configurations. The central C in the middle of the pyramid carries four *different* groups, so that the compounds (a) and (b) are non-superimposable mirror images of each other (the shaded area represents the mirror).➲ This phenomenon of non-superimposable mirror images is called enantiomerism (or optical isomerism) and is of great structural importance for carbohydrate biochemistry and function. The asymmetric carbon atom is said to be a tetrahedral stereocentre.

➲ Enantiomers are molecules that are non-superimposable mirror images of each other.

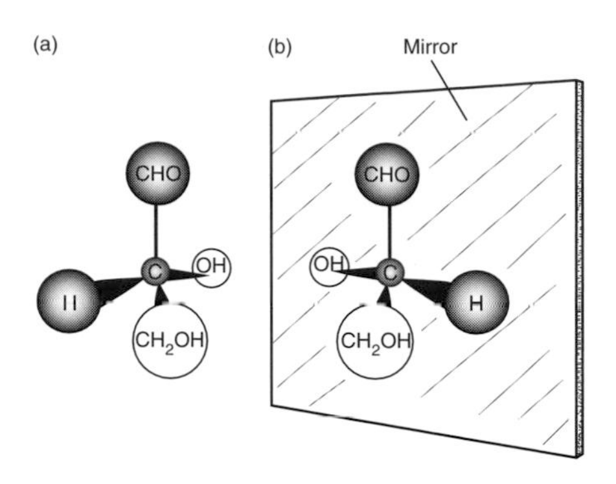

Figure 1.3

The simplest aldose, glyceraldehyde (also called glycerose), contains one chiral carbon atom carrying four different substituents, —CHO, —H, —OH —CH_2OH, and therefore has two different enantiomers (a) and (b).

```
O   H
 \\ /
  C
  |
H—C—OH
  |
HO—C—H
  |
H—C—OH
  |
H—C—OH
  |
H—C—OH
  |
  H
```

Figure 1.4

The six-carbon monosaccharide, glucose, has four different chiral carbon atoms (arrowed), and therefore 2^4, or 16, enantiomers.

All monosaccharides have one or more asymmetric or chiral carbon atoms and thus occur in enantiomers, except the simplest three-carbon monosaccharide, dihydroxyacetone. The simplest aldose, glyceraldehyde (also called glycerose), for example, contains one chiral carbon atom (the central one), carrying four different substituents, —CHO, —H, —OH and —CH_2OH, and therefore has two different enantiomers, as illustrated in *Figure 1.3*. The slightly more complex monosaccharide, glucose, which is a hexose (containing six carbons) has four different chiral carbon atoms, as illustrated in *Figure 1.4*.

For the majority of compounds with n chiral carbons, there are 2^n stereoisometric forms. The number of stereoisomers of aldohexoses like glucose is 2^4 or 16, which occur as eight pairs of enantiomers.

1.5 Monosaccharide 'family trees'

By adding one or more carbon atoms to glyceraldehyde, a 'family tree' of the aldoses can be established (*Figure 1.5*). Similarly, by adding one or more carbon atoms to dihydroxyacetone, a 'family tree' of ketoses can be established (*Figure 1.6*). The relationship of the different members of the 'family trees' are clear: all aldoses carry a terminal carbonyl group and all ketoses carry an interchain carbonyl group. Each member of a 'family' differs from the rest by virtue of its number of carbon atoms, and the configuration at one or more of its carbon atoms.

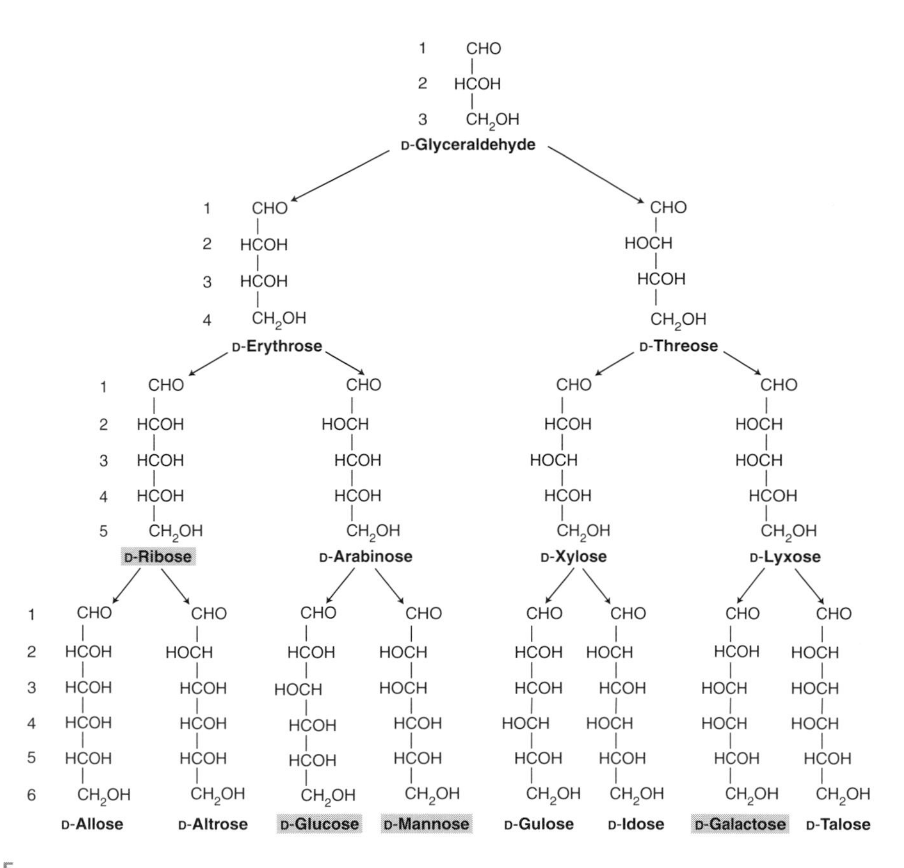

Figure 1.5

The aldose 'family tree'. Members of the 'family' differ from the rest by virtue of their number of carbon atoms, and the configuration at one or more of them. Biologically important monosaccharides are highlighted.

1.6 D-Forms and L-forms of carbohydrates

The way that the structures are drawn in the 'family trees' of *Figures 1.5* and *1.6* are examples of what is termed a 'Fischer projection formula', after the German chemist, Emil Fischer (1852–1919), who won the second Nobel Prize in chemistry in 1902, and who first suggested this type of representation. There are other ways of presenting carbohydrate structures, which will be described later. It is apparent in *Figures 1.5* and *1.6* that the monosaccharides bear the prefix D-, as in, for example, D-glucose and D-galactose. The prefixes D- or L- were designated by Fischer at the beginning of the 20th century. The reason for this designation can be appreciated by consideration of the enantiomers of the triose glyceraldehyde illustrated in the Fischer projection formula in *Figure 1.7*.

Fischer projection formulas simplify chemical structures so that a complex three-dimensional molecular structure can be represented, and appreciated, in two dimensions on the printed page. The formulas are drawn vertically down the page.

A simple way of remembering this is that if the hydroxyl group (OH) is placed on the right-hand side of the asymmetric central carbon atom, the monosaccharide is termed D-glyceraldehyde (from the Latin word *dexter*, which means right). If it is placed on the left side, it becomes L-glyceraldehyde (from the Latin word *laevus*, which means left). The L-form is the mirror image of the D-form.

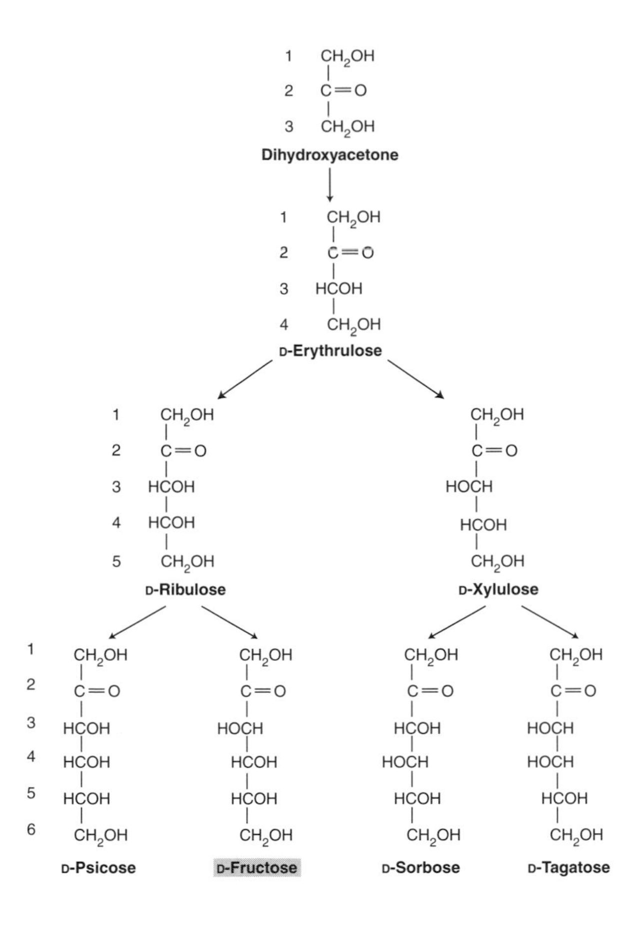

Figure 1.6

The ketose 'family tree'. Members of the 'family' differ from the rest by virtue of their number of carbon atoms, and the configuration at one or more of them. Biologically important monosaccharides are highlighted.

> ➲ L- stands for *laevus* (Latin for left), and D- for *dexter* (Latin for right). In L-forms the OH group on the asymmetric carbon atom lies on the left, and in D-forms on the right.

For monosaccharides with more than three carbon atoms, the convention is that if the hydroxyl group of the carbon *most distant from the carbonyl group* (i.e. the highest numbered chiral carbon atom; *see 1.7*) is on the left in the Fischer projection formula, it is the L-form, and if it is on the right, it is the D-form.➲ It does not matter how the other hydroxyl groups at the other carbons are orientated. By consideration of the 'family trees' in *Figures 1.5* and *1.6* it is now apparent why they carry the prefix D-; the hydroxyl group of the carbon most distant from the carbonyl group is on the right (*dexter*) in all cases. Most hexoses of living organisms occur in the D-isomer form. L-Fucose is a notable exception.

1.7 Numbering the carbon atoms in a monosaccharide molecule

The carbon atoms are numbered within a monosaccharide molecule according to set convention. Carbon atoms are numbered consecutively such that a (potential or actual)

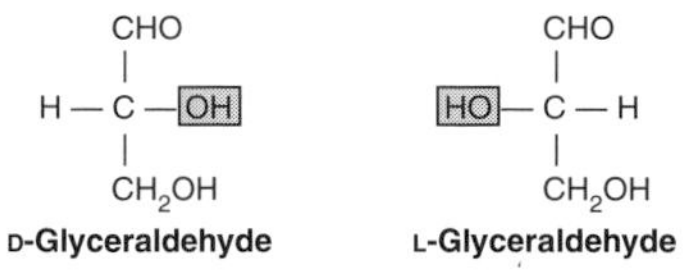

CHO	CHO
H — C — OH	HO — C — H
CH_2OH	CH_2OH
D-Glyceraldehyde	**L-Glyceraldehyde**

Figure 1.7

The L- and D-forms of glyceraldehyde. If the hydroxyl group (OH, shaded) is placed on the right-hand side of the asymmetric central carbon atom, the monosaccharide is termed D-glyceraldehyde (D, *dexter*, is Latin for right). If it is placed on the left side, it is L-glyceraldehyde (L, *laevus*, left). The L-form is the mirror image of the D-form.

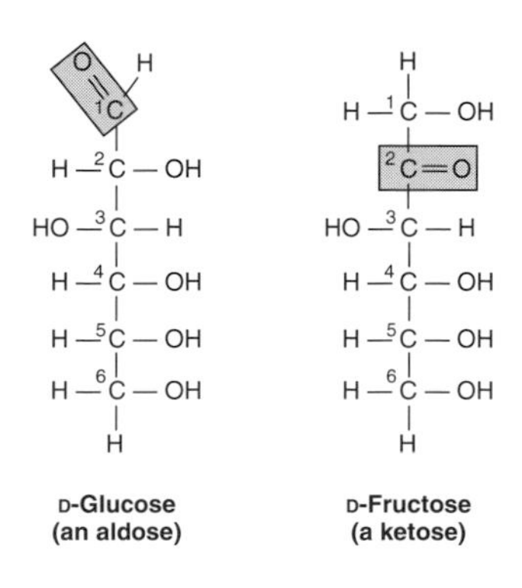

Figure 1.8

Numbering the carbon atoms in monosaccharides. The carbonyl group (C=O, shaded) is placed towards the top of the page in the Fischer projection formula, and the carbon atoms are numbered consecutively from the top downwards. In both glucose and fructose, the highest numbered carbon is C5; the OH group is to the right (*dexter*) in each case, so it is the D-form of glucose and fructose that are shown.

aldehyde group is C1 or the most senior functional group (i.e. carboxylic acid and derivatives are considered 'senior' to potential or actual ketonic carbonyl groups) receive the lowest possible number. So, in the Fischer projection formula, the molecule is drawn with the carbonyl group nearest the top of the page, and the carbons are numbered from the top (C1) downwards. This may be appreciated with reference to the aldose D-glucose and the ketose D-fructose illustrated in *Figure 1.8*. The aldehyde carbon in aldoses is always C1 and the carbonyl group of naturally occurring ketoses is always C2.

The decision as to whether a carbohydrate belongs to the D- or L-configuration is dependent on the asymmetric carbon atom with the highest number (*see 1.6*). In the case of glucose and fructose it is carbon atom number 5, C5, and because the hydroxyl group (OH) is located on the right of the molecules, it is D-glucose and D-fructose which are illustrated in *Figure 1.8*.

In the family trees of monosaccharides shown in *Figures 1.5* and *1.6*, the highest asymmetric carbon atom is the second carbon atom from the bottom in each case. Because all hydroxyl (OH) groups in the figures point to the right (*dexter*) side, all sugars of *Figures 1.5* and *1.6* belong to the D-configuration. If the OH group were to be swapped to the left (*laevus*), the L-forms of the monosaccharides would be presented instead.

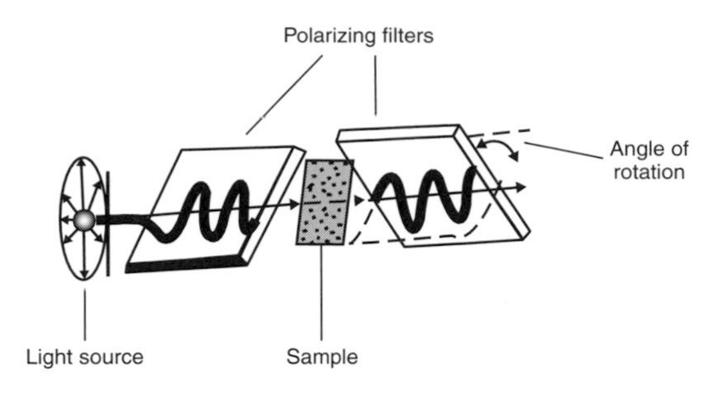

Figure 1.9

Analysis of (+) and (−) forms of monosaccharides by polarimetry. A monosaccharide solution is placed between two polarizing filters. As the molecules are chiral, they rotate the plane of polarization of light. A second filter measures the angle of rotation. It may be necessary to turn the filter clockwise (+) or anticlockwise (−) to different degrees, depending on the wavelength of the light, the concentration and identity of the molecules, and also the path length.

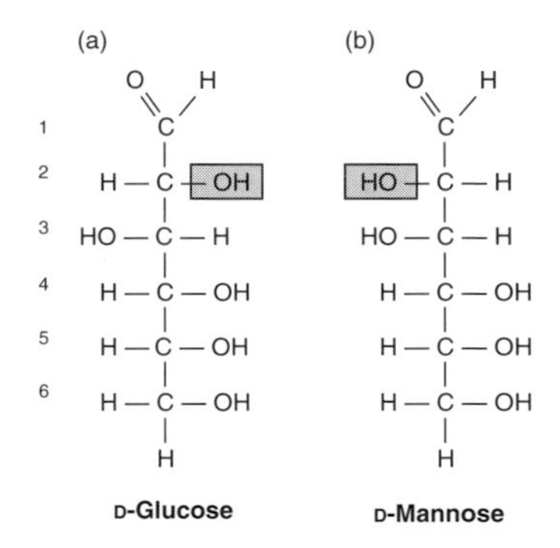

Figure 1.10

Mannose is a C2 epimer of glucose. (a) D-Glucose the OH group at C2 (shaded) lies to the right. (b) D-Mannose the OH group at C2 (shaded) lies to the left.

1.8 (+) and (−) forms of carbohydrates

As carbohydrates are chiral they can be analysed by the technique of polarimetry. To do this, the sample in solution is placed between two polarizing filters, as illustrated in *Figure 1.9*. As the molecules are chiral, they can rotate the plane of polarized light. A second filter can be used to measure the angle of rotation. Depending on the type of monosaccharide in solution, it will either be necessary to turn the filter clockwise (indicated as +) or anticlockwise (indicated as −). Hence a monosaccharide can be either classified with the prefix (+) or (−). This angle of rotation, which depends on the wavelength, concentration and identity of the monosaccharide molecules, and also the path length, must be determined experimentally for each individual molecule.

1.9 The concept of epimerization

Each asymmetric or chiral carbon atom in a monosaccharide molecule has two possible configurations. In *Figure 1.10(a)*, which illustrates D-glucose, the OH groups at carbons C2, C3, C4 and C5 could potentially lie to the left or to the right of the carbon chain.

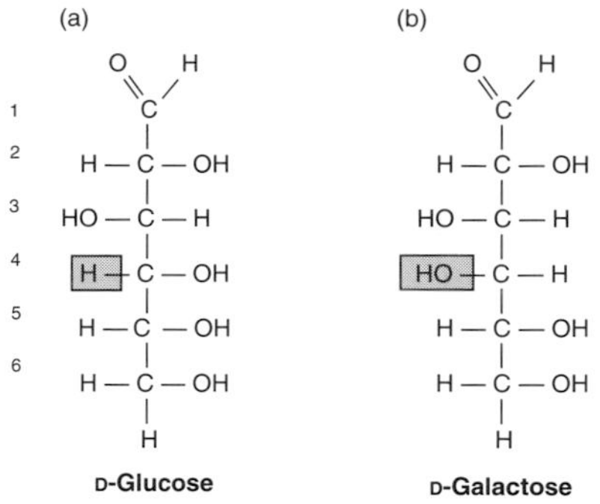

(a)

1 C(=O)H
2 H—C—OH
3 HO—C—H
4 H—C—OH
5 H—C—OH
6 H—C—OH
H

D-Glucose

(b)

1 C(=O)H
2 H—C—OH
3 HO—C—H
4 HO—C—H
5 H—C—OH
6 H—C—OH
H

D-Galactose

Figure 1.11

Galactose is a C4 epimer of glucose. (a) D-Glucose the OH group at C4 (shaded) lies to the right. (b) D-Galactose the OH group at C4 (shaded) lies to the left.

➲ Changing the configuration at any one of these carbons results in the formation of a different molecule, termed an epimer. Changing the positioning of the OH group at the C2 of D-glucose, for example, results in the monosaccharide D-mannose, as illustrated in *Figure 1.10(b)*. D-Glucose and D-mannose are said to be epimers at the C2 carbon. As illustrated in *Figure 1.11*, a similar change in conformation at the C4 of (a) D-glucose results in the formation of (b) D-galactose. D-glucose and D-galactose are thus epimers at C4.

➲ Monosaccharides that differ only in the orientation of the OH group at only one chiral centre are called *epimers*.

1.10 Monosaccharides form ring structures

Although all the above formulas account for many of the structural aspects of monosaccharides correctly, in reality almost all straight chain monosaccharides (with a few exceptions, e.g. glyceraldehyde, tetrulose and dihydroxyacetone) actually exist predominantly in cyclic (ring) forms.

These stable five- or six-sided ring structures occur owing to the formation of intramolecular bonds (*see 1.11*). In biological systems these ring forms predominate because they have lower energies than the straight chain forms. For example, in aqueous solution of glucose at room temperature, only ≈0.0026% of the glucose molecules are found as an acyclic or straight chain form.

These ring structures are most usually drawn as a Haworth projection formula, such as those illustrated in *Figure 1.12*, which shows both the Fischer projection formulae and Haworth projection formulae for the various types of ring structure formed by D-glucose. The Haworth projection formula represents the three-dimensional monosaccharide molecule with the lowest edge of the ring nearest the observer and the upper edge of the ring further away. To emphasize this, the lower edge is often drawn thickened. Substituents are then drawn represented above or below the ring (NB the representation is not meant to imply that some groups lie *within* the ring). Groups that appear to the right in the Fischer formula are below the plane of the ring in the Haworth projection; and groups that lie to the left in the Fischer formula appear above the plane of the ring. The C6 of glucose, which appeared as part of the straight chain in the Fischer formula, now appears as a substituent group above the plane of the ring in the Haworth formula.

The carbon atoms of the ring are often not written in as such, but are taken to lie at the junction of the bonds, as in *Figure 1.12*. It is also common to omit writing in the single

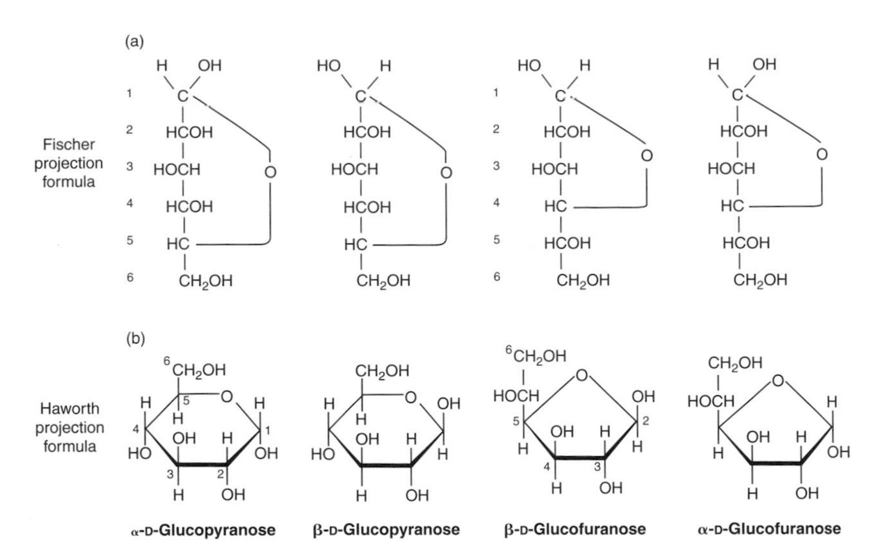

Figure 1.12

Fischer and Haworth formulas for representing the various ring forms of D-glucose. The Haworth projection formula represents the three-dimensional sugar molecule with the lowest edge of the ring nearest the observer, often drawn thickened as here, and the upper edge of the ring further away.

Furan

Figure 1.13

Five-sided ring forms of monosaccharides are related to the heterocyclic compound furan, and the corresponding sugars are therefore called furanoses.

H atoms, which are assumed to appear wherever a bond ends without a specified group. In some ways, this is a little ambiguous as conventionally in organic chemistry a bond without a specified group like this indicates a methyl group. Furthermore, the glycobiology literature is inconsistent and sometimes the bonds to hydrogens are simply omitted altogether.

The Fischer representation of this type of structure, given in *Figure 1.12*, is misleading in that it gives the impression that there is a very long bond between the oxygen of C5 and the C1 atom. In reality, this is not the case, because the C—C chain is bent. For this reason, the Haworth projection of ring forms of monosaccharide molecules is usually considered a more appropriate way of representing them.

The five-membered ring forms can be seen to be related to the heterocyclic (indicating that the ring structure contains atoms other than carbon) compound furan, shown in *Figure 1.13* and the corresponding sugars are therefore called furanoses. The six-membered ring structure can be seen to be related to the heterocyclic compound pyran, shown in *Figure 1.14* and the derived sugars are therefore called pyranoses.

A single monosaccharide may be able to form both pyranose (six-sided) and furanose (five-sided) rings. Glucose, for example, can form glucopyranose and glucofuranose, as illustrated in *Figure 1.12*. The six-sided pyranose ring is more stable than the five-sided furanose ring, so that glucose in solution is predominantly in the glucopyranose, six-sided ring form.

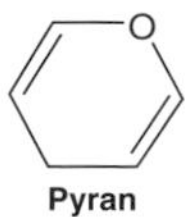

Figure 1.14

Six-sided ring forms of monosaccharides are related to the heterocyclic compound pyran, and the derived sugars are therefore called pyranoses.

The systematic name for the two ring forms of D-glucose are therefore D-glucopyranose and D-glucofuranose.

1.11 How the rings are formed – the formation of hemiacetals and hemiketals

These ring structures form because monosaccharides carry both of the functional groups required to make a hemiacetal/ketal within the same molecule, that is, a hydroxyl (OH) group and a carbonyl (C=O) group. Hemiacetals form from aldehydes by the addition of a hydroxyl group to a carbonyl group as follows:

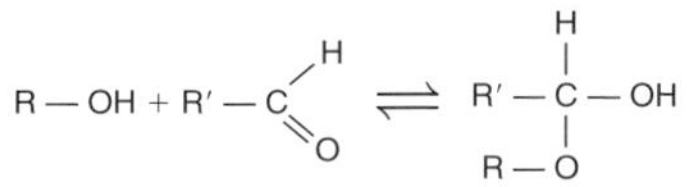

$$\underset{\text{Alcohol}}{R-OH} + \underset{\text{Aldehyde}}{R'-C(H){=}O} \rightleftharpoons \underset{\text{Hemiacetal}}{R'-C(H)(OR)-OH}$$

Similarly, a ketone can react with an alcohol to form a hemiketal as illustrated below:

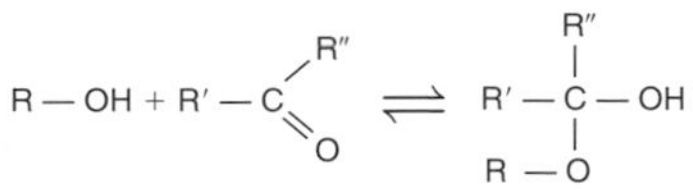

$$\underset{\text{Alcohol}}{R-OH} + \underset{\text{Ketone}}{R'-C(R''){=}O} \rightleftharpoons \underset{\text{Hemiketal}}{R'-C(R'')(OR)-OH}$$

Because monosaccharides contain both OH and C=O groups, these types of reactions can occur between different parts of the *same* molecule. Under acidic conditions both ketoses and aldoses react with alcohols to form hemiketals and hemiacetals, respectively. For example, as illustrated in *Figure 1.15*, in D-glucose, an aldose, the C1 aldehyde group reacts with the C5 hydroxyl group to form an intramolecular hemiacetal. In D-fructose, a ketose, the C2 keto group reacts with the C5 hydroxyl group to form an intramolecular ketal, as illustrated in *Figure 1.16*.

Although the Haworth projection illustrates the ring structures of monosaccharides more clearly than does the Fischer projection, it too can be slightly misleading as it gives the impression that the ring forms are flat with substituents lying either above or below the ring, which is not the case. The angle between bonds in a tetrahedral carbon is 109°, whereas the internal angles of a planar hexagon are 120°. Cyclic compounds composed of rings of six saturated carbon atoms would therefore be under great physical strain if they were actually flat hexagons, as implied by the Haworth projection formula, because the bonds between the carbon atoms would be distorted. In six-membered carbon rings, this strain is relieved by the ring taking up a non-planar shape or conformation in which the bond angles are close to the tetrahedral 109°. Replacement of a carbon by an oxygen, as in pyranose sugar rings, has relatively little effect on the conformation, so pyranoses too take up non-planar conformations. The most stable, and therefore most usual of these is referred to as the 'chair' configuration.

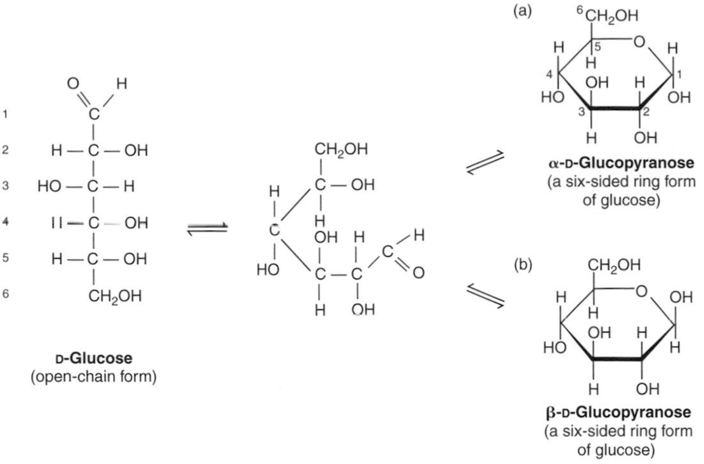

D-Glucose
(open-chain form)

(a) α-D-Glucopyranose
(a six-sided ring form of glucose)

(b) β-D-Glucopyranose
(a six-sided ring form of glucose)

Figure 1.15

Formation of a monosaccharide ring structure in an aldose. The C1 aldehyde of D-glucose reacts with the C5 hydroxyl group of the same molecule to form an intramolecular hemiacetal. Both α-D-glucopyranose and β-D-glucopyranose are possible products.

D-Fructose
(Open-chain form)

α-D-Fructofuranose
(a five-sided ring form of fructose)

β-D-Fructofuranose
(a five-sided ring form of fructose)

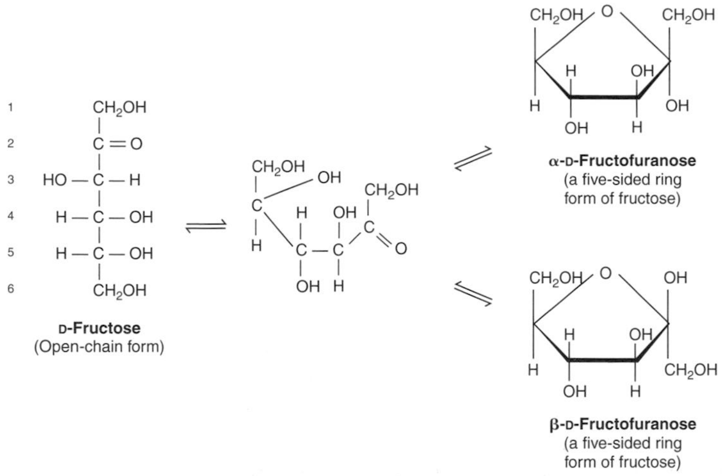

Figure 1.16

Formation of a monosaccharide ring structure in a ketose. The C2 keto group of D-fructose reacts with the C5 hydroxyl group to form an intramolecular ketal. Both α-D-fructofuranose and β-D-fructofuranose are possible products.

This three-dimensional conformation can be represented as illustrated in *Figure 1.17(a)*, which also shows the 'chair' configuration of the two six-sided ring forms of glucose, (*b*) α-D-glucopyranose and (*c*) β-D-glucopyranose.

An alternative configuration, the 'boat' form, is also possible, but is energetically unfavourable. The 'chair' configuration is thus the predominating structure of most hexoses in aqueous solutions. Substituents that lie in approximately the same plane as the ring are said to be in the equatorial position, and their bonds, which accordingly lie approximately in the same

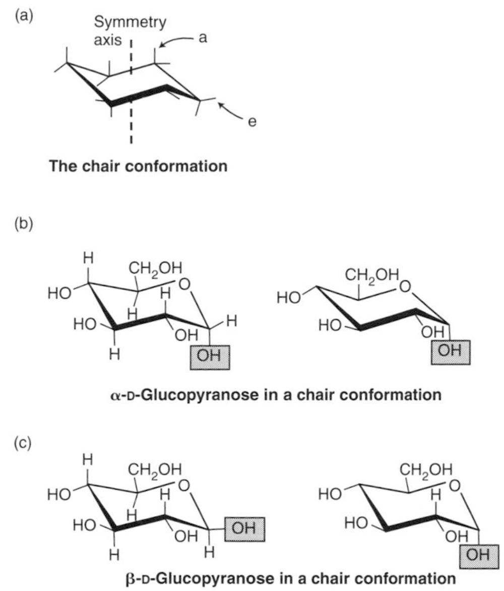

Figure 1.17

(a) The 'chair' form of a pyranose ring. Axial bonds (marked a) project above the plain of the ring, whereas equatorial bonds (marked e) project in the same plane as the ring. (b) The 'chair' conformation of α-D-glucopyranose and (c) β-D-glucopyranose. Both forms of glucopyranose are drawn with all substituents present (left) and with single H groups omitted for simplicity (right). Note that in (b) the α-form, the OH group on C1 atom (shaded) lies axial to the plane of the ring and in (c) the β-form, the OH group on the C1 atom (shaded) lies equatorial to the plane of the ring.

plane as the ring are called equatorial bonds. Substituents that lie roughly vertically above or below the plane of the ring are said to be in the axial position, and their bonds which are also perpendicular to the plane of the ring are called axial bonds. When several large substituents are present in axial positions, they crowd close together and therefore repel each other, whereas if they lie in equatorial positions, they emerge at the periphery of the ring and are therefore less crowded. Thus, the preferred conformation of a molecule is the one in which the large substituents are mostly in equatorial positions. The 'chair' configuration of D-glucopyranose therefore predominates because all axial positions are occupied by the smaller H atoms and the bulkier OH and CH_2OH groups emerge at the periphery, as can be appreciated from *Figure 1.17*.

Changes in conformation do not require the breaking of chemical bonds, only movement (rotation) around them; they therefore occur quite readily in solution. A sugar may be present in a mixture of different conformations, with the overall equilibrium favouring the lowest energy forms.

Because most people find the flat Haworth projection formulas easier to draw than the three-dimensional projection of the 'chair' configuration, the Haworth projection formulas are probably most widely used in the glycobiology literature. They correctly illustrate the configurations of the groups on the ring structure, but do not convey the bond angles. It is important to bear this in mind when considering structures represented in this way.

Furanose (five-sided) rings are not usually represented in the glycobiology literature in the three-dimensional projection form. The reason for this is that several (≈20) different conformations which are energetically very similar are possible, and there is therefore no preferred form.

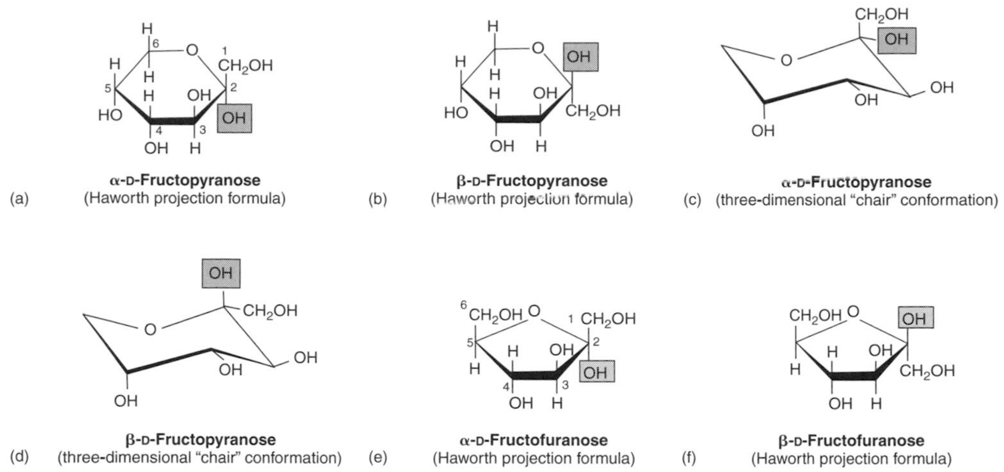

(a) α-D-Fructopyranose (Haworth projection formula)
(b) β-D-Fructopyranose (Haworth projection formula)
(c) α-D-Fructopyranose (three-dimensional "chair" conformation)
(d) β-D-Fructopyranose (three-dimensional "chair" conformation)
(e) α-D-Fructofuranose (Haworth projection formula)
(f) β-D-Fructofuranose (Haworth projection formula)

Figure 1.18

The Haworth projection and three-dimensional chair conformation formulae of the α- and β-anomers of the pyranose (six-sided) ring form of D-fructose, and the Haworth projection formula of the furanose (five-sided) ring form of the same molecule. Note that the three-dimensional chair conformation of furanose rings are not usually represented. In the Haworth projection formula of D-fructopyranose, if the OH group (shaded) is represented to lie below the C2 carbon, the molecule is (a) α-D-fructopyranose and if it lies above the C2 carbon it is (b) β-D-fructopyranose. In the three-dimensional chair conformation of the same molecules, the OH group at C2 (shaded) is represented in the equatorial position in (c) α-D-fructopyranose and in the axial position in (d) β-D-fructopyranose. In the Haworth projection formula of D-fructofuranose, the OH group at C2 (shaded) is represented lying below the C2 carbon in (e) α-D-fructofuranose and below it in (f) β-D-fructofuranose.

1.12 Monosaccharide α- and β-anomers

In some previous figures and discussion, the terms α- and β- have sometimes been used. These designations are now explained. In order for a ring structure to form, the open chain structure has to bend resulting in a ring structure with a new OH group, a hemiacetal OH, at the C1. The C1 now becomes a new chiral centre because it has four different substituents. The OH can emerge at either side of the ring, depending on the way in which the O atom of the C=O group is pointing just before ring closure, and thus two new epimers are created. The numbering of the carbon atoms in the straight chain form of the molecule is retained in the ring structure.

The two forms of D-glucopyranose (the six-sided ring form of D-glucose) which are derived from an epimerization at the position C1 during the closure of the ring structure, are called anomers. The hemiacetal or carbonyl carbon atom, C1, is called the anomeric carbon atom. The term anomers is reserved for epimers created in this way. They are designated α-D-glucopyranose or β-D-glucopyranose depending on whether the OH group is axial (α-) or equatorial (β-) to the plane of the ring. This can best be appreciated in the 'chair' configuration of α- and β-D-glucopyranose given in *Figure 1.17*. Here, the OH group at C1 is axial in (*a*) α-D-glucopyranose, and equatorial in (*b*) β-D-glucopyranose. In the Haworth projection formula, *Figure 1.15*, the same information is conveyed by the OH at C1 being represented as lying below the C1 carbon in (*a*) α-D-glucopyranose and above the C1 carbon in (*b*) β-D-glucopyranose.

The same nomenclature is applied to ketose sugars when they form rings, such as D-fructofuranose the five-sided ring form of fructose and D-fructopyranose the six-sided ring form. Accordingly, the α- and β- refer to hydroxyl groups attached to the C2, the anomeric hemiketal carbon atom, in these structures. This is illustrated in *Figure 1.18* which shows the Haworth

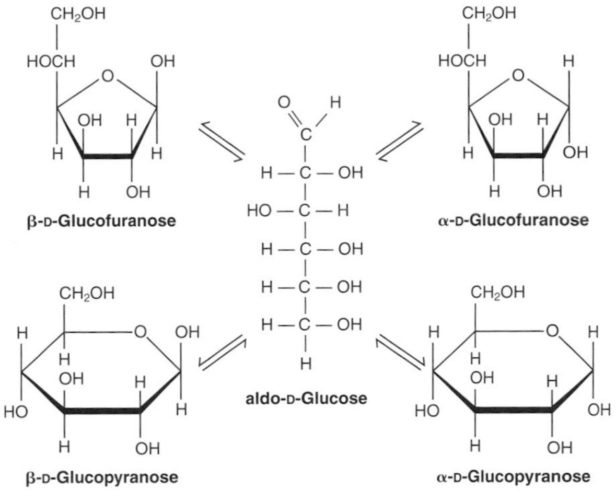

β-D-Glucofuranose

α-D-Glucofuranose

aldo-D-Glucose

β-D-Glucopyranose

α-D-Glucopyranose

Figure 1.19

In a solution of D-glucose, no matter what form of glucose is present in the solution at the beginning, a dynamic equilibrium is established over time in which the open chain form of D-glucose α-D-glucopyranose, α-D-furanose, β-D-glucopyranose and β-D-glucofuranose forms are all present in characteristic proportions.

projection and three-dimensional 'chair' representations of the α- and β-anomers of D-fructopyranose, and the Haworth projection formulae of the α- and β-anomers of D-fructofuranose.

1.13 Equilibrium between different structures

If a monosaccharide such as glucose is dissolved in water, an equilibrium between different chemical forms will develop over a period of a few minutes to a few hours at room temperature. This can happen because the hemiacetal or hemiketal bond is easily broken and reformed. Eventually, no matter what form of glucose is present in the solution at the beginning, a dynamic equilibrium is established in which α-glucose in pyranose and furanose forms and β-glucose in pyranose and furanose forms, and the open chain form of glucose are all present in characteristic proportions, as illustrated in *Figure 1.19*. In the case of glucose, the two pyranose forms predominate because they are the lowest energy forms. At equilibrium at room temperature, the mixture contains ≈66% β-D-glucopyranose, 33% α-D-glucopyranose and traces of the other three forms.

1.14 Examples of some biologically important monosaccharides

In this section, the most common, and therefore biologically important, monosaccharides occurring in mammals are discussed briefly. They are also illustrated, in Haworth projection formulae, in *Figure 1.20*. The topics touched upon in this section are explored in more depth in other parts of this book, as are the rarer monosaccharides of other species not listed here.

1.14.1 D-Ribose and 2-deoxy-D-ribose

The pentoses D-ribose and 2-deoxy-D-ribose form the backbone of the ribonucleic acid (RNA) and the deoxyribonucleic acid (DNA), respectively. Nucleic acids are not dealt with in any detail in this book.

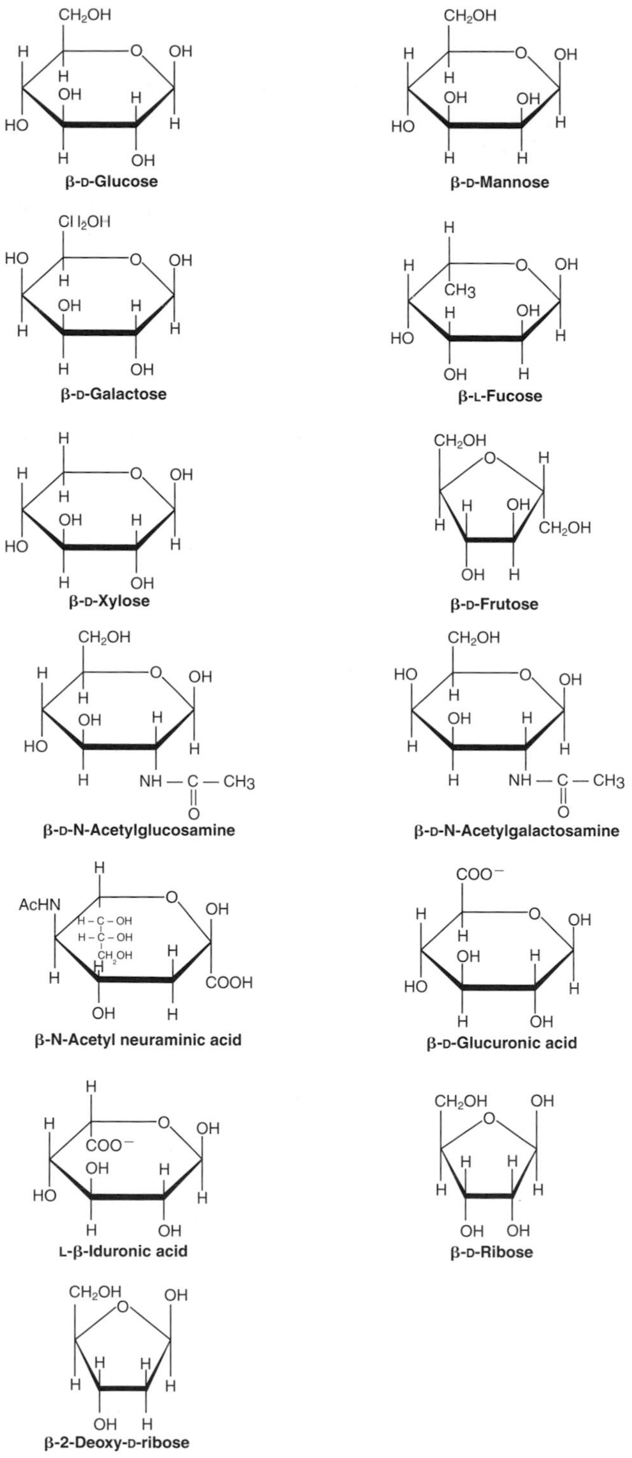

β-D-Glucose

β-D-Mannose

β-D-Galactose

β-L-Fucose

β-D-Xylose

β-D-Frutose

β-D-N-Acetylglucosamine

β-D-N-Acetylgalactosamine

β-N-Acetyl neuraminic acid

β-D-Glucuronic acid

L-β-Iduronic acid

β-D-Ribose

β-2-Deoxy-D-ribose

Figure 1.20

Some common and biologically important monosaccharides.

1.14.2 D-Glucose

Of the hexoses, the aldose D-glucose is the most important biological monosaccharide. It is the final product of photosynthesis and is present in many fruit juices. If many glucose units are covalently linked together, they form either starch (*see 3.4.1*), glycogen (*see 3.4.3*) or cellulose (*see 3.2.1*). Glucose is also part of the disaccharides sucrose (glucose–fructose, cane sugar) and lactose (galactose–glucose, the principal sugar in milk). Glucose occupies a central position in sugar biosynthesis and most other sugars are derived from it.

1.14.3 D-Galactose

The aldose D-galactose is the C4 epimer of glucose (*see 1.9*). It can only be degraded metabolically once it has been transformed into glucose. Together with glucose it forms the disaccharide lactose, which is the main carbohydrate component of human milk. Galactose residues are part of sphingolipids (*Chapter 9*) and the oligosaccharides attached to glycoproteins (*Chapters 4* and *5*), and thus galactose is an important component of biological molecules.

1.14.4 D-Mannose

The aldose D-mannose is the C2 epimer of glucose (*see 1.9*). Again, it can only be metabolized once it has been converted into glucose. It is an important carbohydrate residue in plant and animal glycoproteins (*Chapters 4* and *5*). Its phosphorylated form, mannose-6-phosphate, serves as a molecular 'postcode' for glycoproteins destined for the lysosome.

1.14.5 D-Fructose

Fructose is a ketose. It is usually represented as a furanose, although the pyranose form predominates in simple aqueous solution. Glucose and fructose are the constituents of the disaccharide sucrose (cane sugar) and fructose is present in many fruit juices.

1.14.6 Neuraminic (sialic) acid

The term neuraminic derives from the original isolation of this substance from nervous system tissue. Neuraminic acid is synthesized from mannosamine and pyruvate. If neuraminic acid is N-acetylated to N-acetylneuraminic acid, it is often referred to as 'sialic acid'. However, although strictly speaking this is incorrect, the term neuraminic acid and sialic acid are often used interchangeably. Many different derivatives of neuraminic acid exist, and they are involved in important biological functions (*see 11.8*). As this carbohydrate is negatively charged at neutral pH (because the carboxyl group is ionized) and is also the most abundant terminal carbohydrate residue of human cells, it is the major contributor to their overall negative charge. In erythrocytes this negative charge results in zeta potential and because all negatively charged erythrocytes repel each other, this negative charge prevents the spontaneous agglutination of human erythrocytes. Sialic acids are the topic of *Chapter 11*.

1.15 Examples of some important monosaccharide derivatives

In addition to the simple monosaccharides, such as glucose, galactose and mannose, there are a number of derivatives of monosaccharides in which a hydroxyl group (OH) is replaced with another group, or a carbon atom is oxidized to a carboxyl group. These sugars do not conform to the general formula for carbohydrates $(CH_2O)_n$ (*see 1.2*).

1.15.1 Deoxy-sugars

Monosaccharides in which an alcoholic hydroxy group has been replaced by a hydrogen atom are called deoxy-sugars (because an oxygen has been removed). Examples of deoxy-sugars that

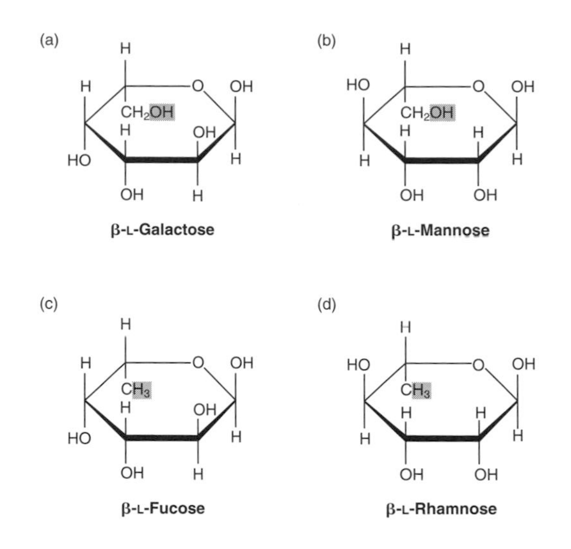

Figure 1.21

The substitution of a hydrogen for the hydroxyl group at C6 (shaded) of L-galactose or L-mannose produces the deoxy-sugars L-fucose or L-rhamnose, respectively.

are common components of glycoproteins, glycolipids and plant polysaccharides include L-fucose or L-rhamnose. These are formed by substitution of a hydrogen for the hydroxyl group at C6 of L-galactose or L-mannose, respectively. This is illustrated in *Figure 1.21*. Fucose is an important constituent of the oligosaccharides found in milk. Although it occurs in its L-form, it is derived from D-galactose and is present in many glycoproteins. Furthermore, fucose is an important terminal carbohydrate residue of human blood group sugars (*see 10.4*).

1.15.2 Sugar acids

Sugar acids are produced when one of the terminal carbons of a sugar is oxidized to a carboxylic acid group. There are two types of sugar acids: uronic acids and aldonic acids.

Uronic acids are formed from aldoses by replacement of the CH_2OH group with a carboxy group. This leaves an aldehyde group unchanged at C1, so uronic acids can still form ring structures through C1. They can also form glycosidic bonds with other sugars. They are important components of polysaccharides (*Chapter 3*), in which they confer an acidic nature to the molecule. They are present in the negative ionized form at physiological pH.

An example of a uronic acid is D-glucuronic acid which is formed from D-glucose. Oxidation of glucose at C6 results in glucuronic acid (*Figure 1.22*), which is structurally important in glycosaminoglycans (*Chapter 8*).

Aldonic acids are formed by oxidizing the aldehyde group of an aldo-sugar to a carboxylic acid group. In aldonic acids, the carboxylic group lies at C1. An example of this is D-gluconic acid, formed from glucose (*Figure 1.23*).

Uronic and aldonic acids can form five- and six-sided ring structures, because C1 or C6 now has a carboxylic acid group attached to it. However, these ring structures are fundamentally different to those formed by monosaccharides; they are cyclic esters called lactones, and are formed by esterification of the carboxylic acid by a hydroxyl group in the same chain. Examples are D-glucono-δ-lactone which is the six-membered ring form of the aldonic acid D-gluconic acid formed from D-glucose, and D-glucourono-δ-lactone, the corresponding lactone formed from D-glucuronic acid. These molecules are illustrated in *Figure 1.24*.

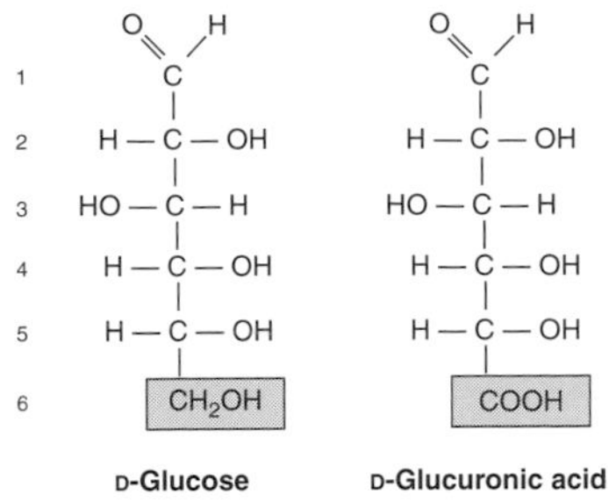

D-Glucose D-Glucuronic acid

Figure 1.22

The uronic acid, D-glucuronic acid, is formed by oxidation of the C6 (shaded) of the aldose D-glucose. The CH_2OH group is replaced by carboxy group.

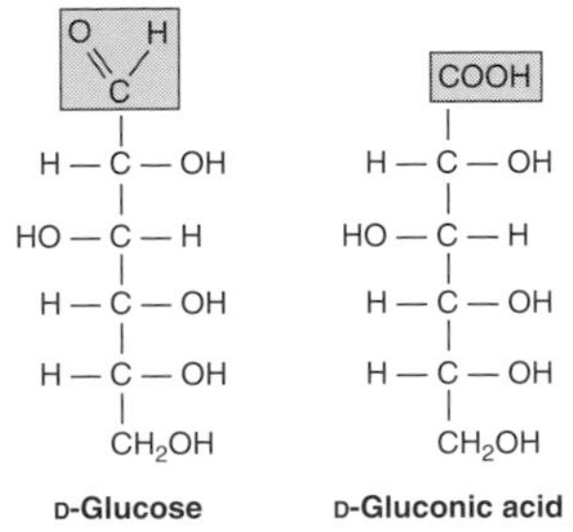

D-Glucose D-Gluconic acid

Figure 1.23

The aldonic acid, D-gluconic acid, is formed by oxidizing the aldehyde group of the aldose D-glucose to a carboxylic acid group. In aldonic acids, the carboxylic group lies at C1 (shaded).

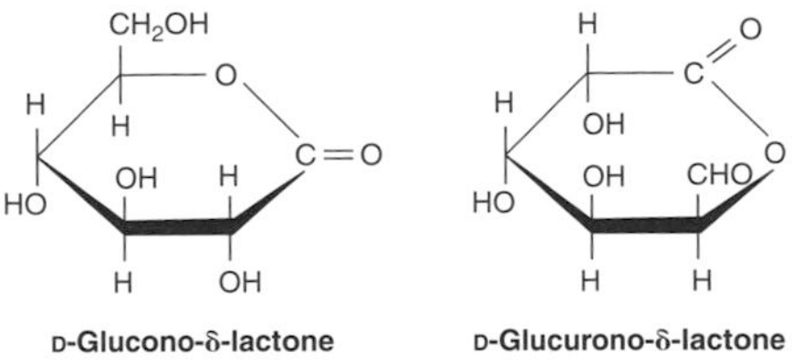

D-Glucono-δ-lactone D-Glucurono-δ-lactone

Figure 1.24

D-Glucono-δ-lactone derived from D-gluconic acid D-glucurono-δ-lactone, derived from D-glucuronic acid.

1.15.3 Amino sugars

Amino sugars are formed when one of the hydroxyl groups is replaced by an amino group. D-Glucosamine, for example, is a modified glucose molecule in which the hydroxyl group of the C2 atom is replaced by an amino group; the same is true for galactosamine. Hence, glucosamine and galactosamine only differ by the position of the hydroxyl group at C4, as illustrated in *Figure 1.25*.

Amino sugars are important components of polysaccharides (*Chapter 3*), in which they often occur as their N-acetyl derivatives, for example N-acetyl-D-glucosamine (*Figure 1.26*),

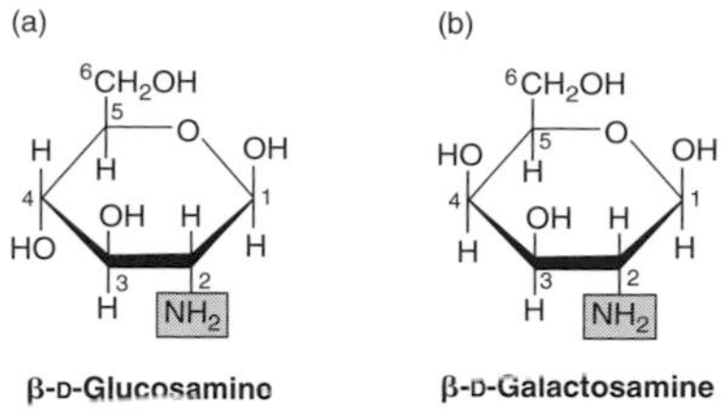

(a) β-D-Glucosamine

(b) β-D-Galactosamine

Figure 1.25

The amino sugars (a) β-D-glucosamine and (b) β-D-galactosamine. The hydroxyl group at C2 of the parent monosaccharide (β-D-glucose and β-D-galactose, respectively) is replaced by an amino group (shaded). β-D-Glucosamine and β-D-galactosamine differ only in the position of the hydroxyl group at C4.

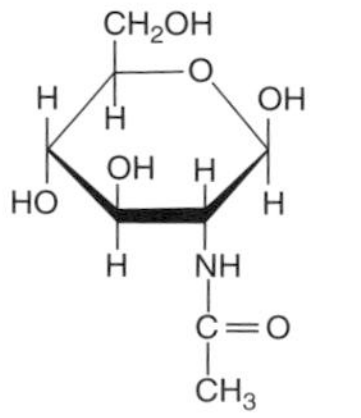

N-Acetyl-β-D-glucosamine

Figure 1.26

The amino sugar N-acetyl-β-D-glucosamine, the N-acetyl derivative of β-D-glucose.

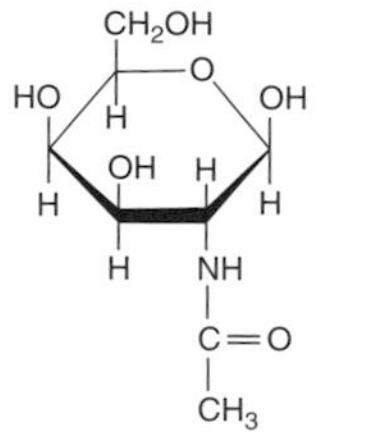

N-Acetyl-β-D-galactosamine

Figure 1.27

The amino sugar N-acetyl-β-D-galactosamine, the N-acetyl derivative of β-D-galactose.

and N-acetyl-D-galactosamine (*Figure 1.27*). N-Acetylmuramic acid (*Figure 1.28*), an important component of bacterial cell walls, is formed when lactic acid is linked to the oxygen at the C3 of N-acetylglucosamine.

1.15.4 Sugar phosphates

In the metabolism of sugars, the first step is usually the formation of a sugar phosphate. Sugar phosphates, for example, are important intermediates in glycolysis. In sugar phosphates, the orthophosphate group may be esterified either to the primary hydroxyl group (e.g. at C6 in glucose-6-phosphate; *Figure 1.29*) or to the anomeric hydroxyl group (e.g. at C1 in glucose-1-phosphate; *Figure 1.30*).

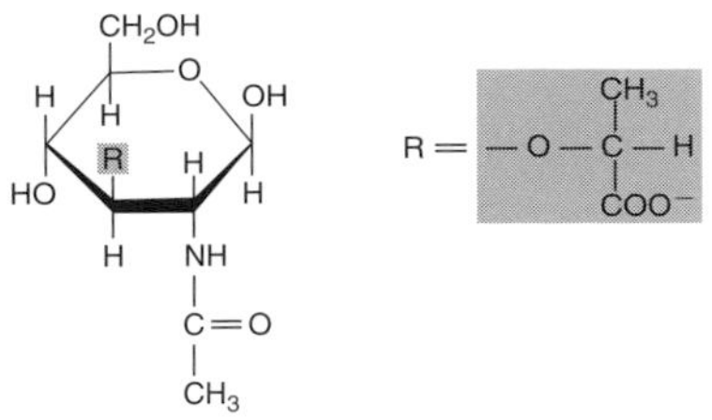

Figure 1.28

The amino sugar N-acetylmuramic acid is lactic acid (shaded) ether linked to the oxygen at C3 of N-acetylglucosamine.

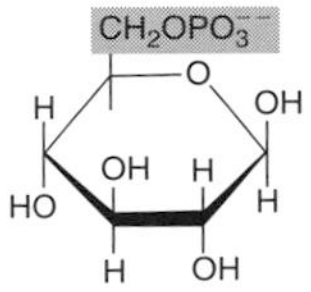

Figure 1.29

The phosphate sugar glucose-6-phosphate. The orthophosphate group is esterified to the primary hydroxyl group (shaded) at C6.

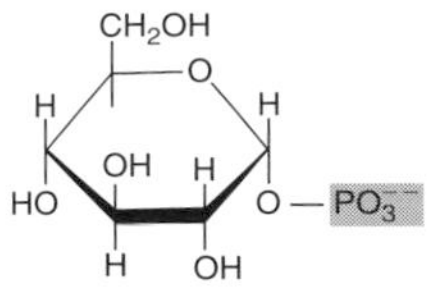

Figure 1.30

The phosphate sugar glucose-1-phosphate. The orthophosphate group is esterified to the anomeric hydroxyl group (shaded) at C1.

1.15.5 Nucleotide sugars

Nucleotide diphosphate sugars, often referred to simply as nucleotide sugars, are important in the interconversion of sugars and in the synthesis of glycosides. In these compounds, which contain two monosaccharide moieties, the monosaccharide to be metabolized is linked through the anomeric position to a nucleotide diphosphate. An example is uridine diphosphate glucose, or UDP-glucose (UDP-Glc) (*Figure 1.31*).

1.15.6 Sugar alcohols

Sugar alcohols are formed by reduction of the carbonyl group of a sugar to a hydroxyl group. They therefore cannot form rings, because they no longer possess a carbonyl group. Examples of this include the formation of glycerol (a sugar alcohol) by the reduction of glyceraldehyde, and D-glucitol (another sugar alcohol, also known as sorbitol) by reduction of D-glucose. Glycerol and D-glucitol are found in large concentrations in the over-wintering stages of some insects and also in some species of fish where they act as antifreeze agents, and glycerol is an important component of lipids. D-Glucitol (usually called sorbitol in this context) is widely used as a 'sugar-free' sweetener in the food industry in diet and diabetic products.

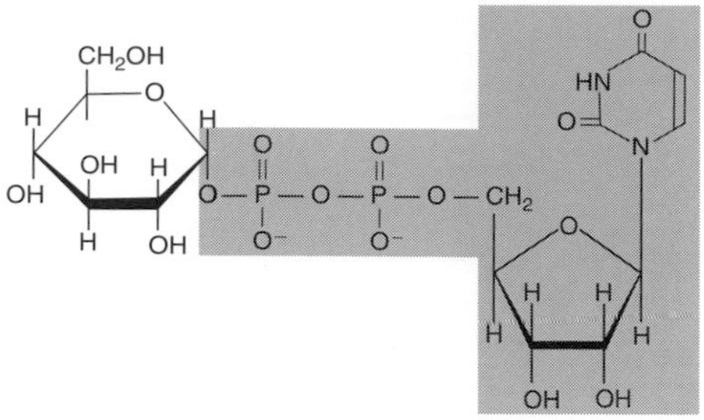

Uridine diphosphate glucose (UDP-Glc)

Figure 1.31

The nucleotide diphosphate sugar uridine diphosphate glucose or UDP-Glc. The monosaccharide glucose is linked through the anomeric position to a nucleotide diphosphate (shaded).

1.16 The linking together of monosaccharides to form chains

1.16.1 Monosaccharides react with alcohol to form acetals

Hemiacetals and hemiketals can react further with alcohols to form acetals and ketals. Monosaccharides in the ring form, which are themselves hemiacetals and hemiketals, can therefore form acetals and ketals by reaction with alcohol.

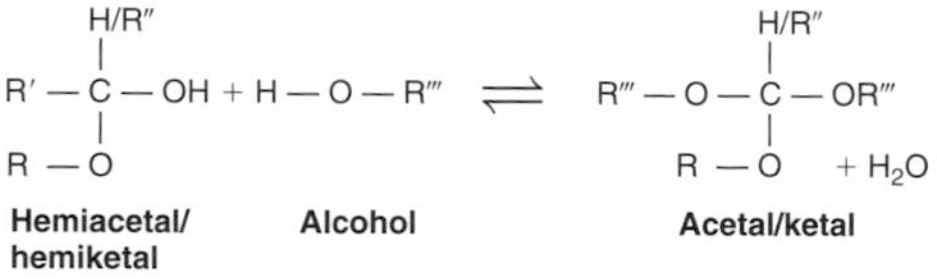

$$\underset{\text{Hemiacetal/hemiketal}}{R'-\overset{H/R''}{\underset{R-O}{C}}-OH} + \underset{\text{Alcohol}}{H-O-R'''} \rightleftharpoons \underset{\text{Acetal/ketal}}{R'''-O-\overset{H/R''}{\underset{R-O}{C}}-OR'''} + H_2O$$

➲ Glycosides are acetals between the hemiacetal or hemiketal OH group of a sugar and an OH group of a second compound. The bond formed is a glycosidic bond. Water is eliminated during the reaction.

Methanol, for example, can react with D-glucose to give an α- or β-acetal, called methyl-α-D-glucoside or methyl-β-D-glucoside, respectively, depending on whether the new OCH_3 group is on the same side of the ring as the reference CH_2OH group (β-form) or on the opposite side (α-form), as illustrated in *Figure 1.32*. Sugar acetals are called glycosides (from the Greek work *glykys* which means sweet). The name is formed by replacing the '-ose' (as in gluc*ose*) by '-oside' (gluc*oside*).➲

1.16.2 Monosaccharides link together to form disaccharides

The same sort of chemical reaction by which monosaccharides react with alcohol to form acetals and ketals (*see 1.16.1*), can occur between two monosaccharides. Monosaccharides may be linked together covalently in this way to form disaccharides. The bond between them is often referred to as an O-glycosidic bond. A glycosidic bond is formed when the OH group of one monosaccharide reacts with the anomeric carbon of the other, with the elimination of a molecule of water. This type of reaction is illustrated in *Figure 1.33*, which shows the reaction of two molecules of α-D-glucose to form maltose. This reaction represents the formation of an acetal from a hemiacetal plus an alcohol, or a ketal from a hemiketal plus an alcohol, described previously, but here the OH group of the second monosaccharide is acting as an alcohol. The formation and breaking of glycosidic bonds do not occur spontaneously, but are

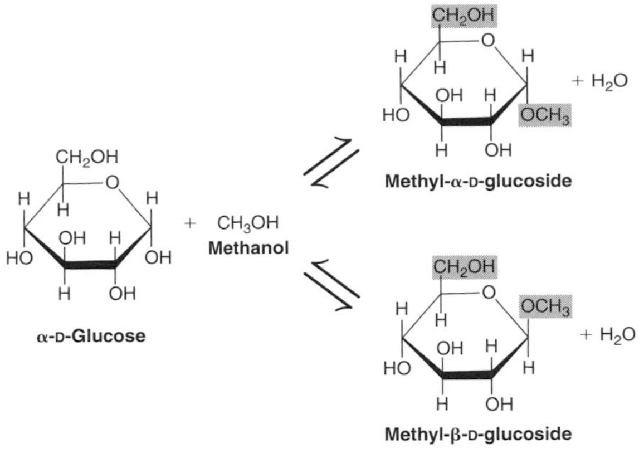

Figure 1.32

α-D-glucose reacts with methanol to form either methyl-α-D-glucoside (if the OCH_3 group is on the opposite side of the ring to the CH_2OH group – both shaded), or methyl-β-D-glucoside (if the OCH_3 group is on the same side of the ring as the CH_2OH group – both shaded).

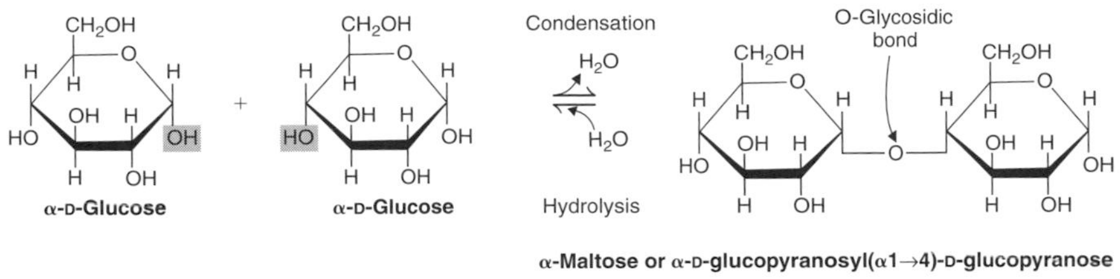

Figure 1.33

Two molecules of α-D-glucose combine in a condensation reaction to form the disaccharide maltose. The hydroxyl group of the right-hand monosaccharide (shaded) acts as an alcohol which reacts with the hemiacetal group (shaded) of the left-hand monosaccharide. The bond formed is an O-glycosidic bond. A molecule of water is produced. As the OH group of the right-hand monosaccharide is in the α-configuration (below the plain of the ring), the disaccharide is α-maltose.

catalysed by glycosyltransferase and glycosidase enzymes, respectively. Glycosyltransferases are the subject of *Chapter 2*.

Following the convention for monosaccharides (*see 1.12*), if the OH group at C1 on the right-hand monosaccharide is below the plane of the ring, the entire disaccharide is said to be in the α-form. If it projects above the plane of the ring, the disaccharide is in the β-form. Thus, the disaccharide illustrated in *Figure 1.33* is α-maltose. This point is further illustrated in *Figure 1.34* which shows both α- and β-forms of maltose in Haworth projection formulae and three-dimensional 'chair' configuration. ➲

An acetal oxygen bridge joins the two monosaccharide units together. This acetal unit, like all acetals, will react readily with water in the presence of acid or an enzyme catalyst (a glycosidase), and this hydrolysis frees the two monosaccharides again.

➲ A disaccharide is a compound in which two monosaccharide units are joined by an O-glycosidic linkage. The reaction is analogous to that in which an acetal is formed from a hemiacetal plus an alcohol, or a ketal from a hemiketal plus an alcohol. One monosaccharide uses the OH group of a second monosaccharide as an alcohol group.

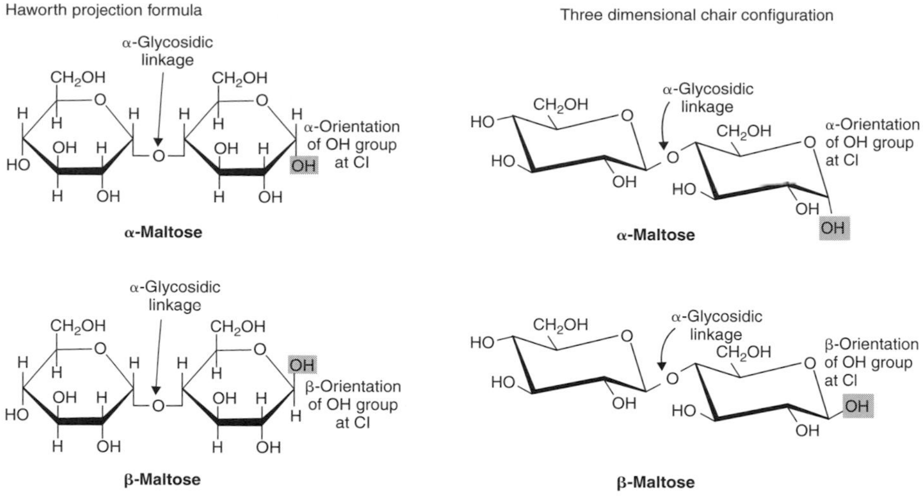

Figure 1.34

The α- and β-forms of the disaccharide maltose differ only in that in α-maltose the OH group at C1 of the right hand monosaccharide (shaded) projects below the plane of the ring, and in the β-form it is above the plane of the ring.

As there are usually several OH groups on the second monosaccharide, there are several different ways of joining two monosaccharides together, and therefore a number of differently linked disaccharides can be produced (*see 1.18*).

1.16.3 Oligosaccharide and polysaccharide chains

The same chemical reaction that links two monosaccharides together to form a disaccharide is responsible for linking together large numbers of monosaccharide units, via glycosidic bonds, in the linear or branching chains of oligosaccharides and polysaccharides. ➲ The oligosaccharides of glycoproteins (*Chapters 4* and *5*), polysaccharides (*Chapter 3*) and glycolipids (*Chapter 9*) are the specific subject of other sections of this book.

➲ The term glycan, strictly speaking, means polysaccharide, but it is widely used in the glycobiology literature to refer to the carbohydrate component of a glycoprotein or glycolipid.

1.17 α- and β-glycosidic bonds

Glycosidic bonds between two monosaccharides can occur in two forms, namely α- or β-forms. In the disaccharide maltose, for example, the glycosidic bond links the C1 of the left-hand monosaccharide with the C4 of the right-hand monosaccharide through an oxygen atom (*Figure 1.34*). Here, the bond linking the bridging oxygen with the C1 of the left-hand monosaccharide is axial, shown below the plane of the ring of the left-hand monosaccharide, so it is an α-glycosidic bond. If it were above the plane of the ring of the left-hand monosaccharide it would be β-glycosidic bond. If the bond was in the β-form rather than the α-form, the disaccharide would be cellobiose, not maltose (*Figure 1.35*). This simple geometrical difference is biologically important (as many of the seemingly trivial differences in sugar linkages and structures are) because humans are only able to digest maltose as an energy source, and are unable to digest cellobiose. Similarly, polymers of glucose units linked in this way by α-glycosidic bonds form amylose (part of starch) and can be degraded by humans, whereas polymers of glucose linked by β-glycosidic bonds form cellulose, which can not.

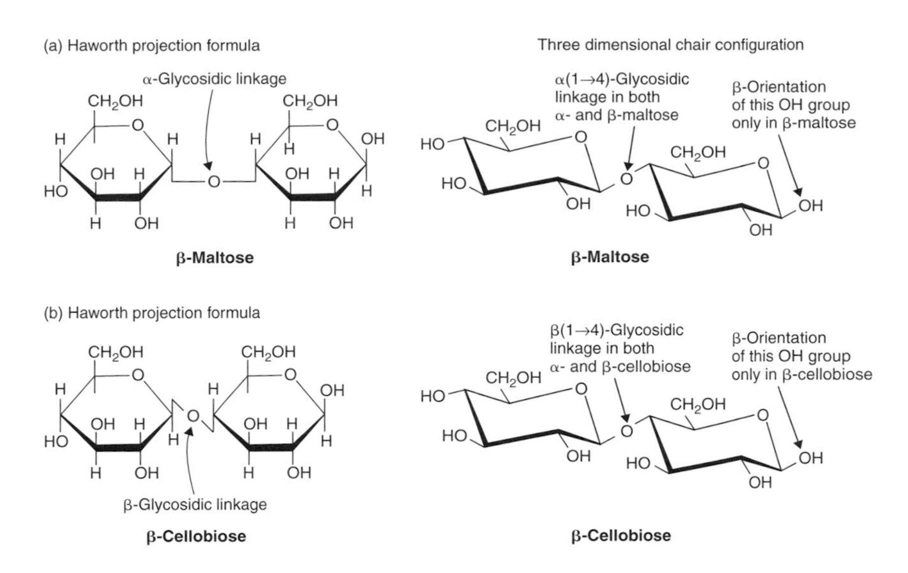

Figure 1.35

(a) β-maltose and (b) β-cellobiose in Haworth projection formulae (left) and 'chair' configuration (right). Both are composed of two molecules of D-glucose. In (a) β-maltose, the glycosidic bond linking the C1 of the left-hand monosaccharide with the C4 of the right-hand monosaccharide is axial, below the plane of the ring of the left-hand monosaccharide, an α-glycosidic bond, making the disaccharide maltose. It is β-maltose because the OH group at C1 of the right-hand monosaccharide is above the plane of the ring. In (b) β-cellobiose, the glycosidic bond linking the left-hand monosaccharide with the C4 of the right-hand monosaccharide is equatorial (the β-form) and the bond is a β-glycosidic bond, making the disaccharide cellobiose. It is β-cellobiose because the OH at C1 of the right hand monosaccharide is above the plane of the ring.

Both the Haworth projection formulae and the three-dimensional 'chair' configuration illustrate the difference between α- and β-forms elegantly. However, in the Haworth projection formula, the chemical bonds that link the oxygen to the two monosaccharides are illustrated as bent lines. This type of representation might be taken to imply, erroneously, that another atom, possibly a hydrogen, lies undrawn at the point of the angle. It is therefore, although a commonly used convention, not ideal. The same bonds are less ambiguously represented in the 'chair' configuration drawings of the same molecules, and for this reason are used to represent the structures of the other disaccharides described in this chapter. In *Figure 1.35*, both maltose and cellobiose are in the β-form (β-maltose and β-cellobiose) because the OH group at C1 of the right monosaccharide lies above the plane of the ring.

1.18 Different linkages

In (di)saccharides, the O-glycosidic bonds can occur between the different carbon atoms. In maltose and cellobiose, illustrated in *Figures 1.34* and *1.35*, the glycosidic bond links the C1 of the left-hand monosaccharide with the C4 of the right-hand monosaccharide, so the bond is said to be a 1→4 bond.

Particularly common in nature are β1→4 (C1 of one monosaccharide linked in a β-configuration to C4 of the next) and β1→6 (C1 of one monosaccharide linked in a β-configuration to C6 of the next) bonds, whereas 1→2 (C1 to C2), 1→3 (C1 to C3) and 1→5 (C1 to C5) bonds occur less often.

The large number of monosaccharide building blocks, the possibility of α- or β-glycosidic bonds, the presence of up to four different monosaccharide residues linked to one core monosaccharide molecule, and the formation of linear and branching chains of virtually any length, lead to the potential formation of a bewildering variety of complex carbohydrate structures. This is in contrast to proteins, which are linear polymers of amino acids, and in which only one type of linkage is possible. Whereas, for example, three amino acids or nucleotide bases can generate six combinations, three hexose sugars could theoretically yield more than 25 000 unique trisaccharides. A chain of six hexose sugars could theoretically yield more than a trillion combinations. However, nature does not make use of all of the possible combinations. This complexity is the major difficulty in determining the molecular formula of carbohydrates, and is thus one of the principle reasons why glycobiology has lagged behind molecular biology. The analysis of carbohydrate structures is the subject of *Chapter 14*.

1.19 Conventions for describing oligosaccharides

Because monosaccharides can be linked together in many different conformations, there are conventions for how these different linkages are described in the scientific literature. This is illustrated by a simple example: the disaccharide maltose contains two D-glucose residues linked by a glycosidic bond between the C1 (the anomeric carbon) of one glucose and the C4 of the other (*see 1.16.2*; *Figure 1.33*). The configuration of the anomeric carbon atom in the glycosidic linkage is α because the bond linking C1 of the left-hand monosaccharide to the oxygen bridge lies below the plane of the ring. Conventionally, this compound is named by considering its non-reducing end (*see 1.20*) to the left, and the name is composed in the following order: (i) the configuration, α or β, at the anomeric carbon joining the first monosaccharide unit to the second is given; (ii) the non-reducing residue is named first, and to distinguish five- and six-membered rings furano- and pyrano- are inserted; (iii) the two carbon atoms joined by the glycosidic bond are indicated in parentheses with an arrow connecting the two numbers, for example, (1→4) indicates that the C1 of the first named sugar residue is joined to the C4 of the second named; (iv) the second monosaccharide is named. If there is a third residue, the second glycosidic bond is named next using the same conventions, and so on. Three-letter abbreviations for monosaccharides, given in *Table 1.2*, are usually used for brevity. The same three-letter abbreviations are frequently used in the

Table 1.2 Common monosaccharides and their derivatives

Monosaccharide/derivative	Abbreviation
Arabinose	Ara
Fructose	Fru
Fucose	Fuc
Galactose	Gal
Galactosamine	GalN
N-acetylgalactosamine	GalNAc
Glucose	Glc
Glucosamine	GlcN
N-acetylglucosamine	GlcNAc
Glucuronic acid	GlcA
Mannose	Man
Muramic acid	Mur
Neuraminic acid	Neu
N-acetylneuraminic acid/sialic acid	Neu5Ac
Rhamnose	Rha
Ribose	Rib
Xylose	Xyl

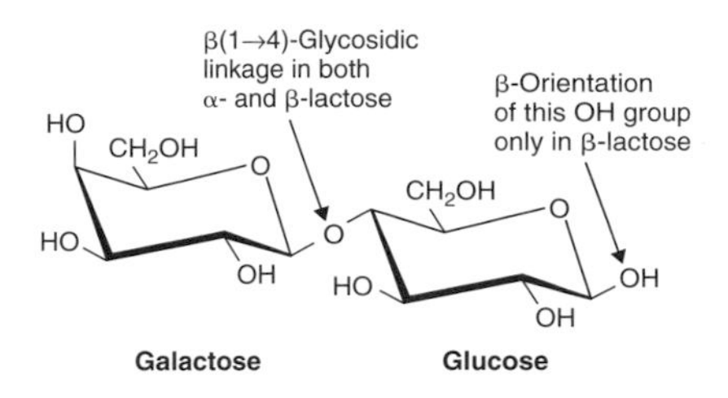

Figure 1.36

β-Lactose. The C1 of the galactose molecule on the left is linked by a β-glycosidic bond to the C4 of the glucose molecule on the right. The disaccharide is β-lactose because the OH group at C1 of the glucose is in the β configuration (above the plane of the ring). β-Lactose is therefore correctly termed β-D-galactopyranosyl-(1→4)-β-D-glucopyranose or Gal (β1→4)Glc.

glycobiology literature to refer to monosaccharides in text. In this book, we have followed this convention, as repetitive use of the full monosaccharide names can be cumbersome.

Following this convention, α-maltose (*Figure 1.33*) is α-D-glucopyranosyl-(1→4)-α-D-glucopyranose or Glc(α1→4)Glc. β-Lactose (*Figure 1.36*), is composed of a galactose on the left linked by a β-glycosidic bond to glucose on the right. The C1 of galactose is linked to the C4 of glucose. The disaccharide is β-lactose because of the β-orientation of the OH group at C1 of the glucose. The correct term for β-lactose is therefore β-D-galactopyranosyl-(1→4)-β-D-glucopyranose or Gal(β1→4)Glc.

1.20 Reducing sugars

In the previous section the concept of disaccharides and oligosaccharides having a 'non-reducing end' was introduced, and this requires explanation. Monosaccharides are all reducing sugars. A common test for this is their reaction with Tollen's and Benedict's reagents. In these reactions, the sugar is oxidized.

In Tollen's test, Tollen's reagent (an alkaline solution of a silver ion in combination with two ammonia molecules $[Ag(NH_3)_2]^+$oxidizes the aldehyde group (RCH=O) to a carboxyl group (in its anionic form RCO_2^-), and silver ion is reduced to metallic silver (Ag), producing a silver mirroring of the reaction vessel. The reaction may be written:

$$RCH{=}O + 2[Ag(NH_3)_2]^+ + 3OH^- \rightarrow RCO_2^- + 2Ag + 2H_2O + 4NH_3$$

Benedict's reagent, which is a basic solution of copper (II) ions and citrate ions, has a brilliant blue colour, due to the presence of copper (II) ions, Cu^{2+}. In Benedict's test, Cu^{2+} ions are reduced to Cu^+, which immediately react with the base OH^- ions to give a brick red precipitate of copper (I) oxide, Cu_2O. The reaction may be written:

$$RCH{=}O + 2Cu^{2+} + 5OH^- \rightarrow RCO_2^- + Cu_2O + 3H_2O + Cu^{2+}$$

In the past, Benedict's test was commonly used to detect glucose in urine in the diagnosis and monitoring of diabetes.

Nearly all disaccharides (apart from sucrose) are also reducing carbohydrates because they have a reducing as well as a non-reducing end, but polysaccharides are not. The reason for this is that reducing sugars must either have a free carbonyl or hemiacetyl group, for example, the free aldehyde or keto group of monosaccharides in the open chain formation, or exist in ring forms that readily give a carbonyl group. Some sugars, such as the disaccharide sucrose and polysaccharides, are non-reducing sugars because they have their potential carbonyl groups substituted in a chemically stable form.

When an anomeric carbon participates in a glycosidic bond, it can no longer be oxidized. The carbon atom can no longer exist in a linear form and thus can no longer act as a reducing agent. In describing disaccharides or polysaccharides, therefore, the end of the chain with the free anomeric carbon is commonly called the reducing end.

1.21 Examples of biologically important disaccharides

A few of the biologically important disaccharides are considered briefly in this section.

1.21.1 Maltose

Maltose is formed by two glucose units linked by an α1→4 glycosidic bond (*see 1.19, Figure 1.33*). Maltose is therefore correctly termed α-D-glucopyranosyl-(1→4)-D-glucopyranose or Glc(α1→4)Glc. It is an intermediary product of starch and glycogen digestion.

1.21.2 Lactose

Lactose (*Figure 1.36*) is the main disaccharide of human milk. Lactose is composed of a galactose linked to a glucose molecule to form a β-D-galactopyranosyl-(1→4)-D-glucopyranose or Gal(β1→4)Glc. Note that the glycosidic bond lies above the plane of the ring of the left hand monosaccharide, and is thus a β-glycosidic bond (*see 1.17*). It is medically important that many humans lack the digestive enzyme lactase necessary for the hydrolysis of the β-glycosidic bond between the galactose and glucose units of lactose, and are thus unable to digest it as a foodstuff. This condition is termed lactose intolerance. It is believed that evolution favoured the lactase gene in humans only after dairy farming became prevalent in human populations.

1.21.3 Sucrose

Sucrose (also sometimes called saccharose; *Figure 1.37*) is β-D-fructofuranosyl-D-glucopyranoside or Fru(α1→2)Glc. It is an example of a disaccharide formed between a pyranose and a furanose. Sucrose is known as cane sugar, and is widely used in foodstuffs for human consumption, and is present in sugar cane, sugar beet and in many fruit juices. The suffix '-ide' as in β-D-fructofuranosyl-α-D-glucopyranos*ide* indicates that sucrose is not a reducing sugar. This is unusual among the disaccharides (*see 1.20*).

1.21.4 Cellobiose

Cellobiose (*Figure 1.35b*) is β-D-glucopyranosyl-(1→4)-D-glucopyranose or Glc(β1→4)Glc. It differs from maltose, Glc(α1→4)Glc only in the nature of the glycosidic bond, which is a β1→4 one in cellobiose and an α1→4 one in maltose. Cellobiose is the basic building block of cellulose (*see 3.2.1*). Cellulose is the main structural component of plant cell walls and as

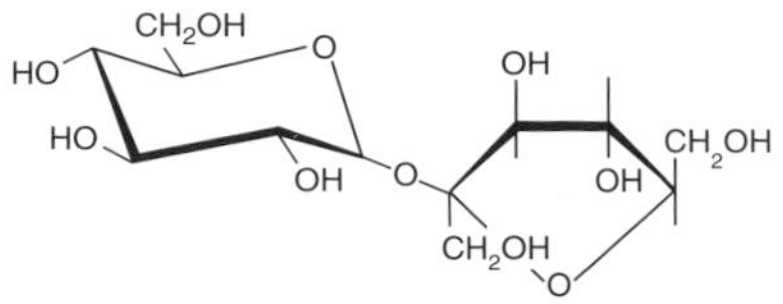

Figure 1.37

Sucrose, or β-D-fructofuranosyl-α-D-glucopyranoside, Fru(α1→2)Glc, is an example of a disaccharide formed between a pyranose (Glc) and a furanose (Fru).

such is consumed in large quantities in plant based foodstuffs. However, it cannot be degraded by mammals. Ruminants have bacteria in their fore-stomach which degrade cellulose to digestible saccharides.

1.22 Ways of illustrating oligosaccharide chains

Oligosaccharide chains are represented in a number of ways in the glycobiology literature. The correct form, as recommended by International Union of Pure and Applied Chemistry and International Union of Biochemistry and Molecular Biology (IUPAC–IUBMB) Joint Commission on Biochemical Nomenclature (JCBN) is to order the monosaccharides with the reducing end of the chain to the left (where there is a glycosyl linkage to a protein, lipid, etc. the monosaccharide residue involved in linkage to the non-carbohydrate appears to the right), use the abbreviation for each monosaccharide, preceded by its anomeric α or β, and its configuration symbol, D- or L-. The ring size should also be indicated by an italic *f* for furanose or *p* for pyranose. The linkage should be given in parentheses between the symbols, with an arrow. A double-headed arrow is used to indicate a linkage between two anomeric positions. For example, cellotriose, a trisaccharide composed of three β(1→4)-linked D-glucopyranose residues would be represented by:

β-D-Glc*p*-(1→4)-β-D-Glc*p*-(1→4)-D-Glc*p*

In a branched structure, the longest chain is regarded as the 'parent'. A branched structure may be represented like this:

β-D-Gal*p*-(1→4)-β-D-Glc*p*NAc-(1→2)-α-D-Man*p*
3
↑
1
α-L-Fuc*p*

or on a single line with the side branch in square brackets, thus:

β-D-Gal*p*-(1→4)[α-L-Fuc*p*]β-(1→3)-D-Glc*p*NAc-(1→2)-α-D-Man*p*

This type of formula may be condensed by omitting the configurational symbol and the letter denoting the ring size. It is usually assumed that the configuration is D- (with the exception of fucose and iduronic acid which commonly occur in L-form), and that the ring will be a pyranose, unless otherwise specified. For convenience, this type of presentation is the one most commonly used in this book. Thus, cellotriose might be represented:

Glc(β1→4)Glc(β1→4)Glc

And the branched oligosaccharide might be represented like this on two lines:

Gal(β1→4)GlcNAc(β1→2)Man
|
Fuc(α1→3)

Or like this on a single line: Gal(β1→4)[Fuc(α1→3)]GlcNAc(β1→2)Man.

For very long sequences, an even shorter form may sometimes be desirable. Here, it is acceptable to omit the locants of the anomeric carbon atoms, the parentheses and the hyphens. Branches may be indicated on the same line by use of round or square parentheses. Configuration symbols and ring size designators can be included if the information would be ambiguous without them. Cellotriose might thus be represented: Glcβ-4Glcβ-4Glc and the branched oligosaccharide: Galβ-4(Fucα-3)GlcNAcβ-2Man.

Table 1.3 Symbols used to indicate monosaccharides in this book

Name of monosaccharide or monosaccharide derivative	Abbreviation	Symbol
Glucose	Glc	▲
Mannose	Man	○
Galactose	Gal	●
N-acetylglucosamine	GlcNAc	■
N-acetylgalactosamine	GalNAc	□
Fucose	Fuc	△
Xylose	Xyl	▽
Neuraminic acid (sialic acid)	NeuAc (SA)	◆
Glucuronic acid	GlcA	⬘
Iduronic acid	IdoA	⬙

In very complex diagrams, it is quite common to see oligosaccharide chains represented as a chain of geometrical symbols. Thus, the branched oligosaccharide in the previous examples might be represented:

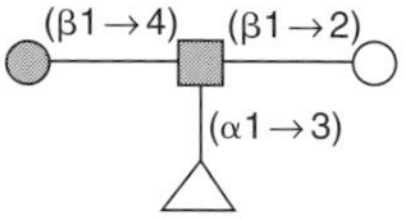

(β1→4) (β1→2)
(α1→3)

There is no official way of doing this, and different glycobiology laboratories have developed their own different conventions. The symbols used in this book are quite commonly found in the glycobiology literature, and are listed in *Table 1.3*.

In this book, for clarity of presentation, symbols have been used to illustrate complex oligosaccharide structures in diagrams. When using this book, the reader is recommended to consult *Plate I* which presents the Fischer projection formulae, Haworth projection formulae, three-dimensional configuration and symbols for common monosaccharides for easy reference.

1.23 The complex carbohydrate structure database (CCSD)

The potentially staggering number of possible carbohydrate structures do not all occur in nature, but an extremely large number do (*see 1.2*). The known structures and accompanying references to the literature are recorded on a computer database called the 'complex carbohydrate structure database' (CCSD), accessed by a software program called CarbBank, which was begun in 1986. Access is free of charge. The database allows searching by author name, key word in title, trivial name, citation, molecular mass, composition, types of residue, type of linkage, complete or fragmentary structure and others. It contains around 50 000 records of more than 22 000 carbohydrate structures, and is an invaluable resource for glycobiologists.

Further reading

International Union of Pure and Applied Chemistry and International Union of Biochemistry and Molecular Biology (IUPAC – IUBMB) Joint Commission on Biochemical Nomenclature (JCBN). Nomenclature of carbohydrates (1996) http://www.chem.qmw.ac.uk/iupac/2carb

The complex carbohydrate structure database (CCSD) CarbBank software from ftp://ncbi.nlm.nih.gov/repository/carbbank

CarbBank website for online searches at http://www.ccrc.uga.edu

Enzymes of glycosylation – the glycosyltransferases

2

2.1 Introduction

Proteins and nucleic acids are essentially linear molecules, made up of building blocks linked together by a series of identical (amide and phosphodiester, respectively) bonds, and are synthesized using a pre-existing template.➲ Carbohydrates are different in several ways. First, oligosaccharides and polysaccharides are often branched molecules, and the monosaccharide building blocks from which they are formed can be linked together in several different ways to give extensive combinatorial diversity (*Chapter 1*). Second, they are not synthesized using a template; instead, the oligosaccharide side chains of glycoproteins and glycolipids are assembled, step by step, as they pass along a cellular processing pathway made up of a series of enzymes located in specific parts of the cell's endomembrane system (endoplasmic reticulum and Golgi apparatus).

> ➲ Glycan synthesis differs from protein synthesis in two main ways:
> (i) glycans are often branching, not linear, molecules; and
> (ii) glycans are not synthesized using a pre-existing template.

Major classes of vertebrate glycoconjugates that are glycosylated in this way include N-linked (*Chapter 4*) and O-linked (*Chapter 5*) oligosaccharides, glycosaminoglycans (*Chapter 8*), glycosphingolipids (*Chapter 9*) and glycosylphosphatidylinositol (GPI) anchors (*Chapter 12*). Specific types of glycosylation also occur in the nucleus and cytoplasm (*Chapter 6*).

2.2 Enzymes of glycosylation – glycosidases and glycosyltransferases

More than 250 different enzymes are involved in the biosynthesis of the elaborate oligosaccharide structures found on glycoconjugates, and more than 30 different enzymes may participate directly in the synthesis of a single complex N-linked oligosaccharide (*Chapter 4*).➲ These enzymes include glycosyltransferases, which are specific sugar-transferring enzymes and glycosidases which are responsible for trimming specific monosaccharides from precursors to form intermediate structures.

> ➲ The enzymes of glycosylation are glycosyltransferases and glycosidases. Glycosyltransferases are sugar-transferring enzymes. Glycosidases are trimming enzymes.

Two major classes of glycosidases are distinguished. The first are endoglycosidases which catalyse the cleavage of internal glycosidic bonds. The second are exoglycosidases which catalyse the stepwise linkage-specific removal of sugars from the non-reducing termini of carbohydrate polymers. Examples of these include sialidases, fucosidases and so on. These are present in lysosomes, where they are involved in the degradation of oligosaccharides (several lysosomal storage diseases result from their defective expression; *see 16.9*) and mannosidases are also expressed in the endoplasmic reticulum (ER) where they take part in the processing of core N-glycosidically linked glycans (*see 4.2.6*). Endo- and exoglycosidases have been used extensively to elucidate oligosaccharide structures (*see 14.6.2*).

Glycosyltransferases are residents of the membranes lining the ER, the Golgi apparatus and the *trans*-Golgi network.

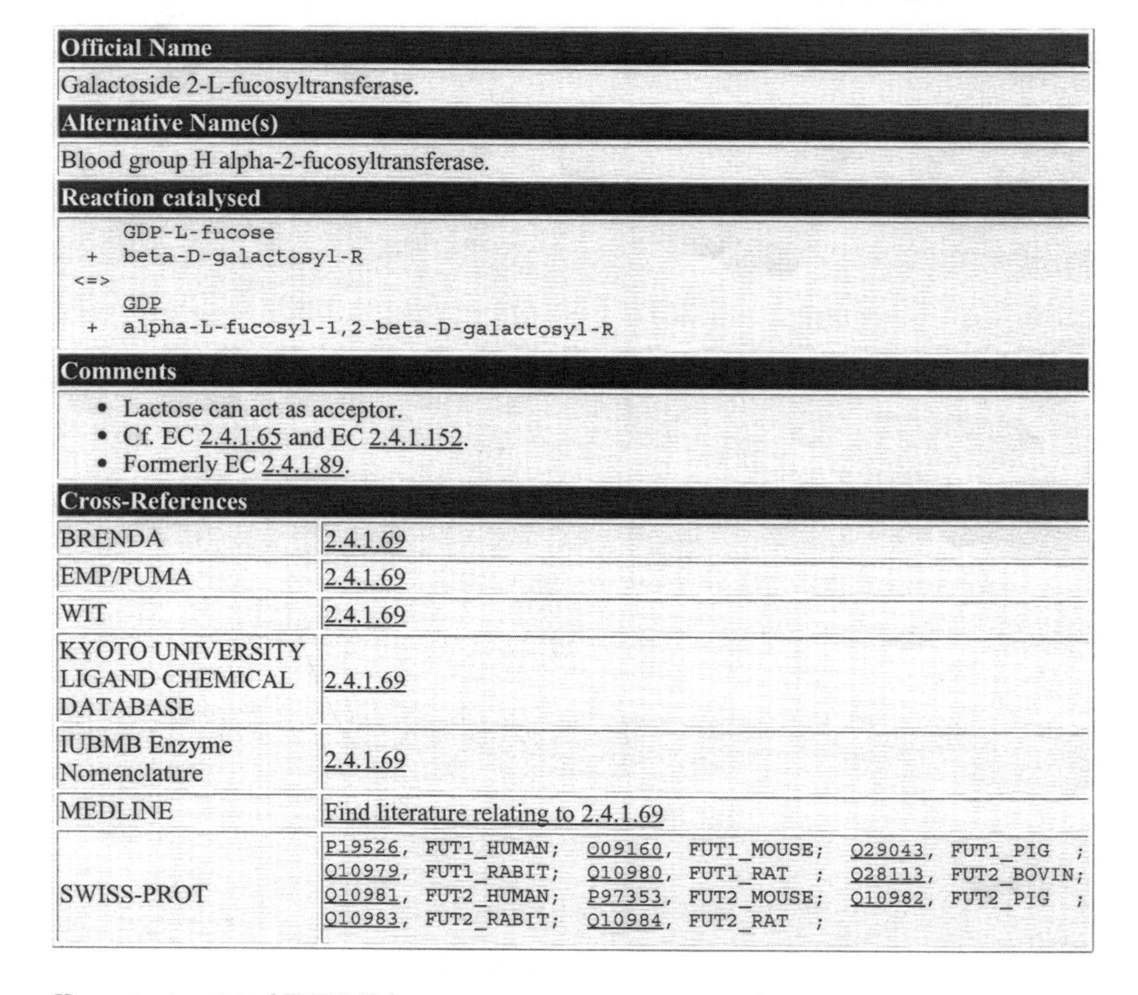

ExPASy Home page	Site Map	Search ExPASy	Contact us	ENZYME

Hosted by SIB Switzerland | Mirror sites: Australia | Canada | China | Korea | Taiwan | USA new

NiceZyme View of ENZYME: EC 2.4.1.69

Official Name

Galactoside 2-L-fucosyltransferase.

Alternative Name(s)

Blood group H alpha-2-fucosyltransferase.

Reaction catalysed

```
    GDP-L-fucose
 +  beta-D-galactosyl-R
<=>
    GDP
 +  alpha-L-fucosyl-1,2-beta-D-galactosyl-R
```

Comments

- Lactose can act as acceptor.
- Cf. EC 2.4.1.65 and EC 2.4.1.152.
- Formerly EC 2.4.1.89.

Cross-References

BRENDA	2.4.1.69
EMP/PUMA	2.4.1.69
WIT	2.4.1.69
KYOTO UNIVERSITY LIGAND CHEMICAL DATABASE	2.4.1.69
IUBMB Enzyme Nomenclature	2.4.1.69
MEDLINE	Find literature relating to 2.4.1.69
SWISS-PROT	P19526, FUT1_HUMAN; O09160, FUT1_MOUSE; Q29043, FUT1_PIG ; Q10979, FUT1_RABIT; Q10980, FUT1_RAT ; Q28113, FUT2_BOVIN; Q10981, FUT2_HUMAN; P97353, FUT2_MOUSE; Q10982, FUT2_PIG ; Q10983, FUT2_RABIT; Q10984, FUT2_RAT ;

View entry in original ENZYME format

If you would like to retrieve all the SWISS-PROT entries referenced in this entry, click here.

ExPASy Home page	Site Map	Search ExPASy	Contact us	ENZYME

Hosted by SIB Switzerland | Mirror sites: Australia | Canada | China | Korea | Taiwan | USA new

Figure 2.1

An example of a page from the ExPASy site giving details of the blood group H α2-fucosyltransferase, EC 2.4.1.69, which catalyses addition of the Fuc residue in an α(1→2) linkage to βGal-R to yield the H blood group antigen α-Fuc(1→2)βGal-R. From www.expasy.org/enzyme/. Reproduced with permission from the Swiss Institute of Bioinformatics.

2.3 Nomenclature of glycosyltransferases

The official names of enzymes are agreed by the Enzyme Commission of the International Union of Biochemistry and Molecular Biology (IUBMB), who also designate a unique Enzyme Commission (EC) number to each enzyme. This information can be accessed at the SwissProt site serviced by the ExPASy molecular biology server at http://www.expasy.ch/enzyme/ which also provides sequence data, links to literature relating to each enzyme, and other useful information, as illustrated in *Figure 2.1*.

In addition to their official name and EC number, glycosyltransferases are described in the glycobiology literature in many ways. It is common to find them described according to the reaction they catalyse.➲ For example, the glycosyltransferase that catalyses transfer of GDP-Fuc in an (α1→2) linkage to β-linked galactose, may be described as GDP-Fuc:Galβ-R α1,2 fucosyltransferase. Because these terms are cumbersome, they are usually abbreviated, in this example most commonly to α1,2 Fuc transferase, or simply α1,2 Fuc-T. Furthermore, many common names are also in wide usage. For example, there are two very similar α1,2 fucosyltransferases one of which is coded for by a gene at the *H* blood group locus and is responsible for addition of the Fuc residue of the H blood group antigen (*see 2.6.1*). It is therefore often referred to as 'the blood group H α-2-fucosyltransferase' or 'blood group H Fuc transferase', or 'blood group H Fuc-T'. This enzyme has the EC number 2.4.1.69 and its entry in the ExPASy site is given in *Figure 2.1*. In this book, for simplicity, we have referred to enzymes by what appear to be their most simple, descriptive and commonly used names.

➲ In addition to the official names and EC numbers for glycosyltransferases, many different descriptions and abbreviations for the same enzyme are found in the glycobiology literature. GDP-Fuc: Galβ-R α1,2 fucosyltransferase is, for example, most usually abbreviated to α1,2 Fuc-T, but may also appear as α2 Fuc-T, α2-FT, α1,2FucT, α1,2FT and other variations.

Several glycosyltransferase 'families' exist (*see 2.11*) in which members often catalyse an almost identical reaction. For example, all nine members of the polypeptide N-acetylgalactosaminyltransferases (or protein-UDP acetylgalactosaminyltransferases, GalNAc-Ts, EC 2.4.1.41; *see 2.11.2*) catalyse the transfer of GalNAc from UDP-GalNAc to serine and threonine residues of the polypeptide backbone at the initiation of O-linked mucin type glycosylation. Different members of the family are believed to operate under different circumstances, for example, in different cell types, or in the glycosylation of different glycoproteins. In the case of such families, the enzymes are usually referred to numerically, for example in the GalNAc-T family, members are referred to in the glycobiology literature as GalNAc-T1 to GalNAc-T9 in the order in which they were first described.

2.4 Initiation of glycosylation by glycosyltransferases

The transfer of sugar residues to proteins and lipids occurs using different mechanisms. Glycolipid biosynthesis (*see 9.4*) begins with a galactosyl- or glucosyltransferase, which links the first monosaccharide using the lipid ceramide as an acceptor. In N-linked glycoprotein synthesis (*see 4.2*), a multimeric enzyme complex called oligosaccharyltransferase (OST) transfers a pre-constructed oligosaccharide block from a dolichol lipid intermediate to the amide nitrogen of an asparagine side chain positioned within the polypeptide. The dolichol intermediate is synthesized on the outer face of the ER until addition of the fifth Man, at which point it is translocated through the lipid bilayer to the lumenal face, via a protein carrier or 'flippase'. Dolichol-dependent glycosyltransferases are otherwise little used by higher eukaryotes, but are responsible for this important step in N-glycosylation. They also play a role in the glycosylation of GPI-anchored proteins (*see 12.4*). In contrast, in mucin-type O-linked glycosylation (*see 5.6*), an initial GalNAc residue is transferred directly onto a serine or threonine residue of the polypeptide chain. Other types of O-glycosylation occur which also do not require a lipid–sugar intermediate are described in *Chapters 6* and *7*.

2.5 Building oligosaccharides – transglycosylation reactions

Once the first monosaccharide has been attached to lipid or protein, or in the case of N-linked glycoproteins, once the oligosaccharide core has been attached and trimmed, the oligosaccharide chain is extended by the sequential action of glycosyltransferases. ➲ Most glycosyltransferases therefore catalyse *transglycosylation* reactions, in which they transfer a monosaccharide from, usually, a high-energy nucleotide-sugar donor to an oligosaccharide acceptor.

> ➲ Most glycosyltransferases catalyse transglycosylation reactions between a high-energy sugar donor and an oligosaccharide receptor.

A generalized scheme for the transglycosylation reaction is:

$$\underset{\text{(donor-sugar)}}{\text{X-sugar}} + \text{acceptor-OH (or -NH)} \xrightarrow[\text{}]{\textit{glycosyl transferase}} \underset{\text{(oligosaccharide, lipid or protein)}}{\text{acceptor-O (or -N)-sugar}} + \text{X}$$

2.6 Sugar donors

Typical nucleotide-sugar donors of eukaryotic cells are listed in *Table 2.1*. Plants, eubacteria and archae bacteria employ different nucleotide-sugar donors. Other sugars donors that are employed by glycosyltransferases less frequently include dolichol phosphosugars (e.g. dolichol phospho-1-glucose, dolichol phospho-1-N-acetylglucosamine, dolichol phospho-1-mannose), sugar phosphates and disaccharides.

The low molecular mass sugar-nucleotide donors involved in glycan synthesis are constructed in the cytosol and transported through the lipid bilayer into the endomembrane system by specific transmembrane nucleotide transporters. ➲ These are usually antiporters which exchange sugar-nucleotides for nucleotides released by previous reactions. This is important, as a high concentration of free nucleotide is potently inhibitory to transglycosylation reactions, and this antiporter mechanism thus keeps levels of inhibitory free nucleotides low. Several putative transporters, based on their ability to restore nucleotide-sugar transport in defective mutants, have recently been cloned.

> ➲ Sugar-nucleotide donors are synthesized in the cytosol and transported into the endomembrane system by an antiporter that reciprocally transports free sugar-nucleotides out, thus keeping their otherwise inhibitory concentrations low.

2.7 The 'one enzyme–one linkage' rule – and exceptions to it

Several hundred different glycosidic linkages have been described, and as a general rule the enzymes that catalyse their formation are specific in their action and, with few exceptions, catalyse the reaction of one specific donor and acceptor to yield a single, specific linkage. Historically, this has been referred to as the 'one enzyme–one linkage' rule, and has been a central dogma of glycobiology for many years.

Table 2.1 Nucleotide sugar donor substrates used by eukaryote glycosyltransferases

Uridine 5′-diphospho-α-D-galactose	UDP-Gal
Uridine 5′-diphospho-α-D-glucose	UDP-Glc
Uridine 5′-diphospho-α-D-N-acetylglucosamine	UDP-GlcNAc
Uridine 5′-diphospho-α-D-N-acetylgalactosamine	UDP-GalNAc
Uridine 5′-diphospho-α-D-glucuronic acid	UDP-GA
Uridine 5′-diphospho-α-D-xylose	UDP-Xyl
Cytidine 5′-monophospho-β-neuraminic acid	CMP-SA
Guanosine 5′-diphospho-α-D-mannose	GDP-Man
Guanosine 5′-diphospho-β-L-fucose	GDP-Fuc

Recently, however, examples have been found of: (i) more than one enzyme, usually from the same enzyme 'family', being capable of catalysing exactly the same reaction; and (ii) rarely, the same enzyme being capable of catalysing the synthesis of more than one linkage.➲ The 'one enzyme–one linkage' rule has therefore been revised to 'one enzyme family–usually one linkage', and it seems likely that this results in a large degree of redundancy as well as a higher level of regulation of the glycoforms synthesized.

➲ The central dogma of glycobiology, the 'one enzyme–one linkage' rule, has been superseded by the 'one enzyme family–usually one linkage' rule.

2.7.1 The *H* and *Se* locus α1,2 fucosyltransferases

An example of a deviation from the 'one enzyme–one linkage' rule is that there are at least two human forms of the GDP-Fuc:Galβ-R α1,2 fucosyltransferase (α1,2 Fuc-T), encoded by different, but spatially very close, genetic loci. One form is encoded by the *H* locus, and the second by the *Se* (secretor) locus, both of which are located on human chromosome 19q13.3, ≈35 kb apart. These two forms of the enzyme are expressed differentially in different tissues. The *H* locus is expressed mostly in tissue derived from ectoderm or mesoderm, such as haemopoietic tissues, and the *Se* locus is expressed mostly in tissues derived from endoderm, such as epithelial cells of the salivary glands and gastrointestinal tract (*see 10.5.7*). This illustrates another important point, that some glycosyltransferases are produced by certain cell types only. About 20% of individuals do not produce the form of the enzyme coded for by the *Se* locus.➲ The α1,2 Fuc-T enzyme itself does not appear to be vital for human survival, but individuals who do not produce it at all have the rare Bombay blood type, which results in difficulty in obtaining compatible blood samples for transfusion (*see 10.5*).

➲ Notable exceptions to the 'one enzyme–one linkage rule' are the two forms of the blood group H determinant synthesizing α1,2 Fuc-T, and the family of GalNAc-Ts that initiate O-glycan chain synthesis.

2.7.2 The GalNAc polypeptide transferase family

Another example of an exception to the 'one enzyme–one linkage' rule is the large family of glycosyltransferases that catalyse the transfer of GalNAc to serine/threonine residues at the initiation of O-linked glycosylation (*see 5.6.1, 5.9*). These are the UDP-GalNAc:polypeptide N-acetylgalactosaminyltransferases (GalNAc-Ts), some of which are expressed ubiquitously, whereas others are highly tissue specific in their distribution. Different members of the family are specific for different regions of the peptide, for example, some members of the family operate only in hydrophobic or in polar regions. This family of glycosyltransferases are described in greater detail below (*see 2.11.2*).

2.8 Cation and pH requirements of glycosyltransferases

To work optimally, many glycosyltransferases require Mn^{2+}, Mg^{2+}, Ca^{2+}, Co^{2+} or other divalent cations as cofactors. They are involved in binding the negatively charged nucleotide-sugar to the protein. The late steps in sialic acid addition utilizing CMP-Sia, and O-acetylation of these residues, are exceptions and do not require the presence of divalent cations.

The optimum pH range of glycosyltransferases is usually between 5 and 7, the pH characteristic of the parts of the endomembrane system in which they function.

2.9 Localization of glycosyltransferases along the secretory pathway

Glycosyltransferases act sequentially, so that the glycosylated product of one reaction becomes the acceptor substrate for the next; in this way, long and complex linear or branching oligosaccharide chains are built up.

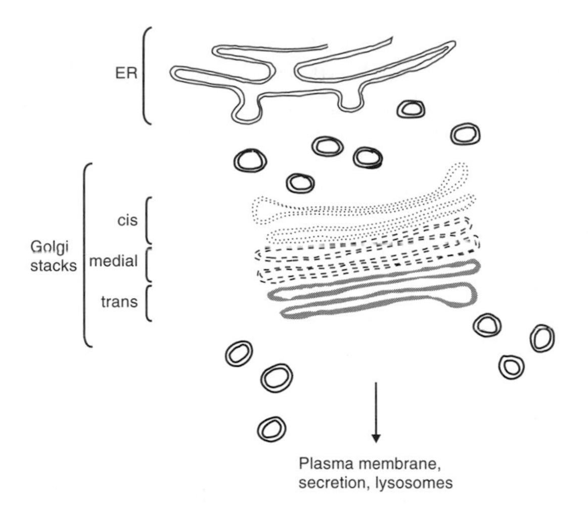

Figure 2.2

Enzymes of glycosylation are located within the Golgi apparatus in accordance with the order in which they operate in glycan synthesis. The glycosidase enzyme α-mannosidase I, involved in initial trimming of the N-linked oligosaccharide core, is located in the *cis*-Golgi (dotted); N-acetylglucosaminyltransferase I, involved in chain elongation, is located in the *medial*-Golgi, (hatched) and β1,4 galactosyltransferase and α2,6 sialyltransferase, involved in terminal glycosylation processes, lie within in the *trans*-Golgi and *trans*-Golgi network (shaded).

➲ The location of glycosyltransferases along the secretory pathway roughly corresponds to the sequence in which they act in building oligosaccharide chains.

To this end, the location of glycosyltransferases within the secretory pathway largely corresponds to the order in which they act; thus in the N-glycosylation of proteins, assembly of the $Glc_3Man_9GlcNAc_2$ oligosaccharide intermediate, its transfer to the asparagine residues of the growing polypeptide chain, and partial trimming take place in the ER and completion of trimming followed by terminal extension takes place in the Golgi apparatus (*see 4.2*).➲

The enzymes responsible for these different processes are located, roughly speaking, accordingly. For example, the glycosidase enzyme α-mannosidase I, involved in trimming, is located in the *cis*-Golgi; N-acetylglucosaminyltransferase I, involved in elongation, is located in the *medial*-Golgi, and β1,4-galactosyltransferase and α2,6-sialyltransferase, involved in terminal glycosylation, are located in the *trans*-Golgi and *trans*-Golgi network. This is illustrated in *Figure 2.2*.

However, there is considerable overlap in localization between different enzymes, and some glycosyltransferases are expressed throughout the Golgi stacks. Furthermore, differences in location may be observed in different cell types, for example, a sialyltransferase may be localized to the *trans*-Golgi alone in goblet cells of the intestine, but be present throughout the Golgi stacks in adjacent absorptive enterocytes. Differences in their localization along the secretory pathway are also observed in disease states. This implies a more complex and subtle regulation of their activity. The mechanism by which glycosyltransferases are positioned within the ER and Golgi apparatus is not well understood (*see 2.12.2*).

2.10 Substrate recognition by glycosyltransferases

Most glycosyltransferases are highly specific for their oligosaccharide receptors through their specific recognition of one or more residues of the underlying oligosaccharide chain. Some, however, are slightly more promiscuous. β1,4 galactosyltransferase (β1,4 Gal-T), for example, recognizes any terminal GlcNAc residue, although some underlying oligosaccharide structures are acted upon preferentially. Mammalian glycosyltransferases are usually very specific in their choice of substrate, whereas the transferases of simpler organisms, like bacteria, are sometimes less selective (*see 18.4.7*).

An intriguing aspect of glycosylation is the fact that certain types of glycan structure are often found at specific glycosylation sites of glycoprotein molecules. How this is regulated is not well understood. To some extent, it may be due simply to spatial accessibility of putative glycosylation sites by particular transferases, but in a few cases there is evidence that the underlying polypeptide chain of a glycoprotein may also be recognized by a particular glycosyltransferase. A good example of this is the glycosylation of Thy-1 (*see 4.4.7*).

2.11 Glycosyltransferase 'families'

Glycosyltransferases belong to several distinct 'families', including sialyltransferases (STs) (*see 11.4*), blood group A and B transferases (*see 10.5*), UDP-GalNAc: polypeptide N-acetylgalactosaminyltransferases (GalNAc-Ts) (*see 5.6.1*), α1,2 fucosyltransferases (α1,2 Fuc-Ts) (*see 10.5*), α1,3 fucosyltransferase (α1,3 Fuc-Ts) (*see 10.6.2*), β1,3 galactosyltransferases (β1,3 Gal-Ts), β1,4 galactosyltransferases (β1,4 Gal-Ts) and others.➲ There is often considerable sequence homology between members of each family, and between similar transferases from different vertebrate and invertebrate species, suggesting very strong conservation through evolution.

➲ Functional features of members of glycosyltransferase families:
- shared specificity for donor nucleotide, but not necessarily donor sugar
- conservation or inversion of the sugar donor linkage to the acceptor (α- or β-anomeric configuration)
- transfer, usually to the same position on the acceptor sugar (notable exception: the sialyltransferase family; *see 11.4*).

The vertebrate genome may contain instructions for making several hundred different glycosyltransferases. Estimates suggest that as much as 2–4% of the active human genome is involved in the process of glycosylation. Glycosyltransferases are generally present in very small amounts in cells, and are also transmembrane proteins which are difficult to extract, hence they are difficult to purify biochemically for sequencing. The first mammalian glycosyltransferase, β1,4 galactosyltransferase (β1,4 Gal-T), was not sequenced until 1986. However, molecular biology methods have been developed that have facilitated the isolation of glycosyltransferase genes without the need for purified protein. Low stringency hybridization and polymerase chain reaction (PCR) cloning with primers derived from conserved sequences from catalytically similar glycosyltransferases have opened up this research area. Since the mid-1980s the number of transferases identified, sequenced and cloned has grown exponentially – although this a rapidly developing field, many glycosyltransferases probably remain to be identified.

2.11.1 The blood group A and B transferases

Most members of a glycosyltransferase family transfer the same type of sugar residue, but may differ in preferred acceptor substrate and sometimes in the linkage formed. However, the blood group A and B transferases are worthy of special mention because they are different.➲ The blood group A and B transferases are very closely related – they differ in only four amino acids. They use the same acceptor substrate and produce the same (α1→3) to Gal linkage, but transfer a different sugar residue: the blood group A transferase transfers UDP-GalNAc and the blood group B transferase transfers UDP-Gal. Functional genes and pseudogenes share sequence homology and belong to the same family (*see 10.5*).

➲ The blood group A and B transferases differ in only four amino acids, but the blood group A transferase catalyses the transfer of UDP-GalNAc and the blood group B transferase catalyses the transfer of UDP-Gal onto the same substrate.

2.11.2 The family of UDP-GalNAc:polypeptide N-acetylgalactosaminyltransferases

➲ The polypeptide GalNAc-T family catalyse the transfer of the first GalNAc to the Ser/Thr residue of the polypeptide backbone at the initiation of mucin type O-linked glycan synthesis.

One family of glycosyltransferases, the UDP-GalNAc:polypeptide N-acetylgalactosaminyltransferases (GalNAc-Ts) will be examined in a little more detail to illustrate some of the general points discussed above.➲ These enzymes are of special interest because they are responsible for the initial key step in mucin-type O-glycosylation (*see 5.6.1*) in which GalNAc is transferred from UDP-GalNAc to serine and threonine residues on the polypeptide backbone of the protein, and there is also evidence that this type of glycosylation is disrupted in some disease states, notably cancer (*see 16.6.2*).

The substrate specificities of these transferases are a major factor in determining the sites of O-glycan attachments and their expression and activity therefore determine the number and position of O-linked glycans on a protein. The enzymes have distinct, but partly overlapping, substrate specificities with short peptide acceptor substrates, are differentially expressed in different tissues, and react at different rates with individual potential glycosylation sites. The regulation of their enzyme activity, substrate specificity and catalytic mechanism are not fully understood.

The rapid rate at which our knowledge of glycosyltransferases is expanding is illustrated by the GalNAc-T family, in that the first member, GalNAc-T1, was cloned independently by two groups from bovine colostrum as recently as 1993. Subsequently, a human equivalent was isolated from human salivary gland in 1995 and a second member of the family, GalNAc-T2, was identified and purified from human placenta at roughly the same time. Speculation that other members of the family existed was rapidly followed by the cloning and expression of GalNAc-T3 and GalNAc-T4, whose discovery was based on a 'GalNAc-T motif' consisting of a conserved stretch of 61 amino acids. At the time of writing, a total of nine members of the mammalian GalNAc-T family has been cloned and characterized. They are termed GalNAc-T1 to GalNAc-T9 respectively. GalNAc-T8 and GalNAc-T9 have been reported within the last 2 years, and it seems probable that further members of this family of enzymes will be discovered.

The members of this family are diverse in their amino acid sequences and genomic organization, raising questions as to their evolutionary origin. That it is a very ancient family, in evolutionary terms, is indicated by the fact that a related family of at least 11 GalNAc transferases is present in organisms as simple a *Caenorhabditis elegans*.➲ It has been suggested that GalNAc-T4, which is coded for by a single exon, may represent the human ancestral gene, and that the other members of the family have arisen by gene duplication and subsequent divergence, although this is far from established. GalNAc-T1, -T2 and -T3 are, in comparison, each coded for by 10–16 small exons. GalNAc-T3 and -T6 are surprisingly similar in sequence throughout their coding regions, and identical in organization for the nine conserved intron/exon boundaries in their coding regions. They also share similar kinetic properties distinct from other GalNAc transferases. All this suggests that they may be closely related and have arisen as the result of a late duplication event.

➲ The polypeptide GalNAc-T family is ancient in evolutionary terms. The ancestral human gene is believed to be GalNAc-T4, from which other members of the family have arisen by gene duplication and divergence.

The GalNAc-T family does not co-localize to a single gene locus. The genes for several family members are located on entirely different chromosomes; for example, GalNAc-T1 is located on chromosome 18q12-q21, GalNAc-T2 on chromosome 1q41-q42, GalNAc-T3 on chromosome 2q24-q31, GalNAc-T4 on chromosome 12q21.3-q22, and GalNAc-T6 on chromosome 12q13. A pseudogene has also been identified on chromosome 3. However, several of the GalNAc-Ts share common intron/exon boundaries and, overall, in spite of their diversity, there seems to be ≈45–50% sequence homology between members of the family, and some similarities in predicted domain structures, sequence motifs and conserved cysteine residues are observed. Around 80% homology is seen within the putative catalytic domain between GalNAc-T1, -T2 and -T3.

Northern blot analysis indicates that the GalNAc-Ts may be classified into two groups on the basis of their tissue expression. GalNAc-T1, -T2, -T4 and -T8 are considered 'housekeeping enzymes', in that their expression appears to be ubiquitous in human tissues, and they are probably involved in the general synthesis of O-linked glycans. GalNAc-T3, -T5, -T6, -T7 and -T9 are tissue-specific enzymes whose distribution is highly regulated. GalNAc-T9, for example, has to date been detected only in brain tissue.

Different members of the family have slightly different, but overlapping, acceptor substrate specificities and they may act cooperatively in a coordinated manner, transferring GalNAc residues to different potential glycosylation sites. GalNAc-T4 and -T7 both selectively catalyse glycosylation of acceptor substrates on glycoconjugates that already exhibit partial glycosylation. These individual differences, coupled with differing patterns of expression and substrate specificity indicate that initiation of O-glycosylation, in which these enzymes are involved, is far more complex than originally thought.

Why several different GalNAc-T enzymes are required is not well understood.➲ It is assumed that each enzyme is required for correct O-glycosylation under different physiological circumstances, and absolute and preferential specificities allow for a high degree of regulation of the initiation process of O-glycosylation. An experimental knockout of a GalNAc transferase results in little obvious phenotypic effect. This suggests that the function of the gene (a GalNAc transferase displaying 93% homology with GalNAc-T1) was null, that its function was dispensable, or, perhaps most likely, that its function was compensated for by other related genes. Naturally occurring alterations in the expression of GalNAc-Ts might therefore be expected to result in subtle changes in biological behaviour. This may be of relevance in disease states such as cancer in which the O-glycosylation of mucins appears to be disrupted.

➲ The diversity of polypeptide GalNAc-Ts is unexplained at present.

Other GalNAc-Ts have been characterized, notably the blood group A-synthesizing enzyme α1,3 GalNAc-T and the β1,4 GalNAc-T that transfers GalNAc to the ganglioside GM_3 and GD_3 to form GM_2 and GD_2. These are evolutionarily unrelated to the UDP-GalNAc: polypeptide N-acetylgalactosaminyltransferase family.

2.11.3 The fucosyltransferases

There are two main families of human fucosyltransferases which are genetically quite distinct. The first consists of the two α1,2 Fuc-Ts coded for by the *FUT1* gene at the blood group *H* locus and the *FUT2* gene at the closely related *Se* locus (*see 2.7.1*). The second consists of five homologous α1,3 fucosyltransferases (α1,3 Fuc-TIII to VII) coded for by the *FUT3, FUT4, FUT5, FUT6* and *FUT7* genes, respectively, and are involved in Lewis sugar synthesis (*see 10.6.2*).

The α1,3 Fuc-T enzymes seem to have arisen through a distant gene duplication to produce, initially, two distinct genes, one located on chromosome 19 and one, *FUT4* (responsible for α1,3 Fuc-TIV) on chromosome 11. Subsequent divergence has generated two very distinct forms of the gene which share ≈50% sequence identity in the catalytic domain. More recent duplication events have resulted in a further two versions of the gene located on chromosome 19p13.3, spatially very close together, and with very high (85–90%) sequence identity in the catalytic domain (α1,3 Fuc-TIII, coded for by the gene *FUT3*; α1,3 Fuc-TV, coded for by the gene *FUT5*; and α1,3 Fuc-TVI, coded for by the gene *FUT6*). *FUT7* is located on chromosome 9, and shares only ≈38–39% homology with the other members of the family.

A number of genes with very high sequence homology to the human α1,3 Fuc-T family have been cloned from other animal species and are summarized in *Table 2.2*.

Table 2.2 Fucosyltransferases of non-human species

Species	Fucosyltransferase/genes	Comments
Chimpanzee	α1,3 Fuc-T	98% homology with human *FUT3*, *FUT5* and *FUT 7* genes
Cow	α1,3 Fuc-T	One common ancestral gene which has about 60% homology with human *FUT3*, *FUT5* and *FUT7* genes
Pig	α1,2 Fuc-T	Two genes equivalent to both human *FUT1* (H blood group antigen) and *FUT2* (secretor) loci
Rabbit	α1,2 Fuc-T *RTF-I*	80% homology with human blood group H α1,2 Fuc-T (*FUT1*)
	α1,2 Fuc-T *RTF-II* α1,2 Fuc-T *RTF-III*	*RTF-II* and *RTF-III* show high homology to each other, and may be equivalent to human *FUT2* (secretor) locus
Rat	Fuc-TIV	Equivalent to human Fuc-TIV
	α1,2 Fuc-T	Two genes equivalent to both human *FUT1* (H blood group antigen) and *FUT2* (secretor) loci
Mouse	Fuc-TIV	Equivalent to human Fuc-TIV
	Fuc-TVII	Equivalent to human Fuc-TIV
	mFuc-TIX	A novel Fuc-T which has a similar substrate specifcity to human α1,3 Fuc-TIV, but is unrelated to the human α1,3 Fuc-T family
	α1,2 Fuc-T	Two genes equivalent to both human *FUT1* (H blood group antigen) and *FUT2* (secretor) loci

2.11.4 The β1,3 galactosyltransferases

β1,3 Galactosyltransferases are responsible for synthesis of a number of key oligosaccharide structures including type 1 chains Gal(β1→3)GlcNAcβ1-R, mucin core 1 Gal(β1→3) GalNAc-R (*see 5.6.3*), the glycosphingolipids GM1 and Gal-Gb4, and the core tetrasaccharide of proteoglycans Gal(β1→3)Gal(β1→4)Xyl (*see 8.4.1*). More than 10 homologous β1,3 galactosyltransferase genes, all of which are typical type II transmembrane proteins containing a conserved DxD motif, have been described. The family includes a β1,3 N-acetylglucosaminyltransferase with substrate specificity for Gal(β1→4)GlcNAc. They may be placed in subgroups according to their structure, as listed in *Table 2.3*. The assignment of β GxT6 to -T9 is tentative as the enzymatic functions of these have yet to be confirmed. The family appears to have evolved by gene duplication. Members of this family show different enzymatic functions using different donor substrates and different acceptor sugars, but all form a common (β1→3) linkage.

The subgroup I transferases are all involved in forming type 1 chains Gal(β1→3) GlcNAcβ1→R. All transfer Gal to the C3 position of βGlcNAc-terminating substrates, but each member of the group shows slightly different substrate preferences. Several of this subgroup of enzymes are expressed mainly or exclusively in the brain. Subgroup III transferases are involved in ganglioside biosynthesis, forming the Gal(β1→3)GalNAc(β1→4)Galβ1→R structure on GA1, GM1 and GD1 (*see 9.4*).

2.11.5 The β1,4 galactosyltransferases

The β1,4 galactosyltransferase gene family currently has seven known members, designated β1,4 Gal-T1 to β1,4 Gal-T7, all with similar function. They can be grouped together in four subgroups on the basis of their structural similarities, as listed in *Table 2.4*. The family has evolved through gene duplication. β1,4 Gal-T1 to -T4 are almost identical in their genomic

Table 2.3 Subgroups of β1,3 galactosyltransferase gene family

Group	Transferases	Chromosome location	Tissue distribution
I	β1,3 Gal-T1	unknown	brain
	β1,3 Gal-T2	1q31	brain, heart
	β1,3 Gal-T3	3q25	brain, pancreas, kidney, reproductive organs
	β1,3 Gal-T5	21q22.3	epithelia of pancreas and intestine
II	β1,3 Gx-T6	12	not documented
	β1,3 Gx-T7	2p12-13	not documented
	β1,3 Gx-T8	19p	not documented
III	β Gal-T4	6p21.3	ubiquitous
IV	β Gn-T	unknown	ubiquitous
V	β Gx-T9	7p14-p13	not documented

x denotes unknown specificity.

Table 2.4 Subgroups of β1,4 galactosyltransferase gene family

Group	Transferases	Chromosome location
I	β1,4 Gal-T1	9p13
	β1,4 Gal-T2	1p32-33
II	β1,4 Gal-T3	1p23
	β1,4 Gal-T4	3p13.3
III	β1,4 Gal-T5	11
	β1,4 Gal-T6	18q
IV	β1,4 Gal-T7	5q35.1-35.3

organization with conservation of the position of five introns within the coding region. All have exclusive specificity for the donor substrate UDP-Gal, and all transfer Gal in a (β1→4) linkage onto GlcNAc, Glc or Xyl. Their expression is regulated in a tissue-dependent manner, but has not been comprehensively mapped. The family is closely related to the polypeptide GalNAc-T family (*see 2.11.2*).

2.11.6 Evolution of glycosyltransferase families

The glycosyltransferase families have arisen through exon shuffling, or by gene duplication and subsequent divergence.➲ Little homology exists between members of different families, indicating that many glycosyltransferases evolved independently. Clues to the genomic organization and protein structure of glycosyltransferases have been elucidated through research by cloning and characterization of glycosyltransferase genes and their cDNA. Sequence analysis suggests that members of each family share common motifs, for example the sialyl-motif of the sialyltransferases (*see 11.4*) involved in sugar-nucleotide recognition. For specific glycosyltransferases, mechanisms such as alternative splicing and alternative promoter usage play a role in the production of multiple isoforms from a single gene, which may differ in enzymatic properties or cellular localization.

Glycosyltransferases are either: (i) single exonic, such as α1,3 fucosyltransferase (α1,3 Fuc-T) and β1,2 N-acetylglucosaminyltransferase (β1,2 GlcNAc-TI) in which the entire coding regions fall within a single exon; or (b) multi-exonic, such as α2,6 sialyltransferase (α2,6 ST), β1,4 galactosyltransferase (β1,4 Gal-T) and α1,3 galactosyltransferase (α1,3 Gal-T) in which the coding region is distributed over several exons.

➲ Shared structural features of members of glycosyltransferase families include:

1. Sequence identity usually limited to a few motifs in the catalytic domain.
2. Conservation of isolated charged and hydrophobic residues between conserved motifs.
3. Conservation of cysteine residues.

➲ Common structural features of glycosyltransferases are:
- type II transmembrane topology
- N-terminal hydrophobic signal anchor sequence
- DxD motif involved in binding to metal-ion cofactor and donor substrate.

2.12 The molecular structure and organization of glycosyltransferases

Most glycosyltransferases are intrinsic membrane proteins, firmly bound residents of specific parts of the ER, Golgi apparatus and secretory pathway where they are present in the membranes of vesicles travelling from ER to the *cis* face of the Golgi apparatus, the cisternae of the Golgi apparatus itself, the *trans*-Golgi network and the membrane bound structures distal to the Golgi apparatus. ➲ In spite of differences in the primary sequence of glycosyltransferases from different families, almost all share a common tertiary structure.

2.12.1 Tertiary structure

The general structure of glycosyltransferases is illustrated in *Figure 2.3*. They are type II transmembrane proteins, with a single transmembrane domain, which is a hydrophobic sequence of 16–19 amino acids containing a proline-rich sequence and of sufficient length to span the lipid bilayer. It operates as a non-cleavable signal-anchor sequence, and contains Golgi targeting or retention signals. It is flanked by a short, 4–27 amino acid, cytoplasmic, N-terminal domain at one side of the membrane, and a longer, generally >325 amino acid residues, C-terminal domain which protrudes into the lumen of the ER–Golgi pathway. This is made up of a 'stem' region of at least 35–37 amino acid residues which act as a flexible tether, to the large, often compact, globular catalytic domain. The positioning of the C-terminal domain, protruding into the lumen, facilitates its interaction with the growing glycan chain attached to glycoproteins and glycolipids during their transit through the secretory pathway. The catalytic and stem regions of the molecule are themselves subject to posttranslational modifications such as glycosylation and phosphorylation.

2.12.2 Localization

Glycosyltransferases are situated in positions along the secretory pathway that correspond roughly to the order in which they are required to act during glycan synthesis (*see 2.9*). Thus, glycosyltransferases involved in the early parts of glycan synthesis are found in the *cis* and *medial* compartments of the Golgi apparatus, whereas those involved in the later stages of glycan synthesis are found in the *trans*-Golgi. The means by which these enzymes are

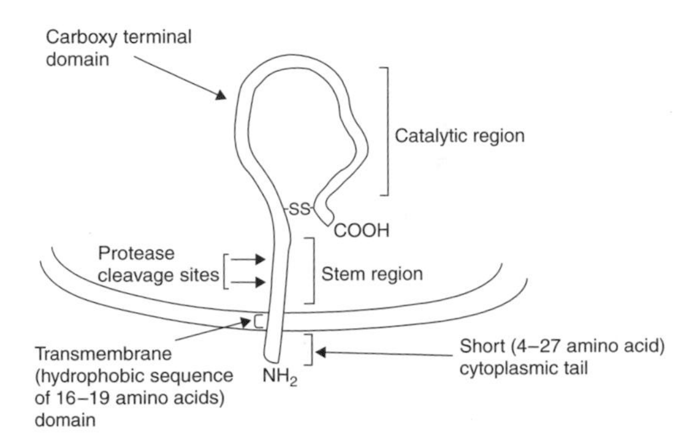

Figure 2.3

Schematic representation of the general structure of a glycosyltransferase enzyme.

directed to their appropriate location within the secretory pathway, and, furthermore, retained there (as opposed to being delivered to the plasma membrane or other parts of the cell as occurs to other transmembrane proteins in the Golgi apparatus), remains poorly understood. ➲

➲ Several theories exist to explain how the glycosyltransferases are located exactly where they are functionally required:
1. The existence of unidentified specific targeting signals has been postulated.
2. The attraction of transferase molecules to parts of the membrane with appropriate thickness and stiffness has been implied.
3. The oligomerization or 'kin recognition' hypothesis postulating formation of large multimeric enzyme complexes by association of their stem regions has been proposed.

2.12.2.1 Targeting signals

The complete lack of homology between sequences of glycosyltransferases from different families makes the concept of a common, linear, amino acid Golgi apparatus retention signal (analogous to the K(H)DEL for ER localization and mannose-6-phosphate for lysosomal targeting) unlikely. There is, however, evidence that some glycosyltransferases, at least, contain sequences in or around their transmembrane region that are involved in location and retention. It is possible that each enzyme carries a unique retention signal, and that there is a corresponding unique membrane receptor for each signal.

2.12.3 Membrane thickness and stiffness

Differing concentrations of cholesterol incorporated into the lipid bilayer in different parts of the secretory pathway result in membranes of differing thickness and stiffness. Membranes get thicker from the ER to the plasma membrane because of higher concentrations of cholesterol, and there is a dramatic gradient from the ER through the Golgi apparatus. It has been proposed that glycosyltransferases find their way to parts of the membrane, lipid microdomains or rafts, that have the correct characteristics for their particular transmembrane segment. Furthermore, the short transmembrane domain of these molecules prevents them from becoming incorporated into cholesterol-rich transport vesicles destined for the plasma membrane.

2.12.4 Kin recognition

There is also evidence that glycosyltransferases that are cooperatively involved in specific parts of the glycobiosynthetic pathways aggregate together through specific interactions between their stem or transmembrane regions (so-called oligomerization or 'kin recognition') to form homo- or hetero-oligomeric complexes that are too large to enter transport vesicles, and are thus retained. Further interactions between cytosolic components of the enzyme complex with Golgi apparatus matrix proteins may additionally anchor these multimeric complexes in place. There have also been suggestions that the pH gradient from the ER through the Golgi apparatus may play a role in the correct localization of glycosyltransferase enzymes, although pH is difficult to assess within a lipid bilayer.

2.13 Cell-surface and secreted forms of glycosyltransferases

Structurally distinct forms of the glycosyltransferases are frequently secreted by cells, sometimes in response to a particular condition such as inflammation, and are detectable, sometimes in quite high concentrations, in body fluids and serum. Their significance remains unknown, as the concentration of donor sugar-nucleotides is too low for effective catalysis of transglycosylation reactions in this environment; it is presumed that their function must be related to their sugar-binding properties, and that they are acting as lectin-like molecules (*Chapter 13*). ➲

➲ Soluble, secreted forms of glycosyltransferases also exist, but their functional significance is poorly understood.

These soluble forms of the enzyme are derived from the corresponding membrane-bound enzymes, but have been cleaved (possibly by the specific action of cathepsin-like proteases in the *trans*-Golgi) within the 'stem' region of the molecule, to release the catalytically active domain, lacking the N-terminal sequence, the transmembrane sequence and part of the 'stem' region. Glycosyltransferases have also been reported to occur at the cell surface where they may play a role in intercellular recognition and/or cell–cell adhesion.

2.14 Regulation of glycosyltransferase expression and activity

Glycosyltransferases act in an ordered and coordinated fashion, and their cellular levels of expression, subcellular localization, substrate specificity, availability of sugar-donor and acceptor, and transit time through Golgi apparatus compartments all determine the glycosylation potential and activity of a cell. Their expression and activity are regulated by the subtle interplay of a number of mechanisms, summarized in *Table 2.5* and described in more detail below.

Glycosyltransferases can compete for a substrate, and the action of one may preclude the action of another. Their expression appears to be predominantly regulated at the level of transcription, but notable exceptions have been described and differential processing of mRNA, alterations in stability of mRNA and differences in the efficiency of translation may also influence control on final enzyme activity. The expression of glycosyltransferases is tissue specific and highly regulated during differentiation and sometimes also cell proliferation. It also appears likely that different substrates may compete for available transferases.

Many glycosyltransferases are constitutively expressed at a low level in all mammalian somatic cells, presumably to provide the everyday machinery of glycosylation for the biosynthesis of cellular glycoconjugates. Cells may also upregulate the expression of specific glycosyltransferases, in response to changes in conditions, to supply a need for higher levels of expression of glycosylated molecules. This seems to be achieved in a number of ways.

2.14.1 Multiple start sites

➲ Many glycosyltransferase genes have multiple transcriptional start sites, suggestive of multiple promoters. The β1,4 Gal-T gene is an example of a glycosyltransferase gene under control of at least two promoters.

Many of the glycosyltransferase genes have multiple transcriptional start sites, suggesting that they have multiple promoters. This is an unusual mechanism requiring special note.➲

Low levels of constitutive expression of the β1,4 galactosyltransferase (β1,4 Gal-T) gene in most cells and tissues is consistent with its role in the assembly of N- and O-linked glycans (*see 4.2*) and glycolipids (*see 9.4*). In the lactating mammary gland, however, levels of expression are increased several fold. Typically, somatic cells carry a single copy of the β1,4 Gal-T gene, but the gene has two different start sites which produce two different transcripts, a 3.9 kb transcript and a longer 4.1 kb form. Two different isoenzymes exist, which differ in primary amino acid sequence by 13 amino acids and are located near the N-terminus, and these exhibit similar

Table 2.5 Methods by which glycosyltransferase activity is regulated

Isoenzymes with different levels of activity or substrate preference, produced by:
use of multiple transcriptional start sites and promoters
alternative splicing
gene duplication
multiple gene copies with different regulatory regions
Availability of sugar donors
Contact time of transferase with substrate
pH
Competition for substrate/sugar donor between different transferases and competition between substrate molecules for available transferase

properties. In somatic cells with low requirements for β1,4 Gal-T, the 4.1 kb form predominates, and in cells with higher requirements, for example, lactating mammary epithelium, the 3.9 kb form is preferred. As the latter is more efficient, being under the regulation of a different promoter, this switch leads to 10× higher levels of mRNA being synthesized.

Expression of β1,4 Gal-T by male germ cells may be under the control of yet another promoter. Using this mechanism, different enzyme isoforms can be generated from a single gene. There is also evidence accumulating that alternative splicing mechanisms are important in generating different enzyme isoforms. In glycosyltransferases coded for by a single exon, gene duplication is a more common method of generating related enzymes with slightly different enzymatic properties.

2.14.2 Alternative splicing mechanisms

A single copy of a glycosyltransferase gene may yield forms of an enzyme with different activity or specificity through alternative splicing. An example of this is the mouse α1,3 galactosyltransferase (α1,3 Gal-T) gene in which alternative splicing of the mRNA results in four different forms of the enzyme molecule that differ in the length of their transmembrane stem region. Generally speaking, there is a direct correlation between the level of mRNA expression and the final glycosyltransferase enzyme activity. However, this does not always seem to hold true and the stability of mRNA, and the rate with which it is degraded *in vivo*, will also influence cellular glycosyltransferase activity, as will the efficiency of translation.

2.14.3 Multiple gene copies with different regulatory regions

An alternative mechanism for regulating cellular glycosyltransferase enzyme activity is the existence of multiple copies of a gene linked to different regulatory regions. This is the case with the family of at least five α1,3 fucosyltransferase (α1,3 Fuc-T) genes (designated α1,3 Fuc-TIII, IV, V VI and VII; *see 2.11.3*), which share considerable sequence homology, and which are involved in the synthesis of the Lewis family of blood group sugars (*see 10.6.2*).

These enzymes exhibit a variable region in their catalytic domain that provides some differences in selectivity of acceptor substrate binding. The most obvious example of this is that of α1,3 Fuc-TIII (sometimes called the Lewis enzyme) which can incorporate Fuc in either a (α1→3) linkage to Gal(β1→4)GlcNAc to make Lewis x, or in a (α1→4) linkage to Gal(β1→3)GlcNAc to make Lewis a. The other is the α1,3 Fuc-T enzyme which can only produce a (α1→3) linkage. The expression of the different forms of α1,3 Fuc-T is tissue specific and also developmentally regulated.

2.14.4 Other regulatory factors

The relative activity of glycosyltransferase enzymes is also controlled by a number of factors, including the levels of donor sugars available, length of time that glycoconjugates are available for glycosylation as they pass through the secretory pathway, intraluminal pH, and competition for sugar donors and acceptors between different glycosyltransferases. Rapid transit of glycoconjugates through the secretory pathway may, for example, limit the amount of time available for some glycosylation reactions to occur. The internal pH of the secretory pathway decreases from the ER to the *trans*-Golgi network, probably owing to the action of one or more ATP-dependent proton pumps, and this probably affects the relative activity of different enzymes which have different pH optima.

2.15 Evidence for the functions of glycans derived from glycosyltransferase mutants

Mutations that alter the function of glycosyltransferases provide an illuminating approach to identifying functions for glycans; both intramolecular functions that rely on the physical or biochemical properties of the glycoconjugate itself and intermolecular functions that involve saccharide–lectin interactions. The consequences of such mutations cover a whole spectrum of effects, from no discernible effect at all to, at its most extreme, death of the cell or organism (*see 15.10*).

2.15.1 Effect of impaired N- and O-glycosylation in cultured cells and on the whole organism

This topic is described very briefly here as it is covered in more depth in *Chapter 15*. Mammalian cells in culture require only simple oligomannosyl N-linked structures for survival. Drastic changes in peripheral sugars of N- and O-linked glycoproteins, GPI anchors, glycolipids and glycosaminoglycans often have relatively little effect on cell viability or growth characteristics, implying that the functions of complex glycosylation lie in processes associated with the whole organism.

Thus, mutation in the N-acetylglucosaminyltransferase I (GlcNAc-T1), for example, results in significantly abnormal N-linked glycoprotein biosynthesis, with N-linked sites having the oligomannosyl structure ($Man_5GlcNAc_2Asn$) instead of complex or hybrid oligosaccharides.➲ This drastic change has surprisingly little effect on the viability or growth characteristics of the mutant cells in culture, and the abnormal glycoproteins exhibit correct folding, exit from the Golgi apparatus and compartmentalization. The function of the abnormally glycosylated glycoproteins is, however, often impaired. Plants lacking functional GlcNAc-TI suffer no obvious abnormalities, whereas the effect in mammals is more drastic – GlcNAc-TI knockout mice die *in utero* (*see 15.10.2.1*).

➲ In cultured cells, drastic alterations in glycosylation have little effect on viability, whereas in the whole organism, very subtle changes in glycosylation can have dramatic effects. Lack of GlcNAc-T1 has much more drastic consequences in the whole organism, resulting in abnormal development and death *in utero*.

2.15.2 Gain-of-function mutants

Gain-of-function mutants also provide an insight into the roles of particular types of glycosylation. Addition of a single monosaccharide residue to oligosaccharides carrying appropriate acceptors may have dramatic effects. For example, the over-expression of GlcNAc-TIII, which transfers bisecting GlcNAc onto N-glycans, renders cells resistant to the Gal-binding lectin, ricin. This may be analogous to the natural situation in which cells may generate or cease to generate ligands for lectin-like molecules, such as selectins, by specific upregulation of expression of glycosyltransferases (*see 16.3*).

Similar results may be obtained by the expression cloning of specific glycosyltransferase genes. This is a rapidly expanding field, and it is envisioned that much progress into our understanding of the functions of glycans, through disruption in the expression or activity of the glycosyltransferase enzymes involved in their synthesis, will be achieved in the near future (*Chapter 15*).

2.15.3 Glycosyltransferase deficiency syndromes

In addition to experimental mutant cell lines and animals there are a number of rare, naturally occurring glycosyltransferase-deficiency syndromes in man, such as I-cell disease, hereditary erythroblastic multinuclearity with a positive acidified serum lysis test (HEMPAS), the carbohydrate-deficient glycoprotein syndromes and others, and these provide an insight into the functions of oligosaccharides on glycoconjugates (*see 16.8*).

Further reading

Amado, M., Almeida, R., Schwientek, T., Clausen, H. (1999) Review: identification of large galactosyltransferase gene families: galactosyltransferases for all functions. *Biochim. Biophys. Acta* **1473:** 35–53.
Colley, K.J. (1997) Mini review: Golgi localization of glycosyltransferases; more questions than answers. *Glycobiology* **7:** 1–13.
Munro, S. (1998) Review: localization of proteins to the Golgi apparatus. *Trends Cell Biol.* **8:** 1–15.
Sears, P., Wong, C.-H. (1998) Enzyme action in glycoprotein synthesis. *Cell. Mol. Life Sci.* **54:** 223–252.
van den Eijnden, D.H., Joziasse, D.H. (1993) Enzymes associated with glycosylation. *Curr. Opin. Struct. Biol.* **3:** 711–721.
Varki, A. (1998) Factors controlling the glycosylation potential of the Golgi apparatus. *Trends Cell Biol.* **8:** 34–40.
www.expasy.org/enzyme/

Polysaccharides

3

3.1 Introduction – how are polysaccharides defined?

Polysaccharides are polymers of monosaccharide units linked by glycosidic bonds into linear or branching chains. Branching is possible because of the diversity in glycosidic bonds that can be formed to link any of the hydroxyl groups of the component monosaccharides (*Chapter 1*). The two main patterns of branching form either comb- or tree-like structures (*Figure 3.1*). There is no rigorously defined division between oligosaccharides and polysaccharides (*see 1.16.3*), but polysaccharides are usually accepted to contain more than 10–20 monosaccharide residues, and in practice are usually much larger than this, typically comprising ≈100 monosaccharide units or more, with cellulose comprising ≈3000 monosaccharide units. Connective tissue polysaccharides which contain one or more amino sugar moieties are usually referred to as glycosaminoglycans, and are considered separately in *Chapter 8*.

3.1.1 Homopolysaccharides and heteropolysaccharides

Polysaccharides are either homopolysaccharides, by far the most common form, composed of repeating units of one type of monosaccharide, or heteropolysaccharides, composed of more than one type of monosaccharide. ➲

➲ *Types of polysaccharide*
(1) homopolysaccharide
(2) heteropolysaccharide
(a) those in which component polysaccharide chains are themselves homopolysaccharides
(b) those in which component polysaccharide chains are themselves heteropolysaccharides

Homopolysaccharides may be classified according to the identity of the monosaccharide of which they are composed, for example, galactans are polymers of galactose, mannans are polymers of mannose, and so on.

Heteropolysaccharides may, in principle, be formed of heterogeneous sequences of monosaccharides (in a way similar to that in which amino acids make up the polypeptide chains of proteins), but in practice they tend to be composed of a limited number of monosaccharide units (no more than about six) ordered in repetitive sequences. Most heteropolysaccharides are composed of two (diheteroglycan) or three (triheteroglycan) different monosaccharide units. Furthermore, heteropolysaccharides are usually either: (i) simple heteropolysaccharides, in which each polysaccharide chain is actually a homopolysaccharide, but two or more different homopolysaccharide chains are present; and (ii) complex heteropolysaccharides in which the polysaccharide chain contains two or more different types of monosaccharide. This is illustrated in *Figure 3.2*.

Heteropolysaccharides are named on the basis of the monosaccharide units of which they are composed. For example, glucomannans are composed of glucose and mannose residues, and arabinoxylans are composed of arabinose and xylose. In simple heteropolysaccharides, it is the convention that the last part of the name (e.g. 'xylose' in arabinoxylose), refers to the principle, core polysaccharide chain, and the first part of the name ('arabino' in arabinoxylose), refers to the composition of the side chains attached to it. In complex heteropolysaccharides,

Functional and Molecular Glycobiology, Susan A. Brooks, Miriam V. Dwek and Udo Schumacher

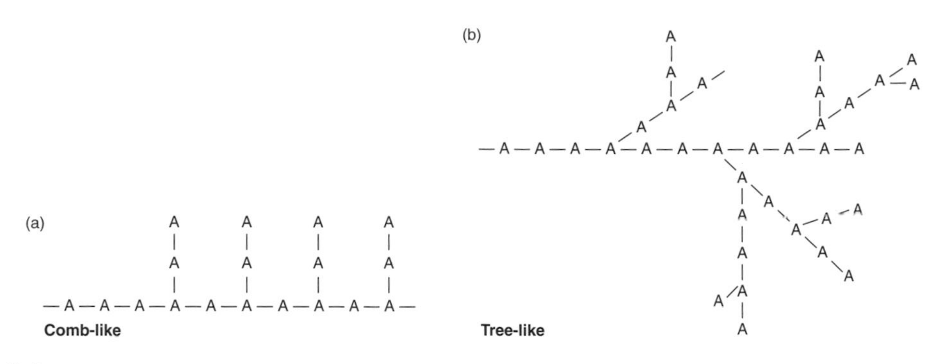

Figure 3.1

The two main patterns of branching in polysaccharides (a) comb-like and (b) tree-like (A represents any monosaccharide residue).

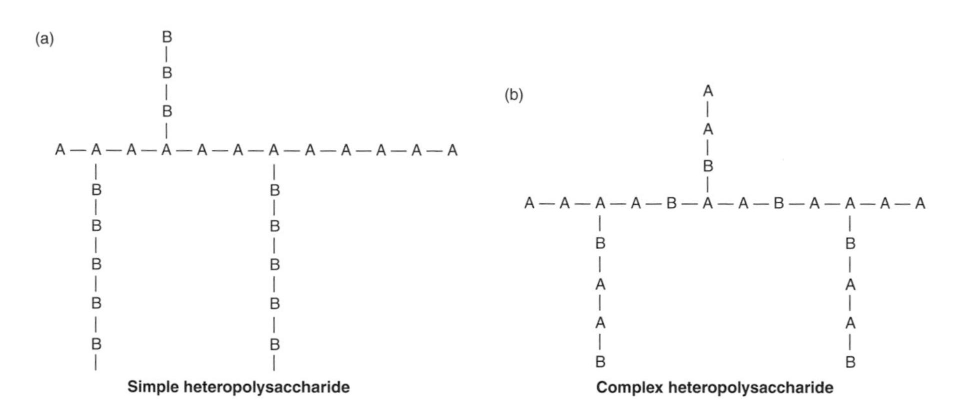

Figure 3.2

Schematic diagram illustrating two different types of heteropolysaccharide: (a) simple heteropolysaccharide: here each polysaccharide chain is actually a homopolysaccharide (composed of only a single monosaccharide residue, indicated by A or B), but two or more different homopolysaccharide chains are present; (b) a complex heteropolysaccharide: here the polysaccharide chain contains a mixture of two different types of monosaccharide (A and B).

Table 3.1 Monosaccharides present in naturally occurring polysaccharides

Pentoses	Hexoses	Modified monosaccharides
D-xylose	D-glucose	D-apiose
L-arabinose	D-mannose	D-mannuronic acid
D-arabinose	D-fructose	D-galactosamine
	D-galactose	D-galacturonic acid
	L-galactose	D-glucosamine
	D-idose	D-glucuronic acid
	D-altrose	L-guluronic acid
	L-fucose	L-iduronic acid
	L-rhamnose	

Table 3.2 Some examples of naturally occurring polysaccharides

Homopolysaccharides

Common name	Monosaccharide	Linkages	Sources
cellulose	glucose	(β1→4)	cell walls of higher plants
amylose	glucose	(α1→4)	starch of higher plants
amylopectin	glucose	(α1→4) and (α1→6)	starch of higher plants
glycogen	glucose	(α1→4) and (α1→6)	microorganisms, animals
dextran	glucose	(α1→6), (α1→3) and others	bacteria
inulin	fructose	(β2→1)	plants including Jerusalem artichokes and dahlias
carrageenan	galactose	(β1→3) and (α1→4)	red seaweeds
fucoidin	fucose	(α1→2) and (α1→4)	brown seaweeds
chitin	N-acetylglucosamine	(β1→4)	insects, crustaceans, spiders, fungi

Heteropolysaccharides

Name	Monosaccharides	Chain type	Sources
arabinoxylan	arabinose, xylose	branched	plant cell walls
glucuronoxylan	glucuronic acid, xylose	branched	plant cell walls
arabinogalactan	arabinose, galactose	branched	coniferous trees
glucomannans	glucose, mannose	linear	seeds, bulbs, some trees
galactomannans	galactose, mannose	branched	seeds of legumes
guluronomannuronan	guluronic acid, mannuronic acid	linear	brown seaweed alginate

it is usual to arrange the monosaccharide names in alphabetical order, for example, arabinogalactoglucuronorhamnoglycan. Only a limited number of monosaccharides are ever present in polysaccharides, and these are listed in *Table 3.1*. Many polysaccharides are commonly referred to by long-standing, traditional names, such as chitin, cellulose, inulin, amylopectin and so on. Some examples of naturally occurring homo- and heteropolysaccharides are listed in *Table 3.2*.

3.1.2 Heterogeneity in composition and structure

The same polysaccharide from different sources may differ in its composition and structure. An example is starch derived from different plants. To clarify matters, it is necessary to indicate the source in addition to the polysaccharide, for example, corn starch or yeast mannan. ➲

Larger polysaccharides are polymolecular, that is they tend not to have a defined molecular mass, as, unlike proteins they are not synthesized from a genetic template. In addition, even in polysaccharides with a highly regular structure, such as cellulose, there are occasional anomalies in their structure, resulting in ≈ 1 in 700 irregular monosaccharide linkages.

➲ Polysaccharides exhibit a high degree of heterogeneity:
- they do not have a defined molecular mass
- polysaccharides from different sources may differ markedly in structure and composition, particularly in chain length.

Table 3.3 Examples of polysaccharides and their functions

	Animals	Plants	Insects	Bacteria	Algae
Energy storage	Glycogen	Starch			Inulin Laminaran
Structural/protective		Cellulose Hemicelluloses Pectic substances Gums Mucilages Callose	Chitin	Levans Dextrins	Cargeenan Agar Alginates

3.1.3 Functions of polysaccharides

➲ The principle functions of polysaccharides are:
- structural support
- protection
- energy storage.

Polysaccharides are produced by virtually all living organisms, where their functions are either energy storage, structural or protective.➲ Glycogen (*see 3.4.3*), starch (*see 3.4.1*), fructans (*see 3.4.2*) and galactomannans (*see 3.3.1*) are important energy reserves. Common polysaccharides with an obvious structural function include cellulose (*see 3.2.1*) and chitin (*see 3.2.3*). Microorganisms and the cells of higher organisms all require mechanical support and protection. In the case of bacteria, this is provided by a thick polysaccharide-rich cell wall which lies outside the cell membrane. Polysaccharides similarly play a very dominant role in the mechanical strength and stability of the plant cell wall. In both cases, the principle is that of mechanically strong, linear threads or fibres embedded or supported by a matrix of more amorphous material. Plant gum polysaccharides and 'callose' act to seal wounds and protect the organism that produces them (*see 3.3*). Some biologically important polysaccharides are described in detail below, and their functions summarized in *Table 3.3*.

3.2 Structural polysaccharides

Structural polysaccharides protect and support unicellular organisms, such as bacteria, and also higher plants. They are often complexed with other materials, such as proteins, and may rely on interactions with other components such as calcium ions. The most usual arrangement is for polysaccharide fibres to be embedded in a supportive matrix.➲ The plant cell wall, for example, comprises a three-dimensional network of cellulose microfibrils and cross-linking hemicellulose polysaccharides embedded in a gel-like matrix rich in pectins.

➲ Structural polysaccharide fibres are usually embedded in a supportive matrix.

3.2.1 Cellulose

Cellulose is the most abundant naturally occurring organic substance. Over half the carbon in the biosphere is incorporated into cellulose at any one time. Cellulose is the principle structural component of the rigid wall of higher plant cells, forms the stiff outer casing of the marine invertebrates tunicates, and is produced by algae and a limited number of bacteria. Its structure was first elucidated in 1923. A subject of intense research activity, a number of approaches has been used to determine its structure, including single-crystal X-ray studies, infrared spectroscopy, high-resolution nuclear magnetic resonance (NMR), dark-field electron microscopy and diffraction electron microscopy.

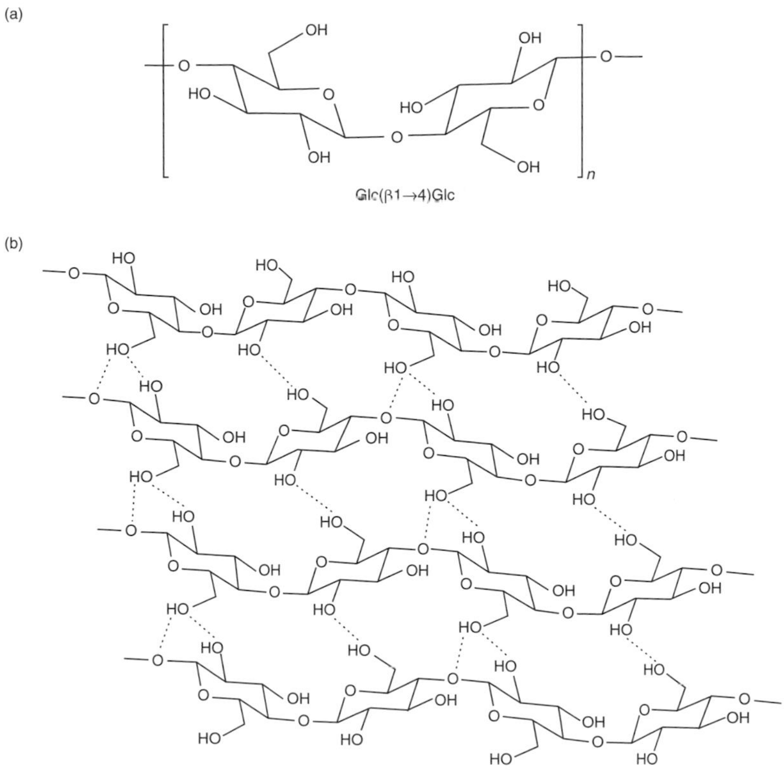

Figure 3.3

Schematic diagram illustrating the structure of cellulose: (a) the molecule is composed of chains of (β1→4)-linked Glc molecules; (b) the (β1→4) linkages mean that each residue is at an angle in relation to its neighbours, giving a bent, ribbon-like formation. The sheets stack vertically, such that they are staggered by half the length of a Glc residue, to form a cellulose fibre. Hydrogen bonding (indicated by dotted lines) between the cellulose glucan chains help to stabilise and form the cellulose fibres.

3.2.1.1 Structure and biosynthesis

Cellulose is a linear homopolysaccharide composed of ≈2500–15,000 repeating (β1→4)-linked D-glucopyranose residues, although there is evidence that other monosaccharides, for example, mannose, may sometimes be integrated into the polysaccharide chain. The biosynthesis of the molecule is complex and in spite of intense research activity over many years, is still not completely understood. There is also some controversy over precisely how the polysaccharide chains are arranged to form cellulose fibres, but it is thought that the glucan chains, which are of a bent, ribbon-like conformation, line up laterally to form sheets with a width of ≈100–250 Å and a thickness of ≈30 Å, which in turn stack vertically to form a cellulose fibre of 35–100 Å in diameter. Hydrogen bonding between the cellulose chains helps to stabilize and form the cellulose fibres. The entire structure consists of ≈40 parallel glucan chains held together by intrachain hydrogen bonds, although structural variations in this model have been proposed. The basic structure of cellulose is illustrated in *Figure 3.3*.

A distinction between at least two structural variations of cellulose is recognized: native cellulose, or cellulose I, is composed of glucan chains arranged in sheets held together by hydrogen bonds, the sheets themselves are stacked into fibres held by van der Waal's forces alone; in cellulose II, or regenerated cellulose, anti-parallel packing of the chains allows stronger, intersheet hydrogen bonding. The structure and organization of cellulose results in a structural material of remarkable strength, as exemplified by wood, used in many of our buildings, or by the resistance of large trees to the potentially devastating effects of storms and high winds.

3.2.1.2 Integration of cellulose into plant cell wall structure

X-Ray diffraction studies and electron microscopy have been crucial in advancing our understanding of plant cell wall structure and organization. Cellulose is usually integrated into a matrix of other polysaccharides (e.g. hemicelluloses and pectic substances; *see 3.2.2*), water, glycoproteins and lignin.

3.2.1.3 Practical and economic importance of cellulose

Cellulose forms the basis of many everyday products including cotton, jute, sisal, coconut matting, hemp, films, sheeting, fibres and wood pulp. Bamboo, papyrus and wood owe their usefulness to their high cellulose content.

3.2.1.4 Cellulose as a foodstuff

The structure of cellulose makes it a challenging foodstuff and living organisms that rely upon it as a source of nutrients have adapted to digesting it. Vertebrates do not produce the enzymes, cellulases, required to hydrolyse the (β1→4) linkages of cellulose, but the digestive tracts of ruminants contain symbiotic microorganisms that do. The densely packed structure of the cellulose fibres makes it resistant to digestion, ruminant animals, such as cattle, have multi-chambered stomachs and chew their cud in order to facilitate the prolonged digestive process. Similarly, the digestive tracts of termites harbour microorganisms that are able to cleave the β(1→4) linkages, enabling termites to use cellulose as a foodstuff. One convenient way of destroying termites is to kill these bacteria; the termites starve to death.

3.2.2 Non-cellulose polysaccharide structural components of plant cell walls – hemicelluloses and pectic substances

The tough walls of most higher plant cells are composed of large amounts of cellulose fibres embedded in a continuous matrix of lignin, pectic substances and, predominantly, hemicelluloses. These substances are all in close association, physically entangled and supported by van der Waal's forces and hydrogen bonds.

Plant growth is accompanied by a loosening of the hydrogen bonds that bind xyloglucan hemicelluloses to cellulose microfibrils and enzyme-mediated cleavage of the xyloglucan chain. A number of enzymes are involved in this process, including xyloglucan-endotransglycosidases, endo-β-glucanases and the so-called expansins which disrupt hydrogen bonds and steric interactions between cellulose and other polysaccharides. ➲

> ➲ Polysaccharides that support plant structure must be disrupted and re-formed to allow plant growth.

Other plant polysaccharides include gums and mucilages (*see 3.3*). The term hemicellulose was first suggested in 1891, at which time the erroneous idea that these substances were an intermediate in cellulose biosynthesis was current.

3.2.2.1 Structure

Most hemicelluloses are relatively small (typically 50–200 monosaccharide residues) hetero-polysaccharides containing typically two to four, but sometimes as many as six types of monosaccharides. The most common monosaccharides are neutral sugars such as the hexoses Glc, Man and Gal and the pentoses Xyl, and Ara. Uronic acids, such as glucuronic acid and galacturonic acid may also be part of the structure. The cell walls of monocots are especially rich in arabinoxylans and the dicots are especially rich in xyloglucans. Most hemicelluloses are branched, and many are partially acetylated.➲

➲ Hemicelluloses are small, usually branching heteropolysaccharides, comprising between two and six different monosaccharide building blocks. Principle hemicelluloses in plants include xyloglucans and xylans.

3.2.2.2 Xyloglucans and xylans

Important groups of hemicelluloses include xyloglucans and xylans which are present in all parts of land plants. D-Xyloglucans (*Figure 3.4*) are composed of a (β1→4)-linked D-glucopyranose backbone to which single xylopyranoses are (α1→6) linked. Both Fuc and Gal residues can be attached via (β1→2) bonds in some types of xyloglucan.

D-Xylans (*Figure 3.5*) are based on a (β1→4)-linked D-xylanopyranosyl core chain composed of, usually, over 200 residues. The principle side chains consist of (α1→2) 4-O-methyl-D-glucopyranosyl uronic acids, but (α1→3) L-arabinofuranosyl residues may also occur. They

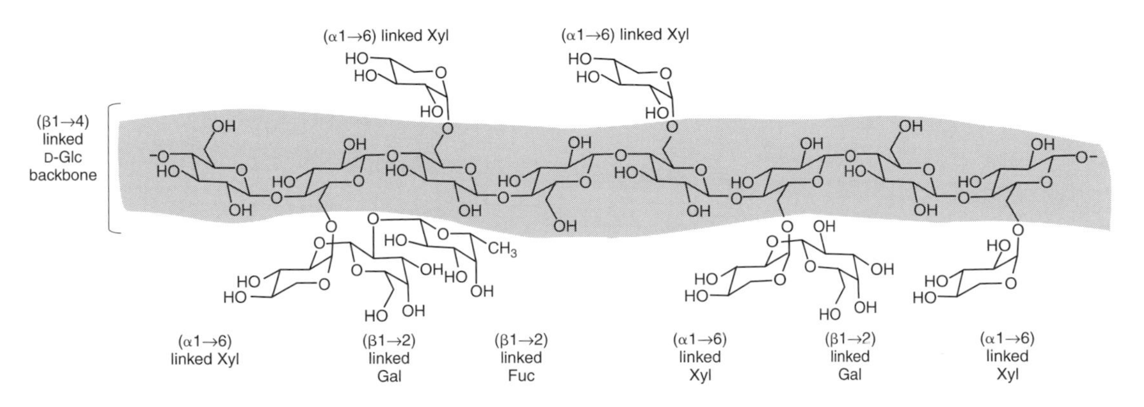

Figure 3.4

The structure of D-xyloglucans. The molecule consists of a backbone of (β1→4)-linked glucopyranose residues (shaded), to which single (α1→6) xylopyranose residues are linked. Both (β1→2)-linked Fuc and Gal residues may also be present.

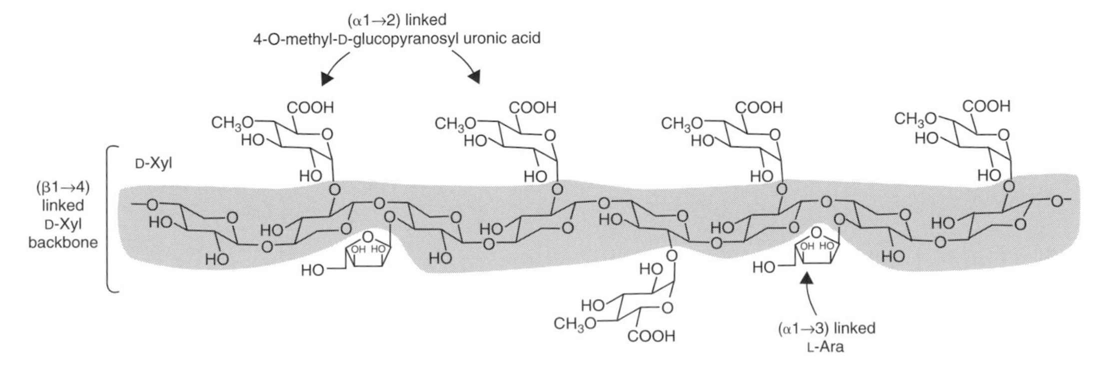

Figure 3.5

The structure of D-xylans. The molecule consists of a backbone of (β1→4)-linked xylopyranose residues (shaded). Side chains are of (α1→2)-linked 4-O-methyl-D-glucopyranosyl uronic acids, and sometimes (α1→3)-linked L-arabinofuranosyl residues.

are amorphous in structure and do not form fibres. Homopolysaccharide D-xylans are unusually found in Esparto grass.

3.2.2.3 Mannans

Mannans (*Figure 3.6*) form similar microfibrils to cellulose, and consist of (β1→4)-linked D-mannopyranosyl units in chains of varying length. Polysaccharides containing predominantly D-mannan homopolysaccharide chains are the principle storage carbohydrate of palm seed endosperm, and have also been isolated from green coffee beans and the tubers of some orchids. The mannan from the ivory nut is one of the best known; it resembles cellulose in its structure and arrangement of chains, and is often referred to as 'vegetable ivory'.

3.2.2.4 Arabinogalactans

Arabinogalactans (*Figure 3.7*) have a backbone of (β1→3)-linked D-galactopyranosyl units to which β-D-galactopyranosyl-(1→6)-D-galactopyranose disaccharides are linked via (β1→6) bonds. A (α1→6)-linked disaccharide (L-arabinofuranosyl-(β1→3)-L-arabinofuranose) can also be linked to the galactan backbone. These molecules are highly branched and, unusually, water soluble. They are present in conifer wood. Some types are of commercial importance in that they are easily extracted and useful as gums (*see 7.4.5.3*).

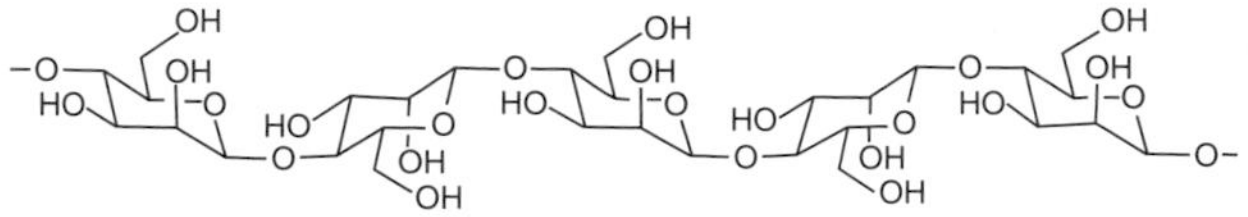

Figure 3.6

The structure of a (β1→4)-linked D-mannan chain.

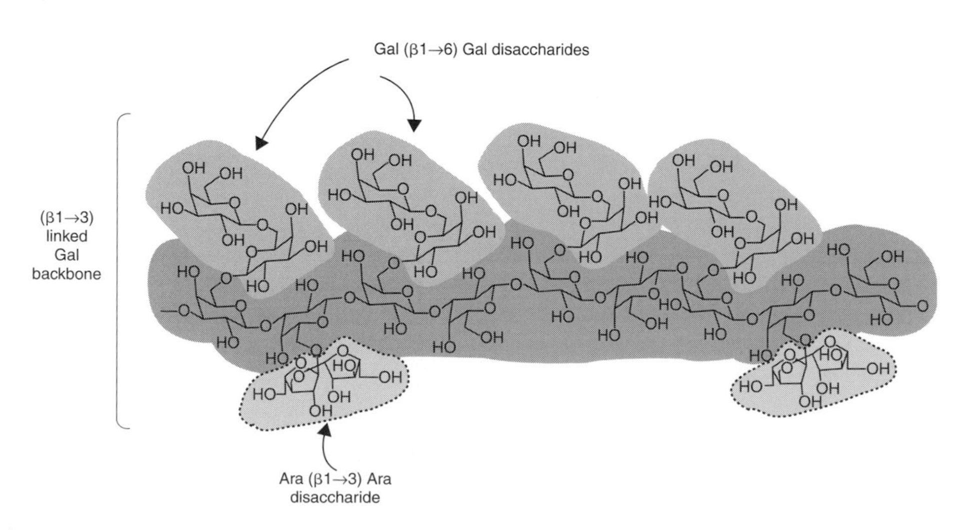

Figure 3.7

The structure of arabinogalactans. The molecule has a backbone of (β1→3)-linked D-galactopyranosyl units (shown as dark shaded area) to which a β-D-galactopyranosyl-(1→6)-D-galactopyranose disaccharides (shown as lightly shaded area) are linked by (β1→6) linkages. (α1→6)-Linked β-L-arabinofuranosyl(1→3)-L-arabinofuranose (shown as dotted outline) can also be attached.

3.2.2.5 Pectic substances

➲ Pectic substances are rhamnogalacturonans and form sticky viscous solutions.

Pectic substances are also important structural components of plant cell walls. They are very diverse and, as a result of being possibly the most complex polysaccharides known in terms of chemistry and biosynthesis, their structure has been difficult to determine. The terms pectin, pectic acids and pectinic acids are frequently used to describe different types of pectic substances, but although in common usage they are not correct terms.➲ Pectic substances are actually a large family of complex, substituted rhamnogalacturonans (*Figure 3.8*). They have a backbone of long sequences of (α1→4)-linked D-galactopyranosyl uronic acid residues, in which a small number of (α1→2)-linked rhamnopyranosyl residues occur. Side branches containing D-xylose, L-fucose, D-glucuronic acid and other monosaccharides are common. They form sticky, viscous solutions (responsible for the setting of jams, for example) which cement adjacent cells together. They are much used as gelling and thickening agents in the food, pharmaceutical and other industries.

3.2.3 Chitin

➲ Chitin is a glucan homopolysaccharide and is the principle structural material of insects and other invertebrates.

Chitin is the principle structural polysaccharide of the exoskeleton of invertebrates such as arthropods, annelids and molluscs and also of many fungi, mycelial yeasts and green algae.➲ Chitin may also occur as a proteoglycan in some animals and fungi. It has been identified in 25 million-year-old fossilized insects. The chitinous exoskeleton of arthropods typically consist of only ≈30% chitin complexed with protein and inorganic salts such as calcium carbonate. Similarly, fungal chitin is found to be associated with other structural materials. Chitin is the second most abundant organic compound in nature (after cellulose). In the oceans, the moulting of cuticles and the death of marine organisms results in a continuous fall of chitinous material to the ocean floor, known as 'marine snow'.

3.2.3.1 Structure

Structurally, chitin is extremely similar to cellulose (*see 3.2.1*), but instead of being built up of Glc residues it is composed of (β1→4)-linked D-GlcNAc residues (*Figure 3.9*). It has, typically, a similar and wide range of molecular masses to that of cellulose. Peptide chains may

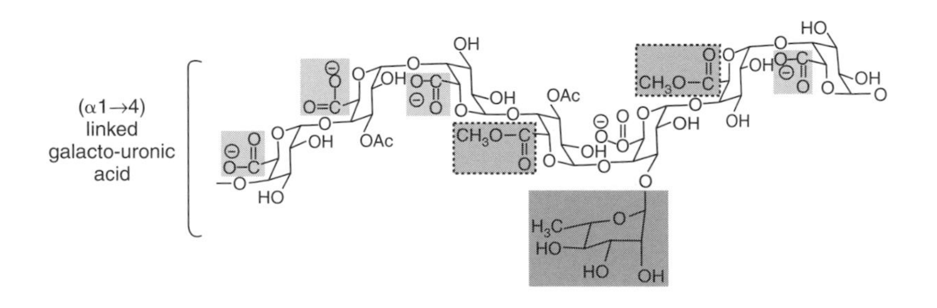

Figure 3.8

The structure of pectic substances, a group of substituted rhamnogalacturonans. They have a backbone of (α1→4)-linked D-galactopyranosyl uronic acid residues (the negatively charged groups are lightly shaded). The uronic acid can be methylated (shown as dotted outline). α-D-Rhamnopyranosyl residues are attached to the backbone by (α1→2) linkages (shown as darkly shaded).

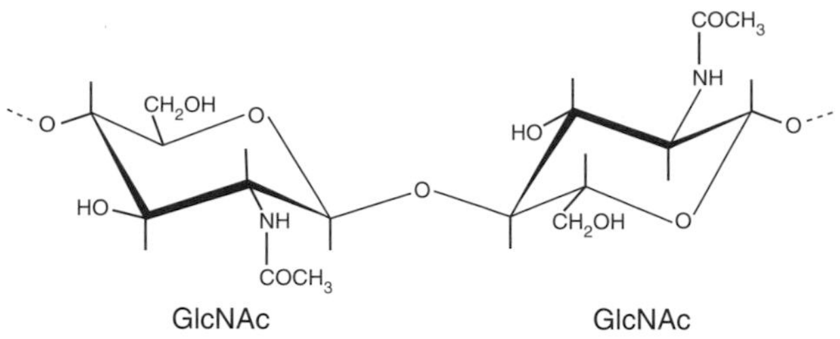

Figure 3.9

The structure of chitin: a linear homopolymer of (β1→4)-linked GlcNAc residues.

be attached to its acetamido groups to form a proteoglycan. The β-linkage causes the polysaccharide chain to lie flat and form a helical structure with two monosaccharide residues per turn, stabilized by hydrogen bonds. Chains are then stacked to form a fibre, again stabilized by hydrogen bonds, and also van der Waal's forces. Like cellulose, chitin appears to be present in at least two forms: in the α-form in which the polarities of neighbouring chains run anti-parallel and in the β-form in which they are parallel. Its biosynthesis is not well understood.

3.2.3.2 Properties

The intermolecular hydrogen bonds make chitin one of the most rigid polysaccharides and, like cellulose, it is a material of remarkable tensile strength, being very insoluble in water. Its unusual properties mean that it has diverse applications in medicine and the pharmaceutical industry.

3.2.3.3 Catabolism

> ➲ Chitin is extremely resistant to degradation, but many organisms produce effective chitinases which break it down.

Chitin is very resistant to degradation.➲ However, enzymes that break down chitin occur widely in microorganisms, as well as some protozoa, earthworms, snails and some mammals. These enzymes allow insectivorous invertebrates to digest their prey and for soil microorganisms to degrade chitinous debris. Marine bacteria, for example, which are responsible for the degradation of most of the estimated 10^{11} metric tonnes of chitin produced in the aquatic biosphere annually, possess complex signal transduction systems for finding chitin, adhering to chitinous substrata, degrading it into oligosaccharides, then transporting and catabolizing the oligosaccharides to fructose-6-phosphate, acetate and NH_3. Marine bacteria also degrade this insoluble material very efficiently; this is illustrated by the fact that the half-life of a crab shell in sea water is only about 2 weeks.

At least two enzymes are required for the digestion of chitin, one, a chitinase, to depolymerize the polysaccharide structure into component $(GlcNAc)_2$ disaccharides, and then a β-N-acetylhexosaminidase (β-GlcNAcidase or chitobiase) to hydrolyse $(GlcNAc)_2$ into GlcNAc monosaccharides. A large number of putative chitin-degrading enzymes have been described and classified into one of these two categories, often based on similarities in amino acid sequence with existing members, rather than activity. The situation may be more complex than originally believed, and several different classes of these enzymes exist, many with overlapping specificities, and each may play unique and important roles in the catabolism of chitin in the organisms that produce them. They may, for example, be involved in linking the catabolism of chitin with other cellular processes such as substrate adhesion or chemotaxis.

3.2.4 Cell wall structural polysaccharides of fungi, lichen and yeasts

The cell walls of fungi, lichen and yeasts are rich in polysaccharides of diverse structure, which may make up as much as 80–90% of the cell wall mass, and stabilize its structure. This is a complex topic owing to the vast number of different species involved and the complexity and diversity of the polysaccharides they produce. Fibrillar cellulose or, frequently, chitin (or other polysaccharides) are embedded in an amorphous mannan (often linked to protein, or sometimes phosphorylated), glucan, galactan (usually as a heteropolymer) or other polysaccharide matrix. Mannans and glucans, and other fungal polysaccharides, are often present in the form of glycoproteins, rather than pure polysaccharides (*see 7.3*).

The taxonomic classification of structural polysaccharides is often based on their constituents. Eight groups of fungi can be distinguished on the basis of whether their cell walls are predominantly composed of:

- cellulose and glycogen
- cellulose and glucan
- cellulose and chitin
- chitosan and chitin
- chitin and glucan (the most common)
- mannan and glucan
- mannan and chitin
- polygalactosamine and galactan.

3.2.4.1 α-D-Glucans

α-D-Glucans are common fungal polysaccharides and can include: amylose; glycogen (sometimes in combination with mannans); pullulan (*Figure 3.10*), which is predominantly composed of repeating (α1→6)- and (α1→4)-linked D-glucopyranosyl units; mycodextran or nigeran, which is made of alternate (α1→3)- and (α1→4)-linked D-glycopyranosyl units;

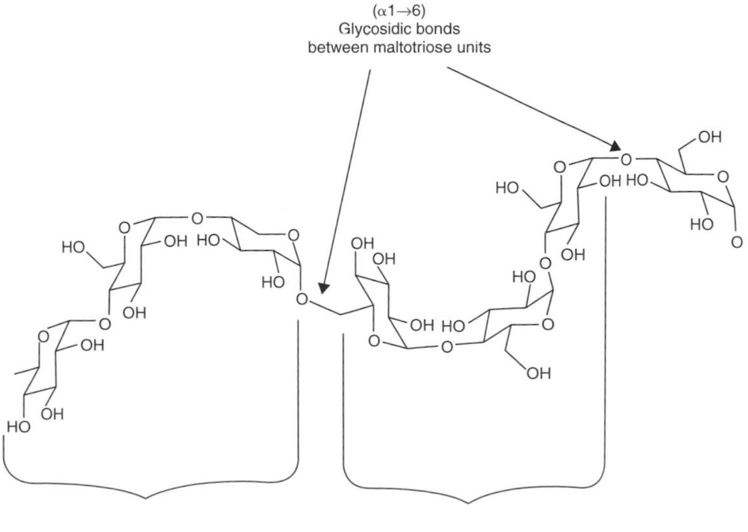

Figure 3.10

The structure of pullulan. The molecule is composed of building blocks of maltotriose units Glc(α1→4)Glc(α1→4)Glc. The maltotriose units are linked together by (α1→6) linkages.

and isolichenan, a common glucan of lichens, composed of (α1→3)- and (α1→4)-linked D-glucopyranosyl residues.

β-D-Glucans, in addition to cellulose, are common in a variety of plants and fungi, and may be composed of linear mixtures of (β1→6) and (β1→3) linkages or (β1→3) and (β1→4) linkages, or may be complex branching structures.

3.2.4.2 Rhamnomannans

Many varieties of rhamnomannans have also been described in various fungal species. Their description goes beyond the scope of this book.

3.2.5 Bacterial polysaccharides

Many bacteria produce polysaccharide capsules with extraordinary diversity in composition and structure; this topic is discussed further in *Section 7.2*.

3.2.5.1 Structure

➲ Bacterial polysaccharides are highly complex compared with those of higher organisms, and include unique bacterial monosaccharides. They are often immunogenic to humans.

Bacterial polysaccharides are typically much more complex than those produced by plants or animals, they are often partially composed of unusual monosaccharides seldom seen elsewhere in nature, and may contain up to 8–10 different monosaccharide types in repeating sequences. Over 100 different monosaccharides are found in bacteria. Bacterial polysaccharides are often immunogenic to humans, and are responsible for the specific immunological properties of a bacterial strain or type.➲

3.2.5.2 Function

In addition to their structural role, bacterial polysaccharides have functions including adhesion receptors, protection from desiccation, resistance to infection by phage and resistance to destruction by their vertebrate hosts. Microbial polysaccharides are used in the food industry as thickeners, emulsifiers, gelling or viscosifying agents.

3.2.5.3 Murein

➲ Up to 50% of the cell wall of gram-positive bacteria is the polysaccharide murein. Many antibiotics disrupt its synthesis.

Bacteria have a rigid cell wall that is composed principally of a peptidoglycan substance unique to bacteria, called murein (*Figure 3.11*). This comprises repeating units of a disaccharide containing (β1→4)-linked N-acetylmuramic acid and GlcNAc, giving a linear structure, which is cross-linked by bridges containing D-amino acids, yielding a rigid polysaccharide with very similar properties to chitin.➲ Gram-positive bacteria utilize considerable amounts of murein (up to 50% of cell wall mass), whereas the cell walls of Gram-negative bacteria contain much less (≈10%). Many antibiotics, for example penicillins, inhibit the synthesis of murein and thus prevent bacterial growth.

3.2.5.4 Teichoic acids

Gram-positive bacteria also have teichoic acids in their cell wall (*teichos* is Greek for wall), and both Gram-positive and Gram-negative bacteria produce extracellular teichoic acids. The structure of teichoic acids is illustrated in *Figure 3.12*. Strictly speaking these are not polysaccharides, but are low molecular mass mono- or oligosaccharides linked by phosphoric

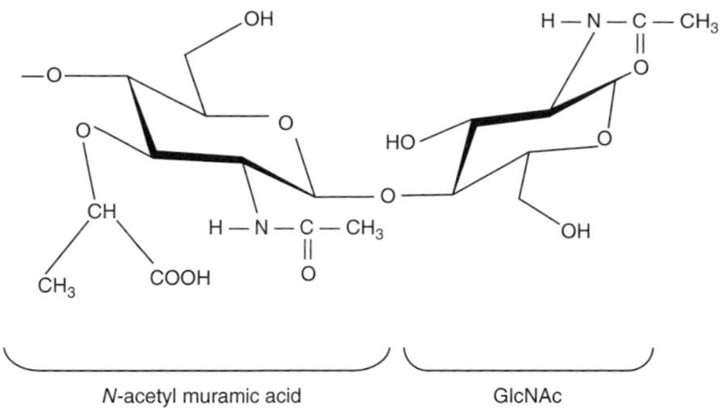

Figure 3.11

The structure of murein. Murein is built of repeating N-acetylmuramic acid (β1→4) GlcNAc disaccharide units.

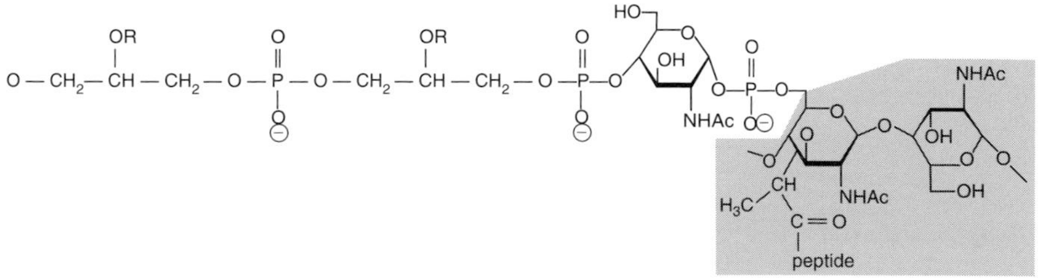

Figure 3.12

The structure of teichoic acid. Teichoic acid may be linked by a phosphodiester linkage to murein (shaded). R can be β-D-glucpopyranose or D-alanine.

diester linkages. Cell wall teichoic acids are covalently linked to the peptidoglycan murein, and cell membrane teichoic acids are linked to lipids to form lipoteichoic acids.

3.2.5.5 Lipopolysaccharides

The cell walls of Gram-negative bacteria are characterized by lipopolysaccharides, which are often antigenic to the host they infect. The polysaccharide chains of these molecules are structurally diverse, and contain monosaccharides that have not been found elsewhere in nature. This may result from evolutionary selection, with the survival of bacteria that have developed protection from immune attack by their host. There are thousands of serologically distinct Gram-negative bacterial species, and consequently, thousands of chemically distinct lipopolysaccharides.

The basic structure of the lipopolysaccharide is illustrated in *Figure 3.13*. It consists of an acylated 'lipid A' anchor portion which is attached to the bacterial cell wall through hydrophobic interactions, linked to a central core region (with inner and outer core), and outer 'O-antigen' side chains (heteropolysaccharides of repeating polysaccharide units). The lipid A anchor and core regions are highly conserved in structure, and are shared by large groups of bacteria, but there are considerable structural variations in the O-antigen chains. Lipid A (*Figure 3.14*) is formed from a framework of (β1→6)-linked GlcNAc. It is one of the most potent biologically active non-protein molecules known, and is an

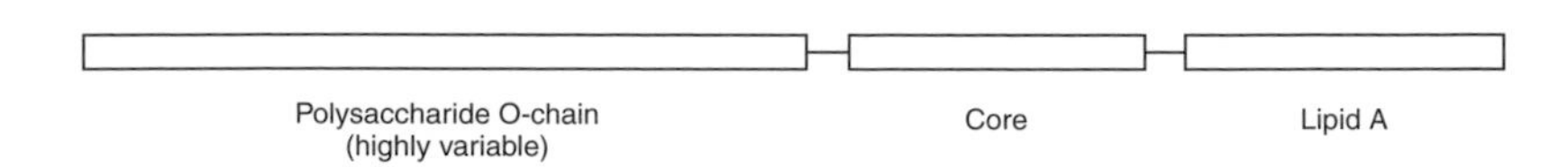

Figure 3.13

The basic structure of the bacterial lipopolysaccharide is an acylated 'lipid A' anchor portion which is attached to the bacterial cell wall via hydrophobic interactions, linked to a central core region (with inner and outer core), and outer 'O-antigen' side chains, heteropolysaccharides of repeating polysaccharide units.

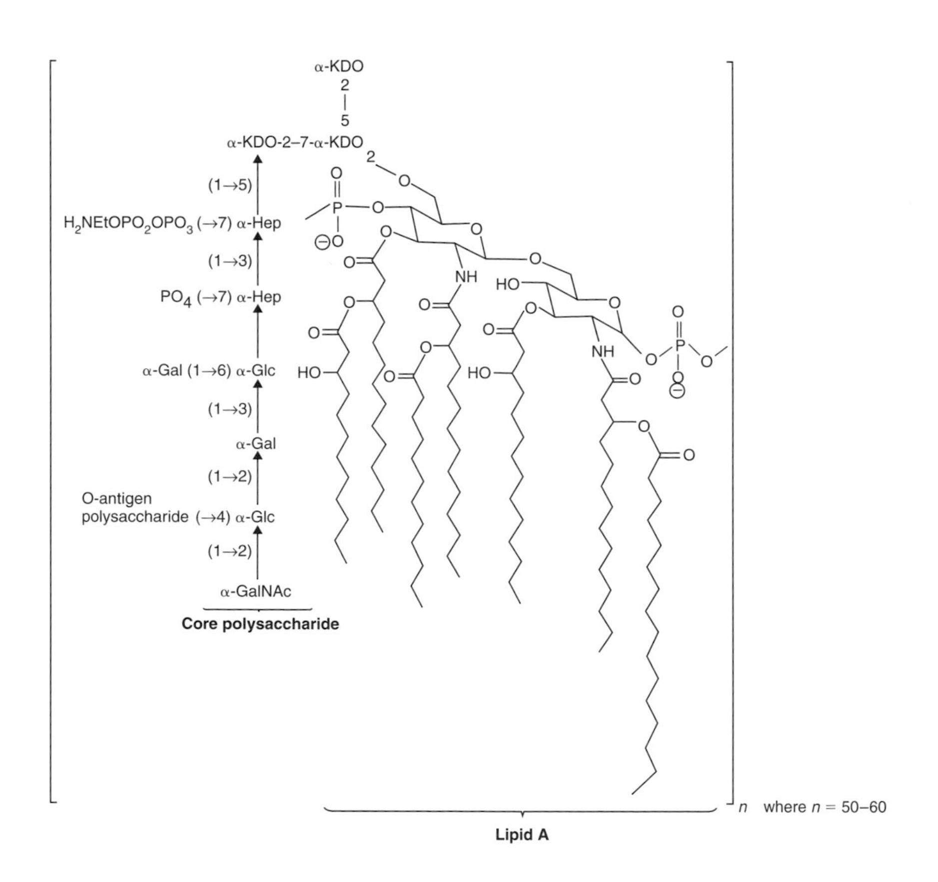

Figure 3.14

The structure of lipid A.

➲ The lipid A portion of bacterial lipopolysaccharides is a potent endotoxin, being one of the most potent biologically active non-protein molecules known.

endotoxin.➲ It is the causative agent of endotoxic shock and is also responsible for the induction of fever in individuals infected by Gram-negative bacteria. It is an amphipathic molecule with a hydrophobic centre of acylated disaccharide and a hydrophilic phosphate peripheral region; it is also amphoteric, carrying both acidic and basic functional groups. The core structure is similar in composition to lipid A but contains the unusual bacterial monosaccharide 3-deoxy-D-manno-octulosonic acid or KDO (*Figure 3.15*; *see 11.1*). The O-antigen varies widely between bacterial types and is typically composed of short (usually fewer than five monosaccharides), repeated polysaccharide sequences.

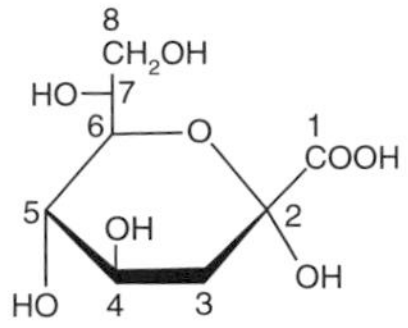

Figure 3.15

The unusual bacterial monosaccharide 3-deoxy-D-manno-octulosonic acid or KDO.

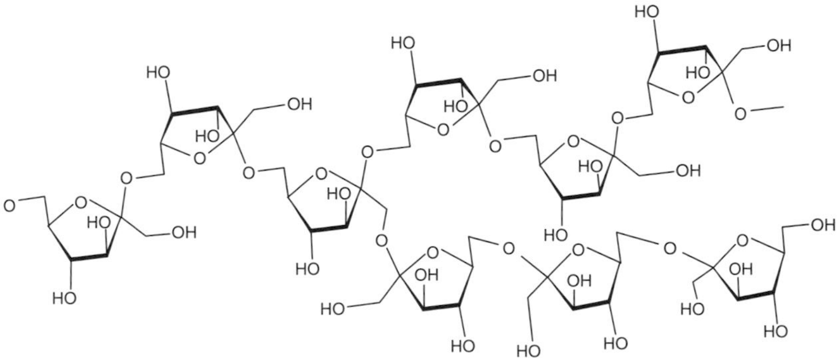

Figure 3.16

The structure of bacterial levans. These molecules consist of long chains of (β2→6)-linked sequences of fructopyranose. Branching occurs via (β2→1) linkages.

3.2.5.6 Extracellular polysaccharides

Bacteria also produce a number of extracellular polysaccharides which are not attached to the bacterial cell wall, but appear as a matrix surrounding them. They include cellulose, levans and dextrins. Bacterial levans (*Figure 3.16*) are mostly (β2→6)-linked sequences of fructopyranose residues with (β2→1)-linked fructopyranose branches. They are very high molecular mass polysaccharides, often with a molecular mass of over 1 million Da (this is in contrast to plant fructans, which are typically no larger than 10 000 Da). Dextrins have a backbone of (α1→6)-linked D-glucopyranosyl residues with (α1→2)-, (α1→4)- or (α1→3)-linked D-glucopyranosyl side branches.

3.2.6 Algal polysaccharides

Algae are one of the first groups of organisms to have evolved. ➲ They show much irregularity in the structure of the polysaccharides they synthesize, and these tend to form slimes or gels rather than structural fibres. These adapt the organism well for life in an aquatic environment, and are essential in, for example, retarding harmful desiccation of seaweeds at low tide, and protecting against damage by violent wave action. There are many types of algae, but the polysaccharides of only four, namely blue–green, red, green and brown are considered briefly here.

> ➲ Algal structural polysaccharides form slimes, gels and mucilages rather than fibres.

3.2.6.1 Blue–green algae

The cell walls of blue–green algae are structurally quite similar to those of bacteria, and are composed of a network of the peptidoglycan murein (*see 3.2.5.3*) coated in an outer envelope of a chemically distinct lipopolysaccharide. The capsule is shrouded in a slimy jelly which, although principally polysaccharide, may be chemically quite heterogeneous in composition. In addition to the extracellular structural polysaccharides, they produce a glycogen-like energy storage polysaccharide inside the cells.

3.2.6.2 Red algae

Red algae make a slightly more sophisticated intracellular granular glycogen storage polysaccharide. They also synthesize a primitive form of cellulose (*see 3.2.1*) plus mannan and xylan analogous to the hemicelluloses of terrestrial plants (*see 3.2.2*) as a structural component of their cell walls. These algae are typified by their production of sulfated polysaccharides including carrageenan and agar, which are a diverse range of polysaccharides that are able to form gels. They are composed of alternating sequences of (β1→3)- and (α1→4)-linked D-galactopyranosyl residues containing various degrees of sulfation. Red algae also produce large amounts of sulfated (α1→3)-linked D-mannans, and some freshwater species synthesize mucilages chemically similar to those of the blue–green algae (*see 3.2.6.1*).

3.2.6.3 Green algae

Green algae contain fibre-forming crystalline cellulose (*see 3.2.1*), mannans or xylans as structural polysaccharides. Glucomannans, sulfated L-arabino-D-xylo-D-galactans, sulfated D-glucurono-D-xylo-L-rhamnans and pectic substances (*see 3.2.2.5*) form water-holding, gel-like matrices and mucilages. They may use starches (*see 3.4.1*), the fructan inulin (*see 3.4.2*) or the (1→3)-linked glucopyranan laminaran as energy reserves. In this large group of organisms, a great heterogeneity in polysaccharide structure is, understandably, seen. O-linked oligosaccharides have also been described from green algae (*see 7.4.6*).

3.2.6.4 Brown algae

Brown algae typically synthesize cellulose (*see 3.2.1*) as their principle structural polysaccharide. They are also rich in laminarans, lichenans and fucoidin, generally covalently bound to protein and are hence not strictly polysaccharides. The most important polysaccharides derived from these organisms are alginates, which make up ≈40% of the dry weight of the plant. They are gelling polysaccharides, linear (1→4)-linked α-L-gulurono-β-D-mannuronans, and are responsible for flexible support and resistance to desiccation in many seaweeds. They are of considerable industrial interest as their unusual properties make them ideal thickening agents and of use in many emulsions as stabilizers.

Alginates are heteropolysaccharides, comprising irregular sequences of D-mannuronic acid and L-guluronic acid, as illustrated in *Figure 3.17*. All linkages are (α1→4). Some regions of the molecule are simple sequences of alternating mannuronic acid and guluronic acid interspersed with irregular blocks of copolymers of the two residues (typically 3–9 copolymers per block). The polysaccharide chain folds into a 3-fold helix along the D-mannuronic acid sequences, and a double ribbon-like structure along the L-guluronic acid sequences which are stabilized by intramolecular hydrogen bonds. Hydrogen bonding also stabilizes adjacent stretches of the polysaccharide chains into a three-dimensional network of fibres. The addition of calcium ions pack the L-guluronic acid sequences into an 'egg box' formation (*Figure 3.18*) in which calcium ions (the 'eggs') lie positioned between crevices in the polysaccharide chains (the 'lid' and 'base' of the 'egg box') supported by ionic interactions between the Ca^{2+} ions and the COO^{-} groups of the L-guluronic acid. The L-guluronic acid sequences therefore form rigid junction zones holding adjacent chains together in a three-dimensional network, and the D-mannuronic acid sequences remain flexible, thus facilitating the trapping of water molecules between the fibres of the polysaccharide.

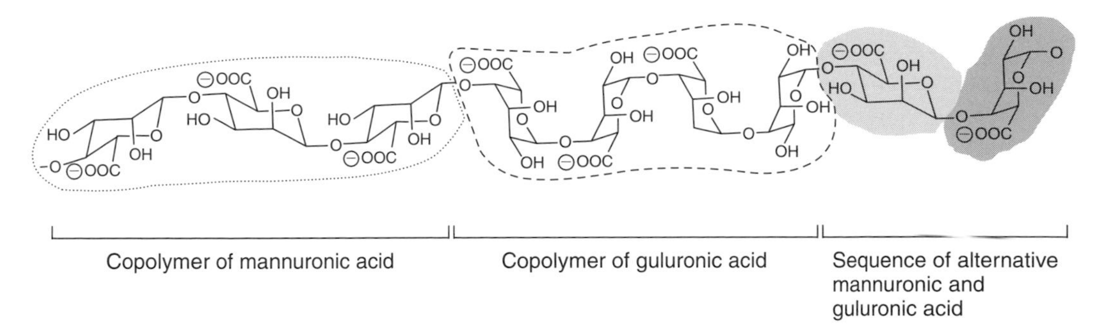

Figure 3.17

The structure of alginic acid. It is composed of β-D-mannuronic acid (dotted outline) and α-L-guluronic acid (dashed outline). All linkages are (α1→4). Some regions of the molecule are simple sequences of alternating mannuronic acid and guluronic acid (dark shading). There are also irregular blocks of copolymers of the two residues, typically 3–9 copolymers per block (lightly shaded).

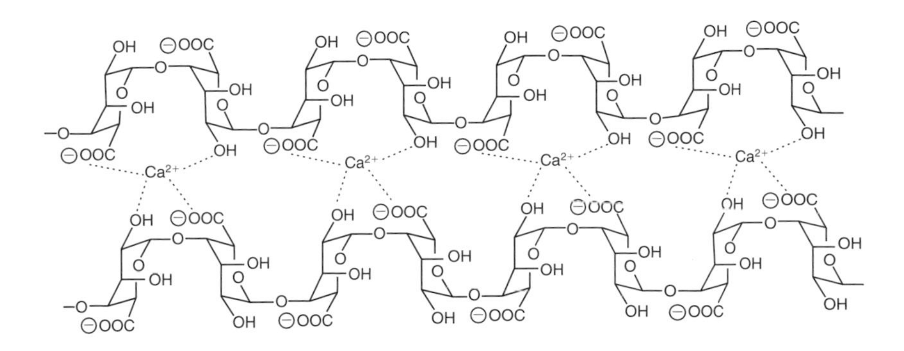

Figure 3.18

The 'egg box' structure of alginates. The 'eggs' are calcium ions which are positioned between crevices in the polysaccharide chains (the 'lid' and 'base' of the 'egg box').

The proportion of L-guluronic acid to D-mannuronic acid sequences therefore determines the properties of the gel: as the L-guluronic acid sequences provide rigidity, the higher the proportion of these in relation to D-mannuronic acid sequences, the firmer the gel. Growing tips of seaweed fronds thus contain alginates with a low L-guluronic acid to D-mannuronic acid ratio yielding a soft and supple structure with few rigid junction zones; in older tissue a higher L-guluronic acid to D-mannuronic acid ratio provides a rigid and firm supportive seaweed stem. ➲

➲ In seaweeds, the ratio of L-guluronic acid to D-mannuronic acid determines the rigidity of the alginate gel, thus yielding firmer support for ageing stems and more flexible support for growing fronds.

3.3 Gums, mucilages and other plant polysaccharides

Because of their physical properties, many plant polysaccharides have historically been classified as gums and mucilages. Callose has a role in sealing wounds and thus protecting the plant from infection (*see 3.3.3*). Amyloid is an important soluble plant polysaccharide (*see 3.3.4*).

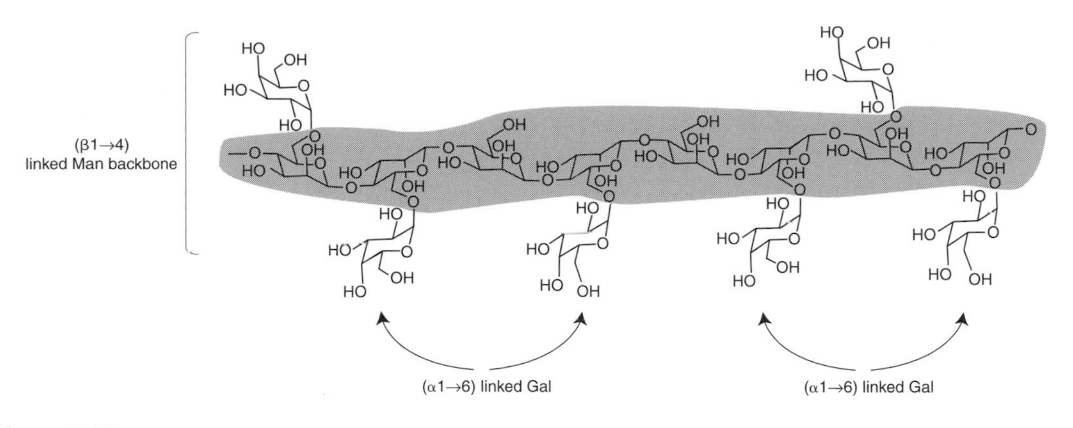

Figure 3.19

The structure of galactomannans. The molecule consists of a backbone of (β1→4)-linked D-mannopyranosyl units (shaded) to which (α1→6) galactopyranoside residues are linked.

3.3.1 Galactomannans

The distinction between mannans (*see 3.2.2.3*) and galactomannans is quite arbitrary. D-galacto-D-mannans are characteristic polysaccharides of some legume seeds and some fungi and yeasts. Galactomannans (*Figure 3.19*) consist of a (β1→4)-linked mannopyranosyl backbone to which (α1→6) galactopyranoside residues are linked. The Gal:Man ratio may vary between 1:1 and 1:5 in galactomannans from different sources. They are soluble in water and are often regarded, like the galactans (*see 3.3.5*), as a type of gum. A good example of this type of polysaccharide is guar gum from *Cyamopsis tetragonaloba*. D-gluco-D-mannans and D-galacto-D-gluco-D-mannans are common hemicelluloses of angiosperm and gymnosperm wood.

3.3.2 Glucomannans and galactoglucomannans

D-gluco-D-mannans are linear chains of (β1→4)-linked D-glucopyranosyl and (β1→4)-linked D-mannopyranosyl residues. The ratio of Glc:Man is usually ≈1:2 in hardwoods (the most common source), but in other sources, for example bulbs of the lily, orchid and iris, it may vary from 1:1 to 1:4. α-D-galactopyranosyl residues may also sometimes be added to the basic structure to give galactoglucomannans.

3.3.3 Glucans (other than cellulose)

Cereal gums, notably from the seeds of barley and oats are linear D-glucan chains of in which (β1→3) and (β1→4) linkages are randomly interspersed. They act as both structural support and as an energy store.

Fungi also produce glucans other than cellulose, which are closely associated with their cell walls. They are typically polymers of (β1→3)-linked Glc with (β1→6) branching points.

Callose is a glucan which is present in plant phloem sieve plates and, in very small quantities, in plant cell walls.➲ Its production is upregulated immediately in response to injury where it forms a hard callous to seal the wound and protect the plant from infection. It is a simple (β1→3)-linked D-glucan polymer with <2% glucuronic acid.

➲ Callose provides structural support for plant sieve plates, and is immediately exuded to form a hard callous at a site of wounding.

3.3.4 Amyloid

Amyloid is an important, water-soluble, component of plant seeds. It is a substituted (β1→4) glucan chain with side chains of, usually, α-D-galactopyranosyl and α-L-fucopyranosyl residues.

3.3.5 Galactans as plant gums

Many plants exude gums in response to injury, in order to seal the wound and prevent infection. These substances are exploited in industrial applications as thickening agents and emulsion stabilizers. One of the most well known plant polysaccharide gums is 'gum arabic', derived from *Acacia* species, which is typical of many plant gums. It consists of a linear core chain of (β1→3)-linked D-galactopyranosyl residues to which side chains of L-arabinofuranosyl, L-rhamnopyranosyl and D-glycopyranosyluronic acid are attached. Different gums have the same basic structure, but exhibit varying degrees of branching and may have slight differences in the nature or the attachment of side chains.

3.3.6 Plant mucilages

These substances, which form a gel when dissolved in water, are thought to act as water-retention reservoirs in seeds to protect against excessive desiccation. They are poorly characterized, but composed of, principally, D-galactopyranosyluronic acids and D-galacturonic acids.

3.4 Storage polysaccharides

Plants, animals and microorganisms store surplus energy as polysaccharides, which can then be utilized at a later time when the energy supply is restricted. The advantage of storing carbohydrate as polysaccharide rather than simple monosaccharide lies in the physical properties of the two types of molecules in that monosaccharides are water soluble and are osmotically active. In contrast, polysaccharides are osmotically inactive as they are relatively insoluble in water. Storage polysaccharides are typically subject to both rapid synthesis and degradation in response to the changing energy balance of the organism.➲ Structurally, these molecules are microheterogeneous. They tend to be roughly spherical structures, excluding water, with many exposed terminal end chains available for enzymatic degradation and also, because they are loosely packed, to facilitate some enzymatic access to their interior.

➲ Storing energy reserves as monosaccharides would cause serious osmotic pressure inside the cell. In contrast, polysaccharides are insoluble and thus osmotically inert, and may be rapidly synthesized or degraded in response to energy requirements.

The biochemical pathways responsible for the metabolism of storage polysaccharides such as glycogen and starch are well understood. Their description goes beyond the scope of this volume and the interested reader is directed to any good biochemistry text.

3.4.1 Starch

Starch is the primary storage polysaccharide synthesized by plants. It is abundant in the seeds, fruits, bulbs, tubers and rhizomes of higher plants, where it forms a long-term energy store, and in the leaves, where it forms a temporary store of carbohydrate produced by photosynthesis during daylight hours. It is also present in some protozoa, bacteria and algae. It is the principle carbohydrate energy source in the diet of humans and many animals. Modified starch is used in paper and glue manufacture, textile production, pharmaceuticals and cosmetics, explosives and building materials.

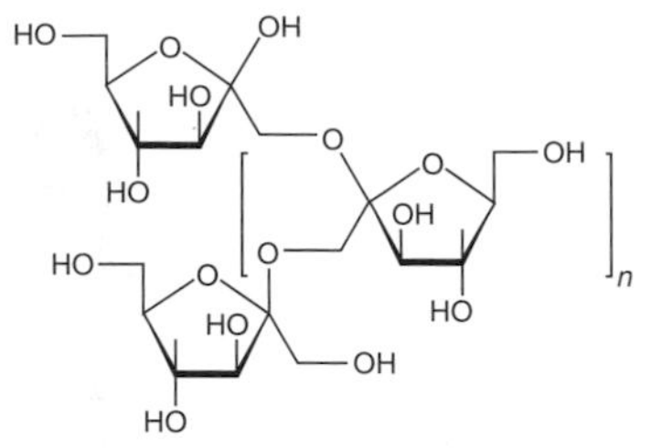

Figure 3.20

The structure of inulin. Inulin is composed of (β2→1)-linked D-fructofuranosyl residues. It is a relatively small polysaccharide comprising only 20–30 monosaccharide units.

3.4.1.1 Structure and synthesis

Starch is a homopolysaccharide, one of the largest known biomolecules, between 10^6 and 10^9 Da in mass depending upon origin, and composed of α-amylose and α-amylopectin (it has also been reported to contain small amounts, perhaps 5–7%, of an 'intermediate material' that appears to have characteristics of both α-amylose and α-amylopectin). Both α-amylose and α-amylopectin are polymers of repeating glucose units.

➲ Cellulose and amylose are isomers of each other, differing only in the linkages between the Glc units, but they have entirely different physical properties.

α-Amylose (*Figure 3.21*) is a mostly linear polymer of typically 1000–6000 repeating (α1→4)-linked Glc units (there may be occasional (α1→6) linkage branching points). It forms a loose randomly coiled structure. α-Amylopectin (*Figure 3.22*) is a highly branched polymer of $\approx 10^6$ mostly (α1→4)-linked Glc units interspersed with (α1→6) branch points typically every 20–30 Glc units, but densely branched sections also appear to exist. The relative proportions of α-amylose and α-amylopectin in starches differ from one source to another, but typically 15–35% of the mass of starch is amylose.➲

Although the chemical structure of the structural polysaccharide cellulose and the storage polysaccharide α-amylose are so similar (they are actually isomers of each other), both composed of homopolymers of repeating Glc units, the differences in the linkages between those units confer very different physical properties. The (β1→4) glycosidic linkages of cellulose cause each successive monosaccharide in the polymer to flip by 180° with respect to its immediate neighbours; this facilitates easy packing of the polymers into a compact and strong fibre. By comparison, the (α1→4) glycosidic linkages of the α-amylose polymer throw it into an irregular, aggregating, helical coil suited to its function as a storage material. This may be appreciated by comparing *Figures 3.3* and *3.21*. These physical differences are relevant to the different biological functions of the molecules, for example, cellulose fibres are well suited to form the complex, rigid and three-dimensional shape of plant pollen tubes, whereas flat amylose molecules form the storage polysaccharide of potato tubers.

3.4.1.2 Biosynthesis

The biosynthesis of starch relies on similar enzymes to those used to synthesize glycogen in animal cells. Starch synthase (α1,4-D-glucan-4-α-D-glucosyltransferase) catalyses the transfer of glucose from either ADP-glucose or UDP-glucose (ADP-glucose is the common donor in plants) to give a chain of Glc monosaccharides linked by (α1→4) glycosidic bonds. Branching points in α-amylopectin are formed by a branching enzyme (α1,4-D-glucan:α1,4-D-glucan-6-glucosyltransferase) which transfers long sections of glucan chain onto the main polysaccharide.

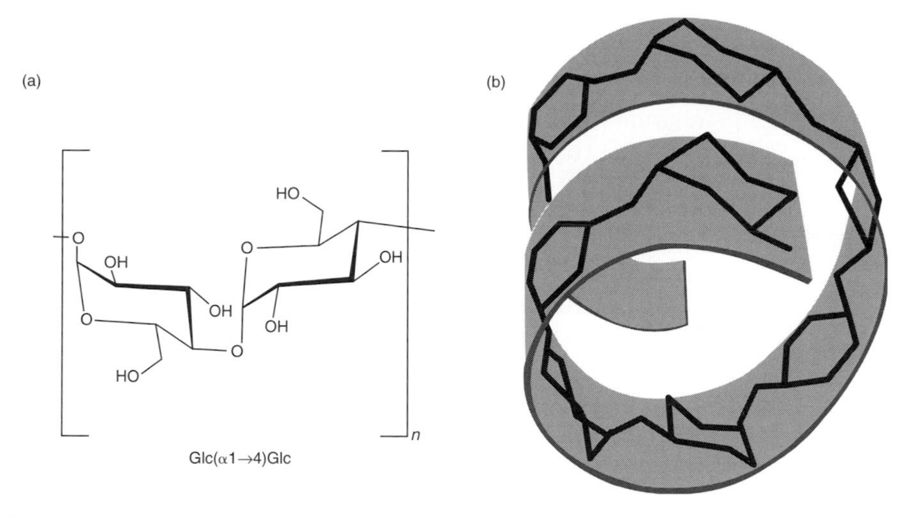

Figure 3.21

The structure of α-amylose. (a) The molecule is composed of chains of (α1→4)-linked Glc molecules. (b) Schematic diagram illustrating the structure of α-amylose: the (α1→4) linkages mean that the molecule forms a loose random coil structure.

3.4.1.3 Storage of starch

Starch is deposited as insoluble storage granules in the plant cell cytoplasm. The insolubility of this material is of considerable importance to the plant in that it is osmotically inactive (*see 3.4*).

3.4.1.4 Metabolism

Plants degrade starch by both phosphorylytic and hydrolytic mechanisms catalysed by phosphorylases and amylases, respectively. It is thought that these different mechanisms are used to different extents in various plants, and in different tissues within individual plants. Starch breakdown in chloroplasts, for example, is predominantly the result of phosphorylytic mechanisms, and it is broken down in germinating seeds by hydrolytic mechanisms. In the germination of cereal grains, cells of the aleurone layer, which surrounds the endosperm, contain α- and β-amylase. α-Amylase is present in an inactive form in the dormant seed and is converted to an active form during germination. β-Amylase is synthesized during germination. α-Amylase hydrolyses the (α1→4) linkages of internal and external polysaccharide chains, yielding a mixture of maltose and low molecular mass oligosaccharides. β-Amylase (α1,4-D-glucan maltohydrolase) hydrolyses alternate (α1→4) linkages of the external chains only to yield disaccharide maltose. (α1→6) Linkages are hydrolysed by amylopectin 6-gluconhydrolase which cleaves small chains of two or more glucose residues from the (α1→6)-linked branch points.

3.4.1.5 Starch as a foodstuff

Starch, eaten in plant-based foodstuffs, forms an important part of the diet of many animals, including humans where it is the most important dietary carbohydrate. Starch digestion, catalysed by the enzyme salivary α-amylase (α1,4-glucan-4-glucohydrolase) begins in the mouth, under alkaline conditions, as the food is chewed. Here, most (α1→4) glycosidic bonds are cleaved and the glucan polymers are reduced to chains of fewer than eight repeating Glc

➲ Starch digestion begins in the mouth where the glucan chains are chopped into short fragments, and completed in the small intestine.

units. Starch digestion cannot proceed in the acidic conditions of the stomach, but is completed in the small intestine.➲ Here, the remaining glucan polymers are broken down to maltose (a disaccharide of two (α1→4)-linked Glc units), maltotriose (a trisaccharide of three (α1→4)-linked Glc units) and dextrins (small oligosaccharides, containing the (α1→6) branches) by pancreatic α-amylase. α-Amylase is unable to break down the (α1→6) linkages of glycogen or the (α1→4) linkages of maltose and maltotriose. Thus, glucosidase enzymes produced by the cells of the brush border of the intestinal mucosa, where rapid monosaccharide adsorption takes place, hydrolyse these small glycans to their constituent Glc units: α-glucosidase breaks down maltose and maltotriose, and α-dextrinase degrades the dextrins.

3.4.2 Fructans

Plant fructans are composed of short chains of fructans terminating in a single Glc residue. They are storage polysaccharides, and include the levans, which are linear (β2→6)-linked fructans and inulin which is a (β2→1)-linked fructan storage polysaccharide characteristic of Jerusalem artichokes and dahlia tubers (*Figure 3.22*). If introduced into the bloodstream, inulin is filtered into the urine and can be used to assess the glomerular filtration rate.

3.4.3 Glycogen

Glycogen is the principle storage polysaccharide in animals, bacteria and some fungi. It is present in all cells, but most of it is stored in the cells of skeletal muscle and liver. As much as 10% of the dry weight of the mammalian liver may be glycogen. The liver stores glycogen when excess energy is present and exports it, as need arises, to maintain blood glucose levels.➲ In other tissues, like skeletal muscle, it is utilized for the specific cellular energy requirements of that tissue. Like starch, glycogen takes the form of cytoplasmic storage granules.

➲ In animals, glycogen is stored mostly in the liver.

3.4.3.1 Structure

The chemical composition of glycogen is remarkably similar to that of starch (*see 3.4.1*) being composed entirely of repeating Glc units. These are linked by (α1→4) glucosidic linkages, with (α1→6) linkages forming branching points around every 10–12 or so Glc residues to form a highly branched 'tree-like' glucan homopolymer. Rat glycogen, as a typical example, ranges in molecular weight from 1 to 5 × 10^8 Da. There is evidence that glycogen may be synthesized in association with a protein backbone, at least under some circumstances.

3.4.3.2 Biosynthesis

Two enzymes are required for glycogen biosynthesis, a synthase (UDP-D-glucose: glycogen 4-α-D-glycosyltransferase) to form the (α1→4) glycosidic linkages, and a branching enzyme (α1,4-D-glucan: α1,4-D-glucan 6-glucosyltransferase) to form the (α1→6) linkages. Glycogen synthase catalyses the transfer of Glc from UDP-Glc onto a non-reducing terminal Glc residue of the glycogen molecule to bring about chain extension and elongation of the polysaccharide. The branching enzyme transfers a side chain onto the C6 position of a mid-chain Glc residue.➲

➲ Glycogen is synthesized by the action of two enzymes: one synthesizes the glucan chains, and the second instigates the branching.

(a)

α-1→6 branch linkage

Glc(α1→4)Glc chains with Glc(β1→6) branch points

(b)

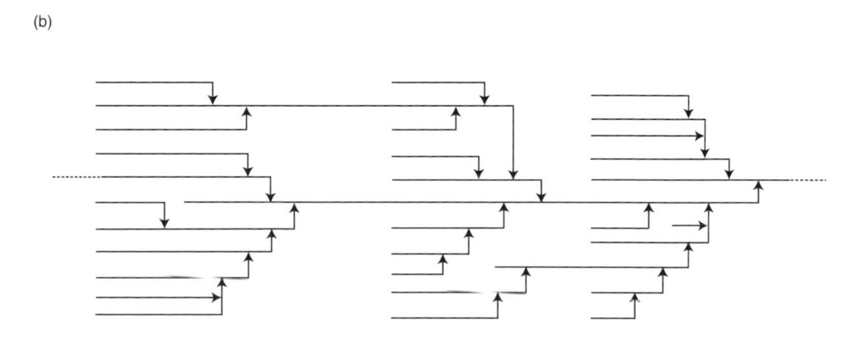

Figure 3.22

The structure of α-amylopectin. (a) The molecule is composed of chains of (α1→4)-linked Glc molecules with (β1→6)-linked branch points to form a bush-like structure. (b) Schematic diagram illustrating the structure of α-amylopectin.

3.4.3.3 Metabolism

Glycogen metabolism is under subtle and complex control.➲ The main hormones involved are glucagon, epinephrine and insulin. Glycogen is broken down in two different situations. The first is extracellular hydrolysis in the gut during digestion of food to yield free glucose which is absorbed, transported around the body in the blood, and taken up by cells. This digestive breakdown of dietary glycogen is identical to that of starch (*see 3.4.1.5*). The second is an intracellular phosphorolysis process by which glycogen is broken down into glucose-1-phosphate in response to the direct energy needs of the cell. A family of glycogen phosphorylase enzymes are involved in this process.

> ➲ Rapid glycogen breakdown in response to the body's energy needs is directed by a family of glycogen phosphorylases.

Glycogen phosphorylase (α1,4-D-glucan:orthophosphate glucosyltransferase) cleaves each (α1→4)-linked Glc residue from a non-reducing end in turn to yield glucose-1-phosphate. The highly branched structure of the polysaccharide means that there are multiple non-reducing ends facilitating rapid degradation by this process, and therefore highly efficient energy release in times of metabolic need. 'Debranching' enzyme action then functions in a two-step process. First an enzyme functioning as a α1,4-D-glucan: α1,4-D-glucan 4-α-D-glucosyltransferase cleaves approximately three Glc residues from the (α1→4)-linked chain from attachment to a (α1→6)-linked branching point Glc and transfers it to an adjacent non-reducing end of the (α1→4)-linked chain, thus exposing the (α1→6)-linked Glc. Then a second enzyme, amylo α1,6-D-glucosidase, hydrolyses the (α1→6) linkage to yield a free Glc monosaccharide.

3.4.3.4 Inherited disorders in glycogen metabolism

Inherited defects in one or more of the enzymes involved in the breakdown of glycogen, results in the devastating effects of rare glycogen storage diseases.

Pompe's disease or glycogenosis type II disease results from defects in the action of lysosomal α-glucosidase, implicating the lysosome in glycogen breakdown in the normal cell, resulting in the massive accumulation of glycogen in swollen vesicles which causes damage to the cells that can be fatal. Sufferers usually die in early childhood (*see 16.9*). Other glycogen storage diseases result in incomplete breakdown of glycogen leading to various metabolic problems resulting from defective glucose metabolism.

Further reading

Calder, P.C. (1991) Glycogen structure and biogenesis. *Int. J. Biochem.* **23:** 1335–1352.

O'Sullivan, A.C. (1997) Cellulose: the structure slowly unravels. *Cellulose* **4:** 173–207.

Ravikumar, M.N.V. (1999) Chitin and chitosan fibres: a review. *Bull. Mat. Sci.* **22:** 905–915.

Smith, A.M., Denyer, K., Martin, C. (1997) The synthesis of the starch granule. *Annu. Rev. Plant Physiol. Plant Mol. Biol.* **48:** 65–87.

N-Linked glycoproteins

4

4.1 What are N-linked glycoproteins?

Glycoproteins are classified according to the type of oligosaccharide chains they carry, and also their site of attachment to the protein molecule.➲ One major type of glycosylation is 'N-linked' glycosylation. An oligosaccharide is N-linked to a protein when it is linked via a GlcNAc molecule in a β-N-glycosidic type bond to a nitrogen (hence N-linked) of the amide group of an asparagine (Asn) amino acid on the polypeptide chain, as illustrated in *Figure 4.1*.

➲ A glycoprotein is a protein containing one or more covalently linked carbohydrate groups. The carbohydrate can be a single monosaccharide or comprise many monosaccharides linked together as branched or linear chains, termed oligosaccharides (or glycans). If an oligosaccharide is attached to the protein, the protein is described as having been 'glycosylated'.

N-Linked glycosylation differs fundamentally from other types of glycosylation, such as O-linked glycosylation (*Chapters 5* and *6*), glycosylation of glycosaminoglycans (*Chapter 8*) and glycolipids (*Chapter 9*), as will become apparent by comparison of the descriptions of these different types of glycoconjugates and their synthesis. Much of our recent understanding of N-linked glycosylation has resulted from studies with carbohydrate processing inhibitors and transgenic animals which lack the enzymes of glycosylation involved in N-linked oligosaccharide synthesis (*Chapter 15*).

4.2 Biosynthesis of N-linked oligosaccharides

N-Linked glycosylation of proteins is a *co-translational* event. This means that it is initiated *during* protein synthesis. This is in contrast to O-linked glycosylation (*Chapter 5*), which begins after protein synthesis is complete, and is therefore said to be *post-translational*. The control and intracellular production of proteins is described in most standard biochemistry textbooks and is referred to only briefly here in the context of protein glycosylation.

Each protein produced in a cell has a corresponding gene sequence, located on the DNA. On receiving a signal that protein production should take place, the gene that encodes the protein is activated. Messenger RNA (mRNA) is produced, and protein synthesis commences – first on ribosomes in the cytoplasm and then, for membrane-bound proteins, the ribosome becomes associated with the rough endoplasmic reticulum (RER) and the growing (nascent) polypeptide chain is inserted through the plasma membrane of this organelle, so that it begins to protrude into the lumen of the ER.

The way in which this happens is illustrated in *Figure 4.2*. The nascent polypeptide chain contains an ER signal sequence and is guided to the ER membrane by a cytosolic signal-recognition particle (SRP) which binds to it. The SRP is then recognized by a SRP receptor embedded in the plasma membrane of the RER. When it has bound to its receptor, and thus positioned the ribosome and the polypeptide chain it is synthesizing correctly on the ER membrane, the SRP is released. Protein synthesis continues with the nascent polypeptide chain being fed through a translocation channel in the RER membrane and through into the

Functional and Molecular Glycobiology, Susan A. Brooks, Miriam V. Dwek and Udo Schumacher

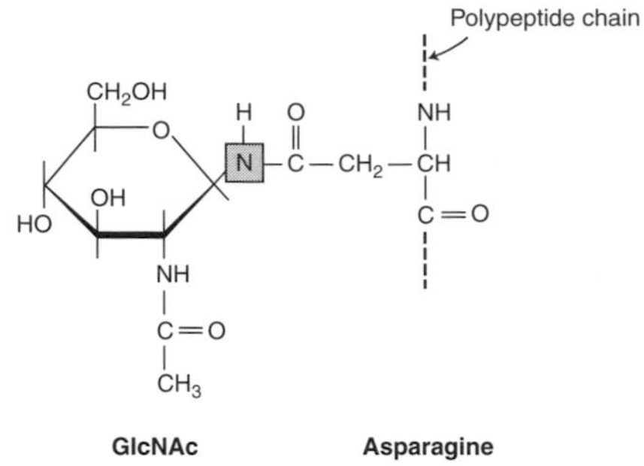

Polypeptide chain

GlcNAc Asparagine

Figure 4.1

N-Linkage. In N-linked glycoproteins, the oligosaccharide is linked via a GlcNAc molecule in a β-N-glycosidic type bond to a nitrogen (shaded) of the amide group of an asparagine amino acid residue on the polypeptide chain.

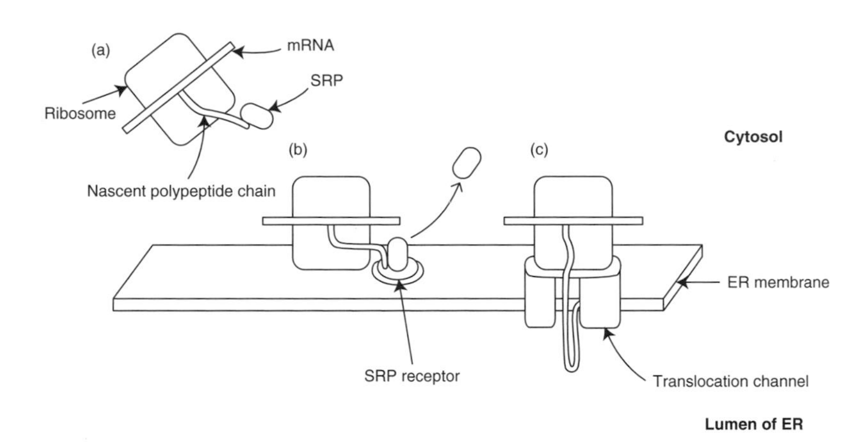

Figure 4.2

During protein synthesis, if the polypeptide chain carries an ER signal sequence (a) then a cytosolic signal-recognition particle (SRP) binds to the ER signal sequence (b) and this is recognized by a SRP receptor in the ER membrane. The SRP is then released, leaving the ribosome in contact with the ER membrane (c). As protein synthesis commences, the nascent polypeptide chain is fed through a translocation channel that spans the ER membrane, and it emerges into the lumen of the ER.

lumen of the organelle. If the conditions are correct (*see 4.2.1*), the addition of N-linked oligosaccharides begins almost immediately (between 12 and 14 amino acids) on the nascent polypeptide chain, commencing as the polypeptide chain enters the lumen of the ER.

N-Linked oligosaccharide synthesis continues as the protein is transported from ER to the Golgi apparatus, and is completed by the time the glycoprotein leaves the *trans*-Golgi network. ➲ The enzymes of glycosylation necessary for this process are transmembrane proteins located along the secretory pathway of ER and Golgi apparatus, and their location corresponds to the sequence in which they act (*see 2.12.2*).

➲ N-Linked glycosylation is a co-translational event that starts in the ER and is completed in the Golgi apparatus.

In some cases, the pattern of glycosylation may be of functional significance to the final destination of the glycoprotein, for example, transport to proteasomes in which the incorrectly folded proteins are degraded (*see 4.2.6.1*) or targeting to the lysosomes (*see 4.2.8*). During the synthesis of proteins destined to reside in the cytoplasm, the ribosome does not become associated with the ER, and thus cytoplasmic proteins are not usually glycosylated. An exception to this general rule is O-GlcNAc-glycosylated cytoplasmic proteins (*Chapter 6*).

4.2.1 The importance of a consensus sequence in N-linked glycosylation

The presence of a consensus amino acid sequence in the nascent polypeptide chain is a prerequisite for N-linked oligosaccharide synthesis. The consensus sequence is Asn-Xaa-Ser/Thr, in which Xaa may be any amino acid other than proline. Occasionally, such as in the leukocyte surface protein (CD69), the amino acid sequence Asn-Xaa-Cys is an acceptable sequon for the addition of N-linked glycans.➲ There is evidence that the consensus sequence allows recognition of the glycosylation site by the first enzyme (the oligosaccharyltransferase, OST; *see 4.2.4*) involved in N-linked oligosaccharide production, ensuring the protein conformation required for enzyme to gain access to the glycosylation site.

➲ The consensus sequence Asn-Xaa-Ser/Thr has two functions:
(i) recognition by the enzyme (OST) that initiates N-linked glycosylation;
(ii) providing the correct protein conformation enabling OST to gain access to the glycosylation site.

4.2.2 Occupation of glycosylation sites with oligosaccharide

The consensus sequence Asn-Xaa-Ser/Thr may occur many times along the polypeptide chain, and each of these potential glycosylation sites may or may not be glycosylated. Several factors influence whether a putative glycosylation site has a covalently linked oligosaccharide. The polypeptide conformation around N-linked glycosylation sites has been modelled using amino acid sequence data and it has been found that successful N-link glycosylation sites are often those in which the consensus amino acid sequence Asn-Xaa-Ser/Thr is in a 'loop' or 'turn' in the polypeptide chain. This is consistent with the idea that the region of the polypeptide needs to be accessible for the correct function of enzymes involved later in N-glycan synthesis.

Glycosylation is cell-, tissue- and site-specific and environmentally sensitive. This is illustrated by observations that only ≈30% of putative N-linked glycosylation sites are glycosylated and when they are, the oligosaccharides present are often a heterogeneous mixture (of the three types of N-linked oligosaccharide described in *4.3.2*). The same glycoprotein may be differently glycosylated under different circumstances. The biological advantage that such heterogeneity in N-linked oligosaccharides may confer is not fully understood. Various models have been proposed to attempt to explain heterogeneity in the glycosylation of proteins. One example of this is the immune recognition of bacteria and yeasts by mannose-binding lectin (MBL; *see 13.12.3.3*). Here, bacteria and yeasts residing in the extracellular space present multiple Man residues on their surfaces and these are recognized by MBL leading to activation of the complement pathway and cell lysis. Other biological reasons for microheterogeneity in glycosylation and the function of N-linked oligosaccharides on proteins are discussed later (*see 4.4*).

4.2.3 Enzymes of glycosylation

In contrast to the production of nucleic acids and proteins, both of which are generated from a DNA 'template', oligosaccharides are not primary gene products; rather they are constructed by the action of 'building' enzymes (glycosyltransferases) and 'trimming' enzymes (glycosidases) (*Chapter 2*). The glycosyltransferases build the oligosaccharide structure, whereas the glycosidases trim it. Hence, many enzymes are involved in oligosaccharide production. Some of the enzyme reactions may not go to completion and this results in a failure of subsequent enzymes to act. This is reflected in the diversity of oligosaccharide structures synthesized. The synthesis of some commonly occurring and biologically important oligosaccharides is considered in detail in *Chapter 10*.

4.2.4 Production of lipid (dolichol)-linked oligosaccharide

N-Linked oligosaccharide biosynthesis occurs as a step-wise process. The first step is the preparation and transfer of a large dolichol-linked oligosaccharide intermediate.

Dolichols are mixtures of saturated isoprenol alcohols (*Figure 4.3a*). They are characterized by a five-carbon repeating isoprene unit, this repeating unit gave rise to their name (the Greek word

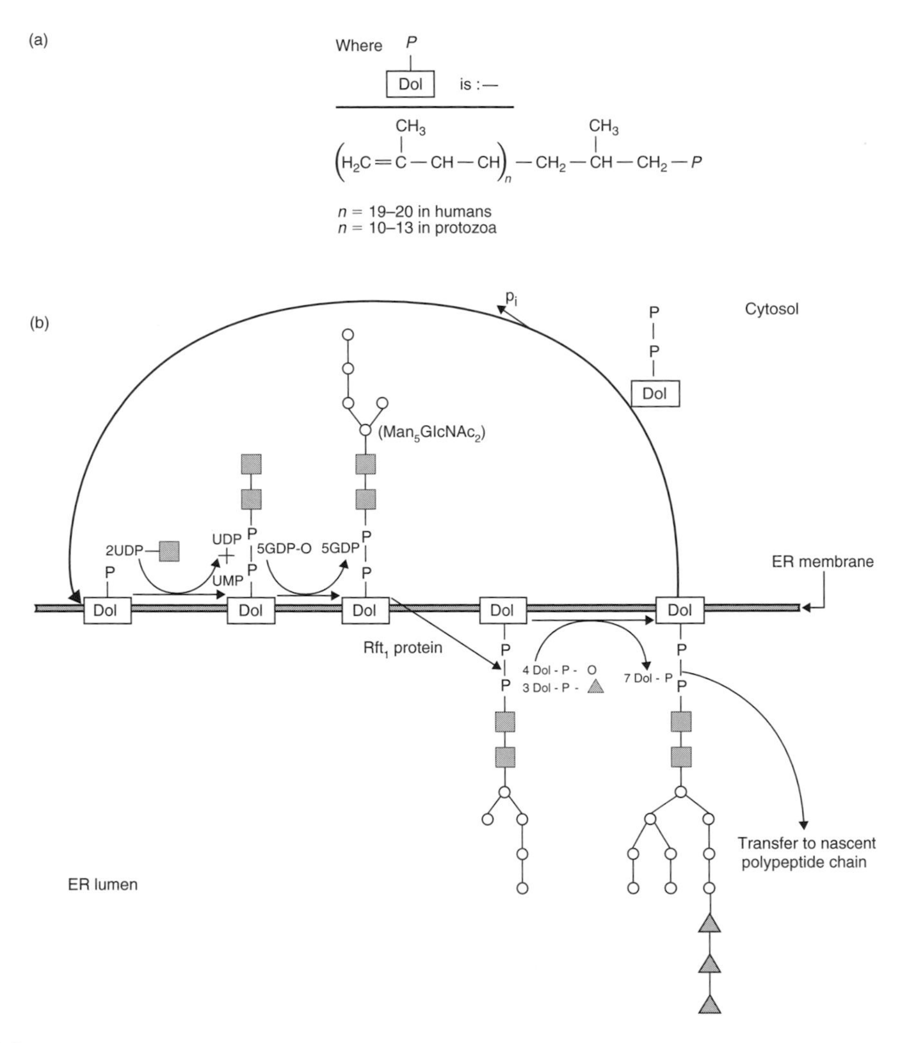

Figure 4.3

The synthesis of N-linked oligosaccharides starts with the attachment of a GlcNAc residue to a dolichol (Dol) phosphate molecule. The Dol phosphate molecule (a) has two GlcNAc molecules and a further five Man molecules attached (b). The $Man_5GlcNAc_2$ molecule is enlarged in the lumen of the ER to form the final dolichol-linked oligosaccharide intermediate ($Glc_3Man_9GlcNAc_2$). Attachment of the monosaccharides to the Dol phosphate and the subsequent formation of the oligosaccharide intermediate require energy input, this takes the form of either nucleotide-activated monosaccharide donor sugars or Dol-P-linked monosaccharide donor sugars as indicated.

dolichol means long) and, indeed, the length of the repeating unit varies according to the species from which they are derived. In general, protozoan dolichols are shorter, for example, in the protozoan parasite *Trypanosomas*, the dolichol repeating unit is typically 10–13, whereas in humans there are typically 19–20 repeating units. The dolichol molecule is located between the two leaflets of the ER membrane and is thought to disrupt the stability of the membrane lipid bilayer.

Formation of the oligosaccharide intermediate molecule on the dolichol phosphate molecule is shown in *Figure 4.3(b)*. Essentially, the structure is formed by the action of first an N-acetylglucosaminylphosphotransferase enzyme which catalyses the reaction Dol-P + UDP GlcNAc to form GlcNAc-P-Dol + UMP and then a further N-acetylglucosaminyltransferase that forms the structure GlcNAc(β1→4)GlcNAc-P-Dol. This structure is also known as the chitobiose core.

The next reaction is mediated by the action of mannosyltransferase I (β1,4 mannosyltransferase) which leads to the formation of the Man(β1→4)GlcNAc(β1→4)GlcNAc-P-Dol structure.

Further reactions occur in the cytoplasm as a result of the action of mannosyltransferases II–V, yielding, finally, a $Man_5GlcNAc_2$-P-Dol structure. All of these reactions require the presence of donor-sugar molecules. These take the form of nucleotide-activated donors: GDP-Man and UDP-GlcNAc. ➲

After the initial formation of the $Man_5GlcNAc_2$ structure in the cytoplasm the entire molecule is 'flipped' into the lumen of the ER.

It is important to reflect on this remarkable sequence of events. For successful N-linked glycosylation of proteins, a very large hydrophilic oligosaccharide needs to be transported ('flipped') across a hydrophobic lipid bilayer of the ER plasma membrane and attached to a growing polypeptide chain. This represents a vital but poorly understood biochemical process that occurs in all eukaryotic cells. A protein called Rft1 has recently been implicated in the 'flipping' process.

> ➲ The process of N-linked glycosylation requires the formation of an intermediate structure linked to a dolichol molecule. The initial formation of the intermediate occurs on the cytoplasmic face of the ER and continues in the ER lumen. First, the entire molecule needs to be 'flipped' into the ER lumen by a 'flippase' called Rft_1.

After the dolichol molecule and partly formed oligosaccharide intermediate ($Man_5GlcNAc_2$) have been flipped into the ER lumen, the partly formed oligosaccharide intermediate is further substituted with four Man and three Glc residues. The donor molecules for this reaction are produced in the cytoplasm (Dol-P-Man or Dol-P-Glc) and flipped into the lumen of the ER. Following attachment of these residues the intermediate oligosaccharide is produced. In its final form it comprises three Glc, nine Man and two GlcNAc residues, conveniently written in the shorthand form $Glc_3Man_9GlcNAc_2$.

4.2.5 Attachment of the oligosaccharide intermediate to the polypeptide chain

The oligosaccharide intermediate is transferred *en bloc* from the dolichol to some, but not all, Asn (if present in the consensus sequence, Asn-Xaa-Ser/Thr; *see 4.2.1*) on the growing (nascent) polypeptide chain. The transfer is catalysed by an enzyme called 'oligosaccharyltransferase' (OST) and the process is illustrated schematically in *Figure 4.4*. ➲ The exact mechanism by which OST achieves this *en bloc* transfer to the growing peptide is not understood, although the structure of the enzyme has been characterized. OST is a complex enzyme (probably the most complex of all the glycosyltransferases). In humans OST contains eight subunits all of which are transmembrane proteins in the RER and are present in association with other proteins, for example DAD1 ('defender of apoptotic cell death 1') as a complex.

> ➲ In N-linked glycosylation, the 'oligosaccharide intermediate' is $Glc_3Man_9GlcNAc_2$ and this intermediate is transferred to the nascent peptide chain by the action of the enzyme oligosaccharyltransferase (OST).

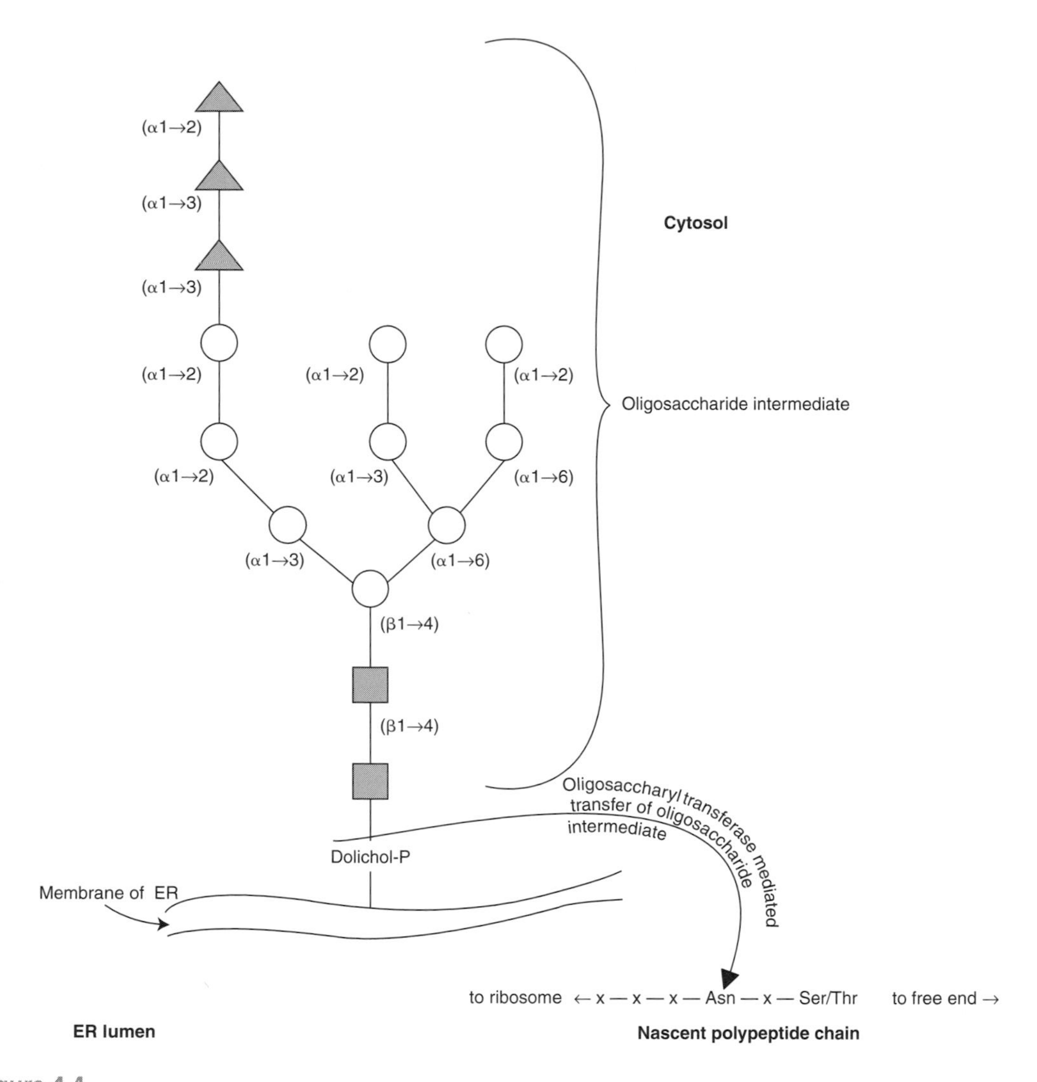

Figure 4.4

As soon as the oligosaccharide intermediate is attached to Dol phosphate, processing of N-linked oligosaccharides begins. First, the oligosaccharide intermediate is transferred from Dol phosphate to an Asn (part of a consensus sequence) on the nascent polypeptide chain. This reaction is catalysed by the oligosaccharyltransferase (OST) enzyme.

Addition of the intermediate molecule to the growing polypeptide chain is the first committed step in the production of an N-linked oligosaccharide.

4.2.6 Trimming the oligosaccharide intermediate in the ER

➲ The removal of two Glc molecules from the oligosaccharide intermediate provides a recognition signal for chaperone molecules that aid the process of protein folding.

The oligosaccharide intermediate is trimmed by glucosidases I and II which, respectively, remove the first and subsequent two Glc residues, and then further trimmed by an α-mannosidase that removes one Man residue. These steps are illustrated in *Figure 4.5*. At first glance, the process of adding monosaccharides to an oligosaccharide and then removing them again appears biologically inefficient.➲ Removal of the Glc residues, however, appears to be an important control mechanism in the subsequent folding of the

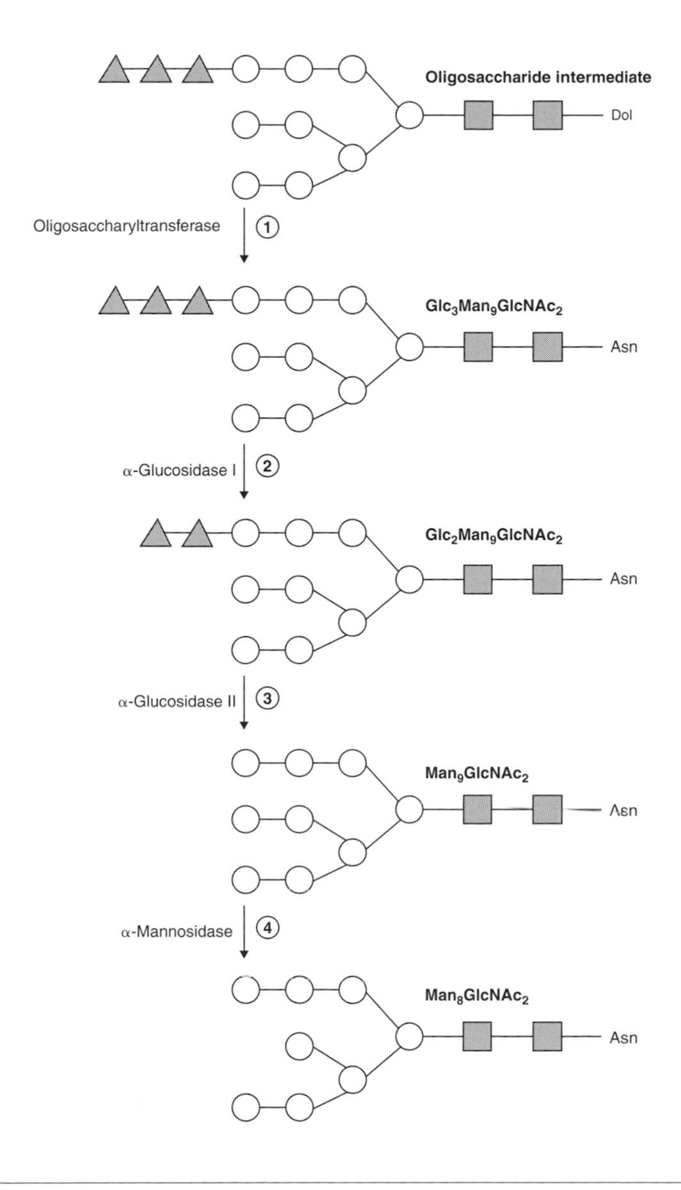

Figure 4.5

After the oligosaccharide intermediate is attached to an Asn on the nascent polypeptide chain (step 1) three glucose residues (▲) are trimmed by α-glucosidases I and II (steps 2 and 3). Finally, an α-mannosidase trims away a mannose residue (○) (step 4).

protein – a step that is accomplished with the aid of chaperone molecules; also, glucosidase I acts while OST is attached, and aids in the release of this enzyme from the growing polypeptide chain.

4.2.6.1 Chaperone molecules

There are many classical chaperone molecules involved in protein folding; an example is protein disulfide isomerase, the molecule responsible for the production of disulfide linkages. Oligosaccharides on proteins, by virtue of their hydrophilic nature, function to keep proteins in solution during and after the folding process. Therefore, indirectly, oligosaccharides aid in the function of classical chaperone molecules. It is notable that proteins expressed in non-glycosylating cells, for example, *Escherichia coli* can aggregate inside the cell.

The structure of calnexin and calreticulin

Two homologous chaperone molecules, calnexin (CNX) and calreticulin (CRT), both with lectin-like properties (*see 13.12.3.4*), function by recognizing the Glc moiety on the oligosaccharide intermediate: $Glc_1Man_9GlcNAc_2$ attached to the polypeptide chain. CNX is a type I membrane protein that resides in the ER, whereas CRT is found in soluble form in the lumen of the ER. Both molecules have a long hydrophilic peptide arm and a globular domain which interacts to form a complex with a thiol oxidoreductase homologue of disulfide isomerase (ERp57). The structure of the luminal domain of CNX has recently been deduced and found to be similar to lectins of the legume family. The role of these molecules in the process of protein folding is becoming better understood.

The calnexin/calreticulin cycle in protein folding

The cycle by which a growing polypeptide chain, or newly produced glycoprotein, binds to CNX/CRT and regulates protein folding/ER-associated protein degradation (ERAD) is shown schematically in *Figure 4.6*.

First, CNX and CRT recognize the (α1→3)-linked Glc moiety of $Glc_1Man_9GlcNAc_2$ on the oligosaccharide intermediate. Upon binding with CNX/CRT, ERp57 catalyses the formation of

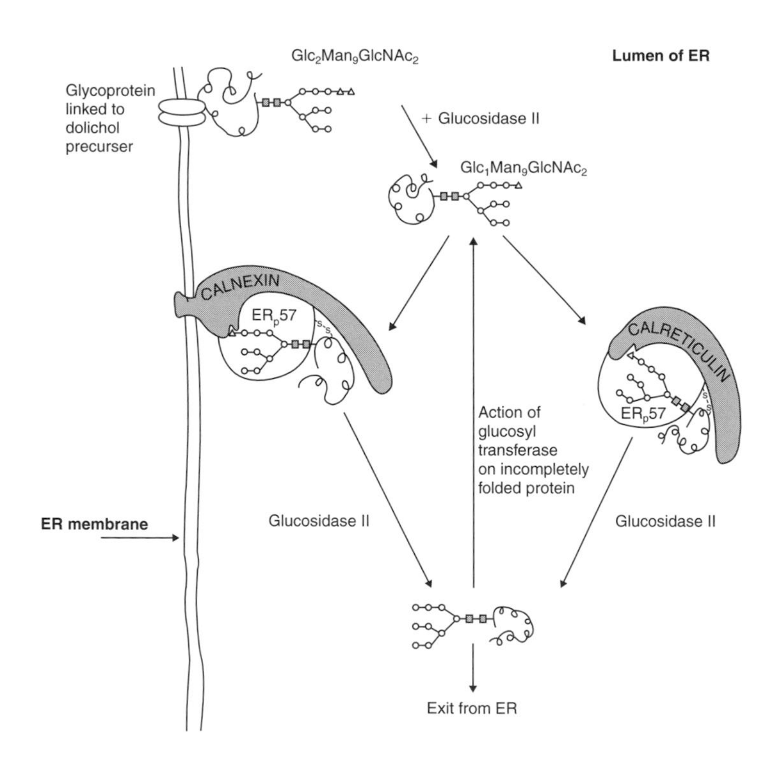

Figure 4.6

The calnexin and calreticulin cycle ensures correct protein folding. Calnexin is a membrane-bound (lectin-like) protein and calreticulin is its soluble homologue. Calnexin and calreticulin are each found complexed with ERp57 (thiol oxidoreductase) a homologue of disulfide isomerase. Calnexin and calreticulin bind the single glucose moiety on the oligosaccharide chain and transient disulfide bonds are formed between the cysteine residues of the protein chain and ERp57. If the protein is properly folded, glucosidase II trims the glucose from the oligosaccharide and releases the protein from the calreticulin/ERp57 or calnexin/ERp57 complex. The protein then exits the ER. If the protein is not properly folded, glucosyltransferase reglucosylates the oligosaccharide and the sequence starts again.

transient disulfide bonds with cysteine residues on the polypeptide chain/glycoprotein. If the protein is now properly folded, the action of glucosidase II results in removal of the (α1→3)-linked glucose from the oligosaccharide intermediate, yielding the product $Man_9GlcNAc_2$, and the concurrent release of the glycoprotein from the CNX/CRT complex. This then signals exit of the glycoprotein from the ER.➲

➲ The CNX/CRT cycle ensures that proteins are correctly folded before they exit the ER.

If, however, the protein is not properly folded, uridine diphosphate glucose:glycoprotein glucosyltransferase (GT) re-glucosylates the protein. The action of GT is triggered by the physical features of the unfolded glycoprotein, for example, the presence of hydrophobic patches. The protein continues through this cycle of re-glucosylation until it is properly folded. If this does not happen, or if proteins are terminally misfolded, they pass into the ER-associated degradation cycle. The importance of the CNX/CRT cycle is highlighted by the fact that transgenic mice without CRT do not survive past day 18 of embryogenesis.

The CNX/CRT cycle and ER-associated degradation of proteins

Misfolded/mutant proteins are transported from the ER into proteasomes in the cytosol for degradation. The trimming of $Glc_{(0-3)}Man_9GlcNAc_2 \rightarrow Glc_{(0-3)}Man_8GlcNAc_2$ by mannosidase I is an important step in this process and it has been suggested that $Glc_1Man_8GlcNAc_2$ is a signal for the ER-associated degradation of proteins.

4.2.7 Processing of N-linked oligosaccharides in the Golgi apparatus

The Golgi apparatus is a series of membranous sacs, comprising the *cis*, *medial* and *trans* cisternae. Each of the sacs contains specific glycosidases and glycosyltransferases (*Chapter 2*) which either trim or add monosaccharides to produce the final N-linked glycoform present on the glycoprotein. This can take the form of any of the following: high mannose type, hybrid type or complex type N-linked glycans.

When membrane-bound and secretory glycoproteins arrive in the Golgi apparatus, the N-linked oligosaccharide intermediate is trimmed by α-mannosidase to $Man_5GlcNAc_2$. This structure is illustrated in *Figure 4.7*. If $Man_5GlcNAc_2$ is not trimmed by α-mannosidase, then it will remain a *high-mannose*-type N-linked oligosaccharide as it terminates in Man residues.➲

➲ $Man_5GlcNAc_2$ forms the basis for hybrid and complex N-linked oligosaccharide chains. It is modified in different ways by a series of trimming and building enzymes.

If up to four Man residues are removed, then the remaining $Man_5GlcNAc_2$ structure will form the basis for all the other N-linked oligosaccharide chains. This structure is modified by a series of

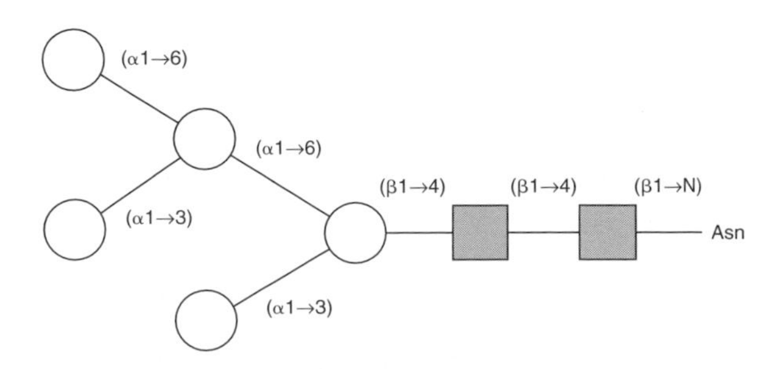

Figure 4.7

When it enters the Golgi apparatus, the N-linked oligosaccharide intermediate is trimmed by a $Man_5GlcNAc_2$ structure.

trimming (glycosidases) and building enzymes (glycosyltransferases) that act as the glycoprotein travels through the Golgi stacks from *cis*- to *medial*- to *trans*-Golgi, to result in many diverse N-linked oligosaccharide structures.

The processing path for one class of N-linked oligosaccharides, the complex N-linked oligosaccharides (*see 4.3.2.2*), is summarized in *Figure 4.8*. In the first step of the process, if the $Man_5GlcNAc_2$ is trimmed by α-mannosidase it will become a suitable substrate for GlcNAc-TI – this enzyme adds (β1→4) GlcNAc to the (α1→3)Man branch. This linkage of GlcNAc then enables a glycosidase (an α1,3/α1,6 mannosidase) to act on the outer-arm Man residues and remove them from the oligosaccharide chain. The resulting structure then acts as a substrate for the enzymes responsible for the formation of the other two types of N-linked glycans: *hybrid* and *complex* types.

As these heterogeneous oligosaccharide structures are produced by glycosyltransferases and glycosidases, in essence, the diversification into the different classes of the N-linked oligosaccharides reflects the levels of the competing glycosyltransferases (*Chapter 2*).

4.2.8 Glycoprotein targeting to the lysosome

One particularly important group of N-linked oligosaccharides includes those found on proteins and enzymes destined for the lysosomal compartment of the cell.

In the late 1970s studies with lysosomal enzymes, for example: β-hexosaminidase, cathepsin D and β-glucoronidase, showed that they were 'capped' with GlcNAc residues. It was subsequently found that on these enzymes the $Man_8GlcNAc_2$ oligosaccharide (illustrated as **1** in *Figure 4.8*) is modified with a GlcNAc-1-phosphotransferase, thereby adding phosphorylated GlcNAc to the sub-terminal Man of the oligosaccharide. The GlcNAc residue is then removed, leaving Man-6-phosphate. This is the recognition determinant for the mannose 6-phosphate receptor in the lysosomes.➲ Mannose 6-phosphate receptor binding is the main event that leads to the targeting of glycoproteins destined for degradation to the lysosomal compartment in the cell.

➲ Proteins that are to be degraded are modified such that their glycans contain mannose-6-phosphate. This is the key structural determinant that enables successful targeting to the lysosomal compartment of the cell.

Inclusion-cell (or I-cell) disease and pseudo-Hurler polydystrophy (mucolipidosis type III) are examples of diseases that occur as a result of congenital defects in glycosylation (CDG; *see 16.8.1*). I-Cell disease and pseudo-Hurler polydystrophy are characterized by a lack of GlcNAc-1-phosphotransferase, the enzyme that is essential for the correct targeting of the proteins for lysosomal degradation. The lysosomal compartments of the cells of affected individuals show enzyme insufficiencies and symptoms that include severe developmental problems, psychomotor problems, hip dislocation, skeletal abnormalities and coarsening of the facial features.

4.3 Common structural features of N-linked oligosaccharides and the generation of structural diversity

Having established the principles of N-linked oligosaccharide synthesis in the previous sections, we examine the structures of N-linked oligosaccharides in more depth.

4.3.1 The common trimannosyl core

➲ N-Linked oligosaccharides all contain the same core glycan. This structure is called the 'trimannosyl' core and has the formula $Man_3GlcNAc_2$.

Common to all N-linked oligosaccharides is the presence of a common pentasaccharide 'core' $Man_3GlcNAc_2$ which may be extended and elongated by many different types of linear and branching oligosaccharide chains.➲ The core comprises two GlcNAc monosaccharides linked together in a (β1→4) configuration to which a Man is attached, again in a (β1→4) configuration, and finally a further

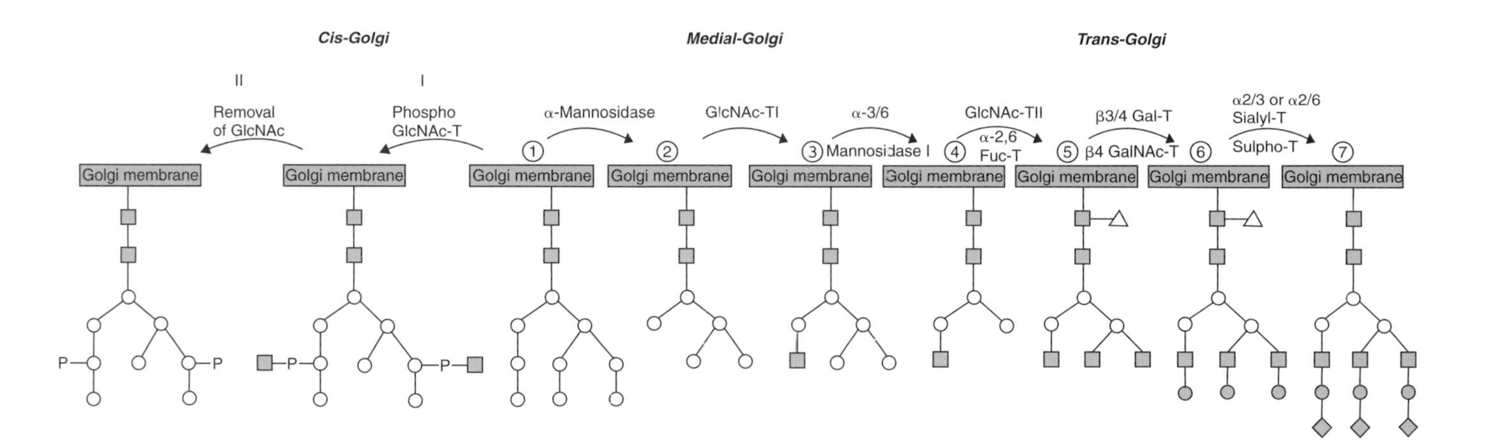

Figure 4.8

The processing path for N-linked oligosaccharides. After entering the Golgi apparatus (1) the N-linked oligosaccharide is modified variably according to the action of a series of glycosyltransferases and glycosidases (shown in steps 1–7). At step 1, the oligosaccharide becomes the high mannose type unless the enzyme GlcNAc-TI acts on the structure. If this occurs, the oligosaccharide becomes either a hybrid or complex type. The removal of mannose by α-mannosidase (steps 3–4) leads to the eventual production of a complex type N-linked oligosaccharide through the sequential actions of GlcNAc-TII (step 4), β1,3/4 Gal-T (step 5) and/or β1,4 GalNAc-T (step 6), α2,3/2,6 ST or sulfotransferase (step 7). The steps shown as I and II are the modification of the oligosaccharide by action of phospho-GlcNAc-T (step I) and an N-acetylglucosaminidase (step II). These modifications enable the recognition of the glycoproteins by the mannose-6-phosphate receptor in the lysosome, resulting in eventual protein degradation.

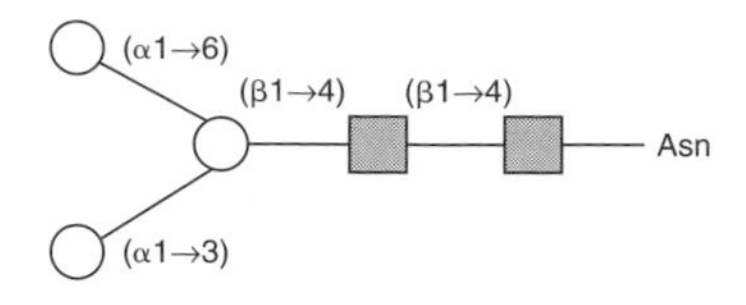

Figure 4.9

The trimannosyl core.

two Man are attached in (α1→3) and (α1→6) linkages to the initial, core Man. This pentasaccharide, illustrated in *Figure 4.9*, is called the 'trimannosyl' core.

4.3.2 Different classes of N-linked oligosaccharides

➲ The three main classes of N-linked oligosaccharides are high mannose, complex and hybrid types. Within each of the classes, there is considerable micro-heterogeneity in glycan composition.

There are three main classes of N-linked oligosaccharides which can be built on the trimannosyl core: (i) high mannose type, (ii) complex type; and (iii) hybrid type.➲

4.3.2.1 High mannose-type oligosaccharides

High mannose oligosaccharides are so called because they contain between five and nine Man molecules attached to the innermost GlcNAc residues of the trimannosyl core. They are therefore often referred to as, for example, Man_7 (containing a total of seven Man residues) or Man_9 (containing a total of nine Man) structures. The high mannose structures always contain a maximum of one branch on the (α1→3)-linked core Man, and a maximum of two branches on the (α1→6)-linked core Man. In the larger structures (Man_7 to Man_9), the outermost Man residues are linked to the proximal residues in an (α1→2) configuration. An example of a high mannose oligosaccharide (a Man 9 oligosaccharide) is illustrated in *Figure 4.10*. By reference to *Figure 4.8*, it is apparent that the $Man_9GlcNAc_2$ shown in step 1 is the starting material for the production of high mannose type N-linked oligosaccharides.

4.3.2.2 Complex-type oligosaccharides

The different branches and the complexity of these structures give rise to their name and classification as complex oligosaccharides. They are the most structurally diverse group of the N-linked oligosaccharides.

➲ The disaccharide GlcNAc(β1→4)Gal is called a lactosamine unit. Many $[GlcNAc(\beta1\rightarrow4)Gal]_n$ units in sequence is termed polylactosamine. These glycans are characteristic of the antennae of complex N-linked oligosaccharides.

Complex oligosaccharides do not contain any additional Man residues other than those found in the trimannosyl core.➲ They are characterized by the disaccharide GlcNAc(β1→4)Gal, also called a lactosamine unit, attached to the trimannosyl core as antennae. The monosaccharides attached as antennae may be found as repeating units, in which case, several GlcNAc(β1→4)Gal lactosamine units appear in sequence, $[GlcNAc(\beta1\rightarrow4)Gal]_n$, and are termed polylactosamine extensions or chains. An example of a complex type N-linked oligosaccharide is illustrated in *Figure 4.11*.

By reference to *Figure 4.8*, it is apparent that the structure shown in step 4 is the starting point for complex-type oligosaccharides. Oligosaccharides with two branches (bi-antennary glycans) are formed as well as glycans with more branches, so-called tri- (three branches) and tetra- (four branches) antennary. These structures are produced by the addition of monosaccharides by glycosyltransferases, as indicated in *Figure 4.11*.

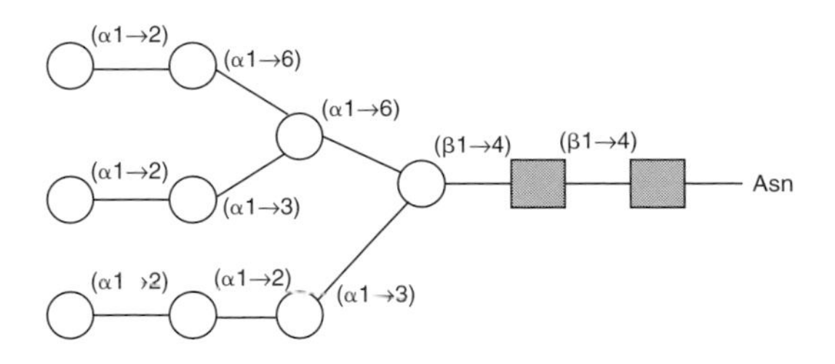

Figure 4.10

An example of a high mannose oligosaccharide. This oligosaccharide has nine mannose residues (○) attached to the two core GlcNAc residues (■), and is a Man_9 oligosaccharide.

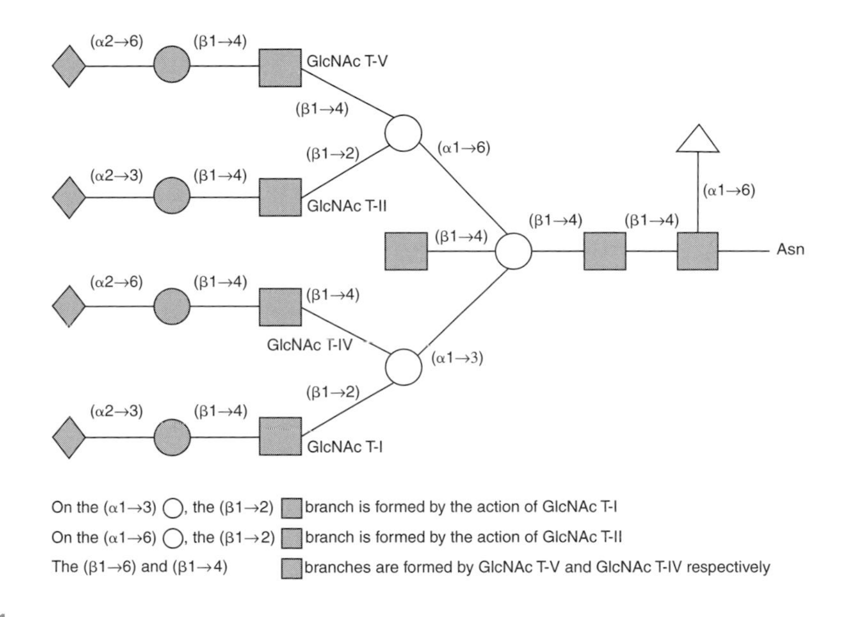

Figure 4.11

An example of a complex oligosaccharide. The only mannose residues (○) involved in this type of structure are the three mannose residues of the trimannosyl core. The core is elongated by lactosamine units (■—●) and the antennae terminate in sialic acid (◆) and fucose (△). The structure is further elaborated by a GlcNAc residue (■) bisecting the core. The enzymes involved in the construction of the branches (antennae) of this oligosaccharide are indicated.

Sialic acids are negatively charged monosaccharides (*Chapter 11*). They are common constituents of complex oligosaccharides and are often the outermost monosaccharide, most distant from the trimannosyl core. Sialic acid is usually in an (α2→6) or an (α2→3) linkage to the adjacent monosaccharide, but other configurations may be found. Occasionally sulfate, α-linked Gal or β-linked GalNAc may be present as the terminal monosaccharides on N-linked oligosaccharide chains.

Further complexity in this group of oligosaccharides occurs if GlcNAc is attached to the trimannosyl core as a 'bisecting GlcNAc' and/or Fuc is attached to the innermost GlcNAc, or on the outer antennae. In the latter case, the monosaccharides found on the outer arms of the complex oligosaccharides are often Lewis antigens (*see 10.6*), saccharide structures common to N- and O-linked glycoproteins and glycolipids.

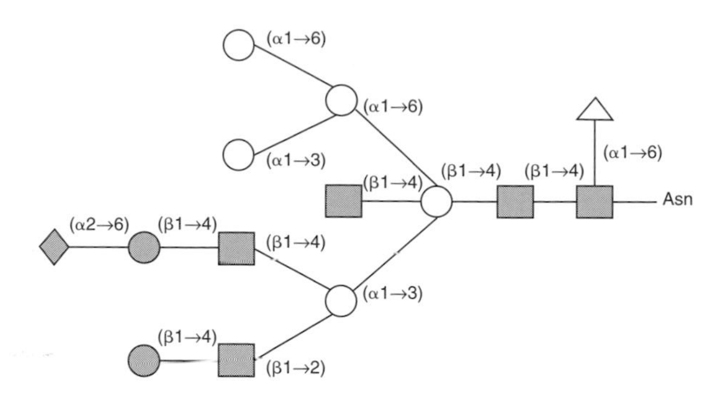

Figure 4.12

An example of a hybrid oligosaccharide. This structure has features of both high mannose and complex oligosaccharides.

4.3.2.3 Hybrid-type oligosaccharides

Hybrid oligosaccharides, as their name implies, possess some of the structural features of both the high mannose and complex oligosaccharides and are therefore best described as a hybrid of these two other classes of N-linked oligosaccharides. They are consequently a diverse group of oligosaccharides. By reference to *Figure 4.8*, it is apparent that the structure in step 3 is the starting material for the production of hybrid type structures. An example of a hybrid oligosaccharide is shown in *Figure 4.12*.

4.4 Functions of N-linked oligosaccharides

4.4.1 Glycosylation and protein conformation

➲ N-Linked oligosaccharides have been shown to have roles in:

Recognition events:
- protein folding
- enzyme function
- serum clearance of proteins
- protein signalling
- as cell adhesion molecules.

Physical functions:
- protease protection/stability
- preventing non-specific interactions
- orientation of cell-surface molecules

The importance of N-linked oligosaccharide structures in relation to correct protein folding has been studied extensively.➲ An example in which N-linked glycosylation is involved in the correct protein conformation is the human cell-surface glycoprotein CD2 (lymphocyte function-related antigen-2, LFA-2). CD2 interacts with its receptor, in humans CD58 (LFA-3), to mediate cell adhesion and signal transduction.

Another example of the importance of N-linked oligosaccharides in correct protein conformation is that of the N-linked oligosaccharides present on each of the constant domains in the hinge region of immunoglobulin G (IgG). In this case, one of the (α1→3) branches interacts with the trimannosyl core of the oligosaccharide on the other IgG constant domain. The other branch of the oligosaccharide (α1→6) then forms a tight interaction with a binding pocket, a lectin-like area on Lys246, Glu258 and Thr260, on the IgG constant domain surface. These, and other data, suggest a role for oligosaccharides in maintaining appropriate presentation of the hinge conformation to result in successful binding to monocytes.

4.4.2 Blood group determinants

The presence of ABO blood group determinants on erythrocytes has been known for over 100 years and that these determinants are oligosaccharides was first discovered over 50 years ago.

ABO blood group sugars can also be found on a range of cell types other than blood cells, including endothelial and epithelial cells, and are present on glycoproteins secreted in body fluids such as saliva, milk and semen in some individuals termed 'secretors' (*see 10.5*). Other blood group antigens, in addition to the ABO system, have been described, and many of these are based on oligosaccharide structures. Examples include Lewis (*see 10.6*), Ii (*see 10.7*) and P-related (*see 10.8*) blood groups. Blood group sugars may form part of both N- and O-linked oligosaccharides.

4.4.3 Serum proteins

The functions of N-linked oligosaccharides on serum proteins include correct folding and conformation of the protein, maintaining protein stability and metabolic turnover. This is an important aspect of glycosylation that is referred to in many other sections of this book.

4.4.4 Glycosylation and enzyme function

4.4.4.1 Tissue plasminogen activator (tPA)

A well-documented case in which N-linked oligosaccharides have been shown to be important to enzyme function is that of tissue plasminogen activator (tPA). tPA is a serine protease, which converts plasminogen to plasmin and induces fibrinolysis (blood clot breakdown). The presence of an N-linked oligosaccharide residue at Asn184 on tPA is important for various properties of the protein including conversion of the single-chain to the double-chain form and its presence reduces fibrinolytic activity by widening the site where tPA binds lysine in fibrin. Plasminogen, the substrate for tPA, is also a glycosylated protein and glycosylation alters the rate of β- to α-conformation changes. These glycosylation patterns have been studied in relation to enzyme/substrate function. Glycosylation modulates the kinetics of association between enzyme and substrate, and, furthermore, glycosylation at Asn184 modulates the enzyme turnover rate. These results illustrate the importance of N-linked oligosaccharides in both enzyme and substrate function.

4.4.4.2 Plasma fibrinogen

Plasma fibrinogen glycosylation is important for the production of clots. Fibrinogen contains heavily sialylated N-linked oligosaccharides involved in calcium binding. Hyperglycosylation may lead to problems in clot formation as illustrated by reduced levels of aggregation of fibrin from fibrinogen in a rare congenital mutation that results in extra N-linked oligosaccharide on fibrinogen.

4.4.4.3 Ribonucleases (RNase) A and B

The ribonucleases (RNase A and B) are responsible for the hydrolysis of 3′- and 5′-phosphodiester bonds in ribonucleic acids. Whereas human pancreatic RNase A is not glycosylated, RNase B has a single glycosylation site variably glycosylated with high mannose N-linked oligosaccharides. Although the oligosaccharides do not dramatically alter enzyme activity, modelling experiments suggest glycosylation may affect the stability of the enzyme and alter the spatial interaction of RNase B with its substrate.

4.4.5 Clearance and catabolism of serum proteins

In the late 1960s it was noted that the copper-transporting protein ceruloplasmin had a dramatically reduced half-life in the circulation if it contained no sialic acid. The removal of other desialylated glycoproteins from rabbit serum via a hepatic asialoglycoprotein receptor, a lectin (*Chapter 13*), was proposed. Since then, the asialoglycoprotein receptor on hepatocytes that binds Gal-containing oligosaccharides has been shown to mediate the lysosomal degradation of many such proteins. In addition to clearance of asialo-glycoproteins, there has been considerable interest in the catabolism of other glycoproteins. Saccharide-binding receptors or lectins (*Chapter 13*) have been found for Fuc-, GlcNAc- and Man-containing oligosaccharides. The glycosylation of serum proteins has been shown to influence their survival, or half-life times, *in vivo*. For example, de-glycosylated erythropoietin has approximately one-tenth the activity of the fully glycosylated molecule, primarily as a result of renal filtration and clearance by hepatocyte receptors. This is particularly important in the biotechnology industry where manufacturing processes have had to be developed to produce correctly glycosylated forms of synthetic erythropoietin in order to ensure that the product is biologically efficacious (*see 18.5.1*).

4.4.6 N-Linked glycosylation and hormone signalling

An increasing body of evidence suggests that the N-linked oligosaccharides found on pituitary hormones and human chorionic gonadotrophin are important signalling molecules (*see 7.6*).

4.4.7 Tissue-specific glycosylation

Tissue-specific alterations in glycosylation of the same protein have been described in many instances. One of the best documented examples is that of Thy-1 extracted from different tissues. This work led to the conclusion that there were no oligosaccharides in common on Thy-1 from brain and Thy-1 from thymus and that the glycosylation of this protein is controlled in a highly tissue-specific manner. The functional reason for tissue-specific glycosylation is not yet well understood.

Further reading

Ashwell, G., Harford, J. (1982) Carbohydrate-specific receptors of the liver. *Annu. Rev. Biochem.* **51:** 531–534.

Helenius, A., Aebi, M. (2001) Intracellular functions of N-linked glycans. *Science* **291:** 2364–2369.

Helenius, J., Ng, D.T., Marolda, C.L., Walter, P., Valvano, M.A., Aebi, M. (2002) Translocation of lipid-linked oligosaccharides across the ER membrane requires Rft1 protein. *Nature* **415(6870):** 447–450.

Hubbard, S.C., Ivatt, R.J. (1981) Synthesis and processing of asparagine-linked oligosaccharides. *Annu. Rev. Biochem.* **50:** 555–583.

Kobata, A., Takasaki, S. (1992) Structure and biosynthesis of cell surface carbohydrates. In M. Fukuda (ed.) *Cell Surface Carbohydrates and Cell Development*, pp. 2–24. CRC Press, London.

Kornfeld, S. (1987) Trafficking of lysosomal enzymes. *FASEB J.* **1:** 462–468.

Rudd, P.M., Elliot, T., Cresswell, P., Wilson, I.A., Dwek, R.A. (2001) Glycosylation and the immune system. *Science* **291:** 2370–2376.

O-Linked (mucin-type) glycoproteins

5

5.1 What are O-linked glycoproteins?

O-Linked-glycoproteins, or O-glycoproteins, are those in which the first monosaccharide of the oligosaccharide chain, usually GalNAc, is attached through an α-O-glycosidic linkage to an oxygen molecule (hence 'O-linked') of an amino acid residue, usually serine or threonine, on the polypeptide chain of a protein (*Figure 5.1*).➲ They differ fundamentally from N-linked glycoproteins (*see Chapter 4*) in which the first monosaccharide is GlcNAc linked by a β-N-glycosidic bond to the amide group of the amino acid asparagine on the polypeptide chain of a protein.

➲ In most O-linked glycoproteins, an initial GalNAc residue is attached in an α-O-glycosidic linkage to a serine or threonine residue of the polypeptide chain.

5.2 The relationship between N-linked and O-linked glycoproteins

Both N- and O-linked glycoproteins share common features, and many, including most mucin-type glycoproteins, carry both N- and O-linked glycans attached to the same protein molecule. Furthermore, many glycosyltransferases, for example those that synthesize blood group ABO sugars, will act on both N- and O-linked glycan chains, so that the terminal saccharides are identical blood group sugars (*see 10.4*). However, N- and O-linked oligosaccharides also have unique features with regard to structure, function and biosynthesis and these will be highlighted in this chapter. O-Glycosylation is less well understood than N-glycosylation.

5.3 Types of O-linked glycoproteins

O-Glycoproteins may be classified into different subgroups according to the nature of the sugar and amino groups involved in the linkage between the first monosaccharide and the peptide. These are listed below:

(a) GalNAc linked to serine or threonine. This is the most common and most variable type of O-glycosylation in vertebrates, and also occurs in plants. Glycoproteins of this type are also classified as 'mucin-type' glycoproteins;
(b) GlcNAc linked to serine or threonine residues that are also sites of phosphorylation (O-GlcNAc type). This is found only in cytoplasmic and nuclear proteins in animals;
(c) Xyl linked to serine or threonine in proteoglycans. This type of O-glycosylation is relatively abundant in the extracellular matrix of animal tissues;
(d) Gal and Glc-Gal linked to the hydroxyl group of hydroxylysine in collagen and proteins containing a collagen-like domain;
(e) Xyl-Glc or Glc linked to serine or threonine in blood clotting factors;
(f) O-linked Fuc, for example, Fuc linked to plasma glycoproteins;
(g) Man O-linked to Ser/Thr in, for example, vertebrate proteoglycans of the brain;
(h) Ara linked to hydroxyproline in plants. Probably analogous to Gal/Glc-Gal linked to hydroxylysine in the collagen of animals;

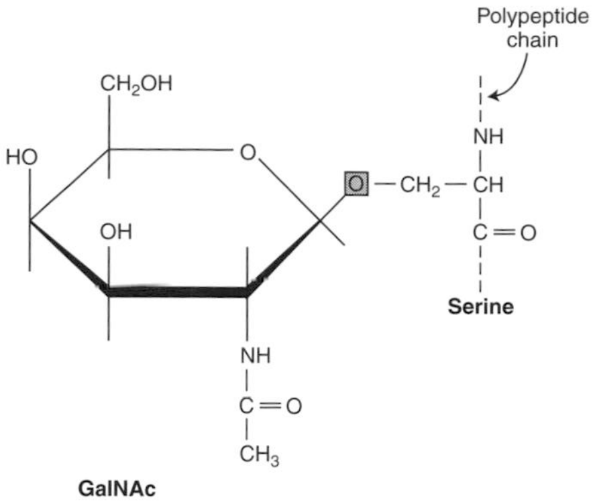

Figure 5.1

O-Linkage. In O-linked glycoproteins, the oligosaccharide is linked via a GalNAc molecule in an α-O-glycosidic type bond to an oxygen (shaded) of (usually) a serine or threonine amino acid residue on the polypeptide chain.

(i) Gal linked to serine in plants;
(j) Man linked to threonine or serine in fungi and yeasts.

In animal tissues, the most common type of O-glycosylation by far is (a) in which the reducing terminal GalNAc is linked to a serine or threonine residue in the polypeptide chain. As this type of glycosylation occurs abundantly in a group of glycoproteins called mucins, glycans O-linked in this way are commonly referred to as 'mucin-type' glycoproteins. They are the topic of this chapter, whereas (b) O-GlcNAc glycosylation is the subject of *Chapter 6*, and some of the other less common types of O-linked glycans, listed above, are described in *Chapter 7*.

5.4 Structure of O-linked mucin type glycoproteins

O-Linked mucin-type glycoproteins may be either secreted or membrane-bound cell-surface molecules. They are very large, usually >200 kDa, with the O-glycan oligosaccharide chains attached at high density and in clusters to give often remarkably heavily glycosylated molecules with as much as 50–80% of the mass as carbohydrate. As many as one in three amino acid residues may be glycosylated. The structure of mucin-type glycoproteins may be likened to that of a 'bottle brush' in which the peptide backbone, represented by the wire handle of the brush, bristles with multiple oligosaccharide chains as illustrated in *Figure 5.2*. The glycans of mucin molecules range in size from a single monosaccharide (a GalNAc residue linked directly to a serine or threonine of the polypeptide backbone of the protein) to complex linear or branched oligosaccharide chains of some 20 residues. There may be several hundred oligosaccharide chains attached to a single protein molecule. Oligosaccharide chains of mucins are predominantly O-linked, but some N-linked oligosaccharides may be present on the same molecule. The O-linked oligosaccharides are composed predominantly of GalNAc, GlcNAc, Gal, Fuc and sialic acid.

O-Linked (mucin-like) glycoproteins may be either secreted or membrane-bound cell-surface molecules.

O-Linked mucin-type glycoproteins are widely distributed in vertebrate epithelia and are heterogeneous in both the structure of the oligosaccharide chains they bear, and the variety of proteins to which they may be attached. Mucin gene expression is species- and tissue-specific, and may be altered in disease states such as cancer.

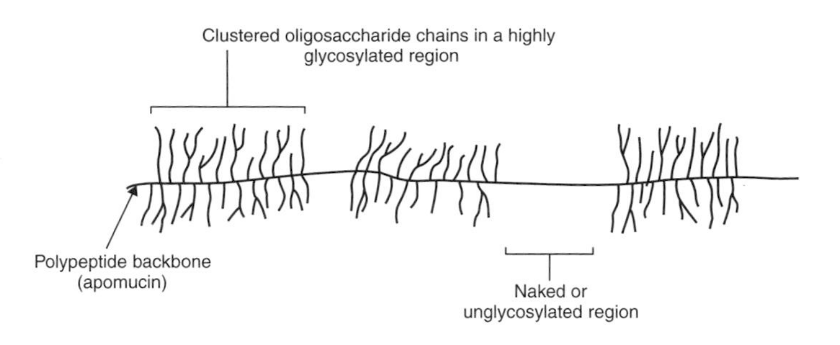

Figure 5.2

Schematic 'bottle brush' diagram of the structure of a typical mucin. Multiple oligosaccharide chains are attached in dense clusters to sequences of the polypeptide backbone or apomucin.

5.4.1 Clustering of oligosaccharides on the mucin backbone

Clustering of the O-linked oligosaccharide chains on the polypeptide backbone of the protein usually occurs where there are many repeating amino acid sequences (tandem repeats) containing multiple potential O-glycosylation sites. These tandem repeat sequences differ in precise amino acid composition and in the genes coding for the different O-glycosylated glycoproteins, but share common features, as they contain many serine and/or threonine amino acid residues in an often proline-rich stretch of the polypeptide chain.➲ Proline-rich regions of the polypeptide chain are favoured because in these regions the side chains of the serine/threonine residues are physically accessible.

➲ Multiple O-glycan attachment sites are clustered within tandem repeat sequences, thus resulting in a high density of oligosaccharide chains attached to stretches of the polypeptide backbone.

Typically, the polypeptide also contains cysteine-rich regions, which are not O-glycosylated, and which are involved in protein folding by means of disulfide bonds. These regions may contain sites of attachment for N-glycans. The peptide part of the mucin molecule is referred to as the 'apomucin'.

5.4.2 Tandem repeat regions on mucin polypeptide chains

O-Linked mucin-like glycoproteins contain densely packed, multiple potential O-linked glycosylation sites. These are provided by multiple, short tandem repeat sequences in the polypeptide chain. There has been speculation that this may have arisen through duplication of exons, but, although the potential glycosylation attachment sites of some O-linked mucin-like glycoproteins, like glycophorin A, are coded for by several exons, this is not always the case. O-Linked mucin sequences may sometimes have arisen by alternative mechanisms. Examination of the coding sequence for a number of other O-glycosylated molecules, for example leukosialin (*see 16.4*), reveals that the entire coding sequence for the molecule lies within a single genomic exon. In submaxillary gland apomucin, for example, there are 25 tandem repeats of an 81 amino acid long sequence. This short tandem repeated region of the polypeptide chain strongly suggests that evolutionary mechanisms other than exon duplication have resulted in the development of polypeptide sequences containing densely packed multiple O-glycosylation sites in these molecules.➲ Alternatively, it is possible that an original exon containing the glycosylation attachment sites was duplicated, and that the new exons were then joined with the elimination of any intervening introns.

➲ The existence of multiple tandem repeats containing densely packed glycosylation sites is most likely to have arisen through exon duplication, although other mechanisms have been postulated in some cases.

➲ Secreted O-linked glycoproteins typically form slippery, viscous mucins, but this is not always the case, and many less heavily glycosylated O-linked glycoproteins do not exhibit mucinous properties.

Mucins are so called because of their mucilaginous nature when secreted by the cell in which they are produced. This is because they cross-link in solution, through disulfide bonds, to produce a slippery, viscous, gelatinous product.➲ Mucins are often produced by specialized secretory epithelial cells, called goblet cells, such as those interspersed between the lining cells of the gastrointestinal and respiratory tracts.

It is important to realize that not all O-linked glycoproteins are actually associated with mucins. Other types of glycoprotein, for example, many serum glycoproteins, carry O-linked oligosaccharide chains. These are usually much less abundant (often only single oligosaccharide chains), are shorter and less complex than those characteristically found on mucins. They are most commonly based on the core 1 and core 2 structures (*see 5.6*).

5.5 An overview of the biosynthesis of O-linked glycoproteins

O-Linked glycoprotein biosynthesis is simpler than N-linked glycoprotein biosynthesis (*see 4.2*) in that the construction of a lipid-linked oligosaccharide precursor to transfer the oligosaccharide to a protein is not required. O-Linked glycosylation is initiated in the Golgi apparatus. Here, monosaccharides are simply added, one at a time, in a stepwise fashion to the growing oligosaccharide chain through the sequential action of glycosyltransferase enzymes (*see Chapter 2*).➲ O-Glycans, unlike N-glycans, are not trimmed by the action of glycosidases. O-Linked oligosaccharides are chains of, principally, the monosaccharides GalNAc, GlcNAc, Gal, Fuc and sialic acid. O-Linked oligosaccharides containing Man and Glc have also been, rarely, reported. They may be sulfated. O-Linked oligosaccharides are found on both glycoproteins and proteoglycans.

➲ O-Linked glycosylation is a posttranslational event and the glycans are simply built by the stepwise addition of one monosaccharide residue at a time to form linear or branching structures.

The addition of the first GalNAc to a serine or threonine of the polypeptide chain of the protein occurs mainly in the *cis*-Golgi (although there remains some controversy over the precise point in the secretory pathway at which it takes place). This is in contrast to N-glycosylation in which the preformed 'oligosaccharide intermediate' (linked to a dolichol pyrophosphate molecule) is linked to the asparagine of the still-growing polypeptide chain as it is processed in the rough endoplasmic reticulum (RER; *see 4.2*). In O-linked glycosylation, following this first step of linking GalNAc to serine or threonine, Glc and GlcNAc residues are subsequently added to the original GalNAc to form characteristic 'core' structures (*see 5.6*).

The core structures may then be terminated by sialic acid or Fuc, or may be elongated to form a longer linear or branching chain, which may itself also be finally terminated by sialic acid or Fuc residues. This type of glycan chain extension is common to O- and N-linked oligosaccharides and glycolipids (*see Chapter 10*).

Like N-glycans, O-glycans are often composed of long chains of monosaccharides, and can have highly variable structures.➲ However, O-glycans can also be very short and simple (as little as a single monosaccharide or disaccharide) and one main difference between N- and O-linked glycans is that O-linked oligosaccharides tend to be less branched than N-linked oligosaccharides. They are based on a simple linear or bi-antennary core (*see 5.6*).

➲ O-Glycans are linear or bi-antennary, and may be as simple as a single monosaccharide or longer oligosaccharides which are based on core structures elongated by polylactosamine units terminated by sialic acid or Fuc.

Elongated O-linked oligosaccharide chains are often composed of a number of repeating Gal(β1→4)GlcNAc(β1→3)Gal, known as 'polylactosamine', units. Fuc or sialic acid are generally found in a terminal position on O-linked oligosaccharide chains and their side branches. Hundreds of different O-glycan chains of varying degrees of complexity are possible.

5.6 Synthesis of O-linked glycans

The structure of O-linked mucin-type oligosaccharides is very heterogeneous, but they can be classified according to their core structures. These core structures are described below and are listed in *Table 5.1* and illustrated in *Figure 5.3*, and the pathways by which they are synthesized are summarized in *Figure 5.4*.

Table 5.1 Mucin-type O-glycan 'core' structures

Core number	Structure	Common name
Core 1	Gal(β1→3)GalNAc-Ser/Thr	Thomsen Friedenreich (TF or T) antigen
Core 2	GlcNAc(β1→6)[Galβ(1→3)]GalNAc-Ser/Thr	
Core 3	GlcNAc(β1→3)GalNAc1→Ser/Thr	
Core 4	GlcNAc(β1→6)[GlcNAc(β1→3)]GalNAcα-Ser/Thr	
Core 5	GalNAc(α1→3)GalNAcα-Ser/Thr	
Core 6	GlcNAc(β1→6)GalNAcα-Ser/Thr	
Core 7	GalNAc(α1→6)GalNAcα-Ser/Thr	
Core 8	Gal(α1→3)GalNAcα-Ser/Thr	

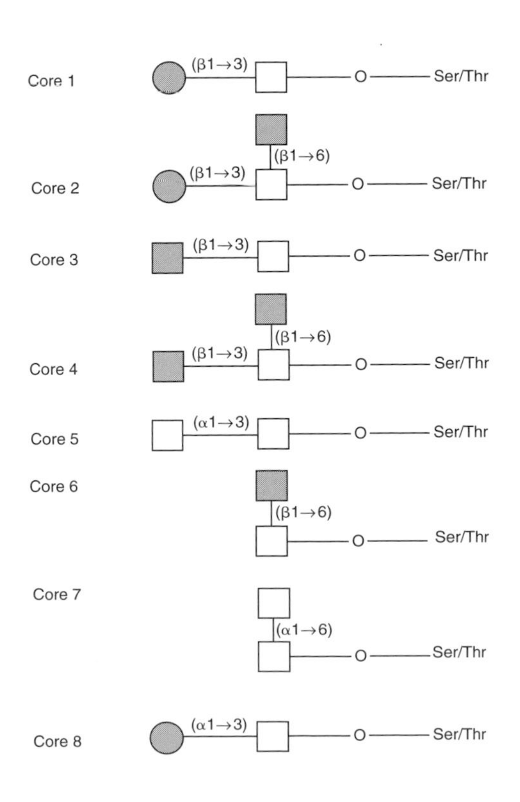

Figure 5.3

Mucin type O-glycan 'core' structures.

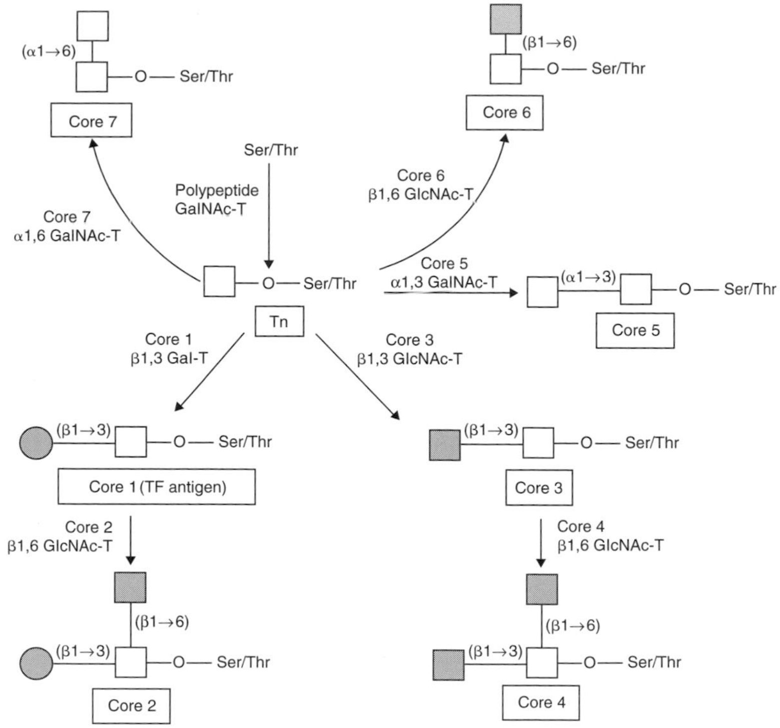

Figure 5.4

Synthesis of core structures 1–7.

5.6.1 Initiation – synthesis of GalNAcα1$\rightarrow$Ser/Thr (also called Tn epitope)

All mucin-type O-glycan synthesis begins with the attachment of GalNAc to a serine or threonine residue on the polypeptide chain, catalysed by N-acetylgalactosaminyltransferase (UDP-GalNAc: polypeptide α1,3-N-acetylgalactosaminyltransferase, polypeptide GalNAc-T). This enzyme is constitutively expressed by all mammalian cells. Attachment of GalNAc to the Ser/Thr of the polypeptide backbone produces the 'Tn epitope', GalNAcα1$\rightarrow$Ser/Thr (*Figure 5.5*).➲

➲ O-Glycan biosynthesis begins with the attachment of a single GalNAc to the serine or threonine of the polypeptide backbone to yield the 'Tn epitope', GalNAcα1$\rightarrow$Ser/Thr.

In N-linked glycosylation a consensus amino acid sequence, Asn-Xaa-Ser/Thr (where Xaa may be any amino acid other than proline), is required at the glycosylation site on the polypeptide backbone (*see 4.2.1*). No comparable consensus sequence has been discovered for O-linked glycosylation. However, some features of O-glycosylation sites have emerged from careful study of the amino acid sequences surrounding sites of O-linked glycosylation. In particular, there is a preponderance of proline residues surrounding sites of attachment of O-glycans, particularly at the -1 and $+3$ positions.➲ The presence of proline disrupts the helical formation of the polypeptide backbone, and instead encourages the formation of β-sheets and β-turns. β-Turns, in particular, favour the attachment of O-glycan chains. There has also been the suggestion that the GalNAc transferase that initiates O-glycosylation may directly recognize the proline residue(s) of the polypeptide chain. In exceptional cases

➲ No consensus sequence for the attachment of O-glycan chains is required, but attachment sites do have characteristic features.

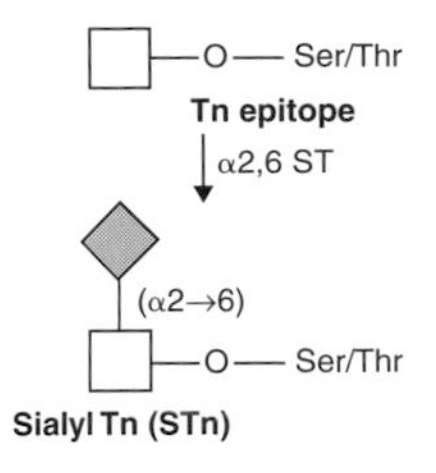

Figure 5.5

The Tn epitope GalNAcα1→Ser/Thr and sialyl Tn (STn) NeuAc(α2→6)GalNAcα1→Ser/Thr.

(e.g. in the glycosylation of glycophorin), O-glycosylation sites are not flanked by proline. Here it is assumed that the polypeptide chain may adopt the required conformation even in their absence. Charged amino acid residues are, in comparison, rarely found at −1 and +3 positions, and charge distribution is important in determining whether glycan attachment takes place. Alanine, serine and threonine are commonly located immediately adjacent to the glycosylated serine or threonine residue.

5.6.2 Formation of NeuAc(α2→6)GalNAcα1→Ser/Thr (sialyl Tn)

The Tn epitope can be acted upon by α-2,6-sialyltransferase (CMP-sialic acid: R_1-GalNAc-R α2,6-sialyltransferase I, ST2,6GalNAcI, ST6GalNAcI) to give NeuAc(α2→6)GalNAcα1→Ser/Thr; 'sialyl Tn' or STn, illustrated in *Figure 5.5*. This disaccharide can no longer be an acceptor for other glycosyltransferases and is always the final product. It is common in submaxillary mucins and some other glycoproteins, and its presence is often increased in certain cancers. Alternatively, the original GalNAcα-Ser/Thr can be extended in several ways to give 'core structures', four of which are common. The O-glycan 'core' structures are illustrated in *Figure 5.3*.➲

➲ The initial GalNAc residue may be sialylated to give 'sialyl Tn', NeuAc(α2→6) GalNAcα1→Ser/Thr, which prevents further chain extension, or may be elaborated to form characteristic O-glycan 'core' structures.

5.6.3 Formation of core 1

The Tn epitope can be extended by an enzyme called core β1,3-galactosyltransferase (UDP-Gal: GalNAc-R β1,3-galactosyltransferase, β1,3 Gal-T or core 1 Gal-T) which adds a (β1→3)-linked Gal to form 'core 1', which is Gal(β1→3)GalNAcα1→Ser/Thr.

There are probably multiple forms of core 1 Gal-T in humans, differentially expressed in various cell and tissue types. This enzyme requires divalent cations such as Mn^{2+} or Cd^{2+} to function, and is inhibited by Zn^{2+}. Its activity is strongly affected by the composition, sequence and length of the polypeptide backbone, the attachment site and number of other sugar residues already present at nearby glycosylation sites.

The reaction by which core 1 is synthesized may be summarized as follows:

UDP-Gal + GalNAcα1→Ser/Thr → Gal(β1→3)GalNAcα1→Ser/Thr + UDP
Tn core 1

The core 1 structure is also known as the Thomsen Friedenreich (TF or T) antigen. The same enzyme that sialylates the Tn structure, ST2,6GalNAcI (*see 5.6.2*) can also act on the TF antigen to yield 'sialyl TF' or 'sialyl T', NeuAc(α2→6)Gal(β1→3)GalNAcα1→Ser/Thr as illustrated in *Figure 5.6*.➲ Substitution of core 1 with sialic acid to yield sialyl TF, or alternatively with Fuc, retards subsequent synthesis of core 2 (*see 5.6.4*).

➲ 'Core 1', Gal(β1→3) GalNAcα1→Ser/Thr, is also known as the Thomsen Friedenreich (TF or T) antigen. It may be sialylated to yield sialyl T antigen, or fucosylated or may be subject to further chain extension.

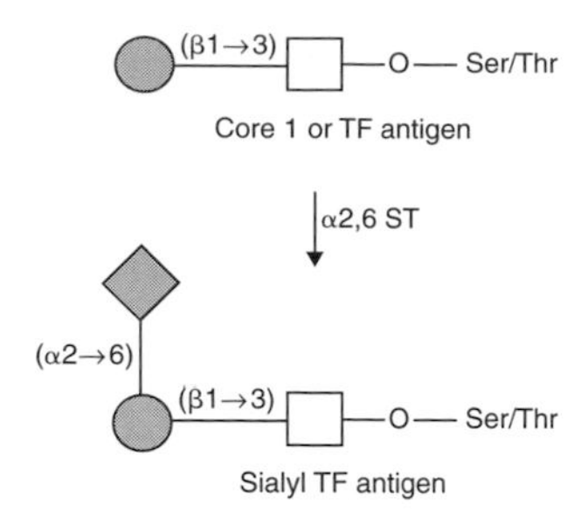

Figure 5.6

Thomsen Friedenreich (TF or T) antigen or 'core 1', Gal(β1→3)GalNAcα1→Ser/Thr and sialyl TF antigen NeuAc(α2→6)Gal(β1→3)GalNAcα1→Ser/Thr.

5.6.4 Formation of core 2

Core 1 can be converted to 'core 2' by the enzyme β1,6-N-acetylglucosaminyltransferase (UDP-GlcNAc: Gal(β1→3)GalNAc-R or (β1→6)GlcNAc transferase, core 2 β1,6 GlcNAc-T or β1,6 GlcNAc-T) which adds a (β1→6)-linked GlcNAc to give a simple bi-antennary structure, GlcNAc(β1→6)[Galβ(1→3)]GalNAcα1→Ser/Thr.

This enzyme does not require the action of a divalent cation for its activity, but may be stimulated or inhibited by various divalent cations in a tissue-dependent manner. Its expression is regulated during growth and differentiation, and is elevated in children with the immunodeficiency disease, Wiskott–Aldrich syndrome (*see 16.4.1*).

The reaction by which core 2 is synthesized from core 1 may be summarized as follows:

UDP-GlcNAc + Gal(β1→3)GalNAcα1→Ser/Thr →
(core 1)

GlcNAc(β1→6)[Galβ(1→3)]GalNAcα1→Ser/Thr + UDP
(core 2)

Core 2-type oligosaccharides are the most common O-linked glycoproteins. Substitution of core 1, for example by the addition of sialic acid or Fuc, prevents core 2 formation.

Unusually for a glycosyltransferase, and as an exception to the 'one enzyme one linkage' rule (*see 2.7*), at least some forms of the core 2 β1,6 GlcNAc-T can also synthesize core 4 (*see 5.6.6*). Furthermore, the same enzyme is able to form the GlcNAc(β1→6)[GlcNAc(β1→3]Gal branch that is part of the I antigen, from the GlcNAc(β1→3)Gal structure that is part of the i antigen (*see 10.7*).

Further elongation of the core 2 occurs in a similar manner to elongation of core 1, but is slightly more complicated in that elongation can occur at both the Gal and the GlcNAc residues of the core 2 structure.

5.6.5 Formation of core 3

The Tn epitope, GalNAcα1→Ser/Thr, can alternatively be acted upon by β1,3-N-acetylglucosaminyltransferase (UDP-GlcNAc: GalNAc-R β1,3-GlcNAc transferase, core 3 β1,3 GlcNAc-T) to give GlcNAc(β1→3)GalNAcα1→Ser/Thr, or 'core 3', *Figures 5.3* and *5.4*. The enzyme requires the presence of Mn^{2+} ions.

The reaction by which core 3 is synthesized from Tn may be summarized as follows:

UDP-GlcNAc + GalNAcα1→Ser/Thr → GlcNAc(β1→3)GalNAcα1→Ser/Thr + UDP
Tn core 3

As both core 1 Gal-T and core 3 GlcNAc-T operate upon the same substrate, Tn (GalNAcα1→Ser/Thr), the preponderance of core 1 and/or core 3 O-glycans depends upon direct competition for substrate between these two glycosyltransferases. The core 3 β1,3 GlcNAc-T is expressed by many mucin-secreting cells and tissues.

The action of core 3 β1,3 GlcNAc-T, to form the core 3 structure, inhibits the subsequent action of core 2 GlcNAc-T, as core 2 GlcNAc-T requires a (β1→3)-linked Gal residue linked to the underlying GalNAc of the GalNAcα1→Ser/Thr. Again, direct competition between the enzymes results in greater or lesser proportions of different core structures synthesized in any given cell or tissue.

In most tissues, synthesis of core 3 precedes that of core 4. In rat colonic mucosa, the core 4-synthesizing β1,6 GlcNAc-T proceeds much faster than the core 3-synthesizing β1,3 GlcNAc-T, thus synthesis of the core 3 structure may be the rate-limiting step in core 4 production.

5.6.6 Formation of core 4

Core 3 can be converted to 'core 4' by the β1,6-N-acetylglucosaminyltransferase (UDP-GlcNAc: GlcNAc(β1→3)GalNAc-R or GlcNAc to GalNAc β1,6 GlcNAc transferase, core 4 β1,6 GlcNAc-T, or β1,6-GlcNAc-T) which adds a (β1→6)-linked GlcNAc to give a second type of simple bi-antennary core structure GlcNAc(β1→6)[GlcNAc(β1→3)GalNAcα1→Ser/Thr. ➲

➲ Core 3 structures can be converted to core 4 structures to give rise to a second type of bi-antennary core O-linked glycan.

The reaction by which core 4 is synthesized from core 3 may be summarized as follows:

UDP-GlcNAc + GlcNAc(β1→3)GalNAcα1→Ser/Thr →
(core 3)

GlcNAc(β1→6)[GlcNAc(β1→3)]GalNAcα1→Ser/Thr + UDP
(core 4)

Core 4 is restricted in its occurrence compared with core 2 owing to the rate-limiting activity and limited distribution of the core 3 transferase. Core 4 cannot be formed after galactosylation of core 3 in which Gal(β1→4)GlcNAc(β1→3)GalNAc is formed. The core 4 transferase does not require the presence of Mn^{2+} ions.

5.6.7 Less common core structures

Although the most common O-linked glycan core structures are types 1–4, other less common core structures have been described. These include:

- *Core 5*: GalNAc(α1→3)GalNAcα1→Ser/Thr has been reported on embryonic gut cells and in human meconium. A sialylated form has been reported on some human adenocarcinomas. This may be synthesized from GalNAcα-Ser/Thr by the action of a core 5 N-acetylgalactosaminyltransferase (UDP-GalNAc: GlcNAc α-N-acetylgalactosaminyltransferase, core 5 α1,3 GalNAc-T), which is Mn^{2+} dependent.
- *Core 6*: GlcNAc(β1→6)GalNAcα1→Ser/Thr has been reported on human embryonic gut, meconium and ovarian cyst mucins, and also in bovine α-casein. This may be synthesized from GalNAcα/→Ser/Thr by the action of a core 6 N-acetylglucosaminyltransferase (core 6 β1,6 GlcNAc-T). Core 6 structures could, alternatively, simply be a degradation product of β-galactosidase cleaving Gal from core 2 glycans during preparation or analysis.
- *Core 7*: GalNAc(α1→6)GalNAcα1→Ser/Thr has been reported on bovine submaxillary mucins. This may be synthesized from GalNAcα-Ser/Thr by the action of a core 7 N-acetylgalactosaminyltransferase (core 7 α1,6 GalNAc-T).
- *Core 8*: Gal(α1→3)GalNAcα1→Ser/Thr has been reported in human respiratory mucin.

The enzymes responsible for the formation of all core structures are restricted in their tissue distribution, so that certain types of O-linked glycosylation are characteristic of certain cell or tissue types. Furthermore, in certain cells or tissues there may be direct competition for substrate between different co-expressed glycosyltransferases resulting in the relative predominance of the synthesis of one or more core structure.

5.6.8 Further elongation of O-glycans

Core structures may be substituted directly by terminating monosaccharides, usually sialic acid or Fuc, or can be further extended by the action of other glycosyltransferases to form more complex, elongated, linear or branched oligosaccharide chains, which themselves may then be terminated, again by sialylation or fucosylation. These processes are described briefly below, and, as they are common to both N- and O-glycans and glycolipids in *Chapter 10*.

Core structures may be elongated to form Gal(β1→3)GlcNAcβ- (called 'type 1 structure' or 'type 1 chain') or Gal(β1→4)GlcNAcβ- (called 'type 2 structure' or 'type 2 chain'). This is illustrated in *Figure 5.7*. As the disaccharide Gal(β1→4)Glc is lactose, the Gal(β1→4) GlcNAc-disaccharide is termed lactosamine or a lactosamine unit. Other 'chain' structures, chains 3–6, occur less commonly, are present on other types of glycoconjugates, and some are not characteristic of O-linked glycans (*see Chapter 10*).

➲ Core structures may be elongated by the addition of (β1→3)-linked Gal to form 'type 1 chains', or (β1→4)-linked Gal to form 'type 2 chains'.

➲ The lactosamine structure, Gal(β1→4)GlcNAc and polylactosamine structure [Gal(β1→4) GlcNAc(β1→3)Gal]$_n$ are common features of O-glycan chains.

Chain structures may be added to any one of the cores, or can be attached to an internal Gal residue of the backbone to give a branched structure. The type 2 structure attached to GlcNAc in (β1→3) linkages results in the i antigen, and a branched type 2 structure in the I antigen (*see 10.7*).➲

The elongation of O-glycans most often involves the addition of repeating units of (β1→3)- or (β1→6)-linked GlcNAc and (β1→3)- and (β1→4)-linked Gal residues to form linear or branched chains. The repeated chains of repeating lactosamine units [Gal(β1→4) GlcNAc(β1→3)Gal]$_n$, termed polylactosamine, often span the core region and the non-reducing terminus.➲ This is a common structure in the elongated backbones of O-linked oligosaccharide chains and is also common in N-linked oligosaccharides and glycosphingolipids. Such oligosaccharide chains may then be terminated by the addition of further monosaccharide residues.

Core structures terminating in Galβ1→ may be elongated to form linear chains including GlcNAc(β1→6)Gal(β→ and Gal(β1→4)GlcNAc(β1→3)Gal(β→ structures, or branched chains

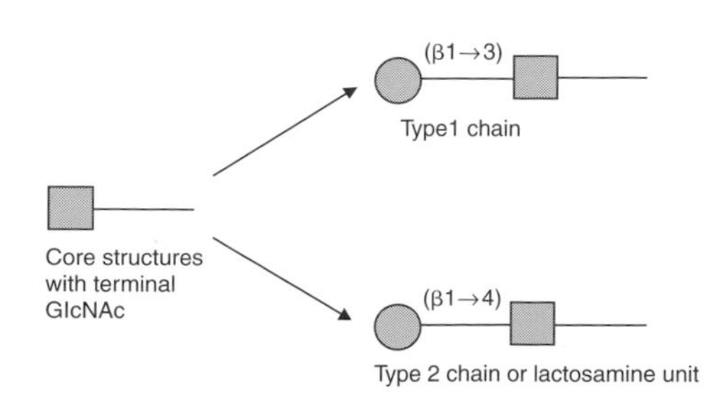

Figure 5.7

Elongation of core structures terminating in GlcNAc to form 'type 1' and 'type 2' chains.

such as Galβ(1→4)GlcNAc(β1→6)[Gal(β1→4)GlcNAc(β1→3)]Gal(β→ (branched polylactosamines). These and some other elongation pathways are illustrated in *Figure 5.8*.

5.6.9 Termination of O-glycan chains – terminal monosaccharides

Elongated O-glycan structures commonly terminate in Gal, GlcNAc or GalNAc or may be substituted with various terminal monosaccharides, typically Fuc or sialic acid, and may be sulfated. Terminal monosaccharides of O-glycan chains are usually α-linked. ➲ Common terminal monosaccharide residues found on O-linked glycans of mammalian glycoproteins, and their linkages, are listed in *Table 5.2*.

> ➲ Terminal monosaccharides of O-linked glycan chains are often α-linked.

Common terminal structures include NeuAc(α2→6)GalNAc- (sialyl Tn), NeuAc(α2→3)Gal(β1→3)GalNAc- and NeuAc(α2→6)[NeuAc(α2→3)Gal(β1→3)]GalNAc- (sialylated core 1). Increased sialylation of O-glycans is a common observation in cancer cells. Sialyl Tn, for example, is rarely seen on normal cells but is commonly synthesized by cancer cells of many types, as is Tn (*see 16.6.6*).

Sialic acids, sulfated groups and Fuc(α1→3) or Fuc(α1→4) residues may also be attached to internal as well as terminal sugar residues. Some examples of typical terminal sequences characteristic of O-glycan chains are given in *Figure 5.9*.

5.6.10 Termination of O-glycan chains – special sugar structures

Many O-linked glycans carry special terminal sugar structures such as those of the ABO and Lewis blood group systems, Cad and S^d blood group antigens. These are illustrated in

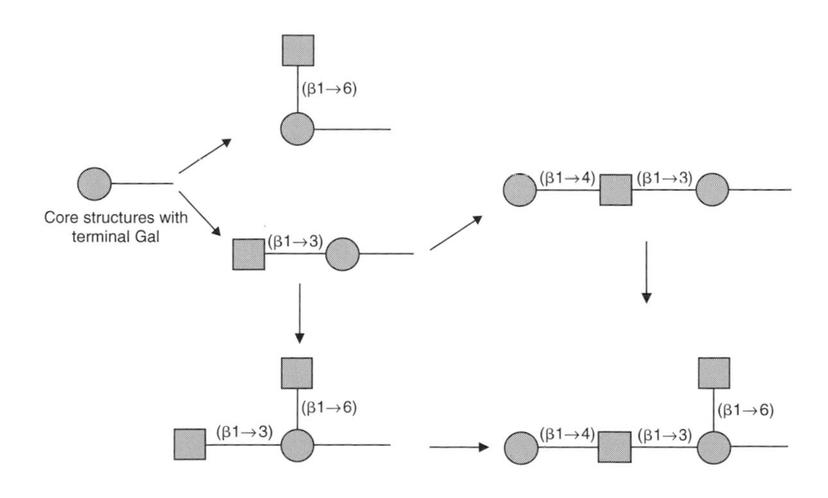

Figure 5.8

Some elongation pathways of core structures.

Table 5.2 Common terminal monosaccharide residues found on O-linked glycans of mammalian glycoproteins, and their linkages and symbols

Sialic acid (◆)	Fuc (△)	GalNAc (□)	GlcNAc (■)	Gal (●)
NeuAc(α2→3) NeuAc(α2→6)	Fuc(α1→2) Fuc(α1→3) Fuc(α1→4)	GalNAc(α1→3) GalNAc(α1→6) GalNAc(β1→4)	GlcNAc(α1→4)	Gal(α1→3)

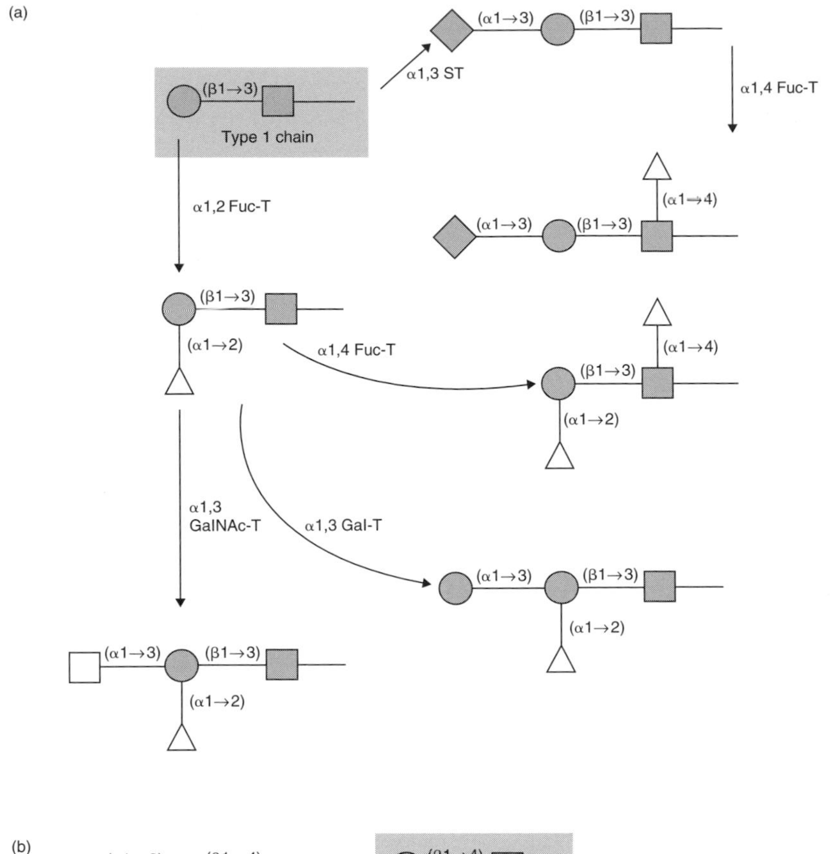

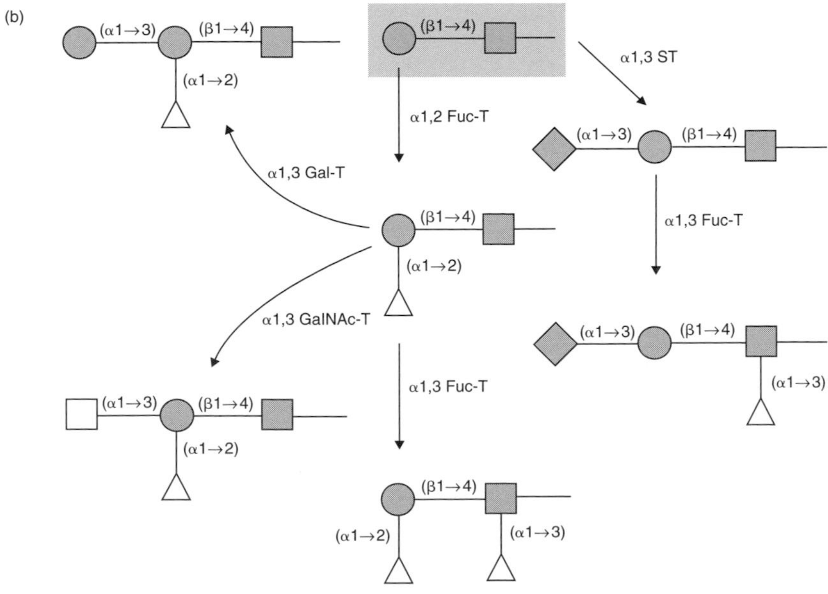

Figure 5.9

Some examples of typical terminal sequences characteristic of O-glycan chains. (a) Termination of type 1 chains by fucosylation and sialylation. (b) Termination of type 2 chains by fucosylation and sialylation.

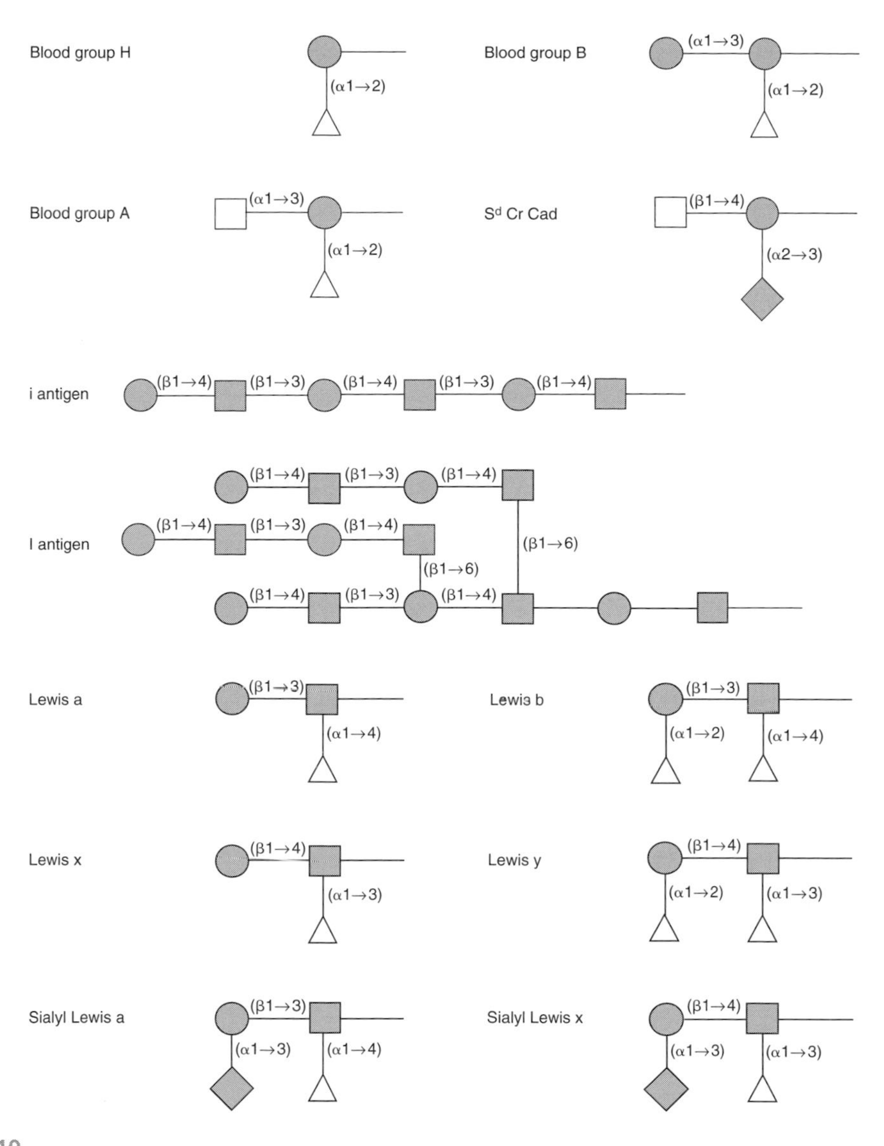

Figure 5.10

Blood group ABO sugars, Cad and S^d antigens, i and I antigens and Lewis family blood group sugars are all commonly found on O-linked oligosaccharide chains.

Figure 5.10. They are common to types of glycoconjugate other than O-linked glycoproteins (*see 10.4*). In this instance, terminal monosaccharides include sialic acid, Fuc, Gal, GalNAc and GlcNAc, usually in the α-linkage, as illustrated in *Table 5.2*. In addition, internal and terminal Gal and GlcNAc residues may be sulfated, although sulfation does not occur on blood group determinants.

5.6.11 Sulfation of O-linked glycans

The Gal, GalNAc and GlcNAc residues of O-linked oligosaccharide chains of most mucins may be sulfated, and many of these oligosaccharides are also sialylated, rendering them highly acidic. Sulfate groups may be attached to terminal and/or internal positions in O-linked glycans. In some cases, the same monosaccharide residue has been reported to be both sulfated and to have sialic acid linked to it. Sulfation and/or sialylation may be protective (*see 11.8*). The degree of both sulfation and sialylation of O-linked glycans is altered in disease states such as inflammation and cancer, although the precise functional significance of this alteration is unknown (*see 16.3, 16.6.3*).

The sulfate donor substrate is 3′-phospho-adenosine-5′-phosphosulfate (PAPS). It is transported into the Golgi apparatus through the action of a transporter molecule in a manner analogous to that by which nucleotide sugars are transported (*see 2.6*). Sulfation of O-linked oligosaccharides is believed to occur through the action of a number of specific sulfotransferases (sulfo-Ts), but little is currently known of their action. They may act relatively late in O-glycosylation pathways, but probably prior to sialylation events.

5.7 Glycosyltransferases involved in O-linked oligosaccharide synthesis

Many of the glycosyltransferases that are involved in the synthesis of N-glycan chains are also involved in O-linked glycan synthesis. An example of this is the glycosyltransferases involved in the synthesis of ABO blood group determinants and of linear and branching polylactosamine chains (*see 2.11, 10.4*). The enzymes that synthesize the common core structures 1–4, and elongate core 1 or 2 and the sialyltransferases that sialylate GalNAc in the ($\alpha 2 \rightarrow 6$) linkage and the Gal of core 1 and 2 in the ($\alpha 2 \rightarrow 3$) linkage appear to be specific to O-glycosylation pathways.

5.8 Regulation of O-glycan synthesis

Mucins contain the greatest diversity of core 1 and core 2 oligosaccharides, whereas secreted glycoproteins and, particularly, glycoproteins synthesized by cancer cells, typically express fewer and simpler O-linked glycan chains. The regulation of O-glycan synthesis is not well understood, but is thought to be influenced by many interrelated factors such as the levels of expression and activity of the relevant glycosyltransferases, the availability of nucleotide-sugar donors and acceptor substrates, cations, and subcellular membrane organization. O-Glycosylation changes during development, differentiation, growth and in disease. ➲ Certain cell or tissue types often express characteristic O-glycans associated with their specialized biological functions.

The rate at which potentially O-linked glycoconjugates are synthesized and the rate of their transit through the secretory pathway may also, theoretically, have a considerable effect; if throughput is high, the machinery of glycosylation may not be able to cope with the demand and incomplete synthesis of oligosaccharides may result.

> ➲ The synthesis of O-glycan structures depends upon the relative activities of competing transferases, and also other factors such as the availability of nucleotide-sugar donors, acceptor substrates, cations, subcellular membrane organization of transferases, and the rate of glycoprotein transit through the secretory pathway.

To a large degree, control of O-glycosylation appears to be a matter of glycosyltransferase availability, activity and specificity, with enzymes competing for available substrates. The glycosylation pathway taken at any point of potential competition is presumed to be dictated by the relative activities of competing transferases. For example, at least three different glycosyltransferases compete for the Tn epitope, GalNAcα1$\rightarrow$Ser/Thr: α2,6 ST, β1,3 Gal-T and β1,3 GlcNAc-T and, depending on their relative availability in any given tissue, relatively greater or lesser amounts of sialyl Tn, core 1 or core 3 may result. This is illustrated in *Figure 5.11(a)*. Similarly, at least five different glycosyltransferase enzymes compete for core 1: α2,6 ST I, α2,3 ST, elongation β1,3 GlcNAc-T, α1,2 Fuc-T and core 2 β1,6 GlcNAc-T, and their levels of activity within a cell determine the

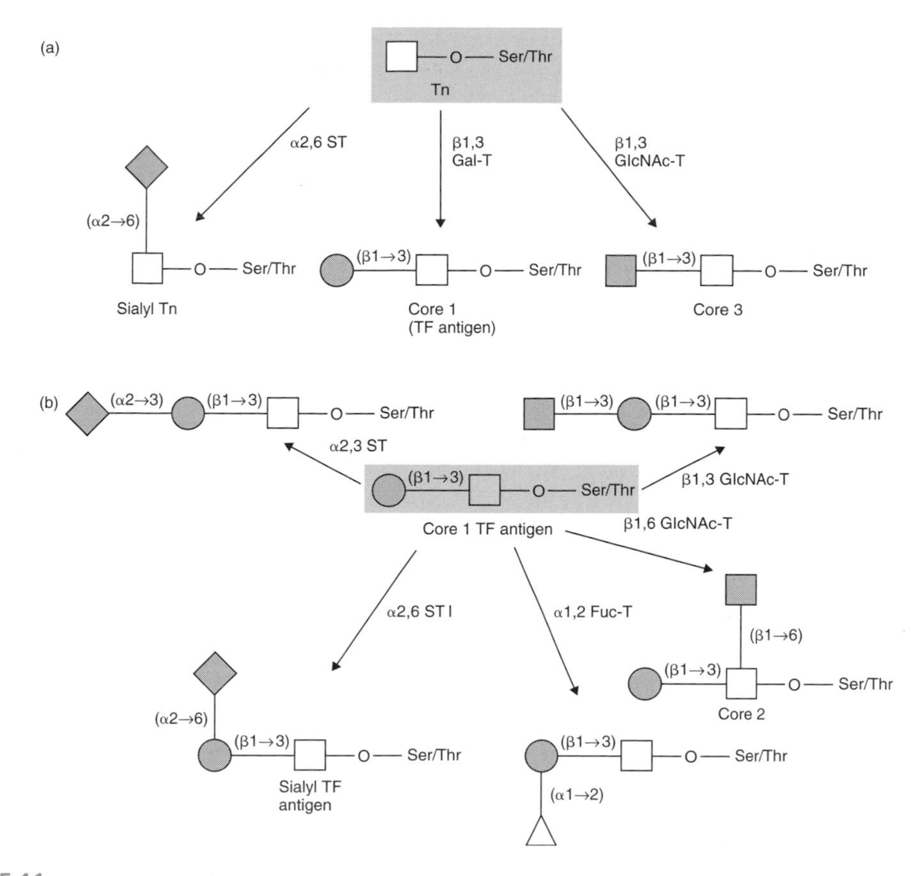

Figure 5.11

(a) At least three different glycosyltransferases compete for the Tn epitope, GalNAcα1→Ser/Thr; and their relative activities within a cell determine the amounts of sialyl Tn, core 1 or core 3 that are synthesized. (b) At least five different glycosyltransferase enzymes compete for core 1, and their relative activities within a cell determine the amounts of core 2 or alternative structures that are synthesized.

relative amounts of core 2 or alternative structures that are synthesized. This is illustrated in *Figure 5.11(b)*.

5.9 Where does O-glycosylation take place?

There is controversy concerning the site of initiation of O-linked glycosylation: a few reports suggest that it is in the RER and the majority suggest that it is within the Golgi apparatus. Currently, it is generally believed that it takes place in the Golgi apparatus, with most evidence indicating that O-glycosylation is initiated in the *cis*-Golgi under normal circumstances, and that elongation and termination of O-glycan synthesis then takes place in later Golgi compartments. As a general principle, there is a gradient from *cis*-Golgi to *trans*-Golgi network of glycosyltransferases involved in early to late O-glycosylation events, as might be predicted. However, there is overlap in expression of transferases involved in early and late events, many glycosyltransferases do not have a fixed localization within the Golgi stacks, and distribution of glycosyltransferases changes with differentiation and maturation, and in disease states.

Initiation of O-linked glycosylation begins with the attachment of GalNAc to a serine or threonine residue of the polypeptide chain by the action of GalNAc-T (*see 5.6.1*). This occurs after the protein has been transported via vesicles from the RER to the *cis*-Golgi apparatus in most cells. There is some experimental evidence to suggest that in some cell types it may occur in the late RER prior to entry into the Golgi apparatus, and it has been suggested that the site at which O-glycosylation is initiated may correspond to the size of the oligosaccharide chains that are synthesized – synthesis of longer or more complex chains beginning earlier in the secretory pathway.

Alternatively, the GalNAc transferase responsible for the initiation of O-glycosylation may reside in different parts of the secretory pathway in different cell types or under different conditions. At least nine different human polypeptide GalNAc-T enzymes are potentially responsible for the initiation of O-linked glycosylation (*see 2.11.2*), and it is possible that each may reside in a different part of the Golgi stacks, and thus, that initiation of O-glycosylation may, under different, as yet undefined, circumstances, occur in different locations. The reason for the existence of so many, apparently very similar, GalNAc-T enzymes is unknown. ➲

➲ At least nine GalNAc-T enzymes potentially responsible for initiation of O-glycan synthesis have been described, but the reason for the existence of so many glycosyltransferases with similar activity is unknown.

Initiation of O-glycosylation occurs later in the secretory pathway than does initiation of N-glycosylation, where synthesis of the asparagine-GlcNAc linkage occurs in the RER by incorporation of a preformed oligosaccharide chain onto the polypeptide backbone immediately after it has been secreted into the lumen of the RER (*see 4.2*).

Subsequent synthesis of core structures and their elongation occurs in the Golgi apparatus. Core 1 β1,3 Gal-T has been localized to the *medial* and *trans*-Golgi compartments, and core 1 oligosaccharide structures are also detectable, for example, through binding of peanut lectin which recognizes Gal(β1→3)GalNAc, in the same locations. In keeping with this idea, core 1 β1,3 Gal-T resides in later compartments of the secretory pathway than the polypeptide GalNAc-T enzyme responsible for the initiation of O-glycosylation, but it shares its location with many of the transferases responsible for further elongation and termination reactions. The GlcNAc-T enzymes responsible for O-glycan synthesis appear to reside mostly in the intermediate and *trans*-Golgi compartments, analogous to the N-glycosylation pathway. However, disruption of these later compartments with the drug monensin (which disrupts trafficking within the Golgi apparatus) results in the absence of galactosylation, but only partial abrogation of the synthesis of core 2 structures, which are also sometimes sialylated, indicating that some GlcNAc-Ts may also reside in earlier Golgi compartments. β1,4 Gal-T has been shown to reside in the *trans*-Golgi.

Sialylation, often the last event in O-glycan biosynthesis, occurs principally in the *trans*-Golgi and the *trans*-Golgi network where several sialyltransferases are localized. ➲ This is also the usual home of the α1,3 GalNAc-T responsible for blood group A sugar synthesis, but this enzyme is found earlier in the Golgi pathway in certain cell types, such as intestinal goblet cells.

➲ O-Glycosylation probably occurs exclusively in the Golgi apparatus. The first GalNAc is attached in the *cis*-Golgi, synthesis of core structures and elongation occurs in the *medial* and *trans*-Golgi compartments, and sialylation in the *trans*-Golgi and *trans*-Golgi network.

Many O-linked glycans are completed without elongation or termination of the core structures, or with only partial elongation. It is assumed that this apparently incomplete synthesis is as a result of the speed or volume of throughput within the processing pathways of the Golgi apparatus leading to an insufficient supply of transferases or available sugar-nucleotides. It is assumed, however, that the synthesis of these apparently ‘incomplete’ glycans has functional significance.

5.10 The *MUC* family of mucin genes

The *MUC* family of mucin genes are worthy of special note. In humans, there are at least eight genes encoding a series of apomucins highly glycosylated by multiple O-linked glycans to form high molecular mass epithelial glycoproteins known as mucins. The genes are designated *MUC1, MUC2, MUC3, MUC4, MUC5AC, MUC5B, MUC6, MUC7* and *MUC8* (conventionally, the acronyms for the genes are printed in italics, for example, *MUC1*, whereas the acronyms for the proteins they produce are printed in plain text, for example, MUC1). The *MUC1* gene was the first mucin gene in this family to be cloned.

Some of the features of the *MUC* family of genes and their products are summarized in *Table 5.3*, and described in more detail below.

With the exception of the MUC1 mucin, which is a membrane-anchored molecule, the other MUC mucins are secreted. The products of *MUC2, MUC3, MUC4, MUC5AC, MUC5B* and *MUC8* are both secreted and gel forming, whereas those of *MUC6* and *MUC7* are secretory and non-gel forming.➲ All are produced by epithelial cells of the gastrointestinal, respiratory and genitourinary (including breast) tracts, and also by the cancers that arise from these tissues. At least four of the *MUC* genes (*MUC2, MUC5AC, MUC5B* and *MUC6*) are located clustered together on chromosome 11p15.5. The others are located on other chromosomes, as indicated in *Table 5.3*. Their O-glycosylation sites, as predicted, have been identified as tandem repeats rich in proline, serine and threonine. Each *MUC* gene seems to have a unique repeat sequence, that is duplicated up to 100 times. Although the transmembrane and cytoplasmic domains of the *MUC1* genes are highly conserved between species, for example, there is ≈80% homology between *MUC1* sequences in humans and mice, the repeat sequences of the glycosylated 'mucin' domain is much less well conserved, and may be characteristic of each mucin, and in addition there is some polymorphism in these genes between

➲ The *MUC* gene family codes for a series of homologous highly O-glycosylated mucins that are produced by epithelial cells. The MUC1 mucin is membrane bound and thus unusual (the others are secreted), however, it is the most intensively studied.

Table 5.3 Properties of the MUC family of mucins

Gene	Gene locus	Membrane-associated or secreted	Mucin produced by
MUC1	1q21–24	Membrane-associated	Glandular epithelial cells including lactating mammary gland, pancreas, bronchus, salivary gland, prostate and uterus, luminal surface of post-capillary venules in the lymph nodes, mesothelium.
MUC2	11p15	Secreted	Small and large intestine, gall bladder and bronchus.
MUC3	7q22	Secreted	Jejunum, ileum, colon, gall bladder, pancreas.
MUC4	3q29	Secreted	Gastrointestinal epithelia (except gall bladder and submaxillary salivary gland) bronchus, prostate and uterine cervix.
MUC5AC	11p15	Secreted	Respiratory and gastrointestinal epithelia (especially stomach), uterine cervix.
MUC5B	11p15	Secreted	Uterine cervix, bronchus, pancreas, acinar cells of the salivary glands.
MUC6	11p15	Secreted	Stomach, gall bladder, Brunner's glands of the duodenum, seminal vesicles.
MUC7	chr 4	Secreted	Submandibular and sublingual salivary gland mucous acinar cells.
MUC8	12q24.3	Secreted	Tracheobronchial and reproductive tract.

individuals, in which different numbers of repeat sequences exist. Glycan chains of between 1 and 20 monosaccharides are attached to serine and threonine of the apomucin by (α1→3) O-glycosidic bonds.

5.10.1 *MUC1*

The *MUC1* gene is located at chromosome 1q21–24. The membrane-bound, polymorphic mucin synthesized by the *MUC1* gene was the first member of the MUC family of mucins to be described and characterized. It is known by several names (*Table 5.4*). It has a transmembrane sequence, and thus, unlike most mucins that are secreted, is normally membrane associated. Alternative splicing of the RNA or protease action both result in a secreted form. MUC1 is produced by mammary epithelium and is secreted in milk, but has been intensively studied largely because it is produced in large amounts by some cancers, notably breast cancers. The molecule has an apparent molecular mass of 300–600 kDa.

MUC1 has a long (69 amino acid) cytoplasmic tail and a structurally variable extracellular domain composed almost entirely of between 20 and 100 tandem repeats of a 20 amino acid motif, which acts as a scaffold for the attachment of oligosaccharides and forms the basis of its polymorphism. The length of the extracellular domain varies between ≈1000 and 2200 amino acids and extends 200–500 nm above the lipid bilayer of the plasma membrane, well above all other membrane-associated proteins. This is illustrated in *Figure 5.12*. The tandem repeat contains five potential glycosylation sites, but two of these, threonine residues situated next to serines, appear to most commonly carry the O-linked oligosaccharides. MUC1 is highly glycosylated, carrying typically between 60 and 200 oligosaccharide side chains. In humans, ≈50% of the mass of the mucin molecule is carbohydrate.

Table 5.4 Alternative names for the MUC1 mucin

PAS-O
Non-penetrating glycoprotein
MAM-6
DF-3 antigen, polymorphic epithelial mucin (PEM)
Epithelial membrane antigen (EMA)
Episialin

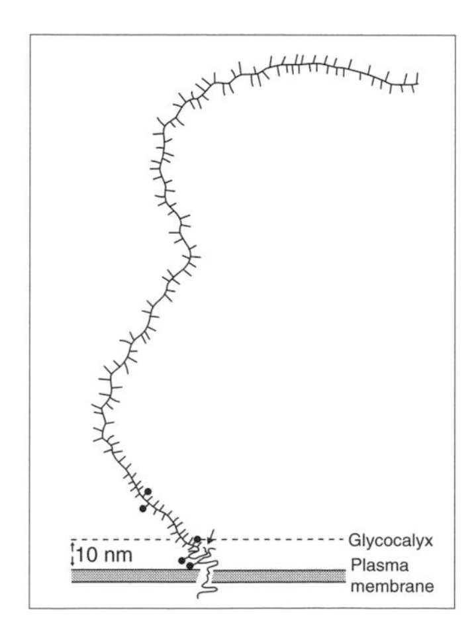

Figure 5.12

The *MUC1* mucin protrudes well beyond the glycocalyx of the cell. N-Linked oligosaccharide chains are indicated by —• and the O-linked oligosaccharides by — The arrow indicates where the precursor is cleaved in the RER to release the secreted mucin. Reproduced from Hilkens J. *et al.* (1992) *Trends Biol. Sci.* **17**: 360, with permission from Elsevier Science.

➲ Physical properties of the MUC1 mucin:
- long, bulky molecule
- protrudes well beyond the glycocalyx at the cell surface
- high density of negatively charged residues at the end of multiple, densely packed oligosaccharide chains.

As few amino acids are hydrophobic, it is unsurprising that the hydrophobic transmembrane domain of the MUC1 apomucin is highly conserved between humans and other mammals. The cytoplasmic tail region is also highly conserved, suggesting an important function. It can interact with actin cytoskeleton filaments, and is also phosphorylated, suggesting that it may be involved in signal transduction events.➲

Up to 50% of the molecular mass of MUC1 is carbohydrate, and most of this is O-linked, although some interspersed N-linked oligosaccharide chains have been described. The oligosaccharides of MUC1 are heterogeneous, but, although changes in the glycosylation pattern are seen, some oligosaccharide sequences are inevitably repeated many times. Most seem to be 2–16 monosaccharides in size. In MUC1 synthesized by normal cells, most oligosaccharides are based on core 2 structures, with polylactosamine branches; this can, however, change in cancer (*see 16.6.10*). The amino acid sequence between the glycan side chains is short and highly immunogenic, and several monoclonal antibodies recognizing this sequence (e.g. the monoclonal antibodies HMFG-1, HMFG-2 and SM-3) have been used extensively in research into this important mucin molecule. It is commonly found at the apical surface of glandular epithelial cells (e.g. lactating mammary gland, pancreas, bronchus, salivary gland, prostate and uterus), and also at the luminal surface of post-capillary venules in the lymph nodes, and on the luminal surface of linings of other body cavities, such as the mesothelium. It is also synthesized by the cells of many types of cancer.

The biosynthesis of MUC1 has been studied extensively. It is synthesized as a single large precursor molecule containing only N-linked glycans; whilst in the RER, it is proteolytically cleaved into a mucin domain and a second large fragment containing the transmembrane domain, the cytoplasmic domain, and an extracellular domain of ≈65 amino acid residues. These two fragments remain associated in a stable, membrane-bound complex. As it passes through the Golgi stacks, the molecule is glycosylated with multiple O-linked oligosaccharide chains. At this stage it is incompletely sialylated, and once it reaches the cell surface, it is internalized and recycled, during which further sialylation takes place to yield the mature molecule. This process of recycling has been described in other O-linked membrane-associated mucins, such as ascites sialoglycoprotein-1 (ASGP-1), and may be necessary to achieve or maintain full sialylation. Cultured breast cancer cells secrete the O-glycosylated part of MUC1 into the medium ≈24 hours after biosynthesis; it is not known whether this is as a result of dissociation of the two parts of the molecule to release the larger, or whether it is as a result of proteolytic cleavage. Expression of the *MUC1* gene may be under hormonal regulation, as the gene carries consensus sequences for binding of the oestrogen–receptor and progesterone–receptor complex and expression in endometrium has been shown to fluctuate with progesterone stimulation.

MUC1 may play a role in anti-adhesion processes. The aggregation and adherence to extracellular matrix components of transfected cells that express high levels of MUC1 are significantly reduced. This may be a function of the position of the molecule protruding well beyond the glycocalyx, where it may sterically hinder underlying adhesion molecules, but the sialylation of the molecule also means that, as well as being very bulky, it carries a heavy negative charge which contributes to its anti-adhesive effect by electrorepulsion. This type of effect has been postulated for other heavily sialylated molecules; for example, the heavily sialylated podocalyxin found on the luminal face of podocytes (the epithelial cells lining the glomerular capillaries in the kidney) maintains the filtration slits between the foot processes of these cells by charge repulsion. On epithelial cells, where MUC1 is found on the apical surface, it may prevent interactions between other molecules located on opposing faces of the membrane preventing adhesion, and maintaining the lumen. In high endothelial venules, it may inhibit adhesion and extravasation of lymphocytes, unless this process is specifically triggered by the selectin–saccharide interaction (*see 13.4.2*).

5.10.2 *MUC2*

➲ MUC2 mucin is the predominant mucin produced by gut goblet cells.

The *MUC2* gene codes for a secretory gel-forming mucin, which is the predominant mucin in human intestinal tissues where it is produced by specialized secretory goblet cells.➲ *MUC2* is located on chromosome 11p15 very close to the genes *MUC5AC*, *MUC5B* and *MUC6*, which are closely related and together form a multi-gene family. It has a central region with a variable number of tandem repeats rich in serine, threonine and proline, like MUC1. It contains two different tandem repeat motifs. One is polymorphic and contains 51–115 repeats of a sequence of 23 amino acid residues with 14 potential O-glycosylation sites. The other is an imperfectly conserved region that consists of 7–40 (average 16) amino acids. It also contains four domains, three in the N-terminal region and one in the C-terminal region, very rich in cysteine which may be involved in formation of intra- and intermolecular disulfide bridges, containing repeated sequences that show high sequence homology with the pre-pro von Willebrand factor domain. This domain is responsible for oligomerization of von Willebrand factor and its storage in specific organelles (the Weibel–Palade bodies), and may play a similar role in the correct folding and oligomerization of biopolymers of MUC2.

MUC2 is synthesized in the RER as an N-glycosylated precursor protein of 600 kDa and forms oligomers with other MUC2 molecules. In the Golgi apparatus multiple O-glycosylated oligosaccharides are attached, and decorated with terminal sialic acid, Fuc and sulfate. The mucin accumulates in vesicles, which mature into storage granules during transport along the secretory pathway to the apical membrane of the cell. Exocytosis releases the mucin into the intestinal lumen. Secretion may be as a result of a basal and unregulated pathway dependent on an interaction with microtubules, or a completely separately stimulated and regulated one which is activated by, for example, acetylcholine stimulation. MUC2 is synthesized by cells of the duodenum, jejunum, ileum, colon, gall bladder and bronchus, but is absent in the stomach.

5.10.3 *MUC3*

The large *MUC3* gene is located on chromosome 7q22. It contains tandem repeats of a 51-nucleotide sequence coding a 17 amino acid peptide. It has an EGF-like domain in its C-terminus cysteine-rich portion. MUC3 is stored in microvesicles (not storage granules like MUC2) and is produced by cells of the jejunum, ileum, colon, gall bladder and pancreas. Like MUC2, it is not secreted by cells of the stomach. In absorptive cells of the intestine, *MUC3* gene expression increases with maturation and migration from the bottom of the crypts of Lieberkuhn to tips of the villi.

5.10.4 *MUC4*

The *MUC4* gene is located on chromosome 3q29 and is expressed by the epithelium lining the gastrointestinal tract (except gall bladder and submaxillary glands), bronchus, prostate and cervix. The gene contains 38 tandem repeats of a 48-bp sequence that codes for a 24 amino acid polypeptide.

5.10.5 *MUC5* genes

The *MUC5* genes, *MUC5AC* and *MUC5B*, are located on chromosome 11p15. The odd nomenclature arose because it was originally believed that there were three closely related genes, *MUC5A*, *B* and *C*, but it was subsequently realized that *MUC5A* and *MUC5C* were actually parts of the same gene, renamed *MUC5AC*.

MUC5AC is expressed by epithelial cells lining the respiratory and gastrointestinal (especially stomach) tract and cervix. *MUC5AC* contains a 24-bp tandem repeat sequence encoding an eight amino acid sequence, plus other threonine and serine rich non-repeated domains, and inserted among these are several cysteine-rich domains implicated in oligomerization. At the 3′-end, there are cysteine-rich clusters and a similar C-terminus domain homologous to the pre-pro von Willebrand factor gene and *MUC2*.

MUC5B is also called MG1 (MG1 and MG2 – the MUC7 mucin – are high and low molecular mass salivary gland mucins). The unusual *MUC5B* gene has a variable length of nucleotide repeat units within the tandem repeat domain. It contains four super-repeats of 528 amino acids (the largest of any mucin gene) each comprising 12 repeats of the irregular tandem repeat sequence of 29 amino acids. It has a cysteine-rich domain similar to that of *MUC5AC* and *MUC2*. It is strongly expressed in the uterine cervix, bronchus, pancreas and acinar cells of the salivary glands.

5.10.6 *MUC6*

The *MUC6* gene is located on chromosome 11p15. The tandem repeat is extremely long (more than five times longer than any other human mucin repeat sequence) and codes for a 169 amino acid polypeptide, making the MUC6 mucin unusually large. The polypeptide backbone alone has a molecular mass of >600 kDa, the largest apomucin to be reported. It contains three cysteine-rich domains, similar to those of MUC5AC. MUC6 is predominantly produced in the stomach, by mucous neck cells of the gastric glands and pyloric gland cells where it may protect the cells from attack by stomach acid or bind to bacteria to be eliminated from the stomach, and also the gall bladder, Brunner's glands of the duodenum, and seminal vesicles.

5.10.7 *MUC7*

MUC7 is also known as MG2. It is a very small molecule with an apomucin of only 377 amino acids, with six 23 amino acid repeat sequences, 54 potential glycosylation sites, and a molecular mass of 39 kDa. The gene is located on chromosome 4. Two cysteine residues (not implicated in oligomerization – the molecule is secreted as a monomer) lie upstream to the tandem repeats region which codes for a 23 amino acid sequence. MUC7 is produced by submandibular and sublingual salivary gland mucous acinar cells.

5.10.8 *MUC8*

The *MUC8* gene is located on chromosome 12q24.3 and codes for a polypeptide with two consensus repeats of 13 amino acids and 41 amino acids. It is highly expressed in the tracheobronchial and reproductive tracts.

5.11 Other membrane-bound mucin-like glycoproteins

Other membrane-bound mucin-like glycoproteins, such as leukosialin, ASGP-1 and epiglycanin, which are genetically unrelated to MUC1, have structural similarities in that they carry long, stiff rod-like, sialylated O-glycan regions that protrude above the lipid bilayer. These have also been implicated in masking epitopes on the cell surface from immune recognition and in serving as anti-adhesion molecules. Membrane-bound mucin-like glycoproteins may represent a class of constitutively expressed anti-adhesion molecules.

5.12 Trefoil factor family (TFF) domain peptides and their relationship with mucins

Trefoil factor family (TFF) domain peptides are mucin-associated molecules produced by the mucin-secreting epithelia of gastrointestinal tract and other tissues. They are co-packaged

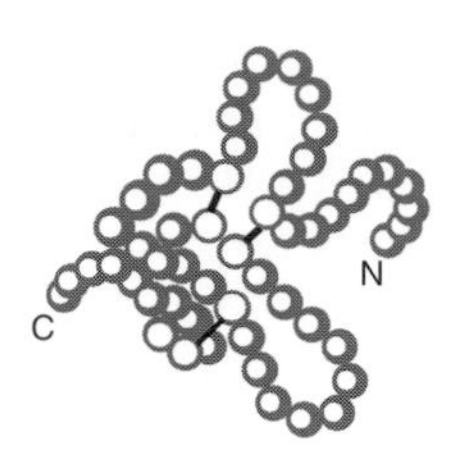

Figure 5.13

Schematic representation of the three-looped trefoil-shaped formation of the polypeptide chain, the trefoil domain.

with mucins by the Golgi apparatus and secreted with them at a high concentration into the protective layer covering mucosal surfaces. Their name derives from the fact the polypeptide chain is folded into a three-looped trefoil-shaped formation in a region where they share a common domain, illustrated in *Figure 5.13*. This is referred to as the trefoil domain. They are highly conserved throughout evolution, and are resistant to degradation by heat, acid and enzymatic attack owing to the way in which the trefoil domain is held tightly in formation by three pairs of disulfide bonds. In humans, three TFFs are known. TFF1 and TFF3 contain one trefoil domain, whereas TFF2 contains two. All three human TFF genes are clustered on chromosome 21q22.3.

The trefoil factor family peptides are implicated in mucosal defence and healing.➲ Expression is associated with gastric ulcers and Crohn's disease, indicating their role in repair of damaged gastric mucosa. They are motogens – that is, they promote cell migration unrelated to cell division, and stimulate the movement of healthy cells into an area of damage, a process called epithelial restitution. They also appear to stimulate cell flattening, again enhancing repair of the defect in the epithelial surface. Recently, a biological role in the protection against gastric tumours has been postulated, as knockout mice lacking expression of these peptides develop gastric tumours more easily. There is also an interest in their role in cancer biology (*see* 16.6.10.2).

> ➲ Trefoil factor family peptides are expressed in association with MUC family mucins, and function with them in mechanisms of mucosal defence and healing.

TFF domain peptides are also produced by tissues other than the epithelia of the gastrointestinal tract; for example, they are normally expressed by the hypothalamus, pituitary gland and breast.

Expression of the TFF domain peptides is strongly correlated with mucin expression. TFF1 and TFF2 are expressed with MUC1, TFF1 is also associated with MUC6 expression, TFF2 with MUC5AC and TFF3 with MUC2. It is likely that there is an interaction between mucins and TFF domain peptides, either through oligosaccharide chains or through the core protein itself. The presence of the TFFs dramatically alters the physical properties of the mucins, leading to an increase in optical density and viscosity, and thus contributes to the protecting effects of mucins on the epithelial surface. TFFs are implicated in the stabilization of the mucus layer covering the epithelia. These peptides hold promise in the development of new therapeutic approaches to ulcerative conditions of the gastrointestinal tract, and possibly in our understanding of at least some gastrointestinal cancers.

5.13 Other cell membrane-associated mucin molecules

In addition to the secreted mucins, large mucin-like domains are also characteristic of a number of cell membrane-bound molecules, including MUC1 (*see 5.10.1*), leukosialin (CD43; *see 5.4.2*), glycophorin (*see 5.4.2*), epiglycan (*see 5.11*) and ASGP-1 (*see 5.10.1*).

The genes coding for several of these molecules have been sequenced and cloned and they share the common characteristic that like the secreted mucins, the polypeptide backbone consists of regions with repetitive sequences of serine and threonine residues which are potential attachment sites for O-glycans. These stretches of amino acids are also rich in proline and other helix-breaking amino acid residues, and the molecule therefore has a rather extended structure with many β-turns and becomes rigid when highly glycosylated. It is thought that in these membrane-bound glycoproteins, this extended rigid structure functions to project the oligosaccharide, which may be a binding ligand for receptors on other cells, high above the glycocalyx of the cell. However, no receptor function has yet been identified for most of these molecules.

5.14 Functions of O-linked mucins

The lack of specific inhibitors of O-glycosylation other than O-benzyl-GalNAc (which terminates O-linked biosynthesis after its initial step; *see 15.7*), has hindered research into the biological function of these heterogeneous oligosaccharides. Nonetheless, there is evidence of a number of important biological roles for the oligosaccharides of these mucins. O-Linked oligosaccharides have been implicated in a range of activities including receptor–ligand interactions in fertilization, infection, immune function and differentiation.

Furthermore, O-glycosylation is often seriously disrupted in disease states, and O-linked mucin-type glycoproteins have been implicated in a pathophysiological role in disease processes. Of particular importance are the following molecules:

- MUC family mucins
- O-linked glycosylation changes in leukosialin during differentiation, immune deficiency and malignancy
- altered glycosylation of O-linked glycoproteins of human chorionic gonadotrophin (hCG) in choriocarcinoma
- the synthesis of truncated tumour-associated glycotopes by cancer cells (*see 16.6.5*)
- increased (β1→6)GlcNAc linkages and polylactosamine synthesis in cancer cells (*see 16.6.4*).

5.14.1 Mucins in protection and lubrication

Internal epithelial covers, such as those lining the gastrointestinal, respiratory and genitourinary tracts contain specialized cells that produce mucin molecules. These mucin molecules form oligomers through disulfide bonds, and are stored within the secretory cell in large granules. Once released, they cross-link through intra- or intermolecular disulfide bonds to form a gelatinous, slimy matrix that binds water and traps ions. This mucus layer covering the cell surface provides a hydrated surface that acts to protect the underlying epithelium from chemical, physical and microbial attack, and also acts as a lubricant.➲ Many mucins are highly sialylated (therefore carboxylated) and/or sulfated and are thus strongly negatively charged. Some examples of the way in which mucins function in protection and lubrication are listed in *Table 5.4*.

> ➲ The physical properties of mucins enable them to protect underlying epithelial cells.

A good example of the protective role of mucins is their function in acting as a physical and chemical barrier to shield gut epithelial cells from body fluids containing potentially harmful digestive proteases. The mucin 'barrier' is very efficient, as digestion of the numerous, heterogeneous and complex oligosaccharide structures present on the mucins would require the action of many different hydrolases. Thus, the oligosaccharides remain intact, and the sticky mucous layer physically prevents the proteases from access to the underlying vulnerable epithelial cells. The mucin layer also forms effective protection for the epithelial cells from the harshly acidic conditions of the stomach; hydrogen carbonate ions are associated with the

Table 5.5 Some examples of the protective and lubricating role of mucins

Anatomical location	Function
Gastrointestinal tract	Physical barrier protecting epithelium from proteases. pH barrier protecting epithelium from acidic conditions in the stomach. Carbohydrate–receptor interaction blocks infection by bacteria in stomach and large intestine.
Respiratory tract	Mucociliary escalator keeps airways clear of particles and sterile.
Salivary mucins	Lubrication during swallowing. Protection against infection.
Lumina	Prevention of adhesion of opposing surfaces.

mucin layer and facilitate the formation of a pH gradient ranging from 1–2 in the gastric lumen to 6–7 at the mucosal surface.

The development of gastric ulcers is also associated with reduced mucin production by goblet cells of the stomach, presumably because a reduction in the protective barrier of mucin results in damage to the underlying epithelium by acids in the stomach. However, there is some evidence that the recognition of gastric mucin oligosaccharides by *Helicobacter pylori*, a microorganism involved in the aetiology of gastric ulcers, may be involved in bacterial adherence and infection (*see 16.2.1*). Altered glycosylation of mucins has been described in patients with inflammatory bowel disorders such as ulcerative colitis and Crohn's disease.

In the respiratory tract, mucins are essential for the normal function of the mucociliary escalator, where solid particles and microorganisms are trapped in the sticky mucus layer, then transported by the beating action of the cilia, up the airway thus keeping the lower regions sterile.

Salivary mucins, in comparison, have a principle role in lubrication to ease the swallowing of food and the glycosylation of salivary mucins also seems to function in inhibiting infection by oral microbes (the role of lectin–saccharide interactions in microbial infection is covered in *13.9*). For example, oral mucins will agglutinate certain *Streptococcal* strains and this is modulated through the O-linked oligosaccharide NeuAc(α1→3)Gal(β1→3)GalNAcα1→Ser/Thr. Individuals suffering from impaired salivary gland function who secrete less mucins are typically prone to chronic low-level inflammation of mucosal surfaces.

5.14.2 Mucins as anti-adhesins

It has been suggested that the repulsion between the glycans of cell membrane-anchored mucin-like glycoproteins, such as MUC1, on opposite walls of lumina, for example, in breast ducts or high endothelial venules, functions in preventing adhesion between membranes and thus keeps the lumen open. This anti-adhesive property may have implications in cancer cell behaviour where expression of these molecules is no longer limited to the apical (i.e. luminal) surface of the cell, but is spread over the entire cell surface. Hence, interactions between adjacent cancer cells and between cancer cells and their normal neighbours may be affected resulting in alterations in cell adhesiveness that may influence metastatic behaviour. Under different circumstances, O-linked glycans of mucin-like glycoproteins may be involved in adhesive, rather than anti-adhesive, mechanisms, possibly as ligands recognized by selectin molecules. These functions are summarized in *Table 5.5*.

5.14.3 Mucin ligands for selectins

Leukocyte homing during the inflammatory response is mediated through recognition of glycan ligands, including sialyl Lewis x (sLe^x) and related structures, by the sugar-binding P- and E-selectins (*see 13.12.3.2*). These glycans may be O-linked, N-linked or part of proteoglycans or glycolipids. However, inhibition of O-glycosylation using O-benzyl-GalNAc (which inhibits O-glycan extension; *see 15.7*), abrogates E-selectin binding, whereas inhibitors of N-glycosylation do not, indicating that O-glycans may be more important E-selectin ligands than N-linked glycans. Deficiency in the glycosyltransferase enzyme Fuc-TVII which mediates attachment of terminal (α1$\rightarrow$3)Fuc results in the failure of normal leukocyte trafficking because leukocytes do not have fucosylated oligosaccharide ligands for selectins. Cancer cells may also have oligosaccharide ligands recognized by selectins, and oligosaccharide–selectin interaction may be implicated in cancer metastasis to distant sites (*see 16.6*).

However, the oligosaccharide ligand alone, for example Lewis x (Le^x), is not sufficient for selectin recognition and binding, as underlying peptide and/or saccharide structures are also of importance. Presentation of the ligand is another important factor in selectin recognition. For example, P-selectin-mediated adhesion of HL60 cells (through recognition of their cell surface sLe^x) can be abrogated by treatment with agents such as O-sialoglycoproteinase which disrupts cell surface sLe^x synthesis. The sLe^x recognized by P-selectin is presented in a particular 'clustered saccharide patch' arrangement, and may be presented on molecules such as leukosialin, glycophorin and MUC1.

5.14.4 Antifreeze glycoproteins

Another function of O-linked glycoproteins is as an antifreeze agent in the serum of some Antarctic fish species. The glycoproteins contain multiple (sometimes > 40) repeat alanine–alanine–threonine sequences in which each threonine carries an O-linked disaccharide Gal(β1$\rightarrow$3)GalNAcα1$\rightarrow$Ser/Thr.

5.14.5 Mucin–microorganism interactions

Many pathogenic microorganisms attach to the epithelial surfaces of the respiratory, genitourinary or gastrointestinal tracts through specific recognition of O-linked mucin oligosaccharides (*see 13.9, 16.2*). Highly glycosylated mucins may act as a trap to prevent bacterial adhesion; the bacteria express lectins (adhesins) that specifically recognize saccharide structures and the highly glycosylated mucins provide a variety of potential binding partners. Infectious microorganisms may thus bind to oligosaccharides of the mucinous coating of epithelial cells, but instead of binding to them directly and thus infecting them, they are shed with the mucin and lost, for example, in stools or by coughing. ➲

➲ Mucins play a powerful role in preventing infection by microorganisms, and, conversely, sometimes in facilitating infection.

Conversely, bacterial recognition of mucin oligosaccharides sometimes enhances infectivity. The binding of two bacteria, *Pseudomonas aeruginosa* and *Pseudomonas cepacia* to O-linked mucin oligosaccharide structures may, for example, play an important role in the infectivity of these microorganisms in the lungs of cystic fibrosis patients and other compromised individuals.

The digestive tract harbours a wide variety of strains of microorganisms, the presence of which are necessary for its normal function. Their presence protects the underlying mucosa from infection by pathological strains, and they are also involved in the breakdown of some components of food and also of mucus (mucolysis). Mucus produced by the cells lining the respiratory tract acts as a trap to particulate matter and microorganisms. The mucus is then transported up the airway by the beating action of the cilia and swallowed, along with salivary mucins. These secretions blend with those produced by the lining of the gastrointestinal tract itself. *Bifidobacterium bifidium* and *Ruminococcus torques* both utilize mucins as their only energy source.

5.14.6 Role of O-linked glycoproteins in sperm–egg recognition

The extracellular matrix coat or zona pellucida of the mammalian egg has a large number of O-glycans, and also some N-glycans. The composition of the zona pellucida is well studied; for example, porcine egg zona pellucida comprises 71% protein, 19% neutral hexose sugars, 2.7% sialic acid and 1.3% sulfate. All zona pellucida proteins are glycosylated (*see 11.8.5.5*).

In the mouse, sperm are believed to bind to O-linked oligosaccharides on the zona pellucida glycoprotein, ZP3, and this interaction is important in fertilization. Oligosaccharide recognition is thought to be through a sperm β-galactosyltransferase which functions like a lectin in this context. Removing the O-glycans by mild alkali treatment results in ablation of sperm binding and ZP3 knockout female mice are infertile, but null GalT sperm still bind to the zona pellucida and achieve low rates of fertilization. The role of oligosaccharide–receptor interaction in fertilization has also been extensively studied in sea urchins, and here too O-linked oligosaccharides have been implicated in sperm–egg recognition and binding. Differences in O-linked glycan synthesis may be partially responsible for species-specific sperm–egg recognition in fertilization. In a novel application, the use of O-linked-type oligosaccharides as contraceptive medication to competitively inhibit sperm–egg binding is being considered. The precise structure of the oligosaccharide(s) involved is undetermined and this complex topic has been the focus of some controversy.

5.14.7 Haemopoietic cells and the immune system

Several cell types of the haemopoietic and immune systems carry characteristic O-glycan chains, and these often change with differentiation. Leukocytes and erythrocytes, for example, have large amounts of cell-surface O-linked oligosaccharides attached to leukosialin and glycophorin molecules.

Leukosialin (CD 43) is the major sialoglycoprotein on the surface of human leukocytes. It has 234 amino acid residues in the extracellular domain, of which 93 are serine or threonine. About 90% of these amino acids are covalently linked to oligosaccharides – one of three amino acids in the extracellular domain is attached to an oligosaccharide. Leukosialin covers the surface of the leukocyte with a large amount of negative charge, thereby preventing leukocytes from adhering to each other, to serum glycoproteins and to endothelial cells while in the blood stream. Leukosialin may also play a role in cell–cell interactions between T- and B lymphocytes. Dramatic changes in the glycosylation of mucin-type oligosaccharides of leukosialin during T-cell maturation and activation are observed. Changes in the glycosylation of leukosialin occur with differentiation, immune deficiency and malignancy (*see 16.4*).

5.14.8 Use of O-glycosylation-deficient mutant cell lines in determining O-glycan function

A mutant Chinese hamster ovary (CHO) cell line, ldlD (so called because it has an incompletely glycosylated, and therefore defective, low density lipoprotein, or LDL, receptor) has proved useful in elucidating some of the functions of O-linked oligosaccharides. ldlD cells lack UDP-Gal/UDP-GalNAc epimerase and are therefore defective in the production of UDP-Gal and UDP-GalNAc, and consequently unable to synthesize O-glycans. If exogenous UDP-GalNAc is provided, these cells synthesize partial O-glycans, and if both exogenous UDP-GalNAc and

Table 5.6 Adhesive and anti-adhesive properties of negatively charged mucins

Anatomical location	Expression	Effect
Epithelia lining lumina	Luminal surface only	Anti-adhesion of opposing surfaces
Cancer cells within a tumour mass	Entire cell surface	Anti-adhesion causing cancer cells to break away and disseminate
Cancer cells in bloodstream	Entire cell surface	Adhesion to endothelium through specific receptor–ligand interaction with selectins

UDP-Gal are provided normal O-glycosylation is fully restored. Experiments with these cells indicate that the correct synthesis of O-glycans may be essential for stability, secretion and folding, as well as intracellular transport and cell surface expression of O-linked glycoproteins.

Further reading

Devine, P.L., McKenzie, I.F.C. (1992) Mucins – structure, function and associations with malignancy. *Bioessays* **14:** 619–625.

Hilkens, J., Ligtenberg, M.J.L., Vos, H.L., Litvinov, S.V. (1992) Cell membrane associated mucins and their adhesion modulating property. *Trends Biol. Sci.* **17:** 359–363.

Hounsell, E.F., Davies, M.J., Renouf, D.V. (1996) O-Linked protein glycosylation structure and function. *Glycoconj. J.* **13:** 19–26.

Seregni, E., Botti, C., Massaron, S., Lombardo, C., Capobianco, A., Bogni, A., Bombardieri, E. (1997) Structure, function and gene expression of epithelial mucins. *Tumori* **83:** 625–632.

Van den Steen, P., Rudd, P.M., Dwek, R.A., Opdenakker, G. (1998) Concepts and principles of O-linked glycosylation. *Crit. Rev. Biochem. Mol. Biol.* **33:** 151–208.

O-GlcNAc glycosylation of nuclear and cytosolic proteins

6

6.1 Introduction – O-GlcNAc protein glycosylation

One of the most recently discovered types of glycosylation is the modification of proteins by the addition of O-linked N-acetylglucosamine (O-GlcNAc). Many proteins in the cytoplasm and nucleus are multiply O-glycosylated at specific serine and/or threonine amino acid residues by the addition of a single β-N-acetylglucosamine (β-GlcNAc) monosaccharide. This structure is not elongated by the addition of further monosaccharides, and is confined almost exclusively to nuclear and cytosolic glycoproteins.

Studies suggesting the existence of this type of glycosylation first appeared in the mid-1970s, but their findings were equivocal and until the mid-1980s it was widely believed that only cell surface and secreted proteins were glycosylated and that proteins within the nucleus and cytoplasm were *not* glycosylated. ➲ Convincing demonstration of the β-O-GlcNAc modification occurred in 1984 when it was shown to be a major type of intracellular glycosylation in mouse lymphocytes. This groundbreaking study used an elegant method to detect O-linked GlcNAc, which is still commonly used today: radiolabelled UDP-[^{3}H]Gal is attached to terminal O-GlcNAc by the action of pure galactosyltransferase enzyme. Radioactive 'tagging' of the cell provided the evidence that terminal O-GlcNAc was available as a substrate for the galactosyltransferase.

➲ Until recently it was believed that nuclear and cytoplasmic proteins were never glycosylated. O-GlcNAc modification of these molecules is now known to be common.

O-GlcNAc modification has since been shown to be a ubiquitous protein modification in virtually all eukaryotes, protozoa, filamentous fungi, plants and animals. It also occurs in viral glycoproteins. It is a particular feature of nuclear pore proteins, which regulate trafficking of molecules in to and out of the nucleus, but O-GlcNAc-modified glycoproteins are present in the cytoplasm and all subcellular organelles except mitochondria. O-GlcNAc-modified proteins include the catalytic subunit of RNA polymerase II, transcription factors, oncogene products, cytoskeletal proteins, nuclear pore proteins, viral proteins, heat shock proteins and enzymes. Some are listed in *Table 6.1*; many hundreds of as yet unidentified O-GlcNAc-modified proteins, mostly of nuclear origin, have been reported and this appears to be a very common modification.

6.2 The functions of O-GlcNAc modification and its interrelationship with phosphorylation

We are only just beginning to understand the functions of this protein modification, but O-GlcNAc-modified proteins share several common features with phosphorylated proteins:

- O-GlcNAc proteins are also phosphorylated, and a close reciprocal relationship between phosphorylation and O-GlcNAc modification often exists in common with phosphorylation, O-GlcNAc modification is a dynamic process, with rapid turnover of O-GlcNAc in response to cell signalling and stages of the cell cycle.

Functional and Molecular Glycobiology, Susan A. Brooks, Miriam V. Dwek and Udo Schumacher

Table 6.1 Some examples of proteins modified by O-GlcNAc glycosylation

Eukaryotic nuclear proteins
Oestrogen receptor
SV40 T antigen
p53 tumour suppressor gene product
Nuclear pore proteins
RNA polymerase II
Transcription factors
c-Myc oncogene product
v-Erb-a oncogene product
Tyrosine phosphatase
Chromatin-associated proteins
DNA-binding proteins
Eukaryotic cytoskeletal proteins
Cytokeratins 8,13 and 18
Neurofilaments H, M and L
Human erythrocyte band 4.1 glycoprotein
Synapsin I
Synaptic vesicle proteins
Microtubule-associated proteins (MAPs)
Tau
Talin
Vinculin
Clathrin assembly protein AP-3
Crystallins
Other eukaryotic proteins
Amyloid precursor protein
p67 translation regulation protein
Viral proteins
Adenovirus fibre
Human CMV UL32
Rotavirus NS26 protein
Baculovirus gp41 tegument protein
Proteins of parasites and other pathogenic microorganisms
Schistosome proteins
Malarial proteins
Trypanosome proteins
Giardia proteins
E. histolytica proteins

- O-GlcNAc modified proteins form reversible multimeric complexes with other polypeptides or structures, and the formation of these complexes is often regulated by phosphorylation.
- O-GlcNAc modification frequently occurs at sites similar to those modified by protein kinases.

➲ A close reciprocal relationship exists between O-GlcNAc modification and phosphorylation of the same protein. Both are dynamic events, and the interplay between them is involved in many cell-regulatory processes.

What is therefore clear, is that the O-GlcNAc modification plays a fundamental role in a wide range of cellular regulatory processes, and disruption in O-GlcNAc modification may thus have profound consequences and is implicated in many disease states.

All O-GlcNAc-modified proteins described to date also occur as O-phosphorylated proteins, and a complex interrelationship between phosphorylation and glycosylation with O-GlcNAc exists.➲ Proteins may carry multiple potential modification sites that can either be glycosylated with O-GlcNAc or phosphorylated, and multiple combinations of O-GlcNAc-modified and phosphorylated forms may be synthesized by the cell in response to cellular signals, and addition or removal of these two differentially regulated modifications may allow for complex and subtle modification of protein function. In some proteins, for example, c-Myc, phosphorylation and O-GlcNAc modification are mapped to the same amino acid residue, where they are mutually exclusive. In other proteins, for example, synapsin and neurofilament H, the phosphorylation sites and O-GlcNAc modification sites lie near to, or flank, each other, but are separate. O-phosphorylation and O-GlcNAc modification occur on different subsets of some cytokeratins, and change with different phases of the cell cycle. Here, O-GlcNAc modification and phosphorylation are not reciprocal, as in most other proteins, but seem to represent functionally different isoforms of the proteins.

Attachment and removal of O-GlcNAc to proteins is a dynamic process, unlike other forms of glycosylation, and, like phosphorylation, is responsive to extracellular signals. The levels of O-GlcNAc modification of many proteins change with the cell cycle, for example, mitogenic stimulation of lymphocytes results in rapid and transient changes in the levels of O-GlcNAc modification.

6.3 Synthesis and cleavage of the O-GlcNAc linkage

The enzymes responsible for the O-GlcNAc modification of proteins may be functionally analogous to the protein kinases and phosphatases that regulate protein phosphorylation.➲ The concept that O-GlcNAc modification and phosphorylation are interdependent and dynamic, changing in response to signals, implies that the O-GlcNAc transferase that adds the O-GlcNAc, the O-GlcNAcase that removes it, and the protein kinases and phosphatases involved in phosphorylation work cooperatively to regulate this process.

➲ Because O-GlcNAc modification and phosphorylation are interrelated and dynamic events, the enzymes regulating both processes must be regulated and function cooperatively.

6.3.1 O-GlcNAc transferase

UDP-N-acetylglucosamine: polypeptide β-N-acetylglucosaminyltransferase (O-GlcNAc transferase, OGT) is responsible for O-GlcNAc modification.➲ It does not belong to any of the established glycosyltransferase gene 'families' (*see 2.11*), but is highly conserved between species with 85% homology in the primary sequence of the enzyme derived from either the nematode *Caenorhabditis elegans*, the rat or humans. Given the diversity of O-GlcNAc attachment sites, it is possible that an, as yet undiscovered, family of transferases responsible for O-GlcNAc attachment exists. The transferase is resident in both nucleus and cytoplasm. Functional OGT is necessary for cell viability.

➲ UDP-N-Acetylglucosamine: polypeptide β-N-acetylglucosaminyltransferase (O-GlcNAc transferase, OGT) is responsible for O-GlcNAc modification.

The O-GlcNAc transferase is composed of two subunits, a 110 kDa α-catalytic subunit and a 78 kDa β-subunit, which may be a result of alternative RNA transcript processing mechanisms or proteolytic digestion. The two subunits undergo posttranslational modification by the addition of tyrosine phosphate and O-GlcNAc, both of which may be involved in the regulation of its catalytic activity. Its activity is also potently modified by changes in the local concentration of UDP-GlcNAc and also UDP (the by-product of the glycosylation reaction), and the relative concentrations of these two substances are also critical in regulating enzyme activity. The importance of O-GlcNAc modification is emphasized by the fact that UDP-GlcNAc synthesis may be responsible for as much as 2–5% of total glucose utilization within the cell, and that cellular concentrations of UDP-GlcNAc may be similar to those of ATP.➲

➲ O-GlcNAc modification is important: as much as 5% of the cell's energy requirements are expended on the synthesis of UDP-GlcNAc, the cellular concentration of which is similar to that of ATP.

6.3.2 Regulation of O-GlcNAc transferase activity

Regulation of OGT is complex, and it has a structure unlike that of other known glycosyltransferases. One striking structural feature is that the enzyme has multiple (11) tandem tetratricopeptide repeat (TPR) domains at the N-terminus which form alternating 'holes' and 'knobs' that may be involved in inter- and intraprotein interactions. TPR domain motifs function in protein–protein interactions, and are important to the functioning of cell cycle, chaperone, transcription and protein transport complexes. The TPR domains of the OGT mediate trimerization of the catalytic subunit, interactions of the enzyme with other proteins, and play a role in substrate selectivity. It has been proposed that numerous tetratricopeptide-binding factors are involved in the regulation of enzyme activity, in a manner analogous to the regulation of RNA polymerase II by transcription factors. Several serine/threonine protein phosphatases contain TPR motifs, and it is possible that O-GlcNAc transferase interacts with one or more of these phosphatases, via the TPRs, to couple phosphorylation and O-GlcNAc modification of proteins directly.

6.3.3 O-GlcNAc modification occurs posttranslationally

Nuclear and cytosolic proteins are synthesized on ribosomes in the cytosol as they lack the signal sequence for direction to the endoplasmic reticulum (ER; *see 4.2*). O-GlcNAc modification, in common with the O-linked glycosylation of mucin-type glycoproteins (*see Chapter 5*) occurs wholly posttranslationally, as distinct from N-linked glycosylation (*see Chapter 4*) which is a co-translational event. ➲ Furthermore, no lipid-linked precursor is involved in its synthesis. A strict primary amino acid sequence requirement does not exist, as it does for N-linked glycans (*see 4.2.1*), but the general pattern (Asp)-Ser/Thr(Xaa)$_n$-Pro is favoured. This sequence is similar to that required for protein phosphorylation, and it seems likely that the transferase competes for serine residues on these proteins.

➲ O-GlcNAc modification of proteins occurs as a posttranslational event.

6.3.4 Cleavage of the O-GlcNAc linkage – the O-GlcNAcase

A cytosolic and a nuclear form of a β-D-N-acetylglucosaminidase selective for O-GlcNAc (O-GlcNAcase) have been identified, which appear to function at a pH optimum of 6.4. Both enzymes have an apparent molecular mass of 106 kDa and consist of two subunits, an α-subunit of molecular mass 54 kDa which carries the active site, and a β-subunit of molecular mass 51 kDa. Together they form an active αβ-heterodimer which is distinct from the lysosomal hexosaminidases. Their cellular regulation is not well understood.

Several analogues of GlcNAc act as inhibitors of β-D-N-acetylglucosaminidase activity, for example, 2-acetamido-2-deoxy-D-gluconhydroxine-1,5-lactone (LOGNAC), O-(2-acetamido-2-deoxy-D-glucopyranosylidene)-amino-N-phenylcarbamate (PUGNAC), 1-amino GlcNAc, and 1-azido GlcNAc. Research using these compounds may increase our understanding of the cellular roles of O-GlcNAc modification.

6.4 Functions of the O-GlcNAc modification

The fact that this is such a common modification would suggest that O-GlcNAc modification may have multiple and important functions, and indeed it is necessary for survival even at the single cell level. It is clear that this is functionally an important modification, possibly involved in diverse aspects of cellular physiology, but its exact functions and the complex mechanisms underlying its precise regulation are not fully understood. Studies have indicated that O-GlcNAc modification is involved in mechanisms as diverse as protection of proteins from degradation, major histocompatibility complex class I antigen presentation, lymphocyte activation and signal transduction pathways. O-GlcNAc modification is a dynamic and sometimes transient phenomenon, with turnover rates similar to those seen for phosphorylation. It seems likely that phosphorylation and O-GlcNAc modification may work in conjunction to regulate biological processes, for example, nuclear trafficking (*see 6.4.1*) and proteins previously believed to be regulated solely by phosphorylation may actually be regulated by dynamic phosphorylation and O-GlcNAc modification together.

6.4.1 Nuclear trafficking

O-GlcNAc modification is integral to the function of nuclear pore glycoproteins, called nucleoporins, which mediate the transport of macromolecules in to and out of the nucleus through the nuclear pore complex.➲ O-GlcNAc may act as an alternative nuclear transport signal on other proteins. Alternatively, it may be involved in direct control of the translocation mechanism.

> ➲ O-GlcNAc modification controls transport into and out of the nucleus via the nuclear pore complex.

O-GlcNAc modification of nuclear pore proteins is directly involved in nuclear transport processes. Nuclear transport of large proteins and RNA through the nuclear pore complex is a specific, ATP-dependent process which requires complex control. Blocking O-GlcNAc on nuclear pore glycoproteins, using either monoclonal antibodies against O-GlcNAc or the lectin wheatgerm agglutinin (WGA) which binds to GlcNAc, abrogates nuclear transport and prevents proteins with a nuclear localization signal from entering the nucleus. This is not simply a case of these substances mechanically blocking nuclear pores, as they affect the energy requiring steps of nuclear transport. Depletion of WGA binding proteins from nuclear envelopes, for example those of reconstituted oocyte nuclei, has the same effect, and reintroducing isolated O-GlcNAc modified nuclear pore proteins restores the transport function. They are not therefore required for assembly of the nuclear membrane, assembly of the nuclear pore itself, or formation of the transport channel, but they do seem to be essential for the recognition of nuclear transport signals. The role of O-GlcNAc in nuclear transport is not a result of their functioning as ligands for lectin-like receptors in nuclear transport because the 'capping' of O-GlcNAc by the addition of Gal does not prevent nuclear transport.

Pores depleted of O-GlcNAc proteins are structurally deficient with spaces where the glycosylated molecules are normally located. At least three O-GlcNAc-modified glycoproteins, p62, p58 and p54, associate to form a macromolecular complex of $\approx$600 kDa that may form the basis of the pore structure required for nuclear transport. p62 is the best studied of these molecules and has at least ten O-GlcNAc modification sites, mostly clustered in the N-terminal part of the molecule. O-GlcNAc is implicated in a sensitive mechanism to detect the integrity of nascent RNA transcripts and prevents nuclear export of defective products.

6.4.2 Chromatin-associated proteins and transcription

Labelling of O-GlcNAc residues on chromosome-associated proteins indicates that O-GlcNAc is concentrated in condensed regions of the chromatin with a dramatic reduction in labelling at sites of puffs where active gene transcription is occurring. O-GlcNAc therefore plays a functional role in transcription.➲

> ➲ O-GlcNAc plays a functional role in transcription.

Most, or all, of the proteins involved in transcriptional regulation are dynamically glycosylated with O-GlcNAc. Transcription factors are synthesized in the cytoplasm, but exert their effects in the nucleus of the cell. Transcription factors including RNA polymerase II, serum response factor (SRF), specificity protein 1 (Sp1), hepatic nuclear factor-1 (Hnf-1) and oestrogen receptor (ER) are extensively modified by O-GlcNAc. A number of specific functions for O-GlcNAc modification of transcription factors have been proposed, including that it is a nuclear targeting signal, that it is involved in regulation of phosphorylation, or in assembly of transcriptional complexes leading to the activation of specific genes. One of the major O-GlcNAc glycosylation sites is adjacent to a regulatory phosphorylation site used by the

34 kDa phosphokinase, p34, which is coded for by the *cdc2* gene and which plays a central role in the transition from G2 to mitosis during the eukaryotic cell cycle.

One example is casein kinase II, a nuclear enzyme implicated in regulating cellular processes including growth, differentiation and cellular proliferation. Both its α- and α′-subunits are multiply glycosylated with O-GlcNAc, but its β-subunit is not. The glycosylation of this type of protein may have consequences for its turnover and transactivation.

The O-GlcNAc modification of Sp1 serves as a nutritional checkpoint: in conditions of starvation, Sp1 becomes markedly hypoglycosylated and is more prone to degradation, thus inducing transcriptional repression. Regulated O-GlcNAc removal from Sp1 may also be necessary for Sp1 binding to specific partners, suggesting that O-GlcNAc modification has multiple roles.

The putative roles of O-GlcNAc modification of the oestrogen receptor (ER) illustrate how it may have differing, and sometimes opposite, functions in different situations. Function may also vary in a protein-specific manner. The ER is a nuclear steroid hormone receptor that regulates gene expression by binding to oestrogen response elements in a ligand dependent manner. O-GlcNAc modification occurs within a sequence (called a PEST sequence) known to be responsible for regulated degradation, and thus may be implicated in controlling ER transcriptional activity via regulated degradation. O-GlcNAc modification has been described in PEST sequences in other proteins, hinting at a similar function. Conversely, in Sp1 it protects against degradation.

O-GlcNAc modification/phosphorylation of RNA polymerase II illustrates how these processes may be involved in subtle mechanisms of transcription control. RNA polymerase II is responsible for mRNA synthesis. It is a complex enzyme composed of at least ten distinct polypeptide subunits. The catalytic subunit contains a seven amino acid sequence (-Tyr-Ser-Pro-Thr-Ser-Pro-Ser-) that is repeated up to fifty times at its C-terminus; this is referred to as the C-terminus domain or CTD, and it is essential for cell viability. The CTD is multiply modified by O-GlcNAc and also phosphorylated. Here O-GlcNAc modification and phosphorylation are reciprocal events; a subtype of RNA polymerase II (type IIA) is not phosphorylated at all, but is heavily glycosylated with O-GlcNAc, whereas type IIO is heavily phosphorylated and is not glycosylated.

Initial O-GlcNAc modification of the CTD induces a turn-like conformation in the molecule which significantly alters its three-dimensional structure. It is not difficult to imagine that this may have profound consequences with regard to interactions with other components of the transcription machinery. The O-GlcNAc residues are subsequently removed, and the molecule is then extensively phosphorylated. This results in transcript elongation and RNA processing. Thus, glycosylation of the CTD functions in: (i) preventing the premature association of the RNA polymerase II elongation factors or mRNA-processing factors, (ii) the O-GlcNAc-modified form may interact with transcriptional inhibitors, or (iii) CTD glycosylation may direct the association of transcription factors during assembly of the transcription initiation complex.

6.4.3 Translation

➲ Dynamic O-GlcNAc modification and phosphorylation play a functional role in the regulation of protein translation.

Reversible O-GlcNAc glycosylation and phosphorylation may play a crucial role in the regulation of protein synthesis. ➲ Elegant research on the interplay of eukaryotic translation initiation factor 2 (eIF-2) and the 67 kDa eIF-2-associated protein, also called serum response factor (p67), illustrates the potential of these modifications in subtle regulation of this process, as illustrated in *Figure 6.1.*

6.4.4 The Rho family of GTP-binding proteins

The Rho proteins are a family of low molecular mass GTP-binding proteins. They modulate changes in the actin cytoskeleton necessary for cell motility, cytokinesis, vesicle trafficking

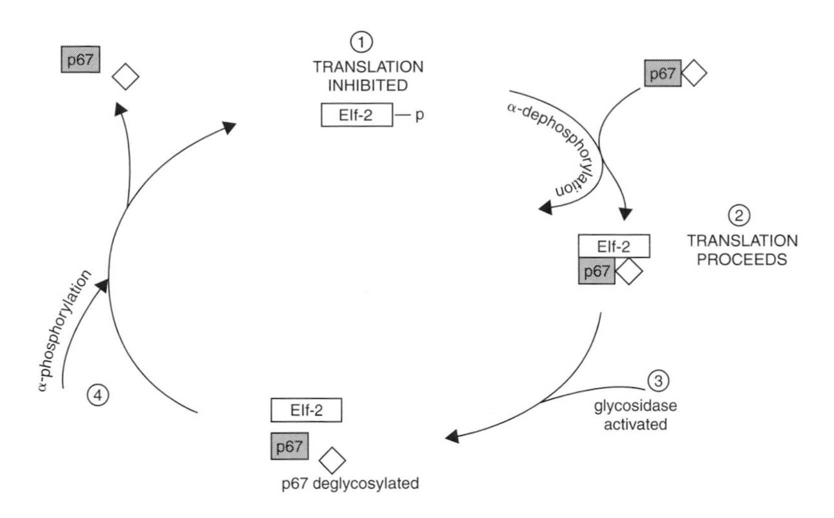

Figure 6.1

The reversible O-GlcNAc modification of p67 and phosphorylation of Elf-2 regulates protein synthesis. (1) The phosphorylated form of Elf-2 inhibits translation, and protein synthesis slows. (2) Only O-GlcNAc (◇) modified p67 can bind to Elf-2. p67 binding to Elf-2 inhibits Elf-2 phosphorylation. The non-phosphorylated form of Elf-2 is not inhibitory to translation and protein synthesis proceeds. (3) Conditions such as serum starvation cause a glycosidase to be activated. The glycosidase removes O-GlcNAc from p67. (4) Deglycosylated p67 is unable to bind to Elf-2, so phosphorylation of Elf-2 is not inhibited. Phosphorylated Elf-2 inhibits translation, and protein synthesis slows.

and pinocytosis. Rho proteins bind to GTP, and this causes activation and signal transduction which affects the action of other proteins downstream, including microtubule-associated protein (MAP) kinases which regulate the association of cytoskeletal proteins. When bound to GDP, Rho proteins are inactive. Some Rho proteins, including Rho, Rac and Cdc42, are the targets of bacterial (*Clostridium* species) glycosyltransferases, which act as toxins. One glycosyltransferase catalyses O-GlcNAc modification of the GTP-binding site of Rho family proteins. This causes functional inactivation, and disruption of the signalling cascade, thus leading to depolymerization of the actin filament system.

6.4.5 Modification of cytoskeletal proteins

Cytosolic proteins may be modified by O-GlcNAc, and many of these are cytoskeletal glycoproteins known to be involved in the phosphorylation-dependent and reversible formation of cytoskeletal structures in cell motility. ➲

➲ O-GlcNAc modification functions in the control of the cytoskeleton.

6.4.5.1 Bridging proteins

Band 4.1, a human erythrocyte cytoskeletal protein involved in maintaining erythrocyte shape by bridging between the spectrin/actin cytoskeleton and the cytosolic tail of the transmembrane glycoprotein, glycophorin, was the first cytoskeletal protein shown to be O-GlcNAc modified. It is now known that many intermediate filament proteins, tubulin- and actin-regulatory proteins, microtubule-associated proteins (such as Tau, MAP1, MAP2 and MAP4), and proteins that bridge the cytoskeleton to transmembrane glycoproteins in a reversible, phosphorylation-dependent manner, such as vinculin, talin and the synapsins, as well as band 4.1 glycoprotein, are modified by O-GlcNAc. The α-crystallins, for example, which act as chaperones in the modulation of the intermediate filament assembly in cell types such as cardiac muscle cells, but are also a major component of eye lens, are dynamically modified by O-GlcNAc many times during the life of the polypeptide.

O-GlcNAc modification is also important in events occurring at the synapse of neurons, once again in relation to bridging to the cytoskeleton. Synapsin I is O-GlcNAc modified, and the glycosylation stabilizes its interactions with the cytoskeleton. Upon depolarization of the neuron, it becomes phosphorylated causing the release of vesicles from the cytoskeleton into a pool that is able to fuse rapidly with the plasma membrane.

Talin is a major cytosolic protein that links the intracellular domains of β1- and β3-integrins to the actin cytoskeleton and is involved in the clustering and activation of these receptors and thus mediates focal adhesion. Talin is glycosylated within its carboxyl tail domain; this interacts directly with vinculin, which is also O-GlcNAc glycosylated. O-GlcNAc modification of vinculin is a dynamic process responsive to thrombin-mediated platelet activation.

6.4.5.2 Intermediate filament proteins

Proteins building intermediate filaments belong to the superfamily of fibre-like molecules that share a common secondary structure. Cytokeratins, which are O-GlcNAc modified, belong to this family, and form the main structural proteins of epithelium. Consistent with the O-GlcNAc modification of other proteins, this glycosylation is dynamic and turns over many times during the life of the protein. Both O-GlcNAc modification and phosphorylation change in relation to phases of the cell cycle. O-GlcNAc modification may be involved in regulating the dynamic assembly/disassembly of the intermediate filaments.

Neurofilaments are the characteristic intermediate filaments of neurons. They are composed of three subunits, all of which are heavily and multiply O-GlcNAc glycosylated and phosphorylated. Glycosylation and phosphorylation are located at different amino acids, and here the glycosylation is involved in neurofilament assembly allowing correct spacing and density of packing of filaments.

6.4.5.3 Microtubule-associated proteins

Microtubule-associated proteins (MAPs; *see 6.4.4, 6.4.5.1*) play important roles in organizing the cytoskeleton. Many are O-GlcNAc modified, including MAP1, MAP2, MAP4 and Tau. Tau is a heavily O-GlcNAc glycosylated neuronal MAP, and organizes microtubules in the axon. Aberrant O-GlcNAc modification/phosphorylation of this protein is implicated in neurodegenerative diseases, such as Alzheimer's disease (*see 6.4.6.2*).

6.4.6 O-GlcNAc modification in disease

6.4.6.1 Oncogenes, tumour suppresser proteins and cancer

A reciprocal relationship between phosphorylation and O-GlcNAc modification exists. Alterations in the phosphorylation of many proteins, particularly those associated with the regulation of cellular function, are of significance in malignancy, and the potential function of O-GlcNAc modification has been little explored. Hitherto, site-directed mutagenesis studies, in which serine or threonine were deleted and alterations in the molecular behaviour were then detected, have been interpreted as an effect of modified protein phosphorylation, as these two amino acids can be phosphorylated. However, it has recently been shown that the same residues are potential sites for O-GlcNAc modification. As this glycosylation effect was not taken into consideration in interpreting the results of the mutagenesis studies, a fresh look at those studies seems to be warranted as the reported effects may be due to glycosylation rather than phosphorylation as originally believed.

> ➲ The involvement of O-GlcNAc modification/phosphorylation in the control of cell-regulatory mechanisms means that disruptions in their control may be of relevance in malignancies.

The interrelationship between phosphorylation and O-GlcNAc modification may be involved in the subtle and exquisite control of cell-regulatory mechanisms.➲ Disturbances in either of these modifications may thus contribute to a transformed phenotype. Many transcription factors are O-GlcNAc glycosylated. Mutations in the genes coding for these molecules, which may sometimes alter their O-GlcNAc glycosylation sites, result in functional effects. This type of alteration in O-GlcNAc modification could be important in malignant transformation.

A number of the transcription factors, such as oncogene products, including p53, c-Myc and v-ErbA are also O-GlcNAc modified. The tumour suppresser and transcription factor gene, *p53*, the most commonly mutated gene in a wide range of human cancers, is O-GlcNAc modified, and the O-GlcNAc moiety has a functional role in DNA binding. The high affinity of *p53* for DNA is dependent on the presence of O-GlcNAc masking a basic region of the C-terminus, which otherwise represses DNA binding. Dynamic O-GlcNAc glycosylation therefore has an important role in modulating DNA binding.

The c-*Myc* oncogene has at least one major phosphorylation/O-GlcNAc modification site, at Thr58, which lies within the transactivation domain. This region is involved in the regulation of gene transcriptional activity, and hierarchical phosphorylation at this (and other) sites is involved in c-*Myc* regulation of the cell cycle. Thr58 appears to be a common site for mutation in some neoplasias, such as AIDS-related lymphomas and Burkitt's lymphoma. Mutation at this site results in enhanced transforming activity and increased tumour-inducing potential of the c-*Myc*-coded protein.

6.4.6.2 Neurodegenerative diseases

O-GlcNAc modification is a common feature of many neuron-associated proteins, for example, neurofilaments, MAPS, clathrin assembly protein and the β-amyloid precursor protein, and it seems likely, therefore, that disruptions in O-GlcNAc modification may be linked to neurodegenerative disorders.➲ Indeed, the OGT gene itself is located on the X chromosome at the same locus as the X-linked Parkinson dystonia gene.

> ➲ Many neuron-associated proteins are O-GlcNAc modified, and disruptions in O-GlcNAc modification are associated with neurodegenerative diseases, such as Alzheimer's disease.

The MAP, Tau, is normally heavily glycosylated by multiple O-GlcNAc residues, carrying up to 12 potential O-glycosylation sites. Tau is involved in the organization of microtubules in the axons. In Alzheimer's disease, Tau is hyperphosphorylated, causing it to dissociate from microtubules and self-assemble into paired helical filaments, which are the major component of neurofibrillary tangles in the brain.

Similarly, the cytoplasmic domain of the β-amyloid precursor protein, which upon proteolysis gives rise to the neurotoxic β-amyloid peptide in Alzheimer's disease, is also O-GlcNAc modified. The similarity of the O-GlcNAc attachment sites and sequences that are known to target proteins for degradation suggest that, here, O-GlcNAc modification is implicated in the degradation of β-amyloid precursor protein, resulting in the release of the neurotoxic β-amyloid peptides. O-GlcNAc modification of the clathrin assembly protein AP-3 is also reduced in Alzheimer's disease, and this is associated with an increase in the density of neurofibrillary tangles. One hypothesis is that some neurodegenerative disorders may arise when decreased glucose metabolism of ageing neurones results in decreased O-GlcNAc modification and the consequent abnormal phosphorylation of key proteins.

6.4.6.3 Diabetes

Hyperglycaemia (high blood glucose levels) is linked to the vascular damage associated with diabetes. A number of studies have provided evidence of an association between disregulation of O-GlcNAc modification and this disorder. Abnormal glycosylation of transcription factors resulting from elevated UDP-GlcNAc levels may be a major factor contributing to insulin resistance and glucose toxicity in diabetes.

Conversion of glucose into glucosamine is an essential prerequisite for the development of insulin-resistant or type 2 diabetes and is catalysed by the enzyme glutamine: fructose-6-phosphate amidotransferase (GFAT). Even low cellular levels of glucosamine result in insulin resistance, but the mechanism by which this occurs is not understood. Glucosamine is about forty times more potent in inducing insulin resistance than glucose. GFAT directs the flow of incoming glucose into glucosamine, which is postulated to serve as a glucose sensor via a negative feedback mechanism. Glucosamine then provides a key metabolic precursor of UDP-GlcNAc, the donor molecule required for O-GlcNAc modification. Although there is as yet little supporting evidence, one hypothesis is that elevated glucosamine levels result in elevated levels of UDP-GlcNAc, which in turn leads to an increase in O-GlcNAc modification of key vesicle-associated proteins, with a resultant decrease in their phosphorylation. Reduced phosphorylation would result in disruption in the release of vesicles from the cytoskeleton and subsequent transport to the plasma membrane.➲ Increased O-GlcNAc modification of transcription factors or signalling molecules involved in response to insulin would also result in insulin resistance. Increased glucosamine levels may also stimulate the transcription of growth factors, for example, TGFα, giving rise to the vascular complications and pathological changes seen in diabetic patients.

➲ High UDP-GlcNAc levels may result in abnormal glycosylation of transcription factors and contribute to insulin resistance and glucose toxicity in diabetes.

6.4.6.4 O-GlcNAc modified viral proteins

As viruses rely on the cellular machinery of their host organisms for glycoprotein synthesis, it is unsurprising that the proteins of many viruses, including some that are responsible for human disease, carry O-GlcNAc modifications.➲ O-GlcNAc modification may be implicated in infection and disease processes. For example, the phosphoprotein UL32, a protein involved in the region between the capsid and the viral envelope (the tegument) of human cytomegalovirus (CMV) is O-GlcNAc modified, and the location of this protein is consistent with it acting as a signal for capsid envelopment during infection. Similarly, fibre proteins, involved in the formation of a trimeric structure necessary for adenovirus attachment to host cell surface are O-GlcNAc modified. Here it is difficult to envisage a function in infection mechanisms as the GlcNAc moieties of the mature trimeric structure are cryptic and hidden within its three-dimensional form. As in many examples, the actual function of this type of glycosylation is not understood.

➲ O-GlcNAc modification of viral proteins may be implicated in infection and disease processes.

Baculovirus proteins have been shown to be O-GlcNAc modified, implying that insect cells are capable of this type of glycosylation. This observation triggered research into the O-GlcNAc glycosylation of many proteins, such as oncogenes and transcription factors, which are normally present in low levels in human cells, but could be over-expressed in baculovirus transfected insect cells. These in turn might secrete the proteins of interest in the O-GlcNAc form.

6.4.6.5 O-GlcNAc-modified parasite proteins

There is increasing evidence that proteins produced by parasites involved in human infection are O-GlcNAc modified, although their function is unknown. Examples include a 92 kDa cytoplasmic protein from *Leishmania major*, a serine-rich protein produced by *Entamoeba histolytica* that is both heavily phosphorylated and O-GlcNAc modified; proteins of *Schistosoma mansoni* and malarial parasites similarly exhibit O-GlcNAc modification.

Further reading

Comer, F.I., Hart, G.W. (1999) O-GlcNAc and the control of gene expression. *Biochim. Biophys. Acta* **1473:** 161–171.

Comer, F.I., Hart, G.W. (2000) Mini review: O-glycosylation of nuclear and cytosolic proteins. Dynamic interplay between O-GlcNAc and O-phosphate. *J. Biol. Chem.* **275:** 29 179–29 182.

Species-specific and unusual types of glycoprotein glycosylation

7

7.1 Introduction

The number of monosaccharides available, and the many different types of linkages between them, mean that a large variety of oligosaccharide and polysaccharide structures can potentially be synthesized by living organisms (*see Chapter 1*). However, the variety that is *actually* produced is nowhere near as large as is theoretically possible, because the enzymes required to synthesize all possible combinations do not exist in nature. In previous chapters (*see Chapters 3–6*), the most common types of glycan structures that have evolved are described. In this chapter, we turn our attention to those glycan structures which are either uncommon or produced by only a limited number of species. Despite their rarity, they often have important biological functions.

7.2 Glycosylation of prokaryotic glycoproteins

Prokaryotes lack internal membrane-bound organelles, including nucleus, endoplasmic reticulum (ER) and Golgi apparatus. As they lack the glycosylation machinery of eukaryotes, they either do not glycosylate their proteins or do so in a very different manner. The other most dramatic difference in the processing of oligosaccharides in prokaryotic species compared with eukaryotes is that it occurs on the outside of the cell membrane, in this case the extracellular space is considered a prokaryotic version of the lumen of the ER of eukaryotes. ➲ The most well-studied examples are archaea bacteria and eubacteria. The need for an understanding of prokaryotic glycosylation pathways is underlined by the importance of bacteria in protein expression systems in biotechnological processes (*see 18.4.7, 18.5.1*), and as pathogens (*see 16.2*). Many antibiotics function by interfering with this glycosylation process.

Many of the reports of glycosylation of proteins in bacteria have focused on the surface coat (S-coat) that surrounds the bacterial cell, the flagellins that assist in the motility of bacteria and the enzymes that bacteria produce. The bacterial cell wall contains many glycoproteins and polysaccharides (*see 3.2.5*) and is difficult to study, as there are numerous technical challenges involved in both protein purification and in detailed oligosaccharide analysis (*see Chapter 14*). Nevertheless, there have been some studies in which the oligosaccharides of bacterial proteins have been chemically characterized and these have led to the identification of both N- and O-linked glycans on the glycoproteins of archaea bacteria and eubacteria. Some of the prokaryotic species whose glycosylation has been characterized are shown in *Table 7.1*.

> ➲ Processing of bacterial oligosaccharides occurs in the extracellular domain.

It is of note that many pathogenic bacteria, including *Chlamydia, Mycobacteria, Neisseria* and *Streptococci*, produce glycosylated proteins, but the information relating to biosynthetic pathways is poorly understood when compared with eukaryotic glycosylation.

Functional and Molecular Glycobiology, Susan A. Brooks, Miriam V. Dwek and Udo Schumacher

Table 7.1 Glycoproteins of prokaryotic species

Species	Glycoprotein
Halobacterium halobium	S-layer
Methanothermus fervidus	
Methanothermus sociabilis	
Bacillus stearothermophilus	
Lactobacillus buchneri	
Halobacterium halobium	Flagellar
Neisseria meningitides	Pili (hair-like), fimbriae (thread-like)
Cellulomonas fimi	β-1-4-glycanase
Chryseobacterium meningosepticum	Endoglycosidases
Mycobacterium tuberculosis	Secreted antigen

$[Glc(\beta1 \rightarrow 4)]_9Glc(\beta1 \rightarrow N)\text{-}Asn$

$SO_4GlcA(\beta1 \rightarrow 4)[(SO_4)GlcA(\beta1 \rightarrow 4)]_2GlcA(\beta1 \rightarrow N)\text{-}Asn$

Figure 7.1

The two most common N-linked oligosaccharides of *H. halobium.*

$Glc(\beta1 \rightarrow N)\text{-}Asn$

$GalNAc(1 \rightarrow N)\text{-}Asn$

$Rha(1 \rightarrow N)\text{-}Asn$

$GlcNAc(\beta1 \rightarrow N)\text{-}Asn$

Figure 7.2

Four common N-linked core monosaccharides found in prokaryotes.

$Me\text{-}O\text{-}3Man(\alpha1 \rightarrow 6)[Me\text{-}O\text{-}3]Man(\alpha1 \rightarrow 2)[Man(\alpha1 \rightarrow 2)]_3GalNAc(1 \rightarrow N)\text{-}Asn$

Figure 7.3

An example of an N-linked oligosaccharide found in *Methanothermus fervidus.*

7.2.1 N-Linked oligosaccharides in prokaryotes

➲ N-Linked oligosaccharides of prokaryotic organisms are often very different to those of eukaryotes.

The first chemical characterization of N-linked oligosaccharides of bacteria was from the glycoprotein of the S-coat of *Halobacterium halobium*. This species of bacteria lives in saturated sodium chloride solutions and, in common with many other prokaryotes, the N-linked oligosaccharides of *H. halobium* are distinctly different to those of eukaryotes (*see 4.2*).➲ The two most common N-linked oligosaccharides of *H. halobium* are shown in *Figure 7.1*.

Examples of other N-linked core monosaccharides found in archaea bacteria and eubacteria are shown in *Figure 7.2*. The core structures are further extended giving rise to unusual oligosaccharides when compared with eukaryotes. An example of this is shown in *Figure 7.3*.

Conversely, some bacterial oligosaccharides are similar to those of eukaryotes, for example, *Chlamydia trachomatis* synthesizes both oligomannose-type N-linked oligosaccharides as well as bi- and tri-antennary complex oligosaccharides. Another example of bacteria synthesizing oligosaccharides in common with mammals is the pathogenic *Helicobacter pylori* which is often found in the stomach of individuals with gastric ulcers (*see 16.2.1*). This pathogenic bacterium contains various oligosaccharides on surface lipopolysaccharides. The structures present on *H. pylori* include Lewis a (Le^a) Gal(β1→3)[Fuc(α1→4)]GlcNAc; Le^b Fuc(α1→2)Gal (β1→3)[Fuc (α1→4)]GlcNAc and their sulfated derivatives (*see 10.6*). The levels of the oligosaccharides on the lipopolysaccharide alter during the process of gastric colonization by *H. pylori* and this may reflect the ability of the bacterium to survive in hosts that have adapted perhaps as a result of previous colonization with *H. pylori.*➲

➲ Sometimes bacteria may synthesize similar N-linked oligosaccharides to eukaryotes. In pathogenic species, this may help them to evade immune attack by the host organism.

7.2.2 Sialic acids

In the past, it was thought that bacteria could not synthesize sialylated oligosaccharides but in the last decade there have been reports indicating that glycoproteins of prokaryotes need not necessarily be asialylated.

The main evidence to suggest that prokaryotes are able to synthesize sialic acids (*see Chapter 11*) has been gleaned from the cloning of genes with a similar sequence to other (eukaryotic) sialyltransferases. An example of this is the gene encoding the enzyme CMP-NeuAc: Gal α2,3 sialyltransferase from *Neisseria meningitides* and *N. gonorrhoeae*. The specificity of the enzyme that this gene encodes was investigated following its expression in *Escherichia coli*, and in this model system was found to be less particular in its substrate specificity than its eukaryotic counterpart:➲ it is able to utilize α- and β-terminal Gal residues and both (β1→4) and (β1→3) linkages of Gal (*see 11.8.6.1*).

➲ Prokaryotes have genes that, when expressed in *E. coli*, enable sialylation of Gal.

A feature of oligosaccharide synthesis in prokaryotes is the ability to produce polysialic acids. Polysialic acids are also found in eukaryotes (*see 11.9*), but in prokaryotes a single enzyme is involved: this adds both the first and subsequent sialic acids to the extended glycan moiety. Another example of 'one enzyme, two functions' is the enzyme CMP-NeuAc: Gal α2,6 sialyltransferase from *Photobacterium damsella* that, when expressed in *E. coli*, successfully sialylates both terminal Gal and GlcNAc. This promiscuity in specificity is quite distinct from the apparent specificity of this sialyltransferase in mammalian systems (*see 11.8.6.2*).➲

➲ Prokaryotes have genes that, when expressed in *E. coli*, enable the synthesis of polysialic acid.

7.2.3 Biosynthesis of N-linked oligosaccharides

In common with eukaryotes, prokaryotic organisms that synthesize N-linked oligosaccharides do so on recognition of a consensus amino acid sequon of Ser/Thr-Xaa-Asn, where Xaa may be any amino acid other than proline.

Other features of the biosynthetic pathway of oligosaccharides are found in cells from both prokaryotic and eukaryotic organisms, but there are also dissimilarities in the biosynthetic processes (*see 4.2*). The first difference is that rather than using a dolichol (lipid)-linked precursor molecule for the initial *en bloc* transfer of an oligosaccharide intermediate, as in eukaryotes, the lipid is a polyisoprenol and the precursor monosaccharides are nucleotide-activated in prokaryotes.

In some bacteria, an oligosaccharide intermediate is prepared, transferred to the protein and then modified further (e.g. in *Methanothermus fervidus)* as in eukaryotes; whereas in others, the oligosaccharide is modified prior to its attachment to the protein (e.g. in *Halobacterium* sp.).

7.2.4 O-Linked oligosaccharides

In eukaryotes, the O-linked oligosaccharides include Gal, Man or Glc linked to hydroxyl groups on serine or threonine (*see Chapter 5*). Serine/threonine is often present in proline-rich domains on the polypeptide chain. In prokaryotes, the O-linked oligosaccharides may be formed by an initial linkage of a monosaccharide to the hydroxyl group of tyrosine. O-linked glycosylation of serine, threonine and tyrosine has been reported in *Mycobacterium tuberculosis, Cellumonas fimi* and *Clostridium thermocellum*. Other prokaryotes may O-glycosylate the same amino acids but these need not be present in proline-rich domains. For example, the bacterium *Chryseobacterium meningosepticum* O-glycosylates serine and threonine when these amino acids are linked to asparagine rather than proline and *Thermoanaerobacter kivui* may O-glycosylate tyrosine when it is linked to valine.

Another common feature of O-linked glycosylation in prokaryotes is the methylation of both Rha and Gal residues. Some of the resulting, diverse, oligosaccharides found in this group of oligosaccharides are illustrated in *Figure 7.4*.

7.2.5 Function of glycoforms on prokaryote glycoproteins

The main functions that have been attributed to the oligosaccharides of prokaryotes are shown in *Table 7.2*.

Thermoanaerobacter thermohydrosulfuricus strain L111-69
Me-O-3-Rha(α1→4){Man(α1→3)Rha(α1→4)}$_{n\text{-}27}$Man(α1→3)-{Rha(α1→3)}$_3$Galβ1-O-Tyr

Bacillus alvie strain CCM 2051
{Gal(β1→4)[Glc(β1→6)]ManNAc(β1→3)}$_n${Rha(α1→3)}$_3$Galβ1-O-Tyr

Thermoanaerobacter thermohydrosulfuricus strain S102-70
Gal*f*(β1→3)Gal(α1→2)Rha(α1→3)Man(α1→3)Rha(α1→3)Glcβ1-O-Tyr

Flavobacterium meningosepticum
Me-O-2-Man(1→4)GlcNAcA(1→4)GlcA(1→4)Glc(1→4)[Me-O-2]GlcA(1→4)[Me-O-2]Rha(1→2)Man1-O-Ser

Figure 7.4

Examples of O-linked oligosaccharides of prokaryotes.

Table 7.2 Function of oligosaccharides on prokaryotic glycoproteins

Function	Example
Molecular mimicry in pathogenic bacteria, enabling the bacteria to evade host recognition	Pathogenic strains of *H. pylori*
Imparting stability to the glycosylated protein	Heat stability of *Bacillus* β-glucanases (when expressed in yeast) is enhanced if the enzyme is properly glycosylated
Structural	*H. halobium* – the glycosylation of S-layer protein is essential for maintaining the cell shape
Cell adhesion	In *N. meningitidis* protein glycosylation is an important aspect of cell adherence to endothelial and epithelial cells

7.3 Glycosylation of yeast and fungal proteins

Yeasts are important expression systems used in the biotechnology industry for the production of recombinant proteins (*see 18.5.5*), and they are also important clinically as pathogens. An example of this is *Candida albicans*, the cause of oral and vaginal thrush.

The entire genome of *Saccharomyces cerevisiae* (brewer's yeast) was recently sequenced and this enabled the identification of glycosyltransferases involved in the synthesis of both N- and O-linked oligosaccharides. A family of yeast strains from *S. cerevisiae* that exhibited defects in glycosylation has been used extensively to elucidate some of the biosynthetic steps in the glycosylation of proteins. As a consequence, the glycosylation pathways in the *S. cerevisiae* species of yeast are the best understood, although other species have also been studied, for example, *Schizosaccharomyces pombe, Pichia pastoris* and *Candida* sp.

The glycosylation of fungal glycoproteins is less well studied than that of yeasts, but glycosylation of *Aspergillus niger* has been examined in some detail.

7.3.1 N-Linked oligosaccharides

Yeasts produce N-linked oligosaccharides that are best described as hypermannosylated; these are large Man-containing oligosaccharides attached to proteins, themselves called mannoproteins or mannans. No complex or hybrid type N-linked oligosaccharides (*see 4.2*) are found in yeasts. ➲

➲ Yeast N-linked oligosaccharides are large extended Man structures.

The N-linked oligosaccharides that yeasts produce all contain a core structure of Man(β1→4)GlcNAc(β1→4)GlcNAc-Asn, as found in higher eukaryotes (*see 4.2*). The other Man residues in yeasts are attached in an α-configuration. Examples of the types of N-linked oligosaccharides found in yeasts are shown in *Figures 7.5* and *7.6*.

In addition to large mannosyl oligosaccharides, galactomannan oligosaccharides have been reported (for example, in *S. pombe*) and novel (β1→2)Man repeating units described (e.g. in *Candida* sp.). Another distinct feature of the oligosaccharides of yeasts is the occurrence of Man-6-phosphate in the inner part of the molecule. This was thought to be a recognition site to allow the trafficking of the glycoproteins into the lysosomal pathway as in the biosynthesis of N-linked oligosaccharides of mammals (*see 4.2.7*). This has not been found to

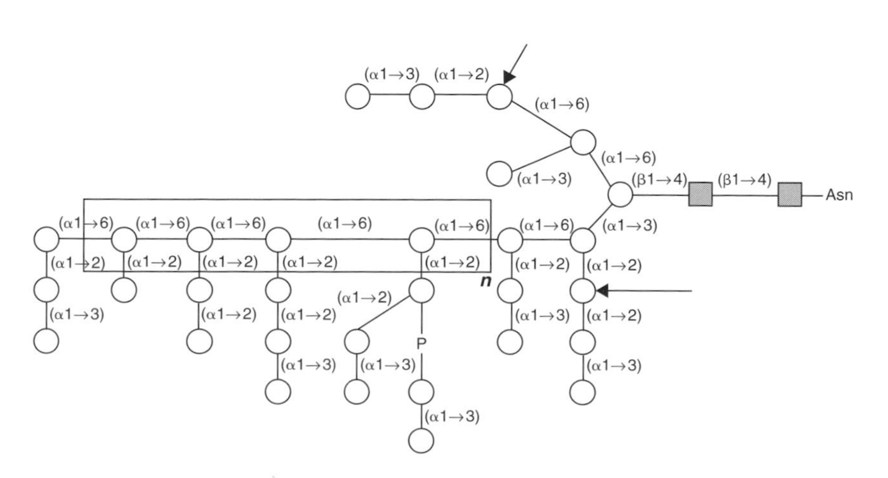

Figures 7.5

N-Linked oligomannose oligosaccharide from *Saccharomyces cerevisiae*, strain X-2180. Arrows show sites for possible attachment of Man-1-6-P.

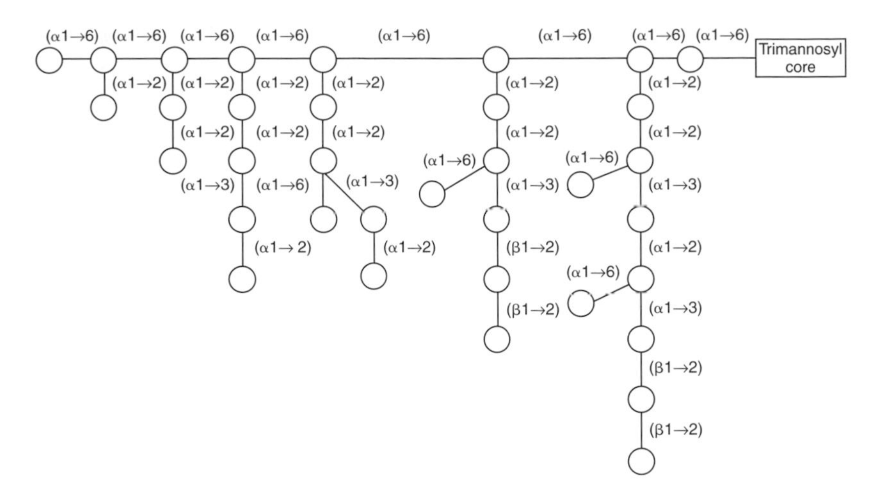

Figure 7.6

The outer chain N-linked oligomannose oligosaccharide from *Candida guillermondii.*

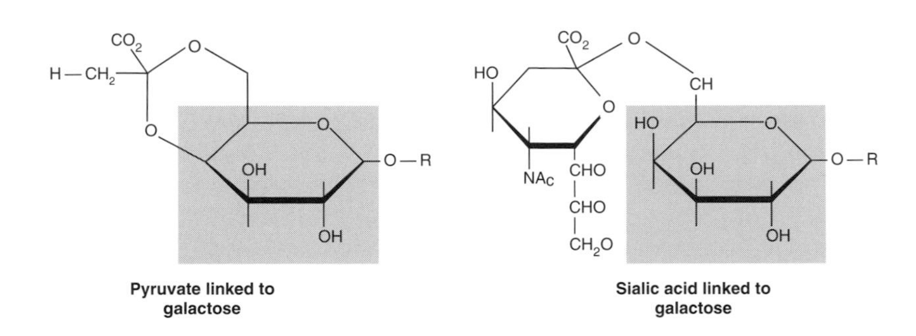

Figure 7.7

The structure of pyruvate galactose of *Saccharomyces pombe* compared with sialic acid linked (α2→6) to galactose. Galactose is highlighted in both cases.

be the case, and Man-6-phosphate-containing oligosaccharides do not appear, at least in *S. cerevisiae,* to be targeted to the yeast equivalent of the lysosome, the vacuole.

The N-linked oligosaccharides of yeasts do not contain sialic acids or sulfate groups, although a terminal pyruvate has been reported in *S. pombe* and this is shown in *Figure 7.7*. The structure of pyruvate is similar to sialic acid (*see 11.2*) and on discovery of this substitution, it was hypothesized that pyruvate may act as a recognition domain, as is often the case with sialic acid, in the N-linked oligosaccharides of higher eukaryotes.

In fungi, the oligomannose N-linked structures found in higher eukaryotes prevail. ➲

➲ Fungi produce oligomannose N-linked oligosaccharides similar to those of higher eukaryotes.

The hypermannosylated structures that are a feature of yeasts are not found in fungi. Another difference in the N-linked oligosaccharides of fungi compared with yeasts is the presence of monosaccharides such as Gal, Glc, galactofuranose and phosphate and sulfate groups. Fungi are similar to yeasts as they neither contain complex or hybrid N-linked glycans; the enzymes that produce these structures are absent. It has been postulated that hybrid and complex oligosaccharides may have evolved in conjunction with animals that possess a circulatory system.

7.3.2 Biosynthesis of N-linked oligosaccharides

The biosynthetic steps involved in the initial stages of production of N-linked oligosaccharides are evolutionarily conserved (*see 4.2*). In yeasts and fungi the steps are essentially as follows:

- an oligosaccharyltransferase transfers the oligosaccharide intermediate *en bloc* to the protein cotranslationally in the lumen of the RER
- glucosidases I and II trim the intermediate
- a Man residue is trimmed to produce $Man_8GlcNAc_2$
- posttranslational modification with Man residues occurs in the Golgi apparatus.

The galactosyltransferase involved in the glycosylation of Man in *S. pombe* has been purified and found to locate to the Golgi apparatus.

7.3.3 O-Linked glycosylation

In common with the higher eukaryotes, yeasts produce O-linked oligosaccharides. The structures are, however, quite different to those found in the higher classes of eukaryotes (*see Chapter 5*) and resemble more closely the N-linked oligosaccharides of yeasts and fungi (*see 7.3.2*). The main feature of O-glycosylation is the formation of linear chains of Man linked to the hydroxyl group of serine or threonine. Typically, up to five Man monosaccharides are linked together. An exception to this glycosylation pathway is found in *A. niger;* this fungus has been reported to produce branched mannosyl O-linked oligosaccharides as well as monomeric and dimeric Man linked to serine/threonine. If yeasts are used to produce recombinant proteins, they are generally found to be O-mannosylated and this O-mannosylation may need to be prevented if yeasts are to be used as expression systems in the biotechnology industry for synthesis of glycoproteins for human use (*see 18.5.5*).

7.4 Glycosylation in plants

Plants synthesize both N-linked (*Chapter 4*) and O-linked (*Chapter 5*) glycans, but it is unusual for both to be produced by the same species; most plant species synthesize either N- or O-linked glycans, but not both.➲ One major difference between animal and plant glycans is that sialic acid (*Chapter 11*), a common terminal sugar of complex mammalian glycans, has not been described on plant glycans. As in other types of organisms, the functional roles of glycosylation in plants are multiple, complex and not well understood. This section outlines some of the ways in which plant glycans differ from the analogous animal glycans, and describes some of the characteristic structures and functions of plant glycans.

➲ Plants synthesize N- or O-linked oligosaccharides, but not both. Plant oligosaccharides do not contain sialic acid.

7.4.1 The plant cell wall

The plant cell wall is made up of a complex three-dimensional meshwork of microfibrils of the structural polysaccharide cellulose and cross-linking hemicelluloses embedded in a matrix of gel-like acidic polysaccharides, called pectic substances (*see 3.2*).➲ These are rich in galacturonic acid, and these molecules are all held together through calcium bridges and cross-linked with aromatic monolignols. Structural proteins and glycoproteins are also involved.

➲ Plant cell wall components include the structural polysaccharides cellulose and hemicellulose embedded in a pectin gel, plus enzymes required for cell wall synthesis and metabolism.

➲ In plants, polysaccharides are important structural components of the cell wall and their composition and structure differ considerably between species.

Also associated with this complex meshwork, by covalent or ionic bonds to hemicellulose or pectin, are a variety of enzymes required for its biosynthesis and modification (such as xyloglucan endotransglycosylase, glucanase and α-fucosidase) and for modification of its metabolites (such as phosphatase, invertase and ascorbic acid oxidase). The two major groupings of flowering plants, the monocots and the dicots, differ significantly in the composition of the hemicelluloses involved in their cell walls. Monocot hemicelluloses are rich in arabinoxylans and dicot hemicelluloses are rich in xyloglucans.

Glycine-, proline- and hydroxyproline-rich structural proteins, similar to collagens in animals (*see 8.2.1*), are also a feature of the plant cell wall.➲ Hydroxyproline-rich proteins are most usually glycosylated with short O-linked arabinofuranosides, made up of between one and four arabinosyl residues. This serves to stabilize the molecule as a rigid, rod-like structure.

The elongation of plant cells during plant cell growth requires considerable remodelling of the structural plant cell wall. The enzymatic breakdown of cell wall components, principally polysaccharides, is achieved by expansins, endo-β-glucanases and xyloglucan-endotransglycosidases.

7.4.2 Defence against pathogens

➲ Glycan signals induce various plant defence strategies against pathogens including:
- strengthening the plant cell wall
- production of enzymes to degrade the pathogen's cell wall
- production of phytoalexins to kill the pathogen.

The plant defence response to attack by pathogens may be triggered by oligosaccharides of the plant or of the pathogen. Defence involves the activation of a number of genes that lead to the production of additional structural cell wall material and cross-linking of cell wall polymers, and also the production of enzymes such as β-glucanases and chitinases that attack and degrade the cell wall of the pathogen, and phytoalexins that kill the pathogen. Degradation of fungal cell walls result in the release of β-glucans, chitin (*see 3.2.3*) and chitosan, degradation of plant cell wall releases oligogalacturonides and fragments of pectic substances (*see 3.2.2*), all of which elicit a defence response.➲

7.4.3 The symbiotic relationship between nitrogen-fixing bacteria and plants

One of the best understood roles of saccharide–lectin interactions is in the establishment of a symbiotic relationship between leguminous plants and the soil-dwelling, nitrogen-fixing bacteria, *Rhizobia*.➲ *Rhizobia* must first invade the roots of the plant and induce the formation of nodules. In these specialized nodules, the *Rhizobia* multiply and fix nitrogen, which is of benefit to the plant, while deriving energy and carbon from their host.

This complex process involves the switching on of numerous genes in both host plant and bacterium. Initially, the plant roots secrete flavonoids, which induce the activation of well-studied bacterial *nod* genes that code for lipochito-oligosaccharides 'Nod factors'. These consist of a pentameric or tetrameric chitooligosaccharide backbone to which a long, unsaturated fatty acid chain is attached at the non-reducing end. The symbiotic relationship between bacterium and plant is species specific and relies on subtle differences in the structure of the Nod factor lipochito-oligosaccharides. In the appropriate species these molecules are potent inducers of root nodule formation and rapidly lead to physiological effects including induction of calcium signals, induction of cell cycle, root hair curling and activation of many plant 'early nodulin' or *ENOD* genes.

➲ Both species-specific lipochito-oligosaccharide 'Nod factors' and lectins are functional in recruiting symbiotic nitrogen-fixing bacteria in leguminous plants.

Lectin–saccharide interactions are also implicated in root nodulation (*see 13.8.6*). Although lectins may not be responsible for the highly species-specific recognition of bacteria and plant, they may play a role in agglutinating bacteria during the nodulation process and experimental evidence strongly supports the idea that lectins are necessary for successful nodulation. The introduction of pea lectin gene into clover roots, for example, results in the host plant recognizing the pea-specific nodulation factor.

7.4.4 The oligosaccharides of honey

In addition to fructose, glucose and maltose, honey contains traces of a number of unusual saccharides, for example, turanose, trehalose, maltotriose, melezitose and erlose, that are not found in other sweet-tasting substances of plant origin, such as cane or beet sugar. Using modern techniques, it is possible to characterize honeys derived from different plant species according to their saccharide composition, and to detect adulteration of honey with sucrose or fructose syrups, but this is a technically difficult specialist field. It may be accomplished using techniques such as high-performance liquid anion-exchange chromatography with pulsed amperometric detection, nuclear magnetic resonance, capillary gas chromatography, mass spectrometry (*see Chapter 14*). Such analysis provides a means of authenticating the source and purity of honey preparations.➲ Ersatz honey is made by inverting glucose.

> ➲ Honey contains many unusual saccharides, analysis of which, although complex, may be useful in authenticating honey and confirming the plant species from which it was derived.

7.4.5 Plant O-glycans

Plant O-glycans are usually linked to serine, threonine or hydroxyproline residues of the polypeptide backbone, and are commonly designated 'hydroxyproline-rich glycoproteins' or HRGPs. They are highly glycosylated, so usually categorized as proteoglycans, and belong to three main categories:

> ➲ Plant O-glycans can be classified as (1) extensins, (2) Solanaceae lectins and (3) soluble arabinogalactans.

1. Extensins, which are insoluble cell wall glycoproteins involved in plant cell wall extension.
2. Lectins (*see Chapter 13*) of the family Solanaceae, for example, potato lectins, which are not only highly glycosylated themselves, but also recognize chito-oligosaccharides.
3. Soluble arabinogalactans (*see 3.2.2.4*), which are present in the cell wall, intercellular spaces, plasma membrane and plant exudates.➲

7.4.5.1 Extensins

Extensins of the plant cell wall are based on a polypeptide backbone rich in hydroxyproline and serine, as much as 60% of which may be the repeat motif SO_4-Ser-Hyp-Hyp-Hyp-Hyp. The polypeptide chain is very heavily glycosylated by short O-linked glycans, consisting of one to four arabinofuranosyl residues β-linked to hydroxyproline, and one or two Gal residues α-linked to serine.➲ The molecule is an extended left-hand helix with three residues per turn, giving an overall rigid, rod-like shape, which is maintained by the glycan side chains and is important structurally. Extensins are secreted as a soluble precursor, then cross-link with other cell wall components to form an insoluble structural matrix. Expression is cell-type specific and highly regulated, being upregulated, for example, in response to wounding and infection.

> ➲ The structurally important extensins are heavily glycosylated by short O-linked oligosaccharides consisting of arabinofuranosyl residues β-linked to hydroxyproline and Gal α-linked to serine.

7.4.5.2 Solanaceae lectins

The best characterized lectin of the Solanaceae family (which also contains the tomato *Lycopersicon esculentum* and the thorn apple *Datura stramonium*), is potato (*Solanum tubersosum*) lectin, which is associated with membranous structures of the plant cell vacuole and cytoplasm. It consists of a short stem-like hydroxyproline and serine-rich domain which is glycosylated with multiple β-linked arabinofuranosyl chains attached to hydroxyproline, and Gal α-linked to serine very similar in structure to that of the extensins, and a chitin-binding domain rich in glycine and disulfide bridges. A similar extensin-like domain has been identified in thorn apple and tomato lectins, and, as in extensins, it is required for structural integrity. The chitin-binding activity of the saccharide-recognition domain in lectins of the Solanaceae has led to speculation that they may have a function in plant defence mechanisms (*see 13.4.1*).➲

➲ Solanaceae lectins share many characteristics with extensins, and may also play structural and defence roles.

7.4.5.3 Soluble arabinogalactans

Arabinogalactans (AGPs; *see 3.2.2.4*) are highly glycosylated, hydroxyproline-rich extracellular proteoglycans with long sugar chains O-linked through Gal to either hydroxyproline or serine residues of the polypeptide backbone. The polypeptide chain is also rich in (unglycosylated) alanine residues. A common amino acid sequence motif is Ala-Hyp-Ala. Common carbohydrate components include the monosaccharides D-galactopyranose, L-arabinofuranose, the uronic acids D-glucuronic acid and D-galacturonic acid and their 4-O-methyl derivatives, L-rhamnopyranose, D-mannopyranose and D-glucopyranose. The most prominent form of glycosylation is the attachment of arabinogalactan chains O-linked through Gal or Glc to hydroxyproline residues. The arabinogalactan chain consists of (1→3) linkages and some (1→6) linkages substituted with terminal arabinofuranose and uronic acids.

AGPs occur as soluble molecules in the cell walls, intercellular spaces, in association with the plasma membrane and in exudates. These proteoglycans are cross-linked by the enzyme extensin peroxidase through the tyrosine residues in Tyr-Lys-Tyr repeat tripeptides to form inter- and intramolecular cross-linking and stabilizing bonds, once again of structural importance to the plant. They may also function in differentiation and embryogenesis as secreted AGPs are capable of stimulating plant cell embryogenesis in culture.

Extracellular and membrane-bound forms of arabinogalactan proteins are very heavily glycosylated with, typically, > 90% of the mass of the molecule being carbohydrate and only as little as 2–10% protein. They are heterogeneous in structure and glycosylation. 'Classical' AGPs have a polypeptide backbone of 130–150 amino acids, consisting of a signal peptide, then a hydroxyproline-, alanine-, serine- and threonine-rich domain, then a membrane-anchoring helical domain. They carry long arabinogalactan chains of 30–50 monosaccharides, mostly consisting of a (β1→3)-linked D-Gal backbone with terminal L-arabinofuranosyl residues on short side chains. The soluble, secreted forms tend to carry longer oligosaccharides than the membrane-bound forms. Some 'non-classical' AGPs are larger molecules with a polypeptide backbone of 170–450 amino acids, and a large, variable C-terminal domain. AGPs may be considered analogous in terms of variability in size and degree of glycosylation to the proteoglycans of animal cells.➲

➲ AGPs may be either extracellular or membrane bound, and are highly glycosylated. They have functions in structure, differentiation and embryogenesis, and are analogous to animal cell proteoglycans.

7.4.6 O-Glycans of green algae

O-glycoproteins have also been characterized in green algae, and resemble those of higher plants. They are rich in glycosylated hydroxyproline and serine residues bearing glycan chains of Ara, Gal and Glc moieties. The glycans are often also sulfated. These molecules function both structurally and also play a role in sexual adhesion.➲

➲ Green algae synthesize oligosaccharides structurally similar to those of higher plants. They are important in maintaining structural integrity and also in adhesion processes involved in fertilization.

One of the better characterized structural O-linked proteoglycans of green algae is that of *Volvox* which has a total molecular mass of 185 kDa, ≈70% of which is carbohydrate. It has a 50 kDa protein core containing an 80 amino acid domain in which almost all amino acids are arabinosylated hydroxyprolines interspersed with occasional threonines, which probably maintain the rod-like structure seen in similar domains in the proteoglycans of higher plants (*see 7.4*). Near the middle of the central rod-like domain, a large 28 kDa sulfated polysaccharide chain consisting of a (1→3)-linked Man backbone with di-arabinosyl side chains is attached. The entire molecule cross-links with others to form a structural honeycomb-like structure.

Under appropriate conditions, the normally unicellular green algae *Chlamydomonas* produces sexual adhesion proteins on the surface of the flagella it normally uses for propulsion. This allows adhesion with another algal cell, and subsequent fusion. The adhesion proteins are high molecular mass (as high as 1300 kDa), long and fibrous. The polypeptide chain is rich in hydroxyproline, serine and glycine and is glycosylated by O-linked chains of Ara, Gal and Glc, which may be sulfated.

7.4.7 Plant N-glycans

7.4.7.1 Structure and biosynthesis

The complex N-glycans of plants differ markedly in structure from those of animals (*see 4.2*), and more closely resemble those of insects and molluscs (*see 7.5*). In addition to the absence of sialylation, plant N-glycans differ from animal N-glycans in that Fuc(α1→6)-linked to the proximal GlcNAc residue and Fuc(α1→3)-linked to the branching GlcNAc residue of N-glycans, both common modifications of animal N-linked oligosaccharide chains, are not usually found in plant glycans.

Plant N-linked glycans are of two main types: high mannose type and complex type. Examples of these are illustrated in *Figure 7.8*.

The high mannose type result from limited processing of the common $Man_9GlcNAc_2$ core by mannosidases. These oligosaccharides consist of a branching trimannosyl core (*Figure 7.8a*), analogous to that of animal N-glycans, and with the general formula $Man_{5-9}GlcNac_2$ (*Figure 7.8b*).

The complex type result from trimming of Man, with subsequent chain extension.➲ Therefore, they feature fewer core Man residues, but have peripheral chain structures involving other monosaccharides including Fuc, Xyl, GlcNAc and Gal. These complex type plant N-glycans may be subdivided further into:

- those having terminal GlcNAc residues or larger antennae (β1→2) linked to the (α1→6)Man or the (α1→3)Man of the common core, usually simply referred to as complex-type N-glycans (*Figure 7.8c*)
- N-glycans having only an (α1→3)-linked Fuc and/or a (β1→2)-linked Xyl linked to the core $Man_3GlcNac_2$ structure. These are called paucimannosidic type N-glycans (*Figure 7.8d*).

➲ Types of plant N-linked glycans:
a. High mannose type, with the general formula $Man_{5-9}GlcNac_2$.
b. Complex type glycans, resulting from trimming of high mannose types and then further chain extension:
(i) complex type
(ii) paucimannosidic type.

They are synthesized and modified in the plant ER and Golgi apparatus in broadly the same way as animal N-glycans (*see 4.2*). The oligosaccharide intermediate $Glc_3Man_9GlcNAc_2$ (*Figure 7.9a*) is transferred from a dolichol-lipid carrier to specific Asn residues, in the sequence Asn-Xaa-Ser/Thr (where Xaa is any amino acid other

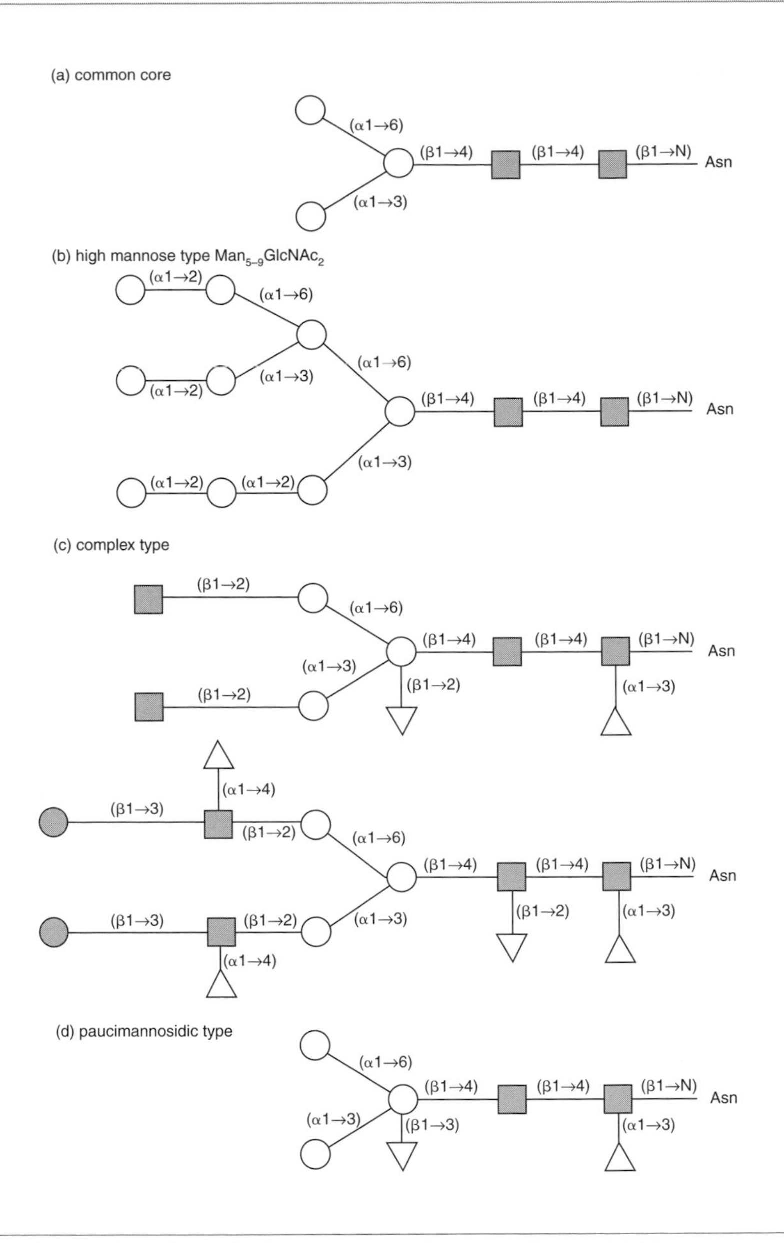

Figure 7.8

N-Linked plant oligosaccharides. (a) The common trimannosyl core, (b) high mannose type, general formula $Man_{5-9}GlcNAc_2$, (c) complex type, (d) paucimannosidic type.

> ➲ The initial steps in plant N-linked glycan synthesis are very similar to those of animal cells using a dolichol-lipid carrier which transfers an oligosaccharide intermediate onto the nascent polypeptide chain.

than proline or aspartic acid), on the nascent polypeptide chain in the ER (*Figure 7.9b*).➲ The three terminal Glc units of $Glc_3Man_9GlcNAc_2$ are first trimmed by glucosidases I and II in the ER (*Figure 7.9c*), and a transient reglucosylation, associated with quality control of unfolded proteins, by an ER UDP-glucose: glycoprotein glucosyltransferase may subsequently occur. In mammals, the next step is cleavage of a single Man residue by an ER mannosidase. An equivalent has not been described in plants, but it cannot necessarily be assumed that this step does not occur.

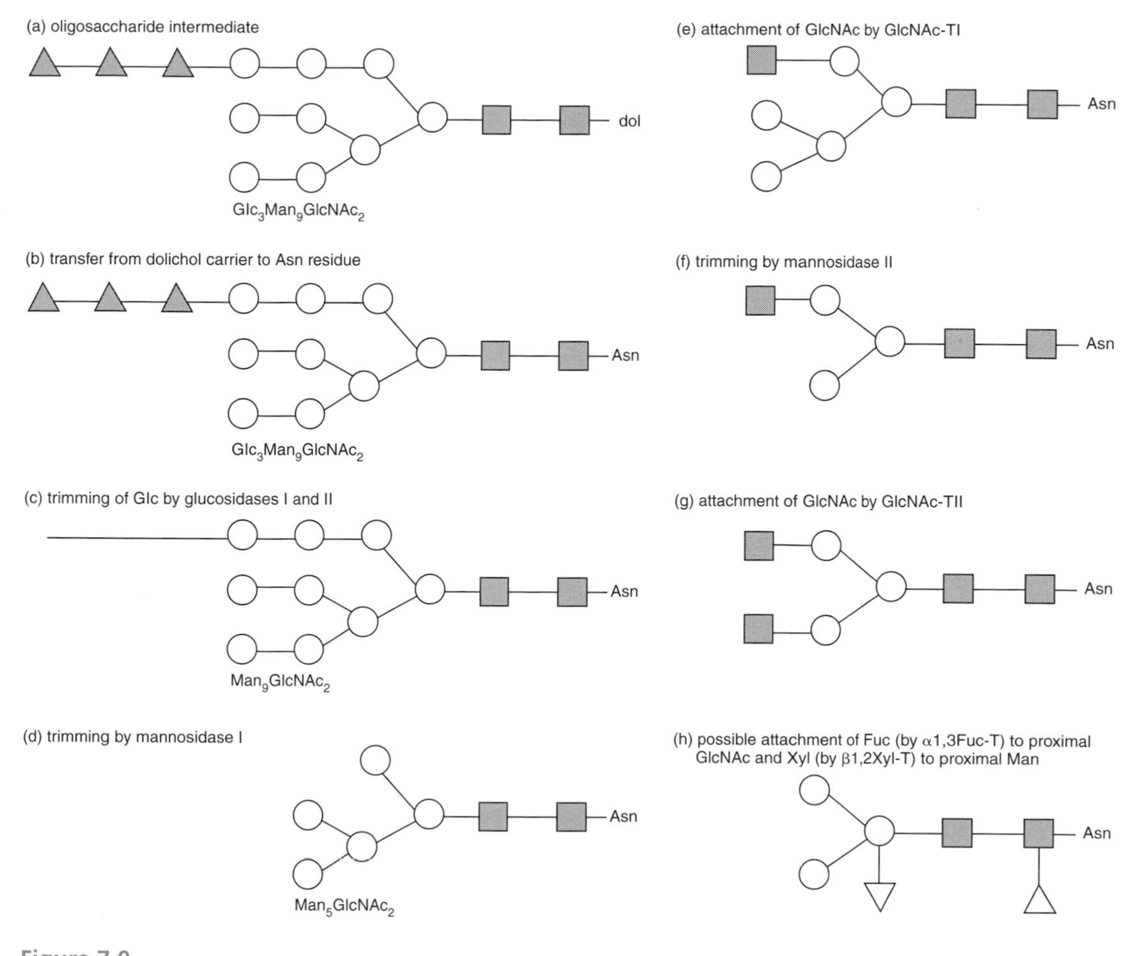

Figure 7.9

N-Glycosylation in plant cells begins with the transfer of a dolichol-linked oligosaccharide intermediate (a) to an Asn residue of the nascent polypeptide chain (b). The three Glc residues are trimmed by glucosidases I and II (c), Man residues are trimmed by mannosidase I (d), GlcNAc is attached to one branch by GlcNAc-TI (e), further trimming of terminal Man residues is accomplished by mannosidase II (f), a further GlcNAc is attached by GlcNAc-TII (g), and finally, a Fuc may be attached to the proximal GlcNAc and a Xyl to the proximal Man, by (α1→3) FucT and (β1→2) XylT respectively (h).

Initial modification in the Golgi apparatus, as in animal N-linked glycans, involves trimming of between one and four (α1→2)-linked Man residues by mannosidase I, and conversion from $Man_9GlcNAc_2$ to $Man_5GlcNAc_2$ (*Figure 7.9d*). Some plant glycoproteins show microheterogeneity in glycosylation, and, for example, a glycoprotein may occur predominantly in the form $Man_7GlcNAc_2$, but variants with six and eight Man residues attached may also be detectable.

The next step is attachment of a terminal GlcNAc to the (α1→3) Man branch of the $Man_5GlcNAc_2$ high mannose type glycan by GlcNAc-TI, to yield $GlcNAcMan_5GlcNAc_2$ (*Figure 7.9e*).

Trimming of two more Man residues by α mannosidase II (α-Man-II) then takes place (*Figure 7.9f*). This is followed by attachment of a second GlcNAc by GlcNAc-TII to the (α1→6) Man branch (*Figure 7.9g*).

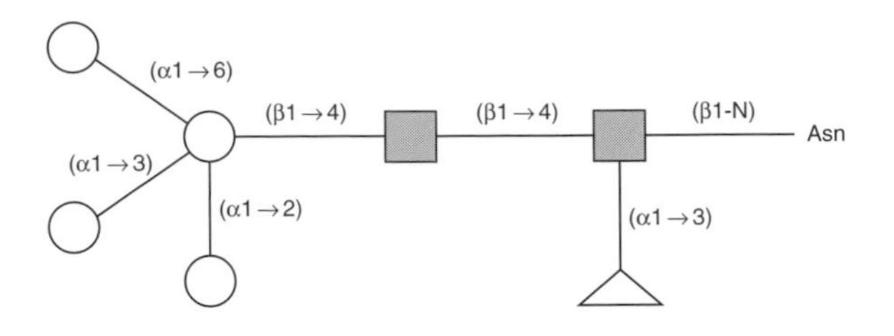

Figure 7.10

A small 'paucimannosidic' N-linked glycan, characteristic of plant vacuolar glycoproteins. The plant vacuole is rich in N-acetylglucosaminidases and mannosidase I, so high Man and terminal GlcNAc residues are trimmed.

At this stage, in plants, an (α1→3) Fuc may be attached to the proximal core GlcNAc, and a (β1→2)-linked Xyl may be attached to the proximal core Man by (α1→3) Fuc-T and (β1→2)Xyl-T respectively (*Figure 7.9h*). This requires the presence of at least one terminal GlcNAc residue, and does not always occur. Xylosylation and fucosylation at this stage are independent events, happening sequentially, and occur in the *medial*- and *trans*-Golgi cisternae, respectively. Plant N-glycans bearing only one (β1→2)-linked Xyl or only one (α1→3)-linked Fuc have been identified.

Further conversion of high mannose type to complex type glycans may take place as proteins move from RER via the Golgi apparatus to their ultimate destination, typically internal or external membranes, the vacuole or the plant cell wall. Modifications are catalysed by glycosidases and glycosyltransferases. These type of structures are illustrated in *Figure 7.8*. Addition of terminal Fuc by (α1→4) fucosyltransferase and Gal by (β1→3) galactosyltransferase to terminal GlcNAc residues yields mono- and bi-antennary plant complex N-glycans, and is a typical modification that occurs late in processing in the *trans*-Golgi. The Gal(β1→3)[Fuc(α1→4)]GlcNAc sequence is found on antennae of these structures and corresponds to the Le^a antigen of mammalian tissues (*see 10.6*), where it is implicated in cell-adhesion and cell-recognition processes.

Many extracellular plant glycoproteins have complex type N-linked glycans with terminal Fuc and Gal residues, such as Le^a (*see 10.6*). Examples of plant glycoproteins glycosylated in this way include lactase, miraculin and the pollen allergen *Cry j* I (from *Cryptomeria japonica*, the Japanese cedar, pollen). In contrast, the plant vacuole is rich in N-acetylglucosaminidases and mannosidase-I and thus, when N-linked glycans are transported to the vacuole, any remaining high mannose residues and terminal GlcNAc residues are trimmed. As a result of this, most plant vacuolar N-glycans are small structures of the paucimannosidic and high mannose core type, as illustrated in *Figure 7.8d* and *Figure 7.10*.

The extent to which the original high mannose structures are modified on different glycoproteins is largely a question of accessibility to the requisite glycan-processing enzymes. Glycans at the surface of folded polypeptide chains are accessible to processing enzymes and thus become extensively modified, whereas those buried in the folds of the polypeptide chain are inaccessible and retain the high mannose form. In addition to this, spatial and temporal differences in expression of modifying enzymes, and other signals directing the formation of different glycan structures may also play a part.

➲ Plant N-linked glycans commonly carry (α1→3) Fuc and (β1→2) Xyl, which are not found on animal glycoproteins, and are thus immunogenic to humans.

(α1→3)-Linked Fuc and (β1→2)-linked Xyl have never been reported in mammalian N-glycans, but are present in the glycoproteins of plants, molluscs, insects and spiders. They are immunogenic in mammals and may act as potent human allergens.➲ This is a stumbling block in the use of transgenic plants to synthesize pharmaceutical glycoprotein compounds for human use (*see*

18.5.8), as they will potentially be glycosylated with these immunogenic (α1→3) Fuc and (β1→2) Xyl residues.

7.4.7.2 Functions of plant N-glycans

N-linked glycans in animal cells have multiple, complex roles in processes such as prevention of proteolytic degradation, induction of correct protein folding or conformation, carrying targeting information, cell–cell recognition and adhesion (*see 4.4*). The conservation of the N-glycosylation pathway in plants, and its complexity, suggests an important function for N-glycosylation of plant proteins as well. However, the function of complex glycans in plants is not well understood. Tellingly, mutant *Arabidopsis* plants that lack the GlcNAc-TI enzyme and are thus unable to convert high mannose glycans to complex type glycans function perfectly normally, at least under laboratory conditions, in contrast to mice lacking the same enzyme in which the mutation is lethal at the embryonic stage.

A number of specific roles for N-glycans in plants have been investigated. One hypothesis is that terminal GlcNAc may have a targeting role in directing glycoproteins to the plant cell vacuole. In many ways the plant cell vacuole is analogous to the animal cell lysosome as it contains a variety of hydrolytic enzymes; it can act as a storage organelle for nutrients and waste products, as a degradative compartment, and it is an important structure for maintaining plant cell turgor. Tunicamycin treatment (which abrogates N-glycan synthesis; *see 15.6.1*) and mutagenesis experiments largely dispute the hypothesis that GlcNAc directs glycoproteins to this organelle. The inactive precursor of the lectin (*see Chapter 13*) Concanavalin A (from *Canavalia ensiformis*, the jack bean) is converted to an active lectin by deglycosylation, suggesting a potential role of N-glycosylation in protein activity. Similarly, N-glycans are implicated in the correct folding of the soyabean (*Glycine max*) lectin molecule, which is essential to its activity. Altering the glycosylation of the lectin ECorL (from *Erythrina corallodendron*) has also been shown to alter its structural conformation, and consequently its saccharide-binding preference (from Man and Glc to Gal).

Tunicamycin treatment and mutation experiments in cultured plant cells have shown that N-glycosylation may be necessary for glycoprotein stability and successful secretion, at least in some cases. Numerous studies have demonstrated that N-glycans protect the protein from proteolytic degradation, and they are responsible for thermal stability, solubility and biological activity.➲ N-Linked glycans may also be implicated in cell-cycle regulation and regulation of plant embryogenesis and development, and as signalling molecules.

➲ Studies with inhibitors of N-linked glycosylation suggest that N-linked glycans may be necessary for protein stability and secretion in plants.

7.5 Protein glycosylation in insects

Insect glycosylation mechanisms are poorly understood but appear to be broadly similar to those of plants and animals, but more limited in scope. This is an area in which our knowledge is likely to increase over the next few years as the increasing applicability of the baculovirus insect expression system and insect cell culture (*see 18.5.7*) to biotechnology stimulates a requirement for greater understanding of the glycosylation pathways of insects.

7.5.1 Insect N-linked glycosylation

In common with plant (*see 7.4*) and animal cells (*see 4.2*), N-linked glycan synthesis proceeds through an intermediate lipid-linked oligosaccharide precursor $Glc_3Man_9GlcNAc_2$, but much evidence supports the idea that insects share only the formative part of the N-glycosylation pathways seen in plants and animals. They are able to synthesize high mannose-type core structures and trim them in a manner analogous to that seen in animal cells, oligosaccharides

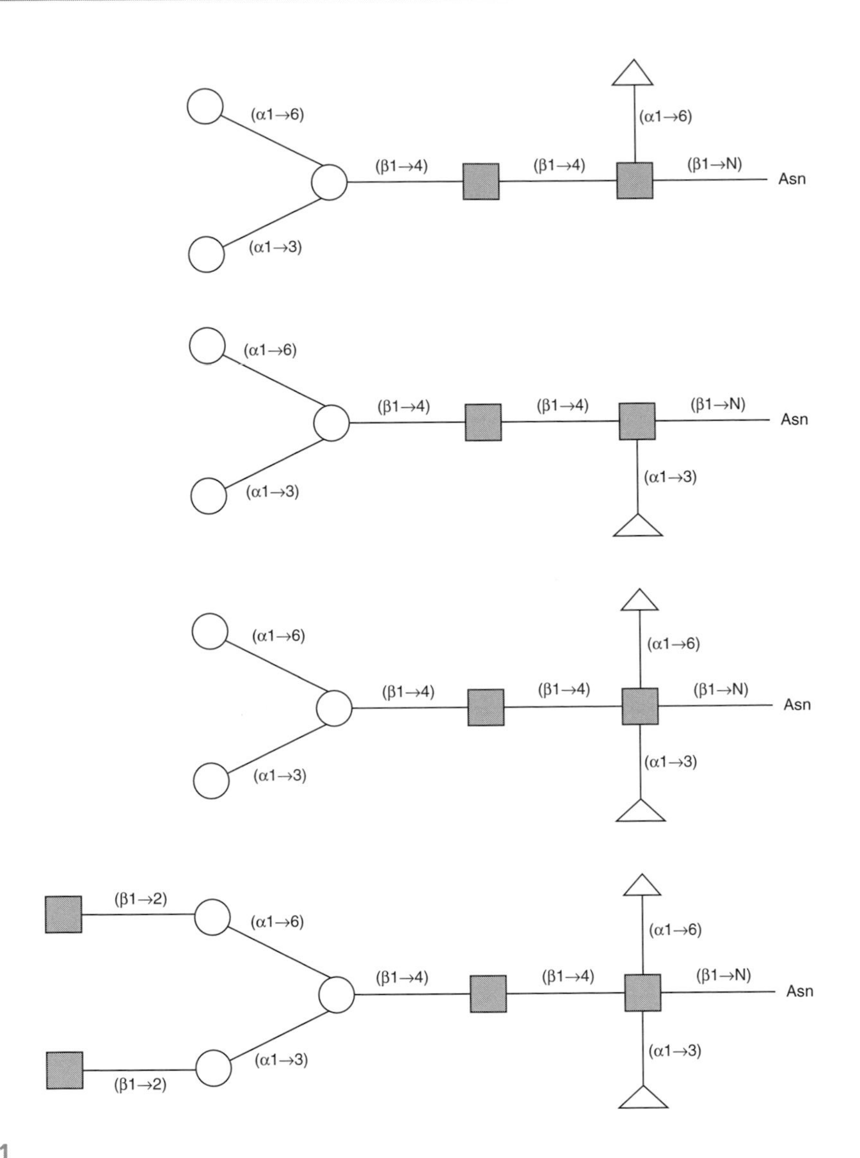

Figure 7.11

N-Glycans characteristic of membrane glycoproteins of cultured insect cells.

based on Man_5, Man_6, Man_7, Man_8 and Man_9 have been reported, but not to elaborate them into complex or hybrid type N-linked oligosaccharides characteristic of animal N-linked glycoproteins. The terms paucimannosidic, truncated or modified are sometimes used to describe these oligosaccharides.

Some N-glycans found on membrane glycoproteins of cultured insect cells are illustrated in *Figure 7.11* as examples, and share a trimannosyl core in which the proximal GlcNAc residue may or may not be fucosylated by either an (α1→3)-linked Fuc, an (α1→6)-linked Fuc or both. The (α1→6) and (α1→3) branches may be terminated by (β1→2)-linked GlcNAc, and variants have been described in which only one of the branches is terminated by GlcNAc in this way. Analogous, more elongated, oligosaccharide chains with the basic formula $Man_{2-9}GlcNAc_2$, either unfucosylated or fucosylated by one or two (α1→6) Fuc and/or (α1→3) Fuc attached to the core GlcNAc have also been described.

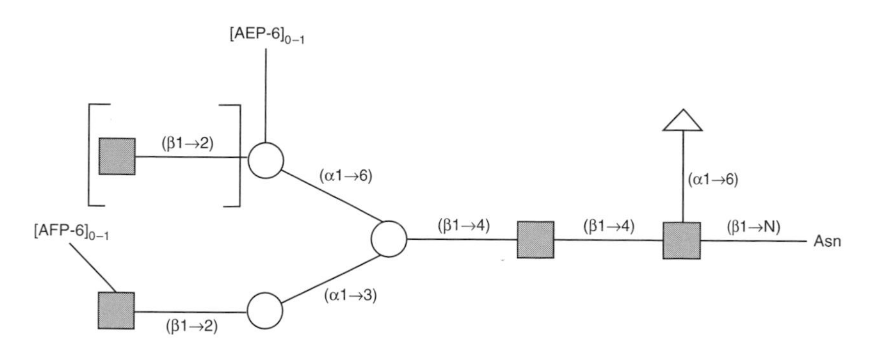

Figure 7.12

The structure of the N-linked glycans of apolipophorin III, from *Locusta migratoria*.

Unusual glycan structures have also been described in insects. Apolipophorin III, a glycoprotein from the haemolymph of the locust, for example, contains a very complex, branched N-glycan with a number of unusual features (*Figure 7.12*). It carries 2-aminoethylphosphonate (AEP) at the 6-position of the (α1→6)-linked branching Man and/or the (α1→3)-linked GlcNAc of the second branch and it has GlcNAc(β1→2) linked to the two branching core Man residues of the trimannosyl core, the product of GlcNAc-TI and -TII.

7.5.2 Insect O-linked glycosylation

Very little is known of insect O-linked glycan synthesis. They are able to synthesize O-linked glycans linked to threonine residues, analogous to O-linked glycan synthesis in animal cells (*see 5.6*), and as in animal cells these glycans may be sulfated.➲

> ➲ Insect cells synthesize N- and O-linked glycans similar to those of animal cells, but glycosylation pathways appear to be more limited; sialylation occurs uncommonly.

7.5.3 Sialylation and galactosylation of insect glycans

Sialylation of insect glycoproteins has not been described, except the unusual example of polysialic acid (*see 11.9*) during *Drosophila* larval development. Galactosylated oligosaccharides have been only seldom reported.

7.5.4 The potential for insect glycoprotein expression systems

In contrast to these findings, complex, sialylated N-glycans have been reported to be produced in transfected insect cells, and it is an intriguing idea that insect cells are able to perform complex glycosylation, but under natural conditions fail to do so. Insect cells may therefore contain the genes coding for transferases required for complex N-glycosylation typical of animal cells, but these are normally silenced. This may have implications for expression systems used to synthesize human glycoproteins (*see 18.5.7*). Very few insect enzymes of glycosylation have yet been sequenced and cloned. Some of the few that are include: α-mannosidase I and -II, and the UDP-Glc:glycoprotein glucosyltransferase responsible for reglucosylation of misfolded glycoproteins. These enzymes show high sequence homology with their mammalian counterparts.

Fucosylation of the proximal GlcNAc in insect glycoproteins acts as a potent allergen in humans; the Fuc(α1→3)GlcNAc moiety is, for example, an essential part of the allergic IgE-reactive determinant of the honeybee sting venom. It is a structure shared by many plant glycoconjugates (*see 7.4*), and may act as an allergen in food, plant and insect glycoconjugates. This may act as a barrier to the use of insect cell-derived recombinant glycoprotein products for human use. The development of insect cell lines that process the complex N-glycans typical of human cells, remains a challenge.

7.6 Unusual vertebrate glycosylation: the glycohormones

➲ Glycohormones contain N-acetyl-galactosaminylated and sulfated N-linked oligosaccharides.

The complexity of glycoforms on glycoproteins has long been thought of as being of biological importance but there are few examples in which oligosaccharides have been shown to play a crucial role. One particularly lucid example is the human pituitary gland hormones. The discovery that oligosaccharides on pituitary gland hormones are important in the function of this class of hormones led to them being termed 'glycoprotein hormones' or 'glycohormones'.

Luteinizing hormone (lupotrophin or LH), follicle-stimulating hormone (follitrophin or FSH), thyroid-stimulating hormone (thyrotrophin or TSH) and chorionic gonadotrophin (CG) are a family of heterodimeric glycoproteins that are important in the ovulatory cycle of vertebrates. Each of the hormones has an α- and β-subunit that contain at least one bi- or tri-antennary N-linked oligosaccharide.➲

7.6.1 Structure of oligosaccharides of LH, FSH, TSH and CG

The N-linked oligosaccharides of these glycohormones contain GalNAc and sulfate groups found only in the same configuration in few other glycoproteins of the submaxillary gland and kidney. *Figure 7.13* illustrates the types of terminal structures that are found on the

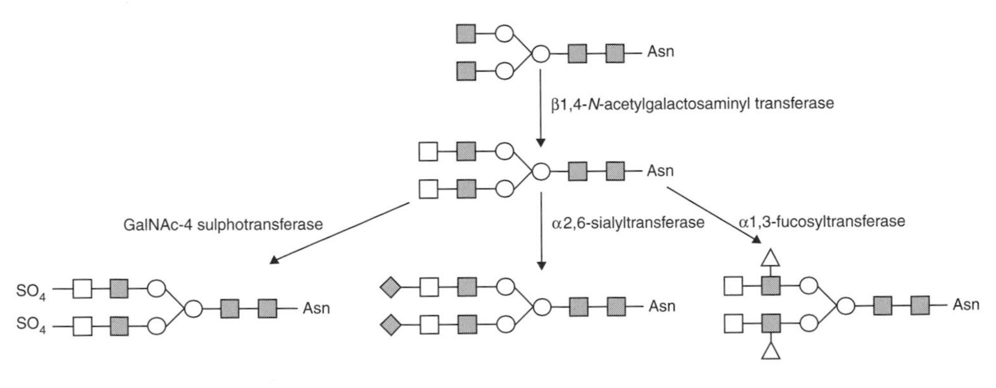

Figure 7.13

Biosynthesis of bi-antennary N-linked oligosaccharides of glycohormones.

Table 7.3 Overall glycosylation of N-linked oligosaccharides of LH, TSH, FSH and CG

Hormone	Pattern of glycosylation
LH	Sulfated and sialylated bi-antennary oligosaccharides
TSH	Sulfated and sialylated bi-antennary oligosaccharides
FSH	Predominantly sialylated bi- and tri-antennary oligosaccharides
CH	Predominantly sialylated bi- and tri-antennary oligosaccharides

oligosaccharides of LH, FSH, TSH and CG, and the glycosyltransferases and pathways involved in their synthesis.➲

➲ Addition of GalNAc to N-linked oligosaccharides by β1,4 GalNAc-T relies on the accessibility of a peptide-recognition domain; this is present on the α-subunit of LH and TSH.

The tissue of origin is important with respect to the glycosylation of the hormone, for example, CG, produced in the placenta, is not N-acetylgalactosaminylated but rather galactosylated. Control of this process is described in greater detail later (*see 7.6.2*). Briefly, the hormones contain either bi- or tri-antennary N-linked oligosaccharide substituted with sub-terminal Gal or GalNAc. The sub-terminal monosaccharide Gal is capped by sialic acid, and GalNAc with sulfate or sialic acid. Their typical patterns of glycosylation are summarized in *Table 7.3*.

7.6.2 Biosynthesis of oligosaccharides of LH, FSH, TSH and CG

The presence of GalNAc linked in a (β1→4) configuration to GlcNAc, the so-called 'LacdiNAc' motif, is unusual, whereas (β1→4)-linked Gal is very commonly found in N-linked oligosaccharides from a wide variety of animal tissues. Because of this, and in light of the fact that β1,4 GalNAc-T and β1,4 Gal-T both use the same oligosaccharides as acceptors, control of the synthesis of (β1→4) GalNAc has been studied in detail. A series of elegant experiments showed that β1,4 GalNAc-T recognizes a peptide sequence motif on the α-subunit of the hormones (this region of the α-subunit is very highly conserved throughout vertebrates). When α- and β-subunits combine, the peptide sequence motif may become inaccessible to β1,4 GalNAc-T. Such a situation arises in the case of FSH and here, the peptide recognition motif is inaccessible to the β1,4 GalNAc-T enzyme, but the oligosaccharides on FSH are available to β1,4 Gal-T, and therefore have Gal rather than GalNAc. In the case of CG, the peptide recognition motif is accessible to β1,4 GalNAc-T but the placenta (the organ in which this hormone is synthesized) does not express β1,4 GalNAc-T, and therefore galactosylated forms of the α-subunits are found *in vivo*.

7.6.3 Sialylation and sulfation of glycohormones

The anionic nature of oligosaccharides is usually due to the presence of sialic acids (*see Chapter 11*). The GalNAc-containing oligosaccharides of LH and TSH may be either sialylated or sulfated and, in the case of the former, sialylation occurs in an (α2→6) linkage to GalNAc, not (α2→3) (*Figure 7.13*). This illustrates the less stringent substrate requirements of the α2,6-sialyltransferase compared with the α2,3-sialyltransferase. In LH and TSH the anionic properties of the oligosaccharides may be due to the presence of sulfate, added to GalNAc by GalNAc-4-sulfotransferase. GalNAc-4-sulfotransferase requires the sequence GalNAc(β1→4)GlcNAc(β1→2)Man and is located in the *trans*-Golgi apparatus, acting in a similar manner to the sialyltransferases.

7.6.4 Structural and biological function of glycosylation of the glycoprotein hormones

The glycosylation of the glycohormones is important for the correct folding of the subunits as well as the appropriate assembly and secretion of these molecules.

Glycosylation relates to the function of the hormones. LH provides a good example for the multiple functional roles that glycosylation can have. The site of production of LH is the gonadotrophic pituitary gland cells. In these cells, the levels of expression of both GalNAc-T and GalNAc-4-sulfotransferase are influenced by the levels of LH, itself modulated in response to the amount of circulating oestrogen. When the level of oestrogen decreases, expression of GalNAc-T and GalNAc-4-sulfotransferase increases, as does the LH synthesized by the gonadotrophic pituitary gland cells. Because of the action of these transferases, the

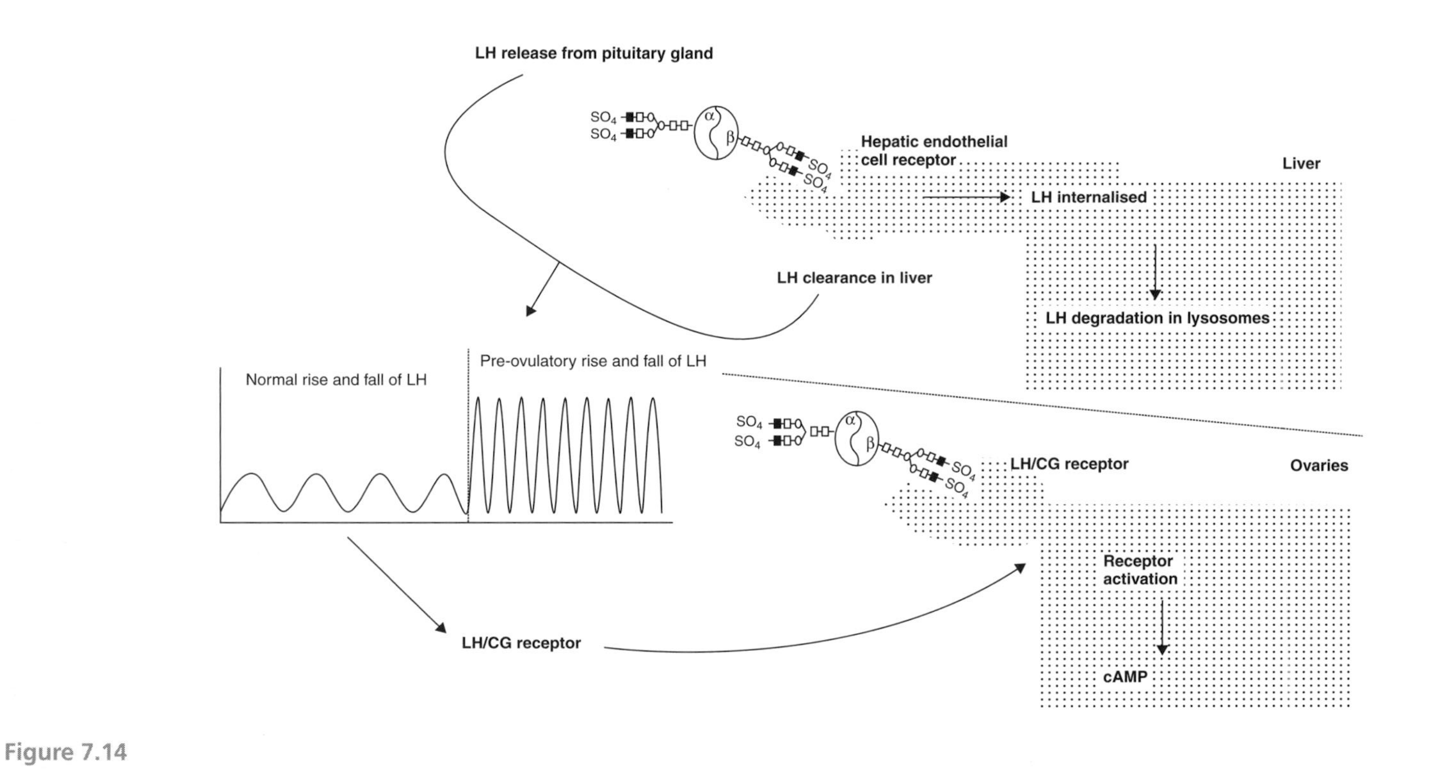

Figure 7.14

Schematic diagram showing the role of oligosaccharides in LH clearance and LH receptor function (for description, see text).

oligosaccharides attached to LH are added specifically when levels of oestrogen fall.➲ This sequence of events is illustrated schematically in *Figure 7.14*.

➲ Function of enzymes responsible for the addition of sulfate and GalNAc to N-linked oligosaccharides occurs in response to levels of circulating oestrogen.

After LH is released from the gonadotrophic cells of the pituitary gland, it has a short half-life. This is because the sulfated oligosaccharides on LH are recognized by a lectin present on hepatic Kupffer cells (*see 13.12.3.1*). In particular, the SO_4-4-GalNAc(β1→4) GlcNAc (β1→2)Man structure is recognized.➲ On LH binding to the GalNAc-4-SO_4 receptor, LH is rapidly internalized and transported to lysosomal compartments for degradation.

➲ Sulfated GalNAc oligosaccharides on LH are recognized by saccharide-binding proteins on Kupffer cells in the liver.

This seemingly contradictory process gives the impression that it should decrease the efficacy of LH to induce ovulation, but the situation is more complex than it appears. As LH is removed from the serum, more LH is produced by the gonadotrophic pituitary gland cells. The level of LH in the serum therefore increases because of the activity of the pituitary gland cells, but then decreases owing to clearance in the liver. The pulsatile nature of this process affects the LH/CG receptor in the ovary. The LH/CG receptor is internalized and inactivated on binding LH. During the process of receptor binding and internalization, cAMP is produced. The amplitude and frequency of LH pulses reach a particular level, as found in the pre-ovulatory phase of the menstrual cycle, the LH/CG receptor becomes maximally stimulated and functionally active.

Other hormones, for example FSH, are released into the serum in the same way as LH, but they do not have the short half-life of LH as they are sialylated rather than sulfated and so serum levels do not have the pulsatile nature of LH.

7.7 Glycosylation of nuclear and cytoplasmic glycoproteins and proteoglycans

The traditional view of glycosylation was that N- and O-linked glycoproteins were assembled in the lumen of the ER and/or Golgi apparatus, and then delivered to the cell surface by the secretory pathway (*see Chapters 3* and *4*). Proteins of the cytoplasm and nucleus were believed to be unglycosylated.➲ In the last few years, it has become apparent that glycoproteins and proteoglycans are also found in the cytoplasm and in the nucleus of the cell.

➲ Until recently it was believed that proteins of the cytoplasm and nucleus were always unglycosylated; this is not the case.

7.7.1 Nuclear high mobility group proteins

High mobility group proteins (HMG) are a large class of non-histone proteins found in the nucleus. They may be modified in a number of ways, including by methylation, acetylation, phosphorylation, ADP-ribosylation and glycosylation. They have been reported to bear N-linked glycans containing GlcNAc, Man, Glc, Gal and Fuc. They are not modified with O-GlcNAc (*see Chapter 6*). Glycosylation has functional significance, as enzymatic deglycosylation results in loss of their binding to nuclear matrix.

7.7.2 Nuclear and cytoplasmic proteoglycans and glycosaminoglycans

In addition to their role as components of extracellular matrix, proteoglycans and glycosaminoglycans, including chondroitin sulfates (*see 8.3.2*), hyaluronic acid (*see 8.3.1*), glucuronic acid and heparan sulfate (*see 8.3.3*), are also to be found in the nucleus and cytoplasm of the cell. It is believed that these molecules are synthesized in the ER/Golgi apparatus in the

➲ Nuclear and cytoplasmic proteoglycans and glycosaminoglycans may be involved in cell cycle regulation.

conventional manner, then specifically endocytosed via a non-lysosomal pathway.

The function(s) of these substances in the nucleus and cytoplasm are largely unknown, but they may be involved in regulation of the cell cycle.➲ Evidence for this comes from a number of sources. Proteoglycans are, for example, able to sequester, and thus inactivate, growth factors, which may have consequences for cell growth and division; furthermore, heparan sulfate and other proteoglycans are known to be involved in cell growth and the regulation of the cell cycle. Heparan sulfate inhibits cell growth by arresting the cell cycle at G1, just prior to the cell's entry into S-phase. Reduced cellular levels of glucose, serum or transforming growth factor β (TGFβ) all cause nuclear levels of heparan sulfate to fall and stimulate cell growth.

7.7.3 How are nuclear glycoproteins synthesized, and how do they enter the nucleus?

N- and O-linked glycosylation occurs in the ER and Golgi apparatus (*see Chapters 4* and *5*). It is still not understood how nuclear glycoproteins reach their destination from their site of synthesis.➲ Several proposals have been put forward:

➲ The mechanism by which glycosylated proteins reach the nucleus is not understood.

1. After processing in the Golgi apparatus, glycoproteins destined for the nucleus may be transported back to the ER, which is a direct extension of the nuclear membrane and is continuous with it, and from there to the nucleus.
2. Following processing, cytoplasmic vesicles may transport glycoproteins from the Golgi apparatus to the nucleus.
3. Glycoproteins may actually be glycosylated in the nucleus itself.

In the case of (1) and (2) the putative mechanism by which glycoproteins are delivered across the nuclear membrane remains obscure.

7.8 Unusual types of O-linked glycosylation

The most common type of O-linked glycosylation, often referred to as mucin-type glycosylation, is that in which the oligosaccharide chain begins with a GalNAc residue attached to the Ser/Thr of the polypeptide backbone (*see 5.6*). A second prevalent type of O-glycosylation is O-linked GlcNAc (*see Chapter 6*). Other, unusual types of O-glycosylation have been described, including O-linked α-Man, O-linked α-Fuc and O-linked β-Glc. They are described briefly in this section.

7.8.1 O-Linked α-mannose

O-linked α-Man was first described in yeasts in the 1950s and is a common glycan of the yeast cell wall. It is also present in the skin collagen of the clam worm. It was first described in mammals in chondroitin sulfate proteoglycan (*see 8.3.2*) preparations from rat brain in 1979. Since then, O-linked Man oligosaccharides have been described in other brain glycopeptides, from the Schwann cell membrane protein, α-dystroglycan, and from the cytoplasmic enzyme phosphoglucomutase. Keratin sulfate may also carry O-linked Man oligosaccharides, in place of the more usual O-linked Xyl, which forms the linking sugar for glycosaminoglycans (*see 8.3.1*).➲

➲ O-linked α-Man oligosaccharides are characteristic of glycoproteins of the nervous system.

The O-linked α-Man core may be elongated by the addition of one or more Man residues by the action of mannosyltransferases which may also function in extending N-linked oligosaccharides. Oligosaccharides with the core structure GlcNAc(β1→3) mannitol

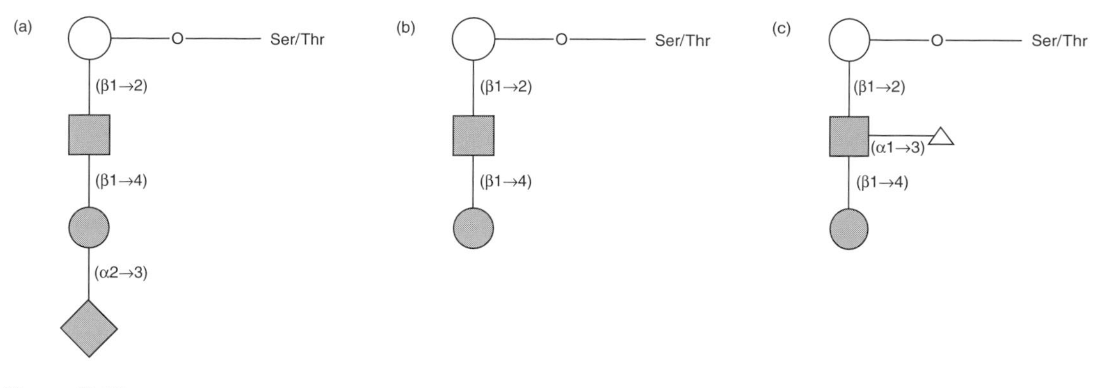

Figure 7.15

O-Linked α-Man carbohydrate structures: (a) Neu5Ac(α2→3)Gal(β1→4)GlcNAc(β1→2)Man-Ser/Thr; (b) Gal(β1→4)GlcNAc(β1→2)Man-Ser/Thr; (c) Gal(β1→4)[Fuc(α1→3)]GlcNAc(β1→2)Man-Ser/Thr.

may also be extended by Gal(β1→4) and Fuc(α1→3) linkages to the GlcNAc residue, and the Gal is frequently substituted with NeuAc(α2→3).

➲ α-Dystroglycan is a classical example of an O-linked α-Man type glycoprotein.

The extracellular peripheral membrane glycoprotein α-dystroglycan contains the structure Neu5Ac(α2→3)Gal(β1→4)GlcNAc(β1→2)Man-Ser/Thr.➲ A non-sialylated version of the Gal(β1→4) GlcNAc (β1→2)Man-Ser/Thr structure and a fucosylated, non-sialylated Gal(β1→4) [Fuc (α1→3)]GlcNAc(β1→2)Man-Ser/Thr have also been isolated from α-dystroglycan from sheep brains and mouse J1/tenascin (a secreted glycoprotein that is part of the extracellular matrix of nervous system and is involved in neuron-astrocyte adhesion). Gal(β1→4 [Fuc(α1→3)]GlcNAc (β1→2)Man-Ser/Thr is the Le^x structure more usual in N-linked and 'typical' GalNAc-O-Ser/Thr linked glycans. These structures are illustrated in *Figure 7.15*.

➲ The O-Man glycosylation pathway in mammals is poorly understood, but a transferase, responsible for adding a branching GlcNAc, is thought to be unique to the synthesis of these types of oligosaccharides.

Very little is known of the O-Man glycosylation pathway in mammals. The transferase responsible for substituting O-Man chains with a second GlcNAc to form a branching oligosaccharide is believed to be unique to the synthesis of O-Man chains, and distinct from the enzymes involved in synthesis of normal N-linked oligosaccharides chains.➲

The function of O-linked α-Man glycans is largely unknown, but they may be implicated in specific cell–cell adhesion processes. α-Dystroglycan links a second protein, β-dystroglycan, to laminin (*see 8.7.2*) in the extracellular matrix. α-Dystroglycan is necessary for normal embryonic development, specifically in development of Reichert's membrane (also known as Bowman's membrane, a thin membrane representing the uppermost layer of the cornea of the eye) and in assembly of extracellular matrix proteins, and glycosylation may be important in these processes. The O-Man sialyl tetrasaccharide Neu5Ac(α2→3) Gal(β1→4)GlcNAc(β1→2)Man-Ser/Thr may be a specific binding ligand for laminin, but this is unproven.

A sulfated O-linked Man structure HSO_3-3GlcA(β1→3)Gal(β1→4)GlcNAc(β1→2)Man-Ser/Thr (a sulfoglucuronyl lactosamine), illustrated in *Figure 7.16*, which was first detected on human natural killer cells and is therefore known as human natural killer epitope 1 (HNK-1, also called CD57), plus substituted forms of it, have also been detected in brain peptides. The HNK-1 epitope and the Le^x structure Gal(β1→4)[Fuc(α1→3)]GlcNAc-R (*see 10.6*) are known to be important recognition molecules and/or cell-adhesion molecules in a number of

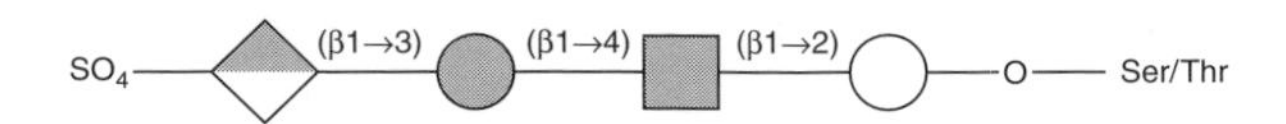

Figure 7.16

The HNK-1 epitope, a terminally sulfated glucuronic acid containing glycan first described on natural killer cells.

situations, and their synthesis is regulated during development and differentiation of the nervous system. They may be implicated in cell–cell adhesion processes, cell migration and neurite extension.

7.8.2 O-Linked α-fucose

Fuc is commonly a terminal monosaccharide residue in N- and O-linked oligosaccharides. However, it can, occasionally, occur as the first monosaccharide, attached in an α-linkage to the Ser or Thr of the polypeptide chain. This modification appears to have been conserved through evolution as it is present in organisms ranging from the primitive eukaryote *Dictylostelium discoideum* to humans. It is found in association with the cysteine-rich consensus sequence Cys_2-Xaa-Xaa-Gly-Gly-Ser/Thr-Cys_3 (the numbers 2 and 3 indicate the second and third conserved cysteines), which is found in epidermal growth factor (EGF) modules. A sequence containing six conserved cysteine residues, which are involved in formation of disulfide bridges, and thus give the molecule a fold, are a characteristic feature of EGF modules. EGF modules are, however, not always glycosylated in this way: tenascin and thrombomodulin, for example, contain multiple EGF modules, but are not O-fucosylated. Attachment of the O-linked α-fucose is catalysed by GDP-fucose: protein O-fucosyltransferase (peptide O-fucosyltransferase, O-FucT-1).

A glycopeptide carrying the glycan Glc(β1→3) Fucα-Thr was first discovered in human urine and later O-linked α-Fuc was discovered on human urokinase and other clotting proteins such as tissue plasminogen activator (tPA), factor XII and factor VII, all of which have EGF modules. ➲ In factor VII, the glycan is extended by a sialyl lactosamine, a glycan common to many conventionally N-linked oligosaccharides, to form Sia(α2→6)Gal (β1→4)GlcNAc (β1→3) Fucα1-O-Ser/Thr. The addition of Gal and Sia is probably achieved by the action of glycosyltransferases that are also involved in synthesizing the structure in N-linked glycosylation pathways, but the attachment of the initial GlcNAc to Fuc in a (β1→3) linkage requires the action of the peptide O-fucosyltransferase and is unique to O-fucosylation. Chinese hamster ovary (CHO) cells, which are used extensively in glycobiology research for the production of glycosylation-defective mutants, synthesize O-linked α-Fuc oligosaccharides.

> ➲ α-O-Fucosylation is a feature of many proteins that contain EGF modules, for example, blood clotting proteins. This type of glycosylation is conserved through evolution, indicating that it has an important function.

The function of O-linked α-fucosylation of most proteins is unknown. However, functions have been elucidated in some cases, for example, in urinary type plasminogen activator (uPA) and Notch proteins.

Binding of uPA to its receptor stimulates mitogenesis in many types of cells. O-linked fucosylation of uPA has been shown to be critical to this signalling pathway: if the uPA is not O-fucosylated (by a single α-Fuc residue) it still binds to its receptor, but the signalling pathway is not activated.

Perhaps the best studied O-linked α-fucosylated proteins are the Notch family. Notch proteins are widely distributed (from *Caenorhabditits elegans* and *Drosophila* to humans), and regulate differentiation (e.g. neurogenesis, angiogenesis and lymphoid tissue development). Notch is a >300 kDa cell-surface receptor and contains 36 tandem EGF modules, which make up most of its extracellular domain. The repeated EGF modules carry potential attachment sites for O-linked α-Fuc (and also O-linked β-glucose, which has been described on other O-fucosylated proteins too, including factor VII and XII; *see 7.8.3*) and are highly conserved across species suggesting an important function. α-O-Fuc and β-O-Glc modification may influence receptor–ligand interactions. Notch was the first membrane-associated protein identified with either O-linked fucose or O-linked glucose modifications, and both are found as monosaccharides and as elongated glycan chains. The elongated form of the Notch O-linked Fuc glycan is the tetrasaccharide Siaα(2→3)Galβ(1→4) GlcNAcβ(1→3) Fucα1-O-Ser/Thr, which has also been described on the human clotting factor protein factor IX. ➲ Ligands that are capable of activating Notch family receptors are broadly expressed in animal development, but their activity is tightly regulated to allow formation of tissue boundaries. Defects in Notch signalling are associated with diseases including T-cell leukaemia. Patients with cerebral autosomal dominant arteriopathy with subcortical infarcts and leukoencephalopathy (CADASIL) a rare inherited disorder resulting from a mutation in the Notch locus, suffer from recurrent heart attacks and strokes. In the inherited disorder, LAD-II (*see 16.8.1.4*), individuals cannot synthesize GDP-fucose, and consequently suffer recurrent infections as the mechanism for selectin-mediated leukocyte recruitment is defective. They also suffer from developmental abnormalities, including short stature and mental retardation. Because Notch signalling is essential for the developmental regulatory mechanisms, it is possible that failure in the O-fucosylation of Notch may contribute to these developmental abnormalities.

> ➲ Notch protein is O-α-fucosylated with both monosaccharide α-Fuc residues and the extended O-linked α-Fuc based tetrasaccharide Siaα(2→3)Galβ(1→4) GlcNAcβ(1→3) Fucα1-O-Ser/Thr.

Notch signalling is induced by its binding to cell surface ligands including Delta and Jagged. The *Fringe* gene family, described in *Drosophila*, have been implicated in limiting Notch activation. The *Fringe* gene codes for a Golgi resident glycosyltransferase enzyme, a fucose-specific β1,3 N-acetylglucosaminyltransferase that catalyses the addition of GlcNAc to the α-Fuc of Notch and alters the ability of Notch to bind its ligands, like Delta. ➲ Three mammalian homologues of Fringe have been identified and given the names Manic, Lunatic and Radical Fringe. This cell type-specific modification of glycosylation may provide a general mechanism to regulate ligand–receptor interactions.

> ➲ Binding of Notch with its ligand Delta is inhibited when the O-linked α-Fuc is capped by GlcNAc as a result of the action of a β1,3 GlcNAc-T coded for by the *Fringe* gene. This may be the way in which the activity on Notch, a protein involved in regulation of differentiation, is controlled.

7.8.3 O-Linked β-glucose

Several glycoproteins carrying oligosaccharides O-linked through a β-Glc residue have been described. They include factor VII, factor IX, protein Z and thrombospondin, which carry Xyl(α1→3)Glcβ-O-Ser and sometimes Xyl(α1→3)Xyl(α1→3)Glcβ-O-Ser, and the human Notch protein. Some of these molecules, including Notch, factor VII and factor XII are also modified by O-linked α-fucose (*see 7.8.2*). The β-Glc is attached to a consensus site comparable with that for O-fucosylation, Cys_1-Xaa-Ser-Xaa-Pro-Cys_2 (the 1 and 2 indicate the first and second conserved cysteines), only nine amino acids from the attachment site of the O-linked α-Fuc glycans (which lies between the second and third conserved cysteines). The Xyl moieties are attached through the action of two different xylosyltransferases.

The function of this unusual type of glycosylation is largely unknown, but may be similar to that of O-linked α-fucose (*see 7.8.2*). Factor VII without O-linked β-Glc glycans has only marginally reduced activity.

7.9 Glycogen synthesis and glycogenin

Glycogen is a major storage polysaccharide composed of repeating Glc units (*see 3.4.3*) and it is synthesized using a 'primer' 37–38 kDa protein called glycogenin. The first Glc molecule is attached to a tyrosine residue at residue 194 to form a glucosyl-1-O-tyrosyl linkage. This is catalysed by glycogen synthase. The 'primed' glycogenin is then elongated by glycogen synthase to form glycogen.

7.10 Glc(α1→2)Galβ-O-hydroxylysine

The disaccharide Glc(α1→2)Galβ- is attached to hydroxylysine in an O-linkage in collagen (*see 8.2.1*) and in some molecules that share a conserved collagen-like domain containing a triple helical structure. These include pulmonary surfactant proteins, complement factor C1q and the rat and human liver mannan-binding proteins. This modification is present in the collagens of organisms as primitive as sponges. Synthesis requires first the conversion of lysine to hydroxylysine (glycosylation appears to occur at a majority of potential sites of attachment modified in this manner), and then the sequential action of two glycosyltransferases. Mannan-binding proteins in the serum form aggregates through disulfide bonds that are dependent upon the presence of hydroxylysine, and its glycosylation.

7.11 N-Linked Glcβ-Asn

Only one unusual type of N-linked Glc glycosylation has been described in vertebrate cells, and that is Glcβ-Asn which exists in basement membrane glycoprotein laminin (*see 8.7.2*), but was first described in *Halobacteria*. Synthesis has much in common with the more usual type of N-linked glycosylation (*see Chapter 4*), in that it involves transfer of a preformed lipid-linked oligosaccharide to the acceptor protein carrying an Asn-Xaa-Ser/Thr sequence. The full structure of the oligosaccharide(s) and their function are unknown.

7.12 C-Mannosylation

Another unusual form of glycan linkage, called C-mannosylation, has been described in RNAse 2 from animal cells. Here the C1 atom of a single Man residue is attached in an α-linkage to the C2 atom of the indole moiety of Trp7. The donor is Dol-P-Man derived from GDP-Man. This is a carbon–carbon linkage, rather than a glycosidic linkage and, as plant cells and bacteria appear to be unable to form this linkage in expressed RNAse 2, it may be a relatively recent development in evolutionary terms. This strange type of glycosylation occurs at a Trp-Xaa-Xaa-Trp sequence, where it is the first tryptophan that is modified, and synthetic peptides containing this sequence are C-mannosylated *in vitro*. At least 300 mammalian proteins carry this sequence and it will be interesting to discover whether they too are C-mannosylated. This type of glycosylation has only recently been discovered, and investigations are still at a very early stage. The β-chain of fibrinogen, for example, contains the sequence, however it appears not to be C-mannosylated. Human IL-12β, however, when expressed in CHO cells is C-mannosylated. The function of C-mannosylation is unknown.

7.13 The Gal(α1→3)Gal antigen and xenotransplantation

One of the principle barriers to xenotransplantation (the transplantation of organs from other species into humans) is the Gal(α1→3)Gal epitope, sometimes called the 'Galili antigen', which is present on the cells and tissues of many animal species, including New World primates, but is largely absent from the cells and tissues of catarrhines, that is, apes, Old World primates (e.g. chimpanzee, orang-utan and gorilla) and man. It is also absent from most

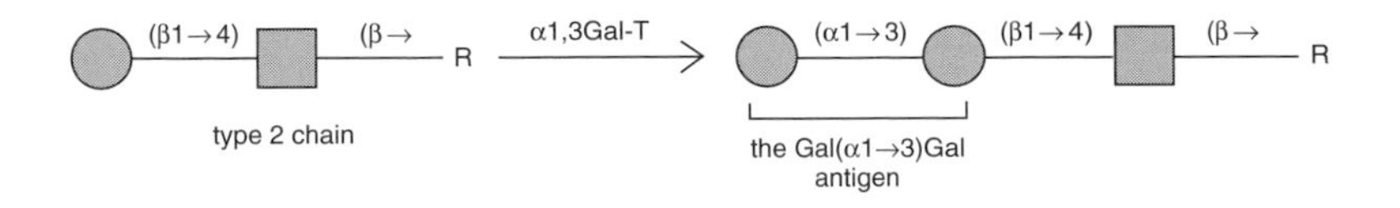

Figure 7.17

Synthesis of the Gal(α1→3)Gal antigen.

non-mammalian species, including fish, amphibians, reptiles and birds. The Gal(α1→3)Gal glycan is synthesized by a α1,3 galactosyltransferase belonging to the same family as the AB blood group transferases (*see 2.11.1*), and uses type 2 chain glycolipid and glycoprotein precursors, as illustrated in *Figure 7.17*. Humans and Gal(α1→3)Gal-negative primates have an inactive gene at the locus coding for the α1,3 galactosyltransferase, and do not elaborate type 2 chain precursors in this way. This may be due to the presence of a processed pseudogene or a frameshift mutation. They do, however, produce naturally occurring antibodies to the structure through exposure to similar glycans, possibly from normal gut microflora. The naturally occurring antibodies to Gal(α1→3)Gal may represent an evolutionary strategy to combat infection. About 1–2% of human antibodies are directed against this epitope.

➲ The Gal(α1→3)Gal structure is not synthesized by humans, but is a common epitope on glycoproteins and glycolipids of most other animals. Its presence is a significant barrier to xenotransplantation. The α1,3 Gal-T enzyme responsible for synthesis of the Gal(α1→3)Gal structure is closely related to the AB blood group transferases.

It is this natural immunization against the Gal(α1→3)Gal epitope that represents a barrier to possible xenotransplantation.➲ The idea of transplanting pig organs, such as heart and kidneys, into human recipients is an attractive one as there is a shortage of human donors and pig organs are of approximately equivalent size to the corresponding human organs. However, human anti-Gal(α1→3)Gal antibodies bind rapidly to exposed Gal(α1→3)Gal epitopes on the vasculature of the transplanted organ resulting in a swift hyperacute rejection. The epitope is abundant on glycosphingolipids, integrins and von Willebrand factor expressed on the endothelium of the blood vessels. This topic is visited in more depth in *Section 17.8.2* where approaches to overcoming this barrier to xenotransplantation, including attempts to produce genetically modified pigs that have reduced or absent Gal(α1→3)Gal, are described. Anti-Gal(α1→3)Gal antibodies have beneficial effects, in that they protect humans from animal retroviruses, which carry Gal(α1→3)Gal residues in their envelope derived from their animal cell hosts.

Leishmania and trypanosome parasites also carry the Gal(α1→3)Gal epitope, and patients infected with these parasites usually have very high titres of antibodies against this structure. It has been suggested that the expression of naturally occurring anti-Gal(α1→3)Gal antibodies in humans and their close relatives may be an evolutionary strategy to protect against infection with these organisms.

Patients with the thyroid autoimmune disorder, Graves disease, have raised titres of naturally occurring anti-Gal(α1→3)Gal antibodies. It has been speculated that aberrant expression of α1,3 Gal-T may result in synthesis of the Gal(α1→3)Gal epitope.➲

➲ The lectin from *Griffonia (Bandeiraea) simplicifolia* IB4 lectin is a useful tool for detecting the Gal(α1→3) Gal epitope.

Further reading

Altmann, F., Staudacher, E., Wilson, B.H., Marz, L. (1999) Insect cells as hosts for the expression of recombinant glycoproteins. *Glycoconj. J.* **16:** 109–123.

Bielinska, M., Boime, I. (1995) The glycoprotein hormone family: structure and function of the carbohydrate chains. In J. Montreuil, H. Schachter & J.F.G. Vliegenthart (eds) *Glycoproteins*, pp 565–587. Elsevier Science, Amsterdam.

Endo, T. (1999) O-Mannosyl glycans in mammals. *Biochim. Biophys. Acta* **1473:** 237–246.

Lehle, L., Tanner, W. (1995) Protein glycosylation in yeast. In J. Montreuil, H. Schachter & J.F.G. Vliegenthart (eds) *Glycoproteins*, pp 475–509. Elsevier Science, Amsterdam.

Messner, P. (1997) Bacterial glycoproteins. *Glycoconj. J.* **14:** 3–11.

Rayon, C., Lerouge, P., Faye, L. (1998) The protein N-glycosylation in plants. *J. Exp. Bot.* **49:** 1463–1472.

Glycoconjugates of the extracellular matrix of animals

8

8.1 Introduction – definition of the extracellular matrix

The term 'tissue' is defined as an association of cells of the same kind plus their surrounding extracellular matrix (ECM). In general, four types of tissues are distinguished:

- epithelia
- connective tissue
- muscle
- nervous tissue.

Epithelia, such as those of the skin which cover our body from the outside, those lining the inside of the body as enterocytes, and the endothelium lining blood vessels, are tightly packed cells. These epithelia are characterized by the almost complete absence of extracellular matrices; a similar arrangement is true for the nervous and muscle tissue. In contrast, connective tissue is often composed of loosely arranged cells, which are characterized by the presence of large amounts of ECM between them. The connective tissue cells which synthesize this matrix are called fibroblasts. The border between the epithelial tissue and the connective tissue is formed by the basal lamina (*Figure 8.1*). The basal lamina is often called basement membrane, but because it is not a typical unit membrane, as in the cell membrane, the term basal lamina is more accurate.

The molecular components of the basal lamina are both secreted by the epithelia and the fibroblasts. Fibroblasts are localized below the basal lamina and are called the fixed cells of the connective tissue, although they have the ability to move around locally. In addition to fibroblasts, several cell types which have migrated from the bloodstream into the connective tissue are present, the mobile cells. They are a diverse population which includes lymphocytes, granulocytes (in particular eosinophils), mast cells and macrophages. ➲ Whereas the epithelia are generally organ specific, the connective tissue is found in every organ as a 'non-specific' component, which is often called the stroma of an organ. Although modifications in the composition of the stroma between organs can occur, its principle composition is the same in every organ. The stroma is generally located between the blood vessels of an organ and its specific epithelia, where it serves as an important transit zone between the blood and the cells specific for the organ. It acts not only as a selective filter, but also as a space for pathological processes. Inflammation, for example, is morphologically characterized by the infiltration of the ECM by inflammatory cells, for example, granulocytes (particularly neutrophils), lymphocytes and macrophages, and is therefore a process which starts within the ECM. The ECM is also important in cancer, where cancer cells have to degrade the ECM around them in order to spread locally and later to disseminate to distant sites. Hence the ECM is an important component of every organ both in health and disease.

➲ The extracellular matrix of multicellular organisms is found in every organ and is often called the stroma. It has important functions in substance exchange between the circulation and the organ-specific cells.

Functional and Molecular Glycobiology, Susan A. Brooks, Miriam V. Dwek and Udo Schumacher

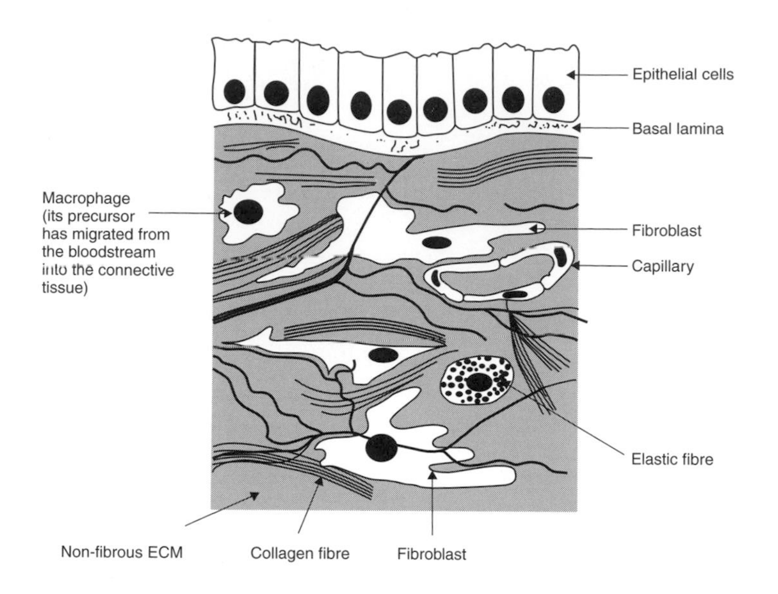

Figure 8.1

The border between epithelia and connective tissue is formed by the basal lamina. Whereas epithelia are a formation of tightly packed cells, the fibroblasts are separated by large extracellular spaces, filled with extracellular matrices. © 1994. From Alberts *et al. Molecular Biology of the Cell*, 3rd Edition. Reproduced by permission of Routledge, Inc., part of the Taylor & Francis Group.

8.2 The fibroblast – producer of the extracellular matrix

➲ The extracellular matrix is rich in glycoproteins and is mainly produced by the fibroblasts.

All cells of the connective tissue can contribute to the composition of the ECM by secreting molecules into it.➲ However, most ECM macromolecules are synthesized by fibroblasts. In common with all secretory proteins, proteins of the ECM are initially synthesized in the rough endoplasmic reticulum (RER) and are transported via vesicles to the Golgi apparatus where they are modified and packed into secretory vesicles. These cellular compartments are the two principal sites for glycosylation processing, so it is not surprising that most ECM components, like most other secretory proteins, are glycoproteins.

Using light microscopy, two components of the ECM can be distinguished: (a) fibres and (b) non-fibrous translucent material between them, often referred to as the ground substance. Both fibres and ground substance are rich in glycoconjugates, mainly produced by fibroblasts. In most extracellular matrices, the fibres predominantly consist of collagen fibrils.

8.2.1 Collagen

The collagens are a large family of fibrous proteins found in all multicellular animals.➲ They are mainly secreted by connective tissue cells (fibroblasts and their relatives the chondrocytes and osteoblasts), but other cells are also able to synthesize collagens, although their contribution to the total collagen of the body is relatively minor. About one-quarter of the protein mass of the human body is collagen. Initially, water-soluble procollagen is synthesized. A pro-peptide sequence is cleaved from procollagen in the ECM by special peptidases. After this cleavage, the combination and aggregation of the collagen begins, which ultimately forms the insoluble collagen fibril. Three α-chains are twisted in a rope-like fashion as illustrated in *Figure 8.2* to form long, stiff, triple-stranded collagen molecules. These in turn form larger aggregates, the collagen fibrils. These fibrils, unlike their precursor procollagen, are insoluble in water.

➲ The collagens are a large family of mainly fibrous proteins of the extracellular matrix.

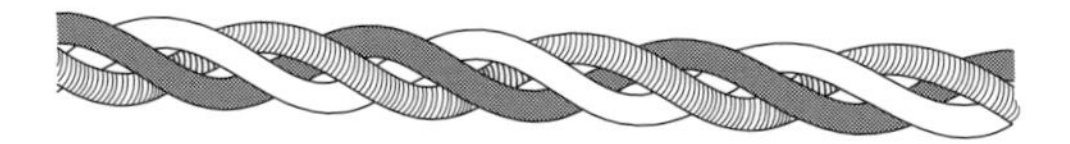

Figure 8.2

The structure of collagen fibres: three long, stiff α-chains are twisted in a rope-like structure. Adapted from Fawcett, D.W. (1994) *A Textbook of Histology*. Lippincott, Williams & Wilkins, New York & London.

Collagens are rich in proline and lysine amino acid residues. Currently, ≈25 different α-chains of collagens have been identified, each coded for by a separate gene.

There are four main types of collagen:

- fibrillar
- non-fibrillar
- fibril-associated
- network forming.

An example of a fibrillar collagen is type I collagen, which is the most abundant (90%) collagen of the human body and is found in skin, tendon, ligaments, bone and internal organs. It forms larger aggregates, the collagen fibril, which can be visualized using a light microscope. Type II collagen also forms fibrils, and is typically found in cartilage. Type VII collagen forms anchoring fibrils which link the basal laminae with the connective tissue fibres below them. Some collagens, such as types IX and XII are fibril associated and cover the surface of the fibrillar collagens. They are presumably involved in linking collagen fibrils to each other and to other molecules of the ECM. In contrast, non-fibrillar, network-forming collagens also exist, for example, type IV collagen. Type IV collagen forms a felt-like network in basal laminae.

Before the three collagen chains form a triple helix, as illustrated in *Figure 8.2*, about half of their proline and a variable amount of their lysine residues must be hydroxylated (*see 7.10*). This hydroxylation is facilitated by prolyl- and lysyl-hydroxylases which are vitamin C dependent. Hydroxylation is a prerequisite for the secretion of collagen molecules into the ECM. The next step after hydroxylation of the collagen molecules is their glycosylation. Only some hydroxylysyl residues are glycosylated. Two enzymes are involved in this glycosylation. UDP-galactosyltransferase and UDP-glucosyltransferase, both of which are manganese dependent. UDP-galactosyltransferase links Gal via a β-glycosidic O-linked bond to the hydroxylysine residues. The subsequent bond between the Glc and the Gal residue, catalysed by UDP-glucosyltransferase, is unusual as it is an (α1→2) linkage (*Figure 8.3*).

This glycosidic bond is relatively stable as it is near to the ε-amino group of the hydroxylysine residue and this is not a configuration recognized by most glucosidases. The degree of glycosylation varies considerably between different anatomical sites: it is low in the fibrillary collagens of the skin and tendons, whereas it is high in the type IV collagen of the basal lamina. This glycosylation takes place within the cell and is restricted to hydroxylysine residues of the collagen triple helix which are located in the non-helical areas.

8.2.2 Reticulin fibres

In addition to the relatively coarse fibres of type I collagen (*see 8.2.1*), networks of delicate (0.5–2 μm) fibres can be found in loose connective tissues. These are not readily distinguishable in normal histological sections, but they are easily recognized in sections stained using silver salt methods. These fibres are also stained by the periodic acid Schiff (PAS) reaction (*see 14.2.1*), which indicates the presence of neutral carbohydrate residues. They consist mainly of type III collagen, and it is not yet clear whether the positive PAS reaction is due to glycosylation of the collagen fibrils or coating of the fibres by proteoglycans and/or glycoproteins.

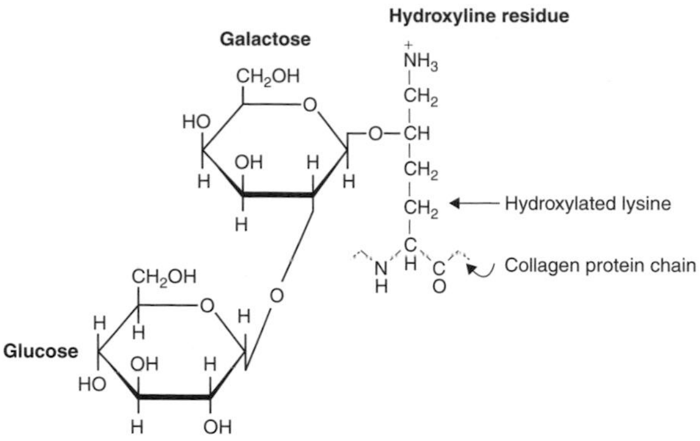

Figure 8.3

Structure of the linkage of the Glc(α1→2)Gal disaccharide unit to the hydroxylysine side chain of the procollagen peptide chain.

8.2.3 Elastic fibres

Many organs require elasticity for their normal function, for example, the large arteries and the lungs. Elasticity is provided by a special class of fibrils, the elastic fibres. About 50% of the aortic tissue consists of elastin, the ECM protein which comprises elastic fibres. In a similar mechanism to that described previously for collagen synthesis (*see 8.2.1*), proelastin molecules are synthesized within the cells, secreted and converted into elastin within the ECM. The elastin molecules then polymerize, bind to collagen, and associate with various glycoproteins to form the elastic fibres. One of these fibre-associated glycoproteins is fibrillin. Two genes code for the homologous proteins fibrillin I and fibrillin II. The other glycoproteins that contribute to the elastic microfibrils are poorly characterized and the functional role of glycosylation of these molecules is completely unknown.

The spaces in the ECM between the collagen, reticulin and elastic fibres are filled by proteoglycans, glycosaminoglycans (*see 8.3*) and glycoproteins (*see 8.7*).

8.3 Proteoglycans and glycosaminoglycans of the extracellular matrix

Proteoglycans are a special type of glycoproteins. They consist of a protein core to which glycan side chains are covalently linked. Whereas the oligosaccharides of glycoproteins are often branched and seldom exceed 20 monosaccharide units, the side chains of proteoglycans consist of long, linear heteroglycan chains. Usually, these chains are formed by repeating disaccharide units. These disaccharide units are composed of a hexosamine monosaccharide plus glucuronic acid (except in keratan sulfate; *see 8.3.4*).

The long chains of disaccharide units alone are called glycosaminoglycans. Only if they are linked to a protein core are they named proteoglycans. ➲ Hyaluronan (*see 8.3.1*) is unusual in that it consists of very long chains of repeating disaccharide units which are not linked to a protein core, and is thus not a proteoglycan.

➲ Proteoglycans are glycosaminoglycans covalently linked to a protein core.

The extensive biological diversity of the proteoglycan family results from three factors:

- proteoglycans differ in their protein backbones.
- the glycan side chain attached to these protein backbones can differ.
- sulfation of the glycan side chains can vary with respect to the position of the sulfated C atom, and the amount of sulfation (i.e. the number of units sulfated).

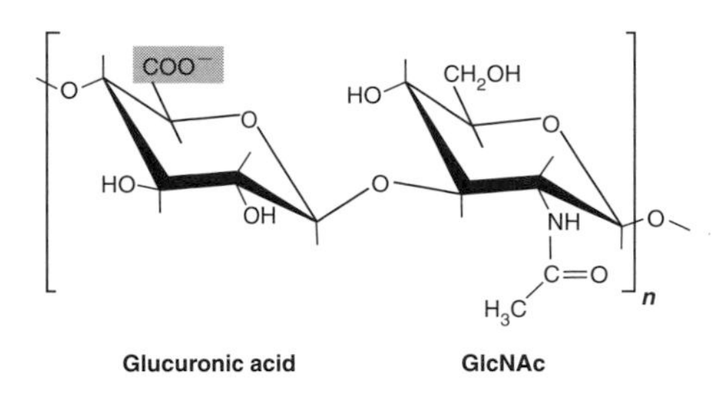

Figure 8.4

Hyaluronan is composed of repeating disaccharide units consisting of glucuronic acid and GlcNAc. The units are [glucuronate(β1→3)GlcNAc(β1→4)]$_n$, where n can be up to 25 000. The carboxylated group is shaded.

Glycosaminoglycans and proteoglycans can be classified into four groups according to their monosaccharide residues, their type of linkage, and the number and location of the sulfate groups. These are:

- hyaluronan
- chondroitin sulfate and dermatan sulfate
- heparan sulfate and heparin
- keratan sulfate.

8.3.1 Hyaluronan

Hyaluronan (also called hyaluronic acid or hyaluronate) lacks a protein core and is hence the simplest of the glycosaminoglycans. It consists of repeating non-sulfated disaccharide units (*Figure 8.4*), each composed of glucuronic acid and GlcNAc, [glucuronate(β1→3)GlcNAc (β1→4)]$_n$, where n can be up to 25 000.

Because of its relatively simple structure, hyaluronan is thought to be the first glycosaminoglycan to have developed during evolution. It is produced by some streptococci bacteria. It is synthesized directly at the plasma membrane by a multi-enzyme complex called hyaluronan synthase. Activated sugar residues enter this multi-enzyme transmembrane complex at the cytoplasmic side. These are linked to the nascent hyaluronan chain, which is injected into the ECM. Here, the large molecule is immediately surrounded by structurally bound water (its large size and hydrophilic nature makes it impossible for this molecule to cross the plasma membrane). The level of hyaluronan varies between tissue types, and is most abundant in embryonic tissues. In connective tissue including cartilage, hyaluronan is interspersed between the collagen fibrils (*Figure 8.5*).

The capacity of hyaluronan to bind large amounts of water means that it forms a jelly-like substance, which has diverse functions, including structural ones. One part hyaluronan binds 100 parts water; thus a hyaluronan molecule occupies a large space within the matrix. It is therefore used in all parts of the body where spaces need to be filled, and is found in high concentrations in the vitreous humour of the eye and it acts as a lubricant in cartilaginous joints. Before fibroblasts divide, they shed hyaluronan into their surroundings to make space for the new cells. Similarly, epithelial cells which migrate during embryonic development secrete large quantities of hyaluronan. Large amounts of hyaluronan are also present during wound healing and, presumably, in cancer metastasis where cancer cells need to create space for migration. After migration is completed, hyaluronate is degraded via the enzyme hyaluronidase.

Hyaluronan is not merely a space filler, however. Cells also attach specifically to hyaluronan via cell membrane-bound receptors, hyaladherins. Sheep brain is rich in a special

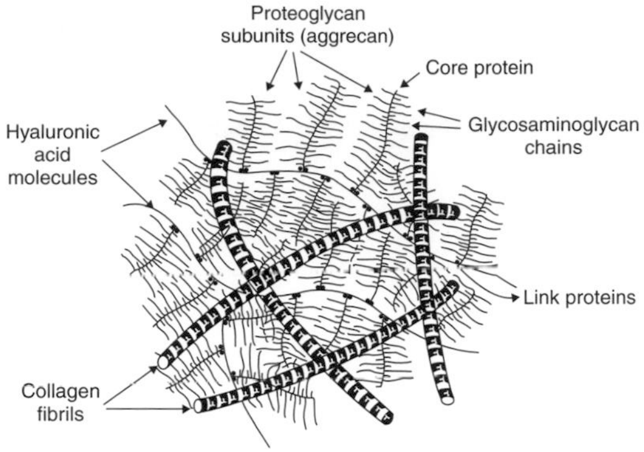

Figure 8.5

The composition of the extracellular matrix in cartilage. The mechanical strength is provided by the collagen fibrils, whereas hyaluronan forms long chains between the collagen fibrils. Link proteins connect other proteoglycans (aggrecan) with the long hyaluronan chain. From Fawcett, D.W. (1994) *A Textbook of Histology*. Lippincott, Williams & Wilkins, New York & London.

➲ Hyaluronan is an important space filler between the collagen fibrils of the extracellular matrix. Many cells express specific receptors for hyaluronan on their cell surface.

hyaluronan-binding molecule called hyaluronectin.➲ Labelled hyaluronectin can be used in the laboratory to demonstrate the presence of hyaluronan in tissue sections. Another important binding protein for hyaluronan is the transmembrane cell-surface receptor, CD44. Much attention has been paid to the interaction between CD44 and hyaluronan in cancer metastasis, as CD44-positive tumours exhibit a more metastatically competent phenotype in comparison with CD44-negative tumours.

Hyaluronan is also an important product in the pharmaceutical industry (*see 17.2.4*). For pharmaceutical purposes, for example, to prevent dehydration of the eye during eye surgery, hyaluronan is either produced by mass culture of hyaluronan-producing streptococci strains, or is isolated from chicken wattle. Because it has to be free from other components to avoid unwanted immunological reactions it is very expensive to produce (1 kg currently costs £100 000 sterling).

Other glycosaminoglycans, considered in the following sections, differ in several ways from hyaluronan:

- glycan side chains are all covalently linked to a protein backbone; hence they are all proteoglycans
- the linkage between the protein and the repeating disaccharide units involves a specially link tetrasaccharide, illustrated in *Figure 8.6*
- glycan side chains are much shorter (< 150 monosaccharides) than in hyaluronan
- the sugar residues are sulfated.

8.3.2 Chondroitin sulfate and dermatan sulfate

8.3.2.1 Chondroitin sulfate

The basic disaccharide unit of chondroitin sulfate is a glucuronate linked via a β-glycosidic bond to GalNAc. As sulfation can take place in position C4 or C6, two types of chondroitin sulfates exist, namely chondroitin-6-sulfate (*Figure 8.7*) and chondroitin-4-sulfate (*Figure 8.8*). They both contain carboxylated and sulfated charged groups within the disaccharide unit.

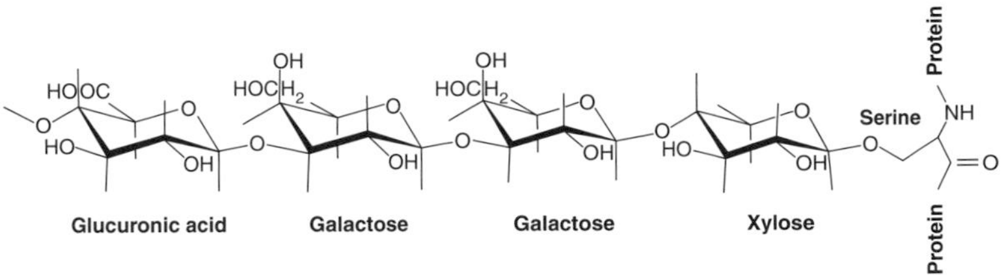

Figure 8.6

The tetrasaccharide that links the protein and the repeating disaccharide units in glycosaminoglycans.

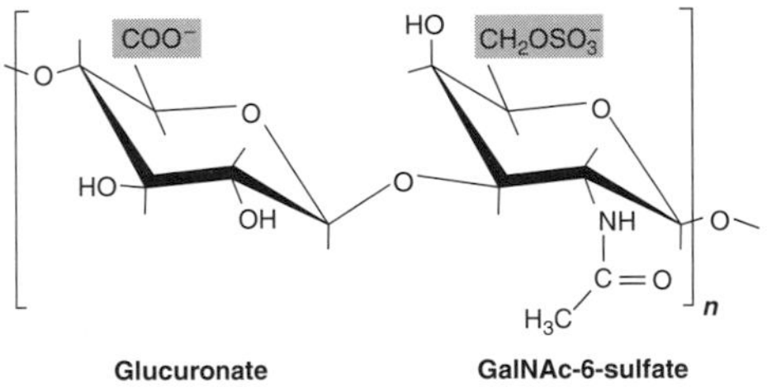

Figure 8.7

Chondroitin-6-sulfate: [glucuronate(β1→3)GalNAc-6-sulfate(β1→4)]$_n$. The charged groups are shaded.

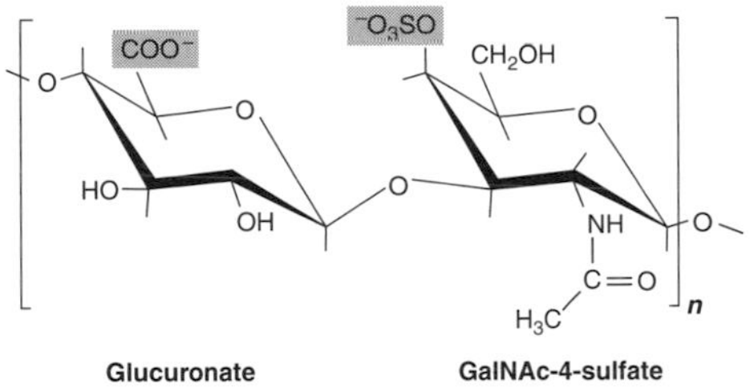

Figure 8.8

Chondroitin-4-sulfate: [glucuronate(β1→3)GalNAc-4-sulfate(β1→4)]$_n$. The charged groups are shaded.

Both types of chondroitin sulfate differ from hyaluronan with respect to their monosaccharides in that the D-glucosamine in hyaluronan is replaced by D-galactosamine which can be O-sulfated. Depending on the tissue, these two chondroitin sulfates occur either separately or as mixtures. The chondroitin sulfates have a molecular mass ranging from 20 to 50 kDa. They are major components of cartilage, where they make up to ≈40% of its dry weight. Other tissues containing chondroitin sulfates are connective tissues, skin and cornea.

8.3.2.2 Dermatan sulfate

Dermatan sulfate is very similar to chondroitin-4-sulfate as it is converted from it via a C5 epimerase (*Figure 8.9*). This epimerase inverts the position of the carboxylate group of the

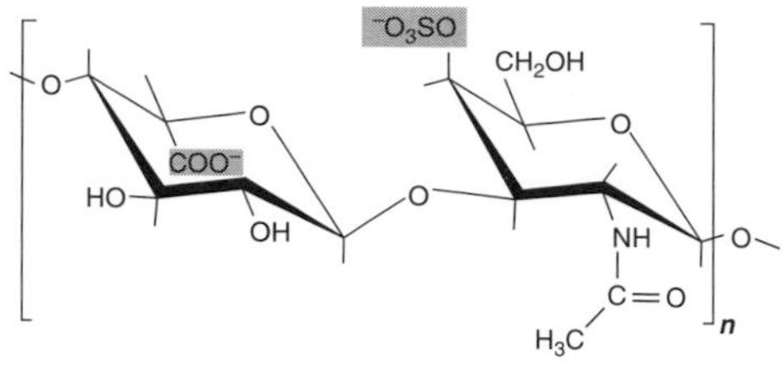

Figure 8.9

Dermatan sulfate [iduronate(α1→3)GalNAc–4-sulfate(β1→4)]$_n$ is an epimerization product of chondroitin-4-sulfate at C5 position of the uronic acid (compare with *Figure 8.7*). The charged groups are shaded.

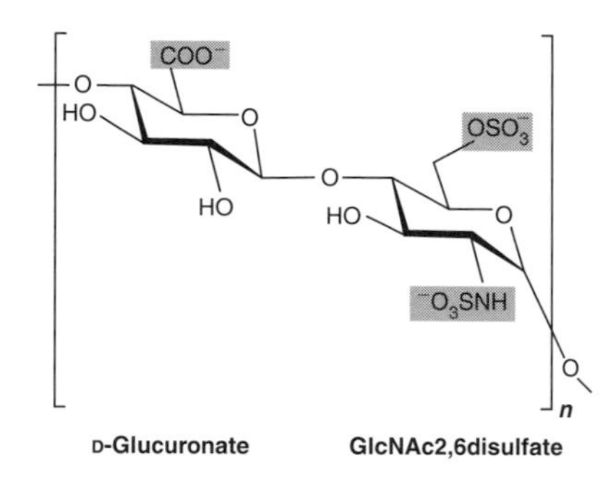

Figure 8.10

Heparan sulfate [D-glucuronate(α1→4)GlcNAc2,6disulfate(β1→4)]$_n$. The charged groups are shaded.

D-glucuronate so that it is converted into iduronate. Dermatan sulfate therefore has the formula [iduronate(α1→3)GalNAc–4-sulfate(β1→4)]$_n$.

The level of conversion of D-glucuronic acid into L-iduronic acid is variable and can range from just a few units to all of them. Furthermore, some of the L-iduronic acid residues can be sulfated at position C2. Dermatan sulfate, as the name suggests, is present in large amounts in the skin (dermis), but can also be found in connective tissue and heart valves.

8.3.3 Heparan sulfate and heparin sulfate

8.3.3.1 Heparan sulfate

The repeating disaccharide unit in heparan sulfate is β-D-glucuronic acid linked to α-D-GlcNAc. The GlcNAc is modified in two ways: (i) N-deacetylation and nitrogen sulfation occurs, which replaces the —NHAc with $NHSO_3H$; (ii) the primary alcohol group at C6 is sulfated. The formula for heparan sulfate is [D-glucuronate(α1→4)GlcNAc2-6-disulfate(β1→4)]$_n$. This is illustrated in *Figure 8.10*.

The polysaccharide is linked to a core protein and is found as a matrix component of many organs, for example, arterial walls, lungs, heart, liver and skin. It is also an important structural part of the basal laminae, particularly so in the kidney, where, at the molecular level, heparan proteoglycan forms the major component of the blood–urine barrier (see 7.7.2).

8.3.3.2 Heparin sulfate

Heparin sulfate (often simply referred to as heparin) is derived from heparan sulfate by the action of a C5 epimerase. The carboxylate group of the D-glucuronic acid residue is

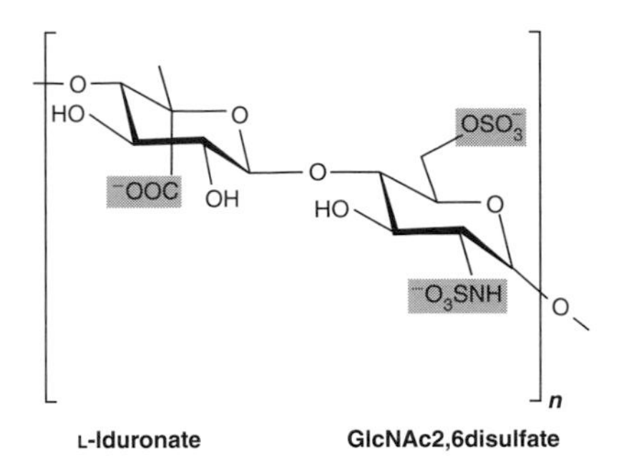

Figure 8.11

Heparin sulfate: [L-iduronate(α1→)GlcNAc2,6disulfate(α1→4)]$_n$. The charged groups are shaded.

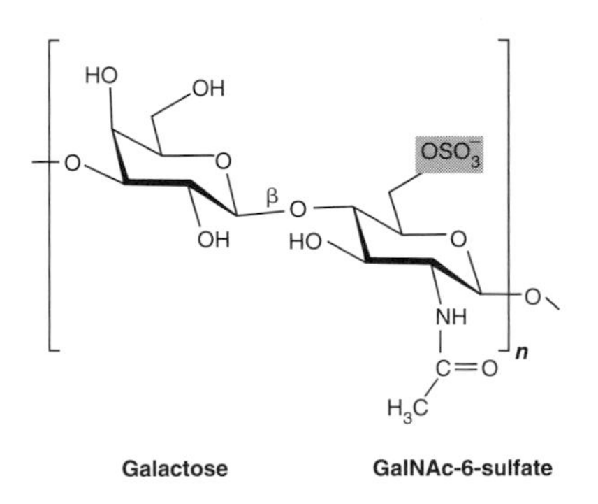

Figure 8.12

Keratan sulfate consists of (β1→3) linked N-acetyllactosamine chains, which are sulfated at the O-6 of GlcNAc. Because the uronic acid is replaced by galactose, only the sulfate group (shaded) carries the negative charge.

inverted, thus forming α-L-iduronic acid. Thus all linkages in heparin are (α1→4) ones (*Figure 8.11*).

The polysaccharide contains a variable number of sulfate groups at C2 of the L-iduronic acid residue, at the C6 hydroxyl and at the 2-amino group of the D-glucosamine residue. ➲ Heparin is stored in the granules of mast cells and when released prevents the clotting of blood. It is widely used in medicine to prevent blood clotting in immobilized patients after surgery.

➲ Heparin is important as an anti-clotting agent in clinical medicine.

8.3.4 Keratan sulfate

This proteoglycan is exceptional as it does not contain uronic acid. Instead, its disaccharide building block consists of β-D-Gal linked (β1→4) to D-GlcNAc, N-acetyllactosamine. Keratan sulfate is therefore often referred to as a poly-N-acetyllactosamine chain. It is sulfated at the O-6 of the GlcNAc. The structural disaccharide is illustrated in *Figure 8.12*.

Of all the glycosaminoglycans, keratan sulfate is the most heterogeneous, as the degree of sulfation is very variable; furthermore it can contain small amounts of Fuc, Man and sialic acid residues. A maximum of ≈30–50 disaccharide units occur. It is present only in cornea, cartilage, bone and on the surface of erythrocytes.

8.4 The protein backbone of glycosaminoglycans

With the exception of hyaluronan, all glycosaminoglycan glycan side chains are linked to a protein backbone (*see 8.3*). Common protein cores of proteoglycans include aggrecan, betaglycan, decorin, syndecan-1, perlecan and serglycin, described in this section. Not all of the glycosaminoglycan structures described in the previous sections occur linked to every protein backbone described here. The number of oligosaccharide side chains associated with a single protein molecule can vary considerably, and ranges from 1 (e.g. betaglycan, decorin) to ≈130 (e.g. aggrecan). The linkage between the repeating disaccharide units and the protein backbone is the same in all proteoglycans, consisting of the tetrasaccharide Xyl(β1→4)Gal(β1→3) Gal(β1→3)GlcA, illustrated in *Figure 8.13*.

> One particular glycan chain of proteoglycans can be linked to various protein backbones giving rise to many different molecules within the proteoglycan family. These have diverse functions.

The formation of the side chain starts with the O-linkage of Xyl to a serine residue. While in N- and O-linked glycoproteins the amino acid sequence determines whether the amino acid side chain will be glycosylated or not, a particular sequence specificity does not seem to exists for proteoglycans. The local conformation of the polypeptide chain therefore determines whether a glycosaminoglycan chain is added.

8.4.1 Aggrecan

The protein core of aggrecan has a molecular mass of ≈210 kDa, to which chondroitin sulfate (*see 8.3.2*) and keratan sulfate (*see 8.3.4*) chains are covalently linked. Two link proteins bind one aggrecan molecule non-covalently to hyaluronic acid. These aggregates can become very large (in excess of 2×10^8 Da) and can occupy a space as large as a bacterium in the ECM (these complexes are illustrated in *Figure 8.5*). It is a characteristic component of articular cartilage.

8.4.2 Betaglycan

The core protein has a molecular mass of ≈36 kDa and is linked covalently to a single side chain of chondroitin sulfate (*see 8.3.2*) or dermatan sulfate (*see 8.3.2*). It is found both in the plasma membrane of cells and in the ECM. It binds to transforming growth factor beta (TGFβ).

8.4.3 Decorin

Decorin has a larger protein core (40 kDa) than betaglycan and is widely distributed in connective tissues. It also carries a single side chain of chondroitin sulfate (*see 8.3.2*) or dermatan

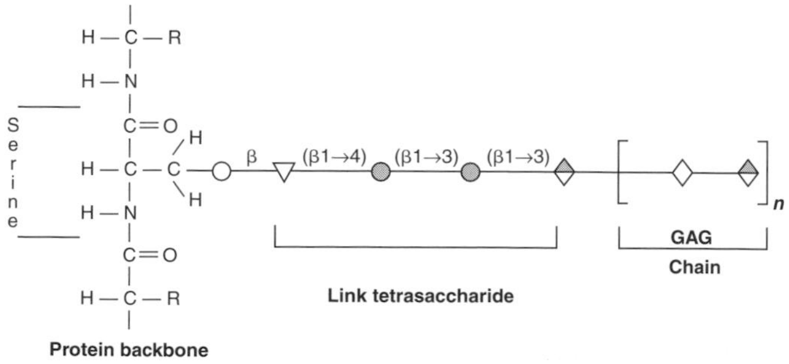

Figure 8.13

The disaccharide units of glycosaminoglycans are linked via a special link tetrasaccharide to the protein backbone to form proteoglycans.

sulfate (*see 8.3.2*). In addition to its TGFβ binding, it binds to (or 'decorates') the surface of type I collagen fibrils, hence its name decorin.

8.4.4 Syndecan-1

Syndecan-1 has a protein core of molecular mass 32 kDa, and carries between one and three side chains of chondroitin sulfate (*see 8.3.2*) and/or heparan sulfate (*see 8.3.3*). It also binds fibroblast growth factor (FGF). It is located on the cell surface of fibroblasts and epithelial cells and mediates cell adhesion.

8.4.5 Perlecan

Perlecan has a protein core with a molecular mass of ≈600 kDa, and carries between 2 and 15 heparan sulfate (*see 8.3.3*) side chains. Perlecan is located in the basal laminae and the charged side chains are important for its filtering function.

8.4.6 Serglycin

This molecule is a small proteoglycan. The molecular mass of the protein core is ≈20 kDa. It carries 10–15 glycosaminoglycan side chains, which are either chondroitin sulfate (*see 8.3.2*) or dermatan sulfate (*see 8.3.2*). It is located in the secretory vesicles of white blood cells and helps to package and store the molecules to be secreted.

8.5 Biosynthesis of proteoglycans

➲ The Golgi apparatus plays the central role in proteoglycan synthesis.

The biosynthesis of proteoglycans is illustrated in *Figure 8.14*. Synthesis of the protein backbone starts in cytoplasmic ribosomes, which attach to the RER as the signal sequence directs the nascent polypeptide chain towards it.➲ The protein chain is then transferred via vesicular transport to the Golgi apparatus, where glycosylation starts with the synthesis of the link tetrasaccharide (*see 8.3.1*).

UDP-Xyl is linked to the serine residue via a xylose transferase. This is followed by the addition of two Gal residues, both also transferred in the activated state by Gal transferases (Gal-TI and Gal-TII, respectively). With the transfer of activated glucuronic acid to the trisaccharide, the synthesis of the link tetrasaccharide (*Figure 8.6*), is completed. For chondroitin sulfate, activated GalNAc and activated glucuronate are successively added until the desired chain length has been reached. As the final step, selected groups are sulfated with the help of sulfotransferases. Two different sulfotransferases are responsible for the sulfation of C4 and C6. Once sulfation is completed, the molecules are packed into secretory vesicles and transported to the cell surface where they are released into the ECM. The synthesis of both proteoglycans and hyaluronate is under hormonal control. High doses of corticosteroids inhibits their synthesis, presumably explaining why high doses of steroids also inhibit wound healing. Other hormones such as thyroid hormone and interleukin-I (IL-I) and interleukin-6 (IL-6) inhibit their synthesis, whereas TGFβ, EGF and platelet derived growth factor (PDGF) stimulate proteoglycan biosynthesis.

8.6 Functions of the glycosaminoglycans

8.6.1 Physicochemical effects

Generally, the glycosaminoglycans of the ECM carry many negatively charged groups (carboxylated and sulfated) and are therefore polyanionic species. Under physiological conditions, the counterions are not protons (H^+) but Na^+, K^+ and Ca^{2+} ions. Hence glycosaminoglycans bind considerable amounts of cations relative to their weight, and are surrounded by structurally bound water. They also interact with each other (hyaluronic acid, link proteins and aggrecan) to

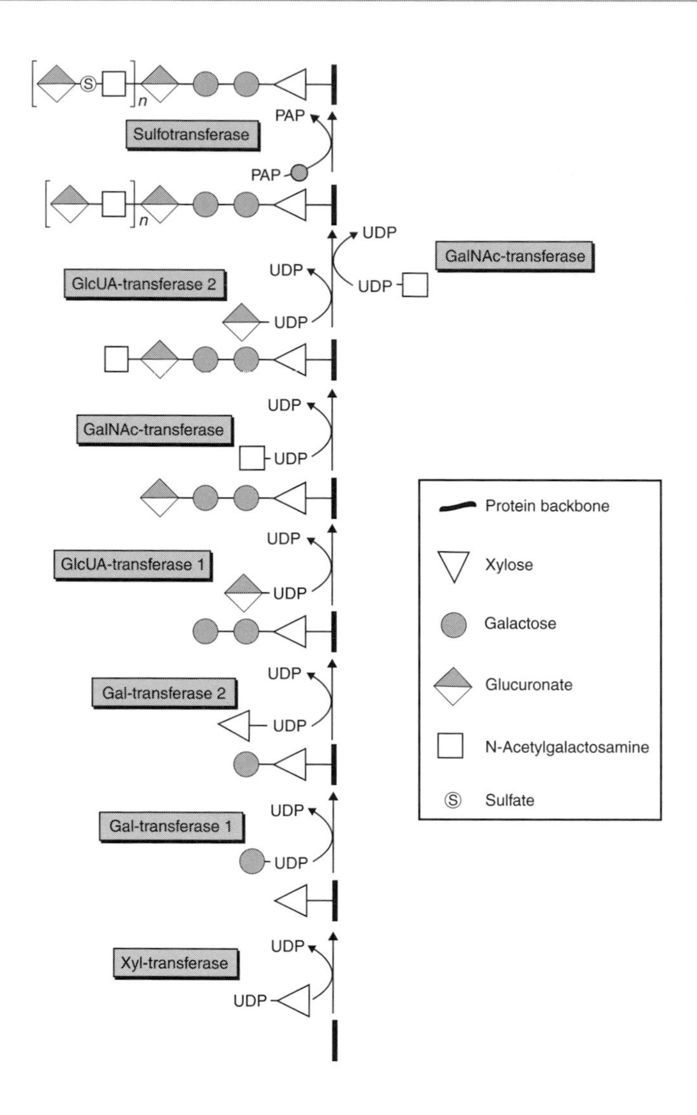

Figure 8.14

Biosynthesis of the glycosaminoglycan side chain of proteoglycans as exemplified with chondroitin-4-sulfate. For details see text (section 8.5). Adapted from figure 26.18 in Löffler, G. and Petrides, P.E. (1997) *Biochemie und Pathobiochemie*, 5th Edition. Springer-Verlag, Germany, with permission from Springer-Verlag.

form large hydrated spaces within the tissue. Because of their charged groups the glycosaminoglycans act as a permeability barrier which influences transport across the extracellular space.

8.6.2 Specific molecular functions

Many proteins interact specifically with glycosaminoglycans. For example, the cell adhesion molecules CD44, TGFβ and basic fibroblast growth factor (bFGF) bind specifically to hyaluronic acid. Glycosaminoglycans also influence the action of proteases and protease inhibitors. Furthermore, proteoglycans can specifically influence the activity of proteins secreted into the ECM. Several mechanisms illustrating how a proteoglycan can influence the activity of a protein have been established. Because of its charge, the proteoglycan immobilizes the protein near its place of secretion, thus limiting its effect, or at least acting as a slow

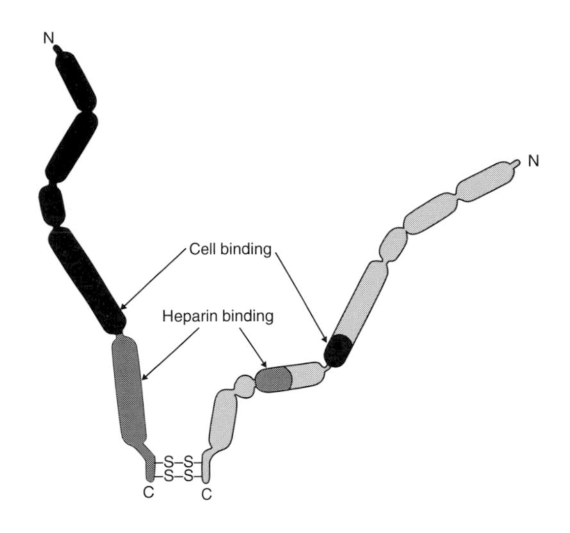

Figure 8.15

Schematic diagram of a fibronectin dimer. © 1994. Adapted from Alberts, B. *et al. Molecular Biology of the Cell*, 3rd Edition. Reproduced by permission of Routledge, Inc., part of the Taylor & Francis Group.

release substrate. Conversely, it can also protect proteins from proteolysis, thus prolonging their biological half-lives. Proteoglycans can sterically block the active site of a protein and therefore alter its biological activity. Proteins may also be concentrated on proteoglycans in order to more effectively present themselves to cell surface receptors.

Several proteoglycans are not constituents of the ECM but are cell membrane bound either by a glycosylphosphatidylinositol (GPI) anchor (*see Chapter 12*) or via a membrane-spanning core protein. An example of the latter are the syndecans, which carry chondroitin or heparan sulfate glycosaminoglycan chains on the extracellular side of the membrane. Together with the integrins, the syndecans act as receptors for matrix proteins, for example, collagens and fibronectin. They also act as co-receptors for the growth factors (TGFβ, FBF) mentioned previously, which collaborate with their specific cell membrane receptors.

8.7 Glycoproteins of the extracellular matrix

The ECM contains not only fibres and non-fibrous proteoglycans filling the space between them, but also a number of glycoproteins.➲ Because these proteins are secretory proteins, they are, in common with most secretory proteins, glycosylated. Therefore, the general principles associated with the various functions of glycosylation apply to these molecules as well. However, there are glycoproteins for which specific functional interaction with saccharides have been described, as outlined below.

> ➲ The ECM is rich in adhesive glycoproteins as well 25 including fibronectin and laminin.

8.7.1 Fibronectin

Fibronectin is composed of two large glycoproteins, which are similar but not identical, as they are the products of differentially spliced mRNA, joined by two disulfide bonds near their C-terminal ends. Each of the two chains contain several distinct binding sites, two of which are highlighted in *Figure 8.15*. The cell-binding site consists of a sequence of only three amino acids: Arg-Gly-Asp or in the single letter code for amino acids, the RGD

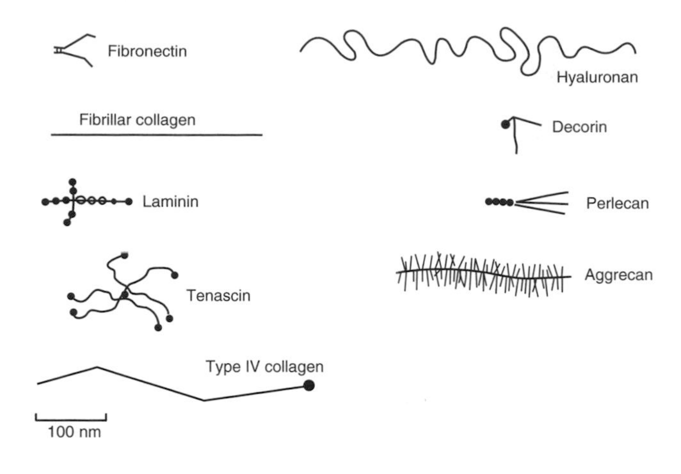

Figure 8.16

Schematic representation of the major components of the extracellular matrix. Note the great variability of the glycosaminoglycan part (one chain in decorin to more than 100 chains in aggrecan). © 1994. From Alberts, B. *et al. Molecular Biology of the Cell*, 3rd Edition. Reproduced by permission of Routledge, Inc., part of the Taylor & Francis Group.

sequence. The RGD sequence is also found on a number of different matrix proteins and is recognized by cell membrane receptors called integrins, which thus attach cells to fibronectin. Fibronectin which has lectin-like activity binds, in turn, to heparin and helps to anchor cells to the ECM. In addition to heparin, fibronectin binds to collagen with binding sites which differ from the heparin binding ones (not shown in *Figure 8.15*) so that an even tighter interaction with the ECM can occur. The interactions of fibronectin with cells is vital for the survival of an organism. Knockout mice in which both copies of the fibronectin gene have been inactivated are not viable and die during early embryogenesis.

In the embryo, an important role for fibronectin is the guidance of cellular movements. This has been shown in experiments with amphibians, in which RGD-blocking peptides blocked the migration of future mesenchymal cells during gastrulation. It is thought that fibronectin stimulates cell migration by attachment of the cells to the ECM. This effect, however, must be balanced against the role fibronectin plays in anchoring cells in position. The exact mechanisms of how these contrary functions are regulated are not clearly understood.

Fibronectin belongs to the class of adhesive glycoproteins called the nectins. Other members of this family include vitronectin, osteonectin, thrombospondin and tenascin. Tenascin is a matrix molecule which plays an important role in guiding cell movements during development. It consists of six polypeptide chains which branch off from the centre like the spokes of a wheel (*see Figure 8.16*). It contains several domains with different binding activities. One domain binds fibronectin, whereas another binds to the cell membrane proteoglycan syndecan (*see 8.4.4*), hence the molecule can be classified as lectin-like (*see Chapter 13*). Tenascin has a dual function in that it can promote *and* inhibit cell adhesion depending on the cell type, presumably regulated by the different protein domains.

8.7.2 Laminins

Another large family of adhesive glycoproteins of the ECM are the laminins, which consist of α-, β- and γ-subunits, ranging in the molecular mass from 150 to 400 kDa. The laminins are an important component of the basal lamina, and several different laminins can be formed by the combination of these subunits. Laminins bind via their RGD epitope to a number of different cells which border basal laminae (e.g. epithelia, neurons, muscle cells and fat cells) on the one hand, and to components of the ECM [including the proteoglycan perlecan

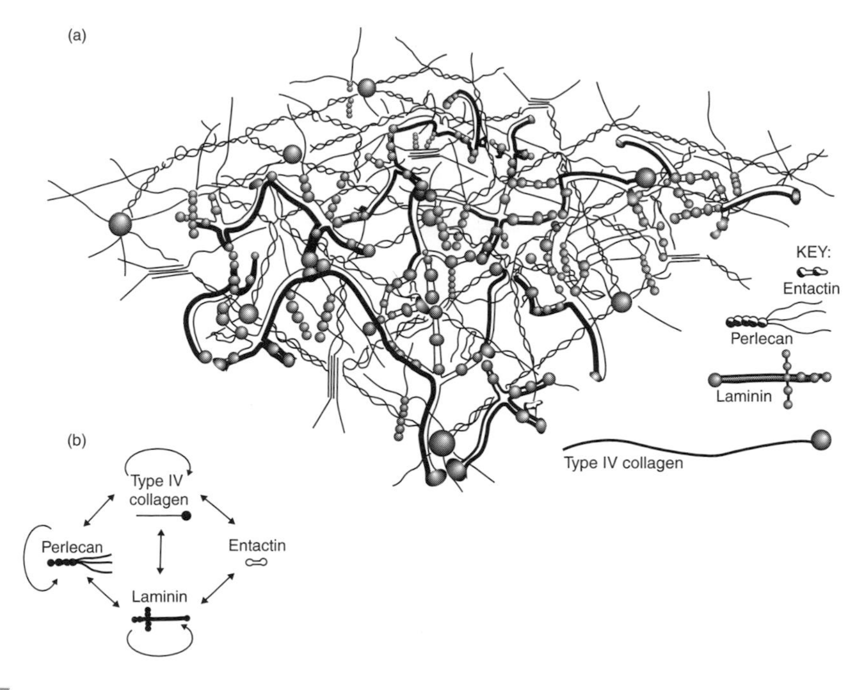

Figure 8.17

Molecular model of the basal lamina. Type IV collagen forms the structural basis and in its meshes the other constituents are anchored (a). The interactions between the different major players are illustrated in (b). © 1994. From Alberts, B. *et al. Molecular Biology of the Cell*, 3rd Edition. Reproduced by permission of Routledge, Inc., part of the Taylor & Francis Group.

(*see 8.4.5*) and type IV collagen (*see 8.2.1*)] on the other hand (*Figure 8.16*). Laminin can contain an unusual type of glycosylation of Glc β-linked to Asn (*see 7.11*).

8.8 Basal lamina

The border between the ECM and epithelial cells is formed by the basal lamina (*Figure 8.17*). Basal laminae are also found around muscle cells and between endothelial cells of blood vessels and the connective tissue surrounding them. The basal lamina between the endothelia of the glomerulus in the kidney and the specialized epithelia of Bowman's capsule (the podocytes) forms an important part of the filtration barrier between blood and urine. Basal laminae form a thin condensed felt-like network of ECM around the cells, which acts not only as an anchoring structure for the epithelia but also presents a diffusion barrier between the cells and the surrounding matrix. They are also important for maintaining cell viability; epithelial cells which loose the contact to the basal lamina undergo a specialized form of programmed cell death called anoikos. Furthermore, basal laminae determine cell polarity, influence the protein composition of the adjacent cell membrane and serve as pathways for cell migration, especially during development. As basal laminae are rich in glycoconjugates they can easily be visualized when tissue sections are stained with the PAS reaction (*see 14.2.1*), which demonstrates the presence of neutral carbohydrates.➲ The main glycoconjugate constituents if the basal laminae are: type IV collagen (*see 8.2.1*), heparan sulfate (*see 8.3.3*), laminin (*see 8.3.2*) and entactin. These molecules are largely synthesized by the epithelia resting on them, but some parts are also contributed by the fibroblasts (especially type IV collagen).

➲ Basal laminae form important tissue borders and are composed of type IV collagen, fibronectin and laminin. These serve as anchoring molecules for epithelial cells.

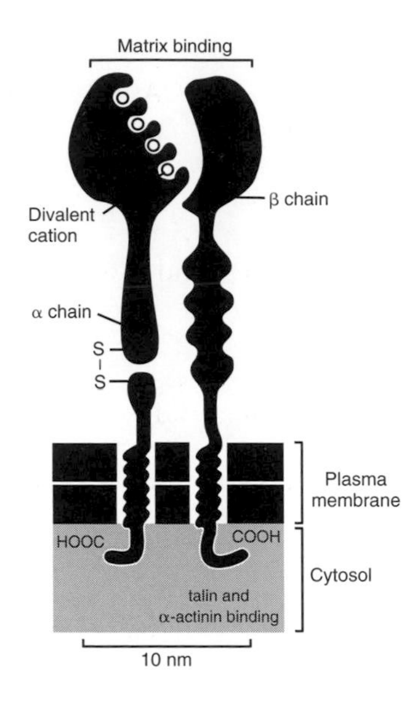

Figure 8.18

Functional integrin receptors are formed by non-covalent association of an α- and a β-subunit. © 1994. From Alberts, B. *et al. Molecular Biology of the Cell*, 3rd Edition. Reproduced by permission of Routledge, Inc., part of the Taylor & Francis Group.

The composition of basal laminae can vary from organ to organ, but type IV collagen can be found in all mature basal laminae. It self-assembles into a felt-like meshwork and can form interactions with all major glycoproteins of the basal lamina. During development, however, the first protein produced in basal laminae is the large glycoprotein laminin (*see 8.7.2*). Functionally and structurally laminin is characterized by the presence of several domains: one binds to heparan sulfate, one to type IV collagen, one to entactin and one to laminin receptors of the cell membrane. The ends of the laminin arms self-assemble, as does type IV collagen; hence both molecules form felt-like sheets. The interaction between laminin and type IV collagen is further strengthened by entactin, which is a short dumbbell-shaped molecule that interacts with both molecules and forms bridges between them.

8.8.1 Receptors for components of the basal lamina – the integrins

The integrins are a large family of glycoproteins, which consist of an α- and a β-subunit, which form non-covalently linked heterodimers (*Figure 8.18*). They act as cellular receptors for basement lamina components fibronectin and laminin, thus attaching cells to the ECM and holding them in place. Integrins function efficiently in this role because they:

- bind to the proteins of the ECM
- span the plasma membrane
- attach to cytoskeletal proteins.

The receptor density of the integrins on cells is much higher than those of other cell membrane receptors, for example, hormone receptors, because the affinity of the integrins to their ligands is much lower than the affinity of hormones to their receptors. Hence many more receptor–ligand interactions are necessary to bind the cells to the matrix. This situation can be compared with the principle by which 'Velcro' operates. This delicate balancing of cells binding to their underlying matrix via weak interactions is necessary to allow for mobility of cells within tissues. In the small intestine, for example, the cells move from the bottom of the crypt to the top of the villi in less than a week. The need to balance mechanical stability with cellular migration therefore has to be finely tuned. As integrins are glycoproteins, the question arises as to whether the glycans have biological function. Several *in vitro* studies using cancer cell lines indicate that the degree of cell–cell adhesion may be influenced by glycosylation. Aberrant glycosylation, as observed in many cancers, may thus downregulate adhesion which contributes to the enhanced mobility of malignant cells. However, the role of integrin glycosylation in cancer progression is not well understood.

Further reading

Bernfield, M., Gotte, M., Park, P.W., Reizes, O., Fitzgerald, M.L., Lincecum, J., Zako, M. (1999) Functions of cell surface heparan sulfate proteoglycans. *Annu. Rev. Biochem.* **68:** 729–777.

Caterson, B., Flannery, C.R., Hughes, C.E., Little, C.B. (2000) Mechanisms involved in cartilage proteoglycan catabolism. *Matrix Biol.* **19:** 333–344.

Fraser, J.R., Laurent, T.C., Laurent, U.B. (1997) Hyaluronan: its nature, distribution, functions and turnover. *J. Intern. Med.* **242:** 27–33.

Funderburgh, J.L. (2000) Keratan sulfate: structure, biosynthesis, and function. *Glycobiology* **10:** 951–958.

Iozzo, R.V. (1998) Matrix proteoglycans: from molecular design to cellular function. *Annu. Rev. Biochem.* **67:** 609–652.

Lander, A.D., Selleck, S.B. (2000) The elusive functions of proteoglycans: *in vivo veritas*. *J. Cell Biol.* **148:** 227–232.

Perrimon, N., Bernfield, M. (2001) Cellular functions of proteoglycans – an overview. *Semin. Cell. Dev. Biol.* **12:** 65–67.

Prydz, K., Dalen, K.T. (2000) Synthesis and sorting of proteoglycans. *J. Cell Sci.* **113:** 193–205.

Scott, J.E. (1995) Extracellular matrix, supramolecular organisation and shape. *J. Anat.* **187:** 259–269.

Glycosphingolipids

9

9.1 Introduction

The term 'glycolipid' describes a glycoconjugate containing one or more monosaccharides bound to a hydrophobic moiety such as a lipid. It is usually used to describe a subgroup of the sphingolipids, the glycosphingolipids (GSL).

GSLs are one of three main types of cell membrane-associated lipids. The others are:

1. *Phospholipids*. These are the most abundant membrane lipids and, as their name suggests, are heavily phosphorylated lipid molecules. A recently described subclass of the phospholipids is the glycophosphatidylinositol (GPI)-anchored proteins (*see Chapter 12*). GPI anchors peripheral proteins within the top layer of the lipid bilayer of the plasma membrane.
2. *Cholesterol*. This is a steroid that fills the spaces between the 'kinks' caused by unsaturated regions in the hydrocarbon tails of neighbouring phospholipid molecules in the plasma membrane, thereby stiffening it.

9.2 Structure of GSL

The lipid part of GSL is sphinganine, previously termed sphingoid. Sphinganine is a long straight-chain aliphatic amino alcohol. The chain length can vary from 14 to 24 carbon atoms with one or more double bonds. Sphinganine is only found in trace quantities *in vivo*, as it is toxic in its unmodified form. In GSL, sphinganine is modified by acetylation to become ceramide (*see Figure 9.1*).

Glycolipids contain two regions, hydrophilic and hydrophobic, and are amphipathic molecules. The hydrophilic region comes from the glycan part of the molecule and the hydrophobic region comes from the lipid. The hydrophilic (polar) part of the molecule prefers an aqueous environment, whereas the hydrophobic (non-polar) region prefers a non-aqueous environment. This structural aspect of glycolipids explains their association with the outer leaflet of cell membranes. The lipid portion is buried in the phospholipid-based, non-aqueous bilayer, and the glycosylated portion extends into the aqueous extracellular space.

Over 200 different GSL have been described and most of the variation in structure of GSL is due to heterogeneity of glycans attached.

If a GSL contains oligosaccharides with sulfate or sialic acid groups it may carry a net negative charge.

The main types of glycolipids and their common names are shown in *Table 9.1*.

9.3 Location of GSL

GSL and cholesterol exist in patches or micro-domains on the surface of the cell. Such micro-domains are proposed to be raft-like structures. One of the functions ascribed to GSL rafts is

Functional and Molecular Glycobiology, Susan A. Brooks, Miriam V. Dwek and Udo Schumacher

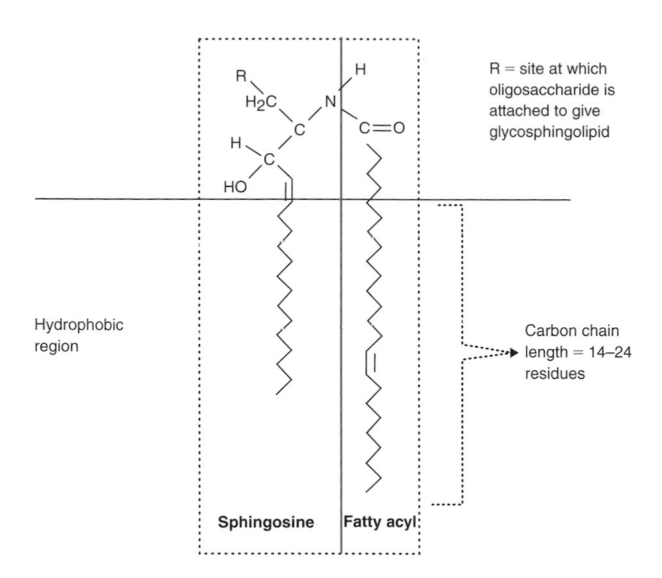

Figure 9.1

Schematic diagram of ceramide. Ceramide is formed by the modification of sphingosine with a fatty acyl group. Variation in the length of the fatty acyl chain can give rise to variability in the ceramide portion of glycolipids. The hydrophobic region orientates the glycolipid in the cell membrane fluid bilayer.

Table 9.1 The two main groups of glycosphingolipids, neutral and acidic

Neutral GSL	glucosyl ceramide galactosyl ceramide
Acidic GSL	gangliosides (sialylated GSL) sulfo glycosphingolipids (sulfated GSL) phospho glycosphingolipids (GSL with phosphate groups)

sorting proteins and lipids into and out of the cell. Much work is still required before the function of GSL rafts become fully understood but it is widely believed that regions of GSL rafts are also specialized areas for signal transduction and receptor-mediated endocytosis. This idea is supported by the observation that GSL rafts are often found on the lumenal (or apical) region of epithelial cell membranes.

GSLs are found in complexes or in micelles with other lipids in body fluids and, intriguingly, incorporated in the membranes of cells other than those in which they were produced. The mechanism by which GSL might be shed from one cell and be taken up by another is not understood, and the significance of this observation remains an area of investigation. GSLs are also located intracellularly following internalization from the plasma membrane.

Most intracellular GSL undergoes degradation in the lysosome but a small amount has been noted in the membranes of intracellular organelles, for example, in the endosome.

9.4 Biosynthesis of GSL

The biosynthesis of GSL starts with the precursor ceramide. This is produced from palmitoyl-CoA and the reader is referred to standard texts for the biosynthesis of this compound.

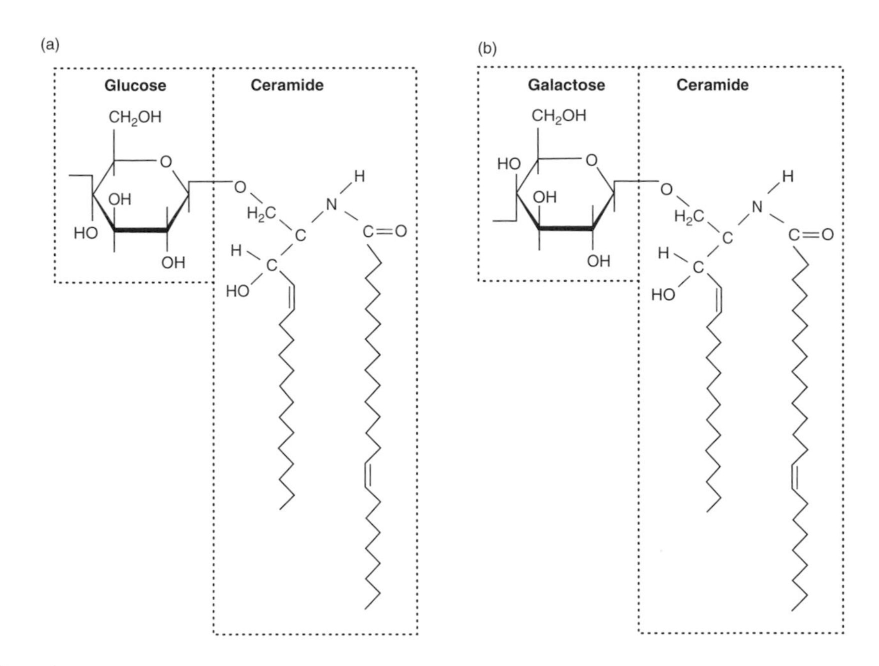

Figure 9.2

In humans, the first monosaccharide found attached to ceramide is either glucose or galactose. The structures that are formed are (a) glucosyl ceramide or (b) galactosyl ceramide.

Ceramide is modified by the addition of either Glc or Gal to yield either glucosyl ceramide (also termed glucocerebroside) or galactosyl ceramide (also termed galactocerebroside), illustrated in *Figure 9.2*. The process requires nucleotide-activated glucose or galactose (UDP-Glc and UDP-Gal) and is mediated by the action of a ceramide-specific glucosyltransferase or ceramide-specific galactosyltransferase. These enzymes are thought to be unique to the GSL biosynthetic pathway as the substrate on which they act is very hydrophobic compared with proteins. This, and subsequent steps in GSL biosynthesis, are illustrated in *Figure 9.3*. There are two biosynthetic pathways, one for the formation of the glucosyl ceramide group of GSL (*Figure 9.3a*), and one for the formation of the galactosyl ceramide group of GSL (*Figure 9.3b*).

The production of glucosyl ceramide starts on the cytoplasmic face of the outer leaflet of the endoplasmic reticulum (ER) membrane. The enzyme glucosyl ceramide synthase adds Glc to ceramide, and by a poorly understood 'flip-flop' mechanism the glycolipid is transferred to the lumenal face of the ER membrane for further processing. This 'flip-flop' transition is probably mediated by a membrane-bound enzyme, termed a 'flippase', which sorts the lipids into the inner and outer leaflet of the lipid bilayer. In contrast, galactosyl ceramide production is initiated on the lumenal face of the ER membrane by the enzyme galactosyl ceramide.

After glucosyl ceramide has been produced Gal is invariably added to give rise to Gal(β1→4)Glc. This disaccharide is also known as lactose and the GSL formed is therefore called lactosylceramide. Lactosylceramide is a precursor for two groups of GSL, the gangliosides and globosides (*Figure 9.3a*).

Glucosyl and galactosyl ceramide are extended to form several characteristic GSL core or root structures, shown schematically in *Figure 9.3*. The glycosyltransferases involved in the construction of these structures are indicated. The core structures of GSL are shown in *Table 9.2* and further illustrated in *Figure 9.4*.

The core structures are extended with oligosaccharide arms to give a large and structurally diverse group of glycolipids. Glycans added to the glycolipid core differ very subtly. An example

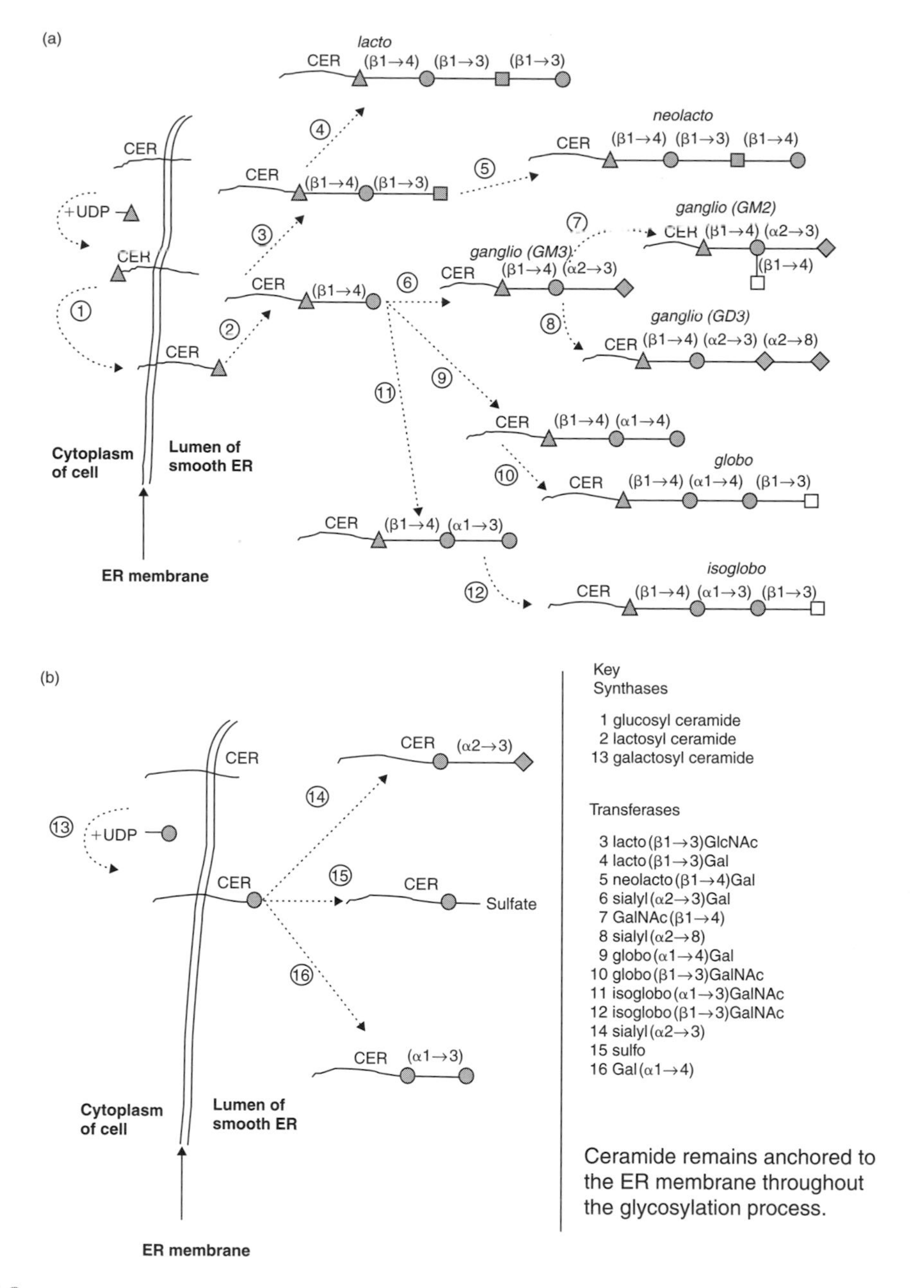

Figure 9.3

Glycosphingolipid biosynthesis involves a complex array of synthases and transferases and activated donor monosaccharides. There are two pathways, one for the formation of the (a) glucosyl ceramide group and the other for the formation of the (b) galactosyl ceramide group. The greatest diversity in structures is found in the glucosyl ceramide group (a: steps 1–12), whereas the galactosyl ceramide group is associated with fewer glycan structures (b: steps 13–16). The formation of glucosyl ceramide commences on the cytoplasmic face of the ER membrane and it is then flipped into the lumen of the ER for further processing. Galactosyl ceramide synthesis commences in the lumenal face of the ER membrane. Ceramide remains anchored to the ER membrane throughout the biosynthesis of the oligosaccharides.

Table 9.2 The core/root structures of the neutral glycosphingolipids

Type	Core structure
Lacto-	Gal(β1→3)GlcNAc(β1→3)Gal(β1→4)Glc(β1→Cer)
Neolacto-	Gal(β1→4)GlcNAc(β1→3)Gal(β1→4)Glc(β1→Cer)
Ganglio-	Gal(β1→3)GalNAc(β1→4)Gal(β1→4)Glc(β1→Cer)
Globo-	GalNAc(β1→3)Gal(α1→4)Gal(β1→4)Glc(β1→Cer)
Isoglobo-	GalNAc(β1→3)Gal(α1→3)Gal(β1→4)Glc(β1→Cer)
Galactosyl-	Gal(α1→4)Gal→Cer
Sulfatidyl-	Sulfo (3→O)Gal→Cer
Muco-	Gal(β1→3)Gal(β1→4)Gal(β1→4)Glc(β1→Cer)
Gala-	GalNAc(α1→3)GlcNAc(β1→3)Gal(α1→4)Gal(α1→Cer)

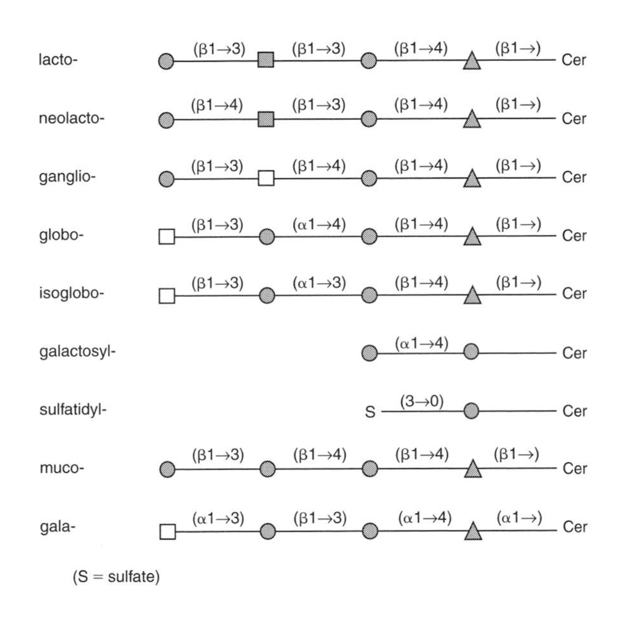

Figure 9.4

Core/root structures of glycosphingolipids. The lacto-, neolacto-, ganglio-, globo-, isoglobo- and muco-core structures all contain Gal(β1→4)Glc(β1→)Cer as they share a common biosynthetic route, as shown in *Figure 9.3*.

of this is the human ABO blood group carbohydrate antigens (*see 10.5*) that may be added to the GSL (*Figure 9.5*). Other types of glycans can adorn the GSL core including the Lewis antigens, for example, sLe^x (*Figure 9.6*; *see 10.6*).

The extension of the GSL core with outer arm oligosaccharide branches dramatically increases the complexity of this group of compounds. Some of the outer arm branched oligosaccharides of GSL are identical to those found on N-linked oligosaccharides (*see Chapter 4*). An example is the neolacto series [GlcNAc(β1→4)Gal]$_n$, which is also commonly found on the outer arm of complex N-linked oligosaccharides of glycoproteins. It is likely that the same glycosyltransferases (*see Chapter 2*) are used for the intracellular construction of glycan linkages common to both glycoproteins and glycolipids. Conversely, some of the oligosaccharides of glycolipids have not been found on N-linked or O-linked glycoproteins. For example, Gal(α2→3)NeuAc(α2→8)NeuAc is found in the ganglio series of GSL and is unique to these glycolipids. Another common substitution of the outer arm oligosaccharides of GSL

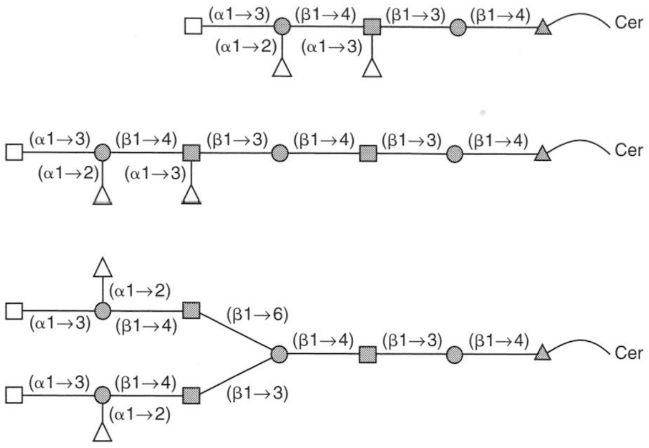

Figure 9.5

The oligosaccharides attached to glycolipids show a vast degree of heterogeneity. Three variants of blood group A type antigens found attached to the glucosyl ceramide core are illustrated.

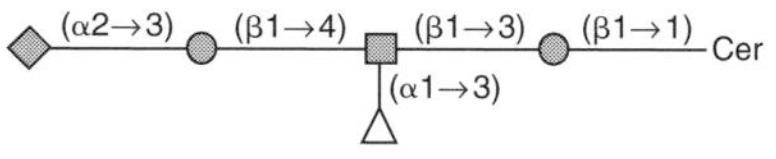

Figure 9.6

The structure of sialyl Lewis x (sLe^{x}). sLe^{x} expression on glycolipids is regulated during development. The aberrant expression of sLe^{x} in many tumours has led to it being described as an oncofetal antigen.

is O-acetylated sialic acid and O-sulfate. These substitutions are unique to GSL and give rise to negatively charged structures.

9.5 Nomenclature

The diversity of oligosaccharides attached to GSL leads to problems with nomenclature. GSLs were originally named trivially as the attached oligosaccharides were identified, and according to the site from which they originated. For example, gangliosides were originally isolated from brain and nerve tissues (*ganglio* means accumulation of nerve cells) and the lacto series of GSL were originally isolated from white blood cells.

The International Union of Pure and Applied Chemistry (IUPAC) guidelines for the nomenclature of glycolipids now exist and, although somewhat cumbersome to follow, describe a logical and comprehensive naming system for this group of molecules.

Under the IUPAC guidelines the GSLs are described first according to the trivial name used for the core (or root) structure of the compound and then by the number of monosaccharides attached. For example, the GSL sometimes referred to by the trivial name lactosylceramide (LacCer) should properly be termed diosylceramide: that is, a ceramide containing two monosaccharides, whose disaccharide structure is Gal(β1→4)Glc(β1→1)Cer.

The complexity of GSLs and the nomenclature for their description means that abbreviated terms, particularly for the brain gangliosides, such as those illustrated in *Figure 9.7*, are in widespread use.

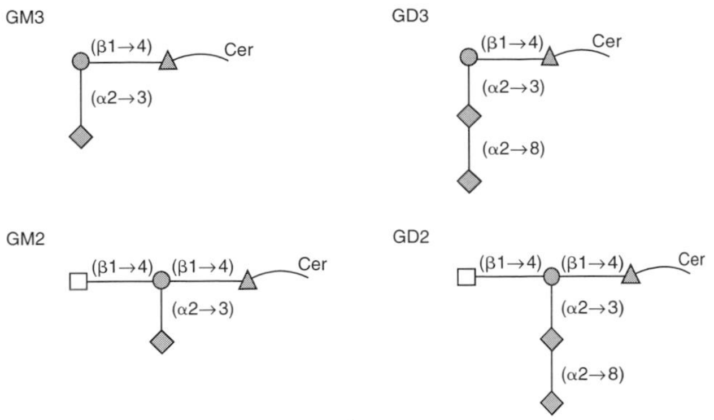

Figure 9.7

The structures of the most common ganglioside glycolipids attached to glucosyl ceramide. The letter 'G' denotes 'ganglioside', whereas 'M' is substituted to illustrate that the structure is 'monosialylated' glycans and 'D' is substituted for 'disialylated' glycans.

9.6 Degradation

GSL shed from the cell membrane into the extracellular space is reincorporated into the cell by internalization within endocytic vesicles formed from invagination of the cell membrane. The endocytic vesicles are transported to the lysosomes and degraded by lysosomal enzymes.

9.7 The distribution of GSL

Understanding of the function and distribution of GSL has been aided by:

- monoclonal antibody technology enabling cell and tissue distribution of lipids to be evaluated
- physicochemical studies of glycolipid and glycosyltransferase expression *in vivo* and *in vitro*
- studies of individuals with glycolipid storage diseases exhibiting an in-born error in the metabolism of glycolipids (*see 16.9*)
- production of knockout mice lacking genes encoding enzymes responsible for the construction of glycolipids.

GSL synthesis is regulated in a cell-specific, tissue-specific, developmental-specific and species-specific manner. ➲

➲ GSL are synthesized in a cell-, tissue-, developmental- and species-specific manner.

9.7.1 Developmental regulation of distribution of GSL

GSL synthesis is developmentally controlled. For example, the trisaccharide Lewis x (Le^x) is linked to glucosyl ceramide and often substituted with sialic acid to give sLe^x. sLe^x is synthesized in a developmentally regulated manner (*see 10.6*). In embryonic mice it is present predominantly at the time of formation of the morula, the solid mass of blastomeres formed after cleavage of a zygote. sLe^x has also been found to be an oncofetal antigen as it is present in fetal tissue and some cancers (particularly adenocarcinoma of the lung) but not in normal tissues. sLe^x is also synthesized by leukocytes and is a ligand for E-selectin on activated endothelium (*see 13.12.3.2*).

Other examples of the developmental regulation of oligosaccharides on GSL include blood group antigens and polylactosamine structures. Blood group carbohydrate antigens are synthesized in a controlled manner during development. A polymeric form of N-acetyllactosamine

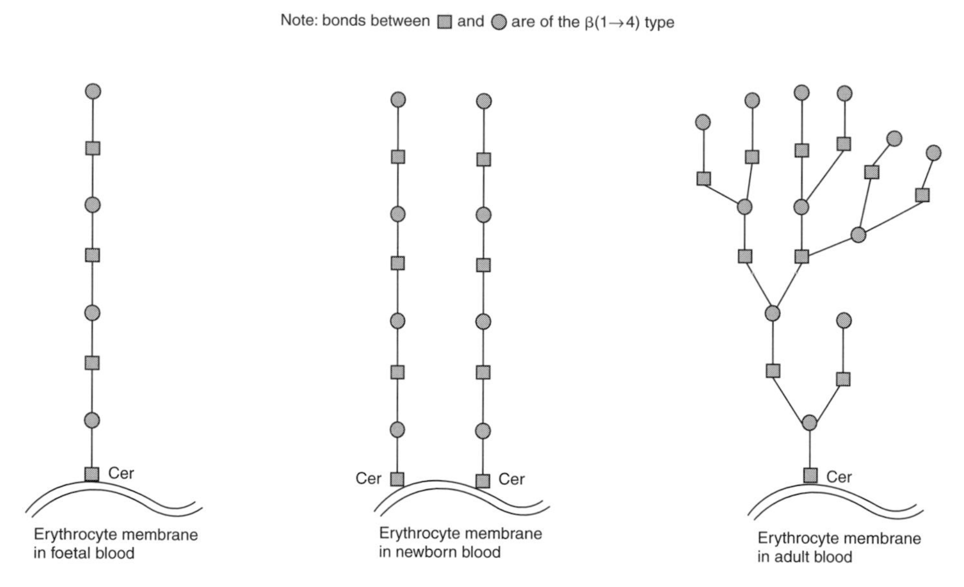

Figure 9.8

Developmental expression of polylactosamine oligosaccharide chains on GSL of the erythrocyte membrane. Note: the relative simplicity in the polylactosamine oligosaccharide chain structure in the fetal erythrocyte membrane and increasing complexity in adulthood.

[repeating GlcNAc(β1→4)Gal units, also known as polylactosamine] is variably synthesized on GSL and glycoproteins during embryogenesis and in development through to adulthood. An example of the differential synthesis of polylactosamine containing oligosaccharides on erythrocyte membranes in fetal, newborn and adult blood is illustrated in *Figure 9.8*.

9.7.2 Cell- and tissue-specific distribution of GSL

9.7.2.1 Galactosyl ceramides

➲ Studies with knockout mice have shown that some GSL (e.g. GalCer) are essential for normal development, whereas others (e.g. GM2 and GD2) are not.

One of the best studied GSL regulated in a tissue-specific manner is the galactosyl ceramides (GalCer). ➲ Galactosyl ceramides are abundant constituents of myelin, and account for ≈20% of its dry weight. The importance of GalCer, and its *in vivo* derivatives, has become better understood following studies of knockout mice lacking the galactosyl ceramide synthase gene. Mice without galactosyl ceramide synthase are live born but most die within 3 months post-parturition. Death is thought to be as a result of gross neurological problems. At 2 weeks, the mice show pronounced tremor even during rest, and as time passes paralysis develops until by 2 months of age most mice are unable to move. The mice have an abnormal distribution of oligodendroglial processes extending along the axon of their neurons in contrast to wild-type mice. The biological effect of knocking-out the galactosyl ceramide synthase gene is that the mice produce myelin containing GlcCer in place of the usual GalCer. The conclusion drawn from these studies is that the GalCer and its derivatives are important for the stability and protective function of myelin.

9.7.2.2 Glucosyl ceramides

Studies of knockout mice lacking GalNAc transferase, GM2 and GD2 synthase, responsible for synthesis of GM2 and GD2 GlcCer-based gangliosides, showed that very low levels of these

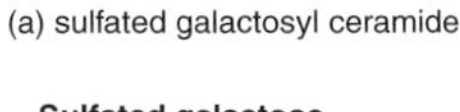
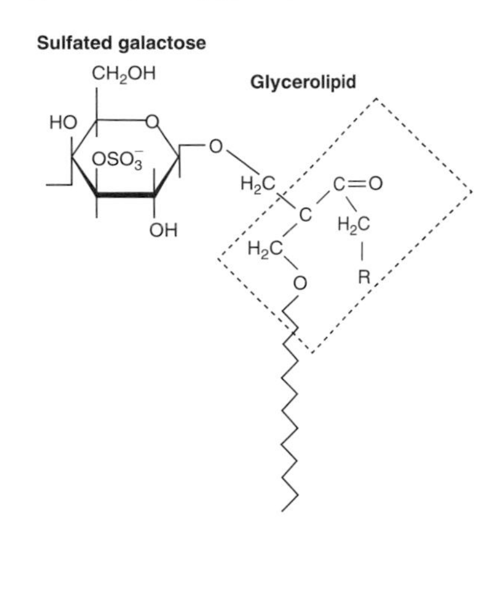

Figure 9.9

The chemical structure of two sulfated glycolipids: (a) sulfated galactosyl ceramide and (b) sulfated galactosyl glycerolipid. The dashed boxes highlight the differences in chemical structure that lies in the lipid portion of these two molecules.

gangliosides are important for the development of the nervous system. This was a surprise finding, given the large amounts of both GM2 and GD2 normally found in the brain and other parts of the nervous system. Detailed histopathological study of the knockout mice brain and nervous system showed GM3 and GD3 to be present in elevated levels suggesting these structures had compensated for the reduced levels of GM2 and GD2. These data showed that GM2 and GD2 are not essential for brain development and that different ganglioside types are, to some extent, interchangeable.

9.7.2.3 Sulfoglycolipids

Sulfated glycolipids are found on GalCer-linked lipids and on other glycolipids, such as glycerolipids. In the case of the former, sulfo-GalCer is detected in the myelin sheath and epithelial cells, as well as in the germ cells of some mammals and lower vertebrates. The closely related structure, sulfogalactosyl glycerolipid, illustrated in *Figure 9.9*, is found in the white matter of the brain, and more recently in male mammalian germ cells where it is mainly localized in the region of the sperm head.

9.7.3 Species-specific distribution

GSL synthesis is conserved throughout the animal kingdom, although the relative distribution of the GSL varies according to the species and tissue studied. For example, the helminth of the genus *Schistosoma* which causes parasitic disease schistosomiasis in humans, has a GSL carrying GalNAc(β1→4)Glc(β1→1)Cer, rather than Gal(β1→4)Glc(β1→1)Cer commonly found in mammalian glycolipids. The eggs of *Schistosoma mansoni* and *Schistosoma japonicum* synthesize an unusual highly fucosylated GalNAc(β1→4)GlcNAc repeating motif on GlcCer, illustrated in *Figure 9.10*.

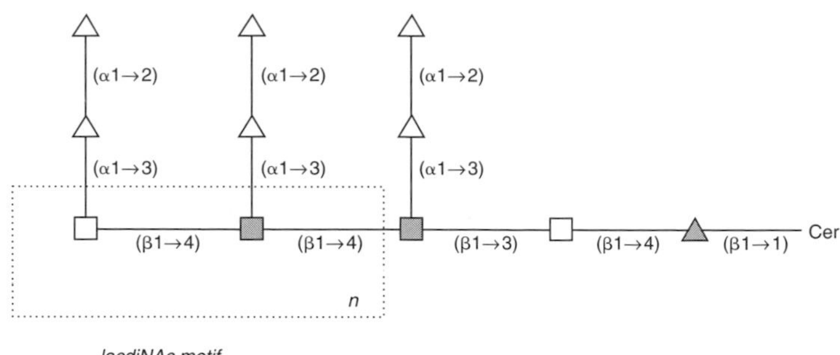

Figure 9.10

GSL of eggs of *Schistosoma mansoni* show some similarities with GSL of eukaryotes although dissimilarities are also seen. For example, GalNAc(□)(β1→4)Glc(▲)(β1→1)Cer compared with Gal(■)(β1→4)Glc (▲)(β1→1)Cer in eukaryotes.

Table 9.3 Some additional, proposed, functions of glycosphingolipids

Proposed function	Supporting data
Signal transduction	GSL form rafts. Rafts associate with signal transduction molecules
Growth factor like	Gangliosides may induce neuron growth
Cell adhesion molecule ligands	Le^x expression in mouse embryo during development

9.8 Function of GSL

Many functions have been postulated for GSL although much remains to be understood. Some functions are described below and others are summarized in *Table 9.3*.

9.8.1 Physical protection

GlcCer are important for the structural integrity of cell membranes and for maintaining the skin permeability barrier. GSL, by virtue of their long acyl chains, tend to cluster together in micro-domains or rafts (*see 9.3*). Such micro-domains are probably important in protection, particularly as they tend to be found at the apex of epithelial cells facing into the lumen of organs such as the gut and other hostile environments.

9.8.2 Signal transduction and growth-factor like function

The recognition of GSL by receptors on other cells suggests they are mediators in cell signalling pathways.

Examples include:

- downregulation of tyrosine phosphorylation of the epidermal growth factor (EGF) receptor by ganglioside GM3
- the high levels of GalCer in myelin and its possible growth factor-like properties and function as neurogenic factors.

9.9 GSL receptors for pathogens

The carbohydrate chains of cell-surface GSL may be specifically recognized as ligands for bacterial adhesins facilitating adhesion and colonisation (*see 13.9*).

Some examples of GSL–pathogen associations include:

- the bacterial toxin from *Vibrio cholerae*, the microorganism that causes cholera, recognizes and binds to ganglioside GM1 allowing the pathogen to adhere to the host cell prior to invasion
- verotoxins, a family of *Escherichia coli* toxins, recognizes and bind the terminal disaccharide of globotriasylceramide
- *Helicobacter pylori*, a microorganism associated with the pathology of gastric ulcers, has adhesins that recognize sialylated polyglycosylceramides.

As with many other aspects of glycobiology, it is expected that as understanding of these complex molecules expands so their functions will be further clarified.

Further reading

Hakomori, S. (1999) Antigen structure and genetic basis of histo-blood groups A, B and O: their changes associated with human cancer. *Biochim. Biophys. Acta* **1473:** 247–266.

Hooper, N.M. (1998) Membrane biology: do glycolipid micro domains really exist? *Curr. Biol.* **8:** R114–R116.

International Union of Applied and Pure Chemistry, available at http://www.chem.qmw.ac.uk/iupac

Lloyd, K.O., Furuwaka, K. (1998) Biosynthesis and function of gangliosides: recent advances. *Glycoconj. J.* **15:** 627–636.

Glycan chain extension and some common and important glycan structures

10

10.1 Introduction

Different types of glycans, including N-linked (*see Chapter 4*) and O-linked (*see Chapter 5*) oligosaccharides and glycolipids (*see Chapter 9*), have been described in previous chapters of this book. They are each characterized by unique core glycans and are synthesized in different ways. These core glycans are often elongated and elaborated, as described in this chapter, which also presents details of some common and biologically important glycan structures.

In contrast to chain precursor synthesis, oligosaccharide chain extension and attachment of terminal glycan structures is highly variable, and is regulated in a cell-lineage and/or tissue-specific manner, and hence may alter with development, differentiation and disease. It is a function of careful regulation of expression and activity of glycosyltransferases (*see Chapter 2*). The alterations are of functional importance, and also make the work of the analytical chemist more challenging.

10.2 Type 1 to 6 chains

The structure of type 1 to 6 chains is illustrated in *Figure 10.1*.

10.2.1 Type 1 chains

The core structures of N-linked (*see Chapter 4*) and O-linked (*see Chapter 5*) glycans and glycolipids (*see Chapter 9*) typically bear terminal GlcNAc residues. Modification by the addition of a (β1→3)-linked Gal to the terminal GlcNAc, by the action of a β1,3 galactosyltransferase,

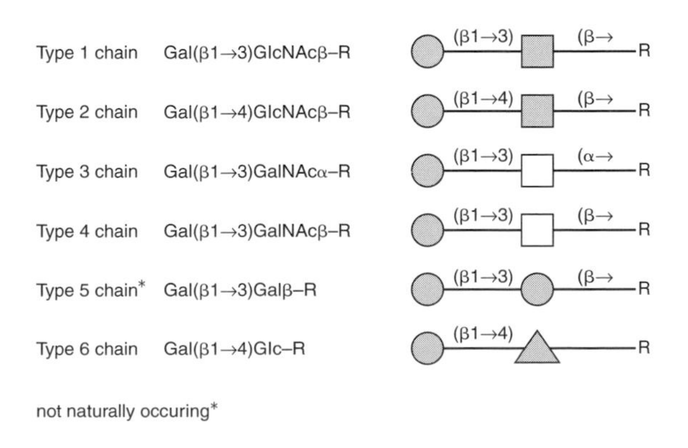

Figure 10.1

The structure of types 1 to 6 chains.

Functional and Molecular Glycobiology, Susan A. Brooks, Miriam V. Dwek and Udo Schumacher

yields the structure Gal(β1→3)GlcNAc, the 'type 1 chain', also sometimes referred to as a 'neolactosamine unit'. It is mostly confined to secretory epithelia of, for example, the respiratory, gut and genitourinary tracts.

10.2.2 Type 2 chains

Type 2 chains are the most common of the six chain types. Core extension and modification most often begins by the addition of a (β1→4)-linked Gal to the terminal GlcNAc residue to yield a 'type 2 chain'. Gal(β1→4)GlcNAc is termed a 'lactosamine unit' and defines a 'type 2 chain'. This type of glycosylation is constitutive, and occurs in all mammalian tissues. It is performed by members of a family of β1,4 galactosyltransferases (*see 2.11*). The lactosamine unit can act as an acceptor substrate for many modifications to both the terminal Gal and the now sub-terminal GlcNAc, to yield many diverse linear or branching structures.

10.2.3 Type 3 chains

The type 3 chain structure, Gal(β1→3)GalNAcα-R, is a simple elaboration of the core GalNAcα-Ser/Thr of O-linked glycans (*see 5.6.1*).

10.2.4 Type 4 chains

The type 4 chain, Gal(β1→3)GalNAcβ-R has been described in glycosphingolipids and is found as an extension of the globoside molecule (*see 9.4*), GalNAc(β1→3)Gal(α1→4)Gal (β1→4)Glc-Cer in which the terminal β-linked GalNAc is substituted with a (β1→3)-linked Gal residue.

10.2.5 Type 5 and 6 chains

Type 5 chain, Gal(β1→3)Galβ-R, does not occur in nature, but is a chemically synthesized disaccharide that functions *in vitro* as an acceptor substrate for the *H* gene encoded α1,2 Fuc-T (*see 2.7.1*). Type 6 chain, which is not commonly found, is Gal(β1→4)Glc-R.

10.3 Polylactosamines

➲ Polylactosamines are chains of repeating lactosamine units Gal(β1→4)GlcNAc (β1→3)Gal.

Glycolipids and glycoproteins, frequently have N- and O-linked glycans containing linear polymers of the type 2 chain lactosamine unit, termed 'polylactosamines'.➲ These are synthesized by the action of one or more β1,4 galactosyltransferases, and one or more β1,3 GlcNAc transferases. O-linked polylactosamine chains tend to be shorter than N-linked ones, and both classes of glycans are often synthesized on (β1→6)-branching structures. Multi-antennary N-linked structures, synthesized by the action of GlcNAc-TV are particularly susceptible to this type of modification.

Polylactosamine chains are often terminated by sialylation or fucosylation. Their length and their hydrophilicity mean that they protrude stiffly some distance from the cell surface. They can protrude different distances from the cell surface depending upon their length. They are thus ideally placed to present terminal glycan structures for interaction with lectin-like receptors, for example, selectins (*see 13.12.3.2*), expressed by other cells. Polylactosamines themselves are recognized by mammalian S-type lectins (*see 13.4.2*).

10.4 Histo-blood group sugars

Type 1 and 2 chain structures (*see 10.2*) may be modified and extended in many ways. One common way is by adding sugars typical of blood group determinants. Several blood group sugar families exist, and they are described below. As 'blood group' sugars are not found exclusively on red blood cells, but are also found on cells from various tissues, and are often present in body fluids and secretions, they are more properly called 'histo-blood group' sugars. ➲

Some human blood group determinants are proteins, whereas others are carbohydrates. Protein blood group determinants include the MNSs, Gerbich, Rhesus, Kell and Duffy determinants, which are beyond the scope of this book, whereas carbohydrate determinants, including the ABO, Lewis, Ii, P-related, T and Tn structures, are described in detail here.

➲ 'Blood group' sugars are found on erythrocytes and cells from other tissues, and are more correctly termed 'histo-blood group' sugars.

The *ABO*, *Hh*, *SEse*, *Ii* and *Lele* loci are genetically independent of each other, but the saccharide structures arising from the action of the glycosyltransferases for which they code are closely related, as will become clear.

At its most simple, a carbohydrate blood group determinant may be a single monosaccharide, linked in an α- or β-glycosidic linkage to a sub-terminal core structure. This may be part of a linear or branched chain, and can be attached to either a polypeptide via a GlcNAc-β-N-asparagine (N-linkage; *see 4.2*), a GalNAc-O-Ser/Thr (O-linkage; *see 5.6*) or can be attached to Glc-β-ceramide (as in a glycolipid; *see 9.4*). Most blood group determinants are part of glycoproteins or proteins with GPI anchors (*see Chapter 12*) found on the erythrocyte membrane.

10.5 The ABO blood group system

Landsteiner first described the ABO blood grouping system in 1900, and its discovery revolutionized blood transfusion in that it became possible to transfuse blood between individuals without the complications of blood group incompatibility. Briefly, the reason for this is that individuals can be separated into blood groups A, B, AB and O, and their sera contain naturally occurring antibodies to the other blood group antigens: blood group A individuals have anti-B antibodies, blood group B individuals have anti-A antibodies, blood group O individuals have anti-A and anti-B antibodies, and AB individuals produce antibodies to neither. Thus, blood can be safely transfused between individuals of the same ABO blood group (e.g. blood from a group B donor into a group B recipient). Furthermore, as group AB individuals do not produce antibodies to other ABO blood group antigens, they can act as 'universal recipients' and safely receive blood transfusions from donors of any ABO blood type. As neither group A, B, AB or O individuals produce anti-blood group O antigen antibodies, blood group O individuals are 'universal donors' and their blood may be safely transfused into recipients of any ABO blood group.

These naturally occurring antibodies to the A and B blood group antigens are of the IgM class and are produced in response to saccharides on intestinal bacteria and perhaps ingested food recognized by the immune system as 'foreign'. Blood group glycotopes are the cause of the severe problems when blood of an incompatible type is transfused; the recipient recognizes the transfused blood cells as carrying a 'foreign' antigen and their immune system therefore seeks to destroy the transfused cells.

10.5.1 The genetic basis of ABO blood groups

ABO blood group is determined by the allelic genes *A*, *B* (which are co-dominant) and a null or inactive allele *O*. They are inherited according to classical Mendelian principles, and six genotypes are therefore possible: *AA*, *AB*, *AO*, *BB*, *BO* and *OO*, which afford four phenotypes: A, B, AB and O.

The synthesis of ABO blood group sugars is dependent on the activity of genes at three separate loci, the *ABO*, *H* and secretor (*SE*) loci (*see 2.7.1, 2.11.1*). The *A*, *B* and *O* alleles are

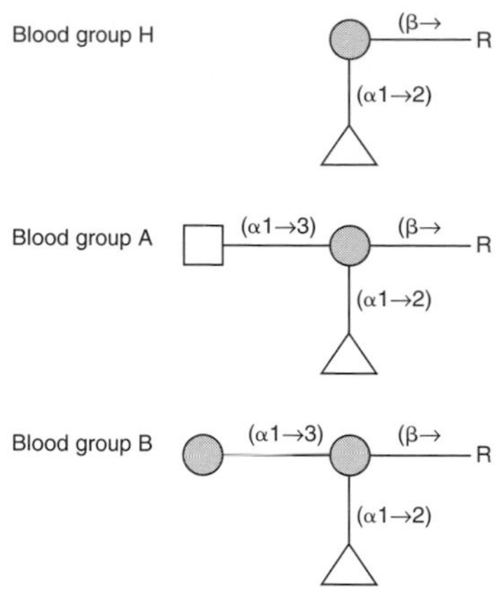

Blood group H (β→ R (α1→2)

Blood group A (α1→3) (β→ R (α1→2)

Blood group B (α1→3) (β→ R (α1→2)

Figure 10.2

The histo-blood group O/H, A and B determinants.

located on human chromosome 9 at q34.2. The H blood group determinant is coded for by a gene at a completely separate locus to that occupied by the A and B blood group determinant genes, on the long arm of chromosome 19. The *SE* locus lies close to it. The blood group A, B and H determinants are illustrated in *Figure 10.2*.

10.5.2 Synthesis of the H determinant

The H determinant is a disaccharide Fuc(α1→2)Galβ-. It is synthesized on various core chain structures by the action of a α1,2 fucosyltransferase (α1,2 Fuc-T). The reaction may be summarized as:

GDP-Fuc + Galβ-R → Fuc(α1→2)Galβ→R + GDP

10.5.3 Synthesis of A and B determinants

In blood group A, B or AB individuals, the blood group H determinant then acts as the acceptor substrate upon which the A and B determinants are synthesized. ➲ The O allele is a null allele that encodes a functionally inert polypeptide that does not further modify the H precursor. About 45% of Caucasians have a null allele at the *ABO* locus (the proportion varies in different races) and are thus blood group O. Blood group A determinant is (α1→3)GalNAc linked to the H determinant disaccharide Fuc(α1→2)Galβ-, and is synthesized using the H determinant as an acceptor substrate through the action of a blood group A α1,3 GalNAc transferase (α1,3 GalNAc-T) as follows:

➲ In group A or B individuals, the blood group H determinant is modified by the addition of a (α1→3)-linked GalNAc (blood group A) or (α1→3)-linked Gal (blood group B). In O individuals, the H determinant is unmodified.

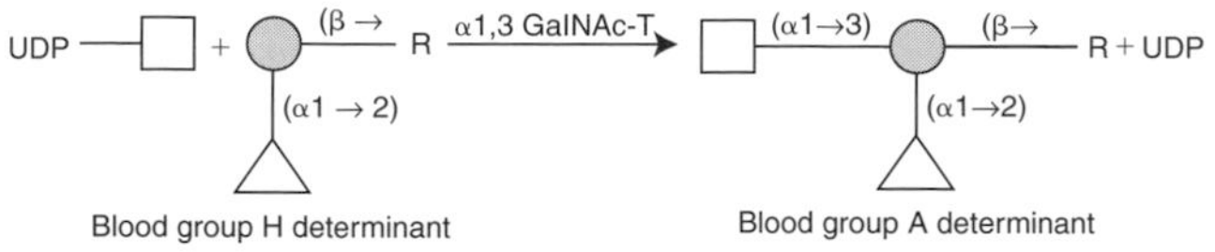

UDP + (β → R (α1 → 2) α1,3 GalNAc-T (α1→3) (β→ R + UDP (α1→2)

Blood group H determinant Blood group A determinant

Blood group B determinant is Gal (α1→3)-linked to the H disaccharide, and is synthesized using the H determinant as an acceptor substrate through the action of the blood group B α1,3 galactosyltransferase (α1,3 Gal-T). The reaction is as follows:

UDP — + (β→ R (α1→2) —α1,3 Gal-T→ (α1→3) (β→ R + UDP (α1→2)

Blood group H determinant

Blood group B determinant

The synthesis of H, A and B structures on a type 2 (*see 10.2*) is illustrated in *Figure 10.3*.

10.5.4 The A and B transferases

The genes for the A and B blood group transferases differ in only four single bases which result in enzymes with four amino acid substitutions at residues 176, 235, 266 and 268 in the polypeptide chain. ➲ Residue 266 which codes for leucine in A individuals and methionine in B individuals, and 268 which codes for glycine in A individuals and alanine in B individuals, are both conserved in primates, whereas residues 176 and 235 are not. This implies that residues 266 and 268 are crucial for the difference in sugar specificity between the A-transferase and B-transferase, whereas residues 176 and 235 are not. The difference between the two types of transferase is thus extremely

> ➲ The genes coding for the blood group A and B transferases are closely related and differ in only four amino acids.

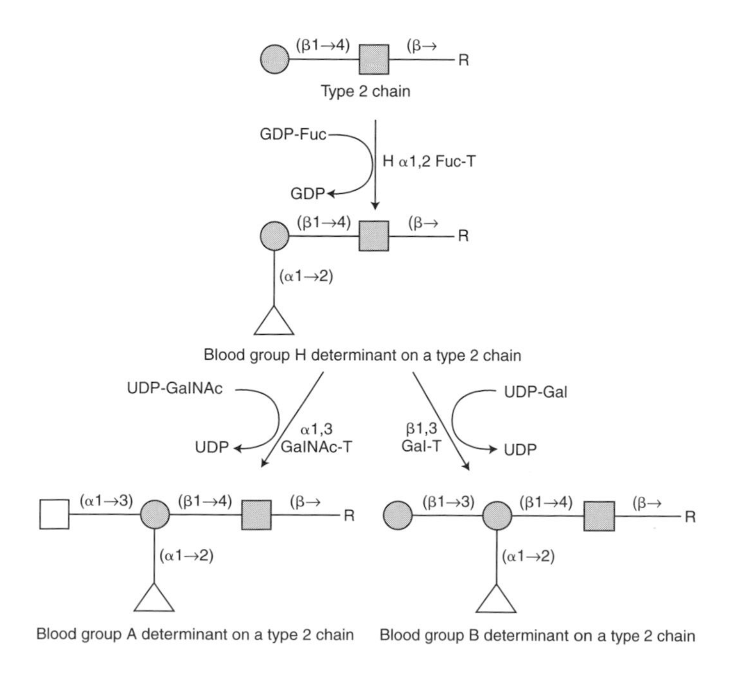

Figure 10.3

The synthesis of the H, A and B blood group determinants on a type 2 chain.

subtle, and indeed the B transferase can utilize UDP-GalNAc as a donor, and the A transferase, although preferring UDP-GalNAc as a donor, is inhibited by UDP-Gal.

10.5.5 Blood group A subtypes

There are two major subtypes of A individuals, A_1 and A_2, and as many as nine different, and rarely occurring, subtypes (A_{int}, A_3, A_x, A_m, A_{end}, A_{el}, A_{bantu}, A_{lae}, A_{fin}) in which the A determinant may be coded for by slightly modified glycosyltransferase enzymes with differential affinities for nucleotide sugar substrates. A_1 and A_2 determinants are coded for by different GalNAc-Ts with different pI, pH optima and other requirements. The genes differ by a single base substitution and a single base deletion in the coding sequence of the last exon. The deletion of nucleotide 1060 near the 3′-end of the coding region results in an additional 21 amino acids added to the C-terminus of the A_2 transferase. The A_1 GalNAc-T appears to be more efficient at transferring GalNAc onto the blood group H disaccharide, and A_1 erythrocytes therefore carry more blood group A determinant than do A_2 erythrocytes. There may also be qualitative differences in the amount of A determinant transferred onto the different determinants on glycoproteins and glycolipids in individuals of different A subtypes reflecting the slightly different substrate preference of the transferases. The relatively weaker activity of the A_2 transferase may be a result of steric hindrance owing to its elongated catalytic domain. The blood group O gene is structurally identical to the A and B genes, but contains a base pair deletion causing a shift in the reading frame which leads to the production of an entirely inactive protein lacking transferase activity. Thus the H determinant remains unmodified in group O individuals.

10.5.6 Synthesis of A, B and O(H) blood group sugars

A, B and O(H) blood group sugars are synthesized using type 1, 2, 3 and 4 chains (*see 10.2*). Most erythrocyte ABO sugars are synthesized on type 2 chains, whereas in most other tissues, for example, gastrointestinal epithelia, they are synthesized predominantly on type 1 and type 2 chains.

Blood group sugars are found in abundance on the surface of erythrocytes. About 75–80% of the ABO sugars present on these cells are associated with the anion transport protein, 'Band 3', which displays ≈1–2 million ABO determinants per erythrocyte. The erythrocyte glucose transport protein, 'Band 4.5', displays about 5×10^5 ABO determinants per erythrocyte.➲ In these two integral membrane glycoproteins ABO determinants are displayed on the multiple terminal branches of a single N-linked polylactosamine. They are also found on other red cell glycoproteins and glycolipids. Furthermore, they are found on the endothelial cells lining the blood vessels and on a variety of epithelial cells where they are displayed by integral membrane glycoproteins and glycolipids.

➲ On erythrocytes, the glycoproteins Band 3 and Band 4.5 carry large numbers of oligosaccharide chains bearing blood group determinants.

10.5.7 Secretors

Body fluids and secretions such as ovarian and breast cyst fluids, saliva, seminal fluid, gastric mucins, urine and milk can also contain ABO sugars, but only in individuals designated 'secretors'.➲

About 80% of the population are secretors. Whether or not a person is a secretor is determined by the expression of alleles at the secretor (*SEse*) gene locus, which lies close to the *H* gene on chromosome 19 (*see 2.7.1*). Secretors carry the dominant allele *SE* (i.e. genotypes *SE*/*SE* or *SE*/*se*), whereas non-secretors are homozygous for the inactive non-secretor allele *se* (i.e. genotype *se*/*se*). Both the *H* and *SE* genes code

➲ Secretors have blood group sugars in their body fluids and secretions.

for an α1,2 Fuc-T, but the enzymes are, although closely related, different.➲ The H α1,2 Fuc-T is expressed in skin and haemopoietic cells such as those in the bone marrow, erythroblasts and platelets and preferentially utilizes type 2 and type 4 chains (*see 10.2*), which, therefore, mostly carry ABO blood group sugars on red blood cells. The gene encoding for the blood group H fucosyltransferase is called *FUT1*. The *SE* α1,2 Fuc-T is expressed in epithelial secretory cells and preferentially utilizes type 1 and type 3 chains (*see 10.2*) which thus, in secretors, mostly carry ABO blood group sugars synthesized by epithelia of the gastrointestinal, genitourinary and respiratory tracts.➲ The gene encoding for the secretor fucosyltransferase is called *FUT2*. The product of each transferase is a similar H determinant disaccharide Fuc(α1→2) Galβ-, but the two products actually differ significantly in chromatographic properties, heat stability and pH activity. The two genes probably arose from a point mutation in a glycosyltransferase gene that gave rise to subtle glycan structural polymorphisms. The H α1,2 Fuc-T does not appear to belong to any of the established glycosyltransferase families.

➲ The blood group *H* locus α1,2 Fuc-T gene is called *FUT1* and the secretor locus α1,2 Fuc-T gene is called *FUT2*.

➲ The H α1,2 Fuc-T gene *FUT1* is expressed in tissues derived from mesoderm, for example, erythroblasts. The secretor α1,2 Fuc-T gene *FUT2* is expressed in tissues derived from endoderm, for example, gut and salivary glands.

10.5.8 The Bombay phenotype

The Bombay phenotype, named after the city in which the first case was reported, results from a null or inactive H transferase gene *h/h* (and *se/se*), and is very rare (≈1 in 10 000 individuals). Bombay individuals lack H α1,2 Fuc-T activity owing to a catalytically inactive mutant allele lacking 50 amino acids at the C-terminus. The enzyme is normally responsible for transferring Fuc in a (α1→2) linkage to a terminal β-linked Gal residue to form the H disaccharide Fuc(α1→2)Galβ- (*see 10.5.2*). As this structure is required as an acceptor for the blood group A glycosyltransferase α1,3 N-acetylgalactosaminyltransferase (α1,3 GalNAc-T) and the blood group B glycosyltransferase α1,3 galactosyltransferase (α1,3 Gal-T), even though Bombay individuals may be able to synthesize active blood group A and B glycosyltransferases, they are unable to make either A or B blood group determinants (*see 2.7.1*). These individuals have anti-H, anti-A and anti-B antibodies in their serum, and are therefore cross-match incompatible with all donors, except other Bombay individuals. Para-Bombay individuals have undetectable blood group H determinants on their erythrocytes, but secrete normal H substance in their saliva.

Bombay individuals have no obvious pathological phenotype associated with their condition.➲ This raises questions as to the purpose of blood group ABO sugars, and the function of the polymorphism at the *ABO* locus. The most likely explanation is that at some point in the distant past, this may have provided some selective advantage for protection against infection by pathogenic organisms (although there is no direct evidence for this speculation) but that this is no longer of relevance in modern times.

➲ Rare Bombay individuals do not synthesize ABO blood group sugars.

10.6 The Lewis system

Antibodies to the 'Lewis antigen' (Lewis a, Le^a) were discovered in the serum of a patient, a Mrs Lewis, in 1946. The Lewis system is closely related to the *ABO/H/SEse* systems. Lewis antigens are not present on the erythrocytes of new-born babies, but appear at ≈3–6 months of age, and are present in their final concentration by 3–6 years of age. Lewis antigens are not synthesized on the erythroblasts themselves, but are synthesized elsewhere – possibly predominantly by gut epithelium where they are shed into the digestive tract, digested, reabsorbed and their glycolipids transported in the plasma, then adsorbed onto the red blood cells. Lewis sugars are detectable on glycoproteins in saliva and other secretions, are present as free oligosaccharides in milk and urine, form part of N- and O-linked glycoproteins, and are found on sphingolipids in plasma on high density lipoproteins, and on erythrocytes.

10.6.1 The Lewis sugars

The Lewis system sugars are a family of fucosylated glycans, illustrated in *Figure 10.4*. Lewis a (Le^a) and Lewis b (Le^b) antigens are fucosylated oligosaccharides based on the type 1 chain structure (*see 10.2*). Lewis c (Le^c) and Lewis d (Le^d) antigens have also been described as part of glycolipids. Lewis x (Le^x) and Lewis y (Le^y) are based on type 2 chains (*see 10.2*). Le^x (also called CD15 and stage-specific embryonic antigen 1 or SSEA-1) is a trisaccharide, fucosyl

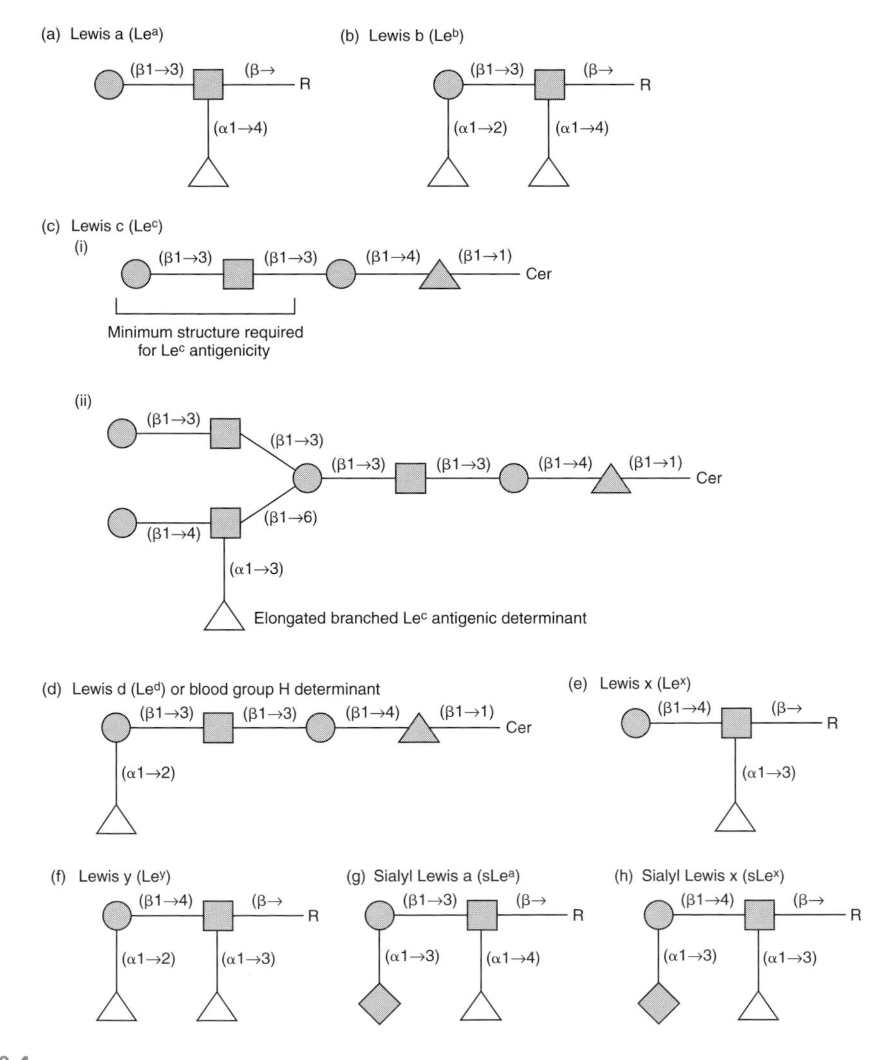

Figure 10.4

The Lewis system determinants. Lewis a, Lewis b, Lewis x and Lewis y are all expressed on glycoproteins; the R indicates where the carbohydrate is linked to the polpeptide chain. Lewis c and Lewis d are expressed on glycolipids, and are therefore linked to ceramide (Cer).

lactosamine; it is not found on erythrocytes, but is present on granulocytes and ovarian cyst glycoproteins. It is closely related to the mouse pre-implantation antigen, SSEA-1, as both have a straight core polylactosaminyl sequence known as the i antigen (*see 10.7*). Le^x is found on glycoproteins and glycolipids and is frequently polyfucosylated with the terminal sequence Gal(β1→4)[Fuc(α1→3)]GlcNAc. ➲ Le^y is a difucosylated structure made up of both Le^x and the H antigen (*see 10.5*).

➲ The Lewis sugars are a family of fucosylated glycans.

10.6.2 The fucosyltransferases responsible for Lewis sugar assembly

Lewis sugars Le^a, Le^b, Le^x and Le^y are synthesized by a number of α-L-fucosyltransferases which are differentially expressed in tissues. The genes responsible for their synthesis are called *FUT1* to *FUT7* in the order in which they were described. *FUT1* encodes the blood group H α1,2 Fuc-T and *FUT2* the secretor locus α1,2 Fuc-T (*see 10.5.7*). *FUT3* to *FUT7* are five homologous α1,3 fucosyltransferase genes encoding Fuc-TIII (*FUT3*), Fuc-TIV (*FUT4*), Fuc-TV (*FUT5*), Fuc-VI (*FUT6*) and Fuc-T-VII (*FUT7*). The blood group H and secretor locus α1,2 Fuc-T genes *FUT1* and *FUT2* are genetically unrelated to the α1,3 Fuc-T genes *FUT3* to *FUT7*. The α1,3 Fuc-T gene family is described in more detail in *Section 2.11.3*.

10.6.3 Synthesis of Lewis a

In the synthesis of Le^a, a fucosyltransferase adds an (α1→4)-linked Fuc to the sub-terminal GlcNAc of the type 1 chain (*see 10.2*) Gal(β1→3)GlcNAc- as follows:

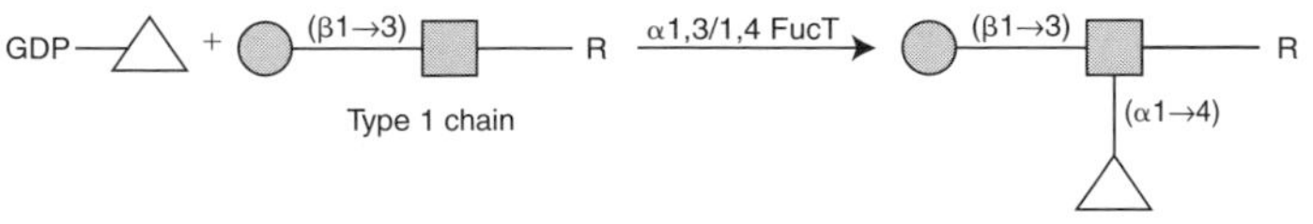

GDP + (β1→3) R — α1,3/1,4 FucT → (β1→3) R (α1→4)
Type 1 chain
Le^a

It is a α1,3/1,4 Fuc-T (GDP-L-fucose: N-acetyl-D-glucosaminyl α-fucosyltransferase, Fuc-TIII), often called the 'Lewis transferase' and is coded for by the *FUT3* gene at the Lewis (*Le*) locus. ➲ This has been mapped to the short arm of chromosome 19 in the region p13.3–p13.2.

Lewis a may be sialylated in an (α2→3) linkage to form sialyl Lewis a (sLe^a).

➲ The Lewis transferase α1,3/1,4 Fuc-T adds an (α1→4)-linked Fuc to the sub-terminal GlcNAc of the type 1 chain to yield Le^a.

10.6.4 Synthesis of Lewis b

The addition of an (α1→4)-linked Fuc to the sub-terminal GlcNAc of the H blood group determinant (*see 10.5*) attached to a type 1 chain (*see 10.2*) results in the difucosylated structure Le^b, as follows:

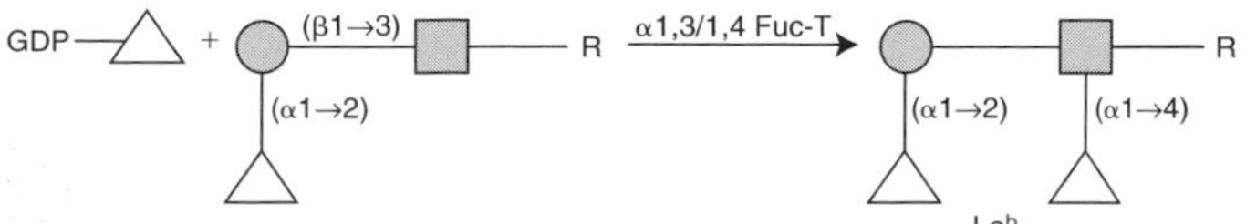

GDP + (β1→3) R (α1→2) — α1,3/1,4 Fuc-T → R (α1→2) (α1→4)
Blood group H on a type 1 chain
Le^b

Le^b synthesis therefore requires both the action of the *Se* locus α1,2 Fuc-T *and* the Lewis α1,3/1,4 Fuc-T on the original type 1 chain (*see 10.2*). Expression of the Lewis α1,3/1,4 Fuc-T, and thus Le^a and Le^b sugars, is largely restricted to the same epithelia as those expressing the

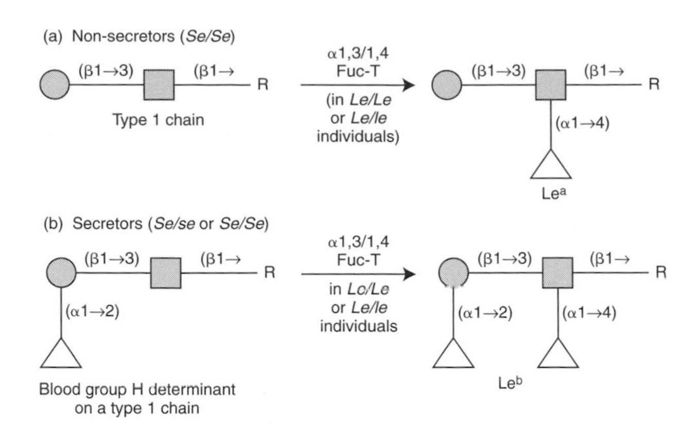

Figure 10.5

The relationship between the Lewis and *Se* locus. Non-secretors (*se/se*) do not synthesize H blood group sugar on type 1 chains. Core 1 chains are therefore available for synthesis of Le^a by the action of the Lewis locus α1,3/1,4 Fuc-T. Therefore, individuals who synthesize the Lewis enzyme (*Le/Le* or *Le/le*) are of the phenotype Le$^{(a+b-)}$, whereas individuals who do not (*le/le*) simply express unmodified type 1 chains that are converted to neither H or Lewis structures, phenotype Le$^{(a-b-)}$. Secretors (*Se/Se* or *Se/se*) synthesize H blood group sugar on type 1 chains. The resulting structure is then further modified to produce Le^b by the action of the Lewis locus α1,3/1,4 Fuc-T. Therefore, individuals who synthesize the Lewis enzyme (*Le/Le* or *Le/le*) are of the phenotype Le$^{(a-b+)}$, whereas individuals who do not (*le/le*) simply express unmodified express H antigen, unconverted to Le^b, the Le$^{(a-b)}$.

> ➲ The addition of an (α1→4)-linked Fuc to the sub-terminal GalNAc of the H determinant yields Le^b.

> ➲ Individuals can be classified into three Lewis/ABH types: Le$^{(a+b-)}$, Le$^{(a-b+)}$ and Le$^{(a-b-)}$. This results from interaction at the genetically independent Lewis (*Le*) locus and the ABH secretor (*Se*) locus.

Se α1,2 fucosyltransferase (*see 10.5.7*). Soluble Le^a and Le^b glycoconjugates are released into secretions from these cell types. ➲

10.6.5 Relationship of Lewis and secretor locus

There is thus a close relationship between the Lewis and *Se* locus and a strong interaction between the *Le* gene and the genes of the ABO, H and SE systems (*see 2.7.1, 2.11.1*). This is further illustrated in *Figure 10.5*. Adult humans therefore fall into three groups on the basis of their Lewis sugar synthesis: Le$^{(a+b-)}$, Le$^{(a-b+)}$ and Le$^{(a-b-)}$, as illustrated in *Table 10.1*. ➲ The proportions of individuals belonging to each group differs between races, and in Caucasians is about 20%, 75% and 5%, respectively.

Le^a and Le^b sugars are also detectable on erythrocytes. They are not synthesized by red blood cells, instead they are acquired from Lewis sugar positive glycosphingolipid molecules circulating in the bloodstream.

10.6.6 Synthesis of Lewis x and Lewis y

The other main members of the Lewis sugar family are Lewis x (Le^x) and Lewis y (Le^y). They are synthesized by a number of tissue-specific α1,3 Fuc-Ts.

Fuc-TV uses sialylated and unsialylated N-acetyllactosamine as a substrate and is thus responsible for synthesis of both Le^x and sialyl Le^x (sLe^x). It catalyses the synthesis of Le^x by adding (α1,3)-linked Fuc to sub-terminal GlcNAc on type 2 chains [Gal(β1→4)GlcNAc] or sLe^x by adding (α1→3)-linked Fuc to sub-terminal GlcNAc on sialylated type 2 chains. It also

Table 10.1 The relationship between the secretor and Lewis loci, and the resulting phenotypes

Genotype and status				Carbohydrate structures synthesized	Lewis phenotype
Secretor locus	Secretor status	Lewis locus	Lewis status		
se/se	non-secretor	*Le/Le* or *Le/le*	Lewis sugar producer	Le^a on type 1 chain	$Le^{(a+b-)}$
se/se	non-secretor	*le/le*	null	unmodified type 1 chains	$Le^{(a-b-)}$
Se/Se or *Se/se*	secretor	*Le/Le* or *Le/le*	Lewis sugar producer	Le^b on H blood group sugar	$Le^{(a-b+)}$
Se/Se or *Se/se*	secretor	*le/le*	null	unmodified blood group H sugar	$Le^{(a-b-)}$

catalyses the synthesis of Le^y by adding (α1→2)-linked Fuc to sub-terminal GlcNAc on 2′-fucosyl-N-acetyllactosamine. Conformational studies on the Le^x saccharide indicate that the Fuc and Gal residues have unusual nuclear Overhauser effect (NOE) distance constraints holding that part of the structure more rigidly than otherwise might be expected.

Fuc-TIV is mostly expressed in leukocytes and myeloid cells and acts preferentially on type 2 precursors [Gal(β1→4)GlcNAc] to synthesize Le^x.

10.6.7 Sialylation of Lewis sugars

A number of sialyltransferases are responsible for sialylation of Lewis sugars.

These include the Galβ1,3/1,4 GlcNAc α2,3 sialyltransferase which has a high substrate preference for type 2 chains as precursors [Gal(β1→4)GlcNAc] over type 1 chains [Gal(β1→3)GlcNAc] and thus contributes to the synthesis of sLe^x. Another Galβ1,3/1,4 GlcNAc α2,3 sialyltransferase has a higher substrate preference for type 1 chains as a precursor [Gal(β1→3)GlcNAc] over type 2 chains [Gal(β1→4)GlcNAc] and thus is involved in the synthesis of sLe^a. A third α2,3 sialyltransferase uses oligosaccharide substrates with terminal Gal attached to [Gal(β1→3)GalNAc] type 1 chain and [Gal(β1→4)GlcNAc] type 2 chain sequences to yield NeuAc(α2→3)Gal(β1→3)GalNAc-R and NeuAc(α2→3)Gal(β1→4)GlcNAc-R.

10.6.8 Sulfation of Lewis sugars

In addition to sialylation, Le^x and Le^a are often further modified by sulfation to yield sulfo-Le^x, sulfo-sialyl Le^x and sulfo Le^a. Sulfation occurs after sialylation but before fucosylation. Sulfation is catalysed by a GlcNAc-6-O-sulfotransferase.

10.6.9 Lewis c and Lewis d sugars

Lewis c (Le^c) and Lewis d (Le^d) determinants are also considered part of the Lewis family of blood group sugars.

The type 1 precursor chain Gal(β1→3)GlcNAc, lactotetraosylceramide, is the minimum structure required for Le^c antigenicity. It is referred to as a Lewis family antigen although it is not synthesized by the action of Lewis locus genes. It is found on the erythrocytes of Lewis-negative ABH non-secretors (*le/le*, *se/se*), and is also present in their plasma and saliva. Polyclonal antisera prepared against Le^c preferentially recognize an elongated branched structure based on the lactotetraosylceramide core. Both structures are illustrated in *Figure 10.4(c)*.

The Le^c Gal(β1→3)GlcNAc disaccharide is the precursor for synthesis of the blood group H structure, which is also referred to as Le^d : Fuc(α1→2)Gal(β1→3)GlcNAc. It is formed from the Gal(β1→3)GlcNAc type 1 chain by the action of the secretor transferase (*see 10.5.2*). It is the main antigen found on erythrocytes and in plasma of group O/H, Lewis-negative secretors (*le/le, Se/Se* or *Se/se*). It is the precursor for synthesis of blood group A and B antigens.

10.7 The I and i antigens

Polylactosamine chains, consisting of repeating Gal(β1→4)GlcNAc units (*see 10.3*), form the basis of the I- and i-blood group system. The I and i antigens are also structurally related to the ABO/H and Lewis systems in that they are potential substrates for further conversion to H, A and/or B structures. They are present on polyglycosylceramide molecules, on the oligosaccharide chains of the erythrocyte Band 3 and Band 4.5 glycoproteins, and on the neo-lacto series of glycolipids. They are found on the surface of erythrocytes, leukocytes, macrophages and some epithelial glycoproteins. The structures of the I and i determinants are illustrated in *Figure 10.6*.

10.7.1 Synthesis of the i antigen

The synthesis of the i antigen is catalysed by an i-specific β1,3 N-acetylglucosaminyltransferase (β1,3 Gn-T) which attaches a (β1→3)-linked GlcNAc to the type 2 chain or lactosamine unit, Gal(β1→4)GlcNAc to yield the i antigenic determinant which is represented by GlcNAc(β1→3)Gal(β1→4)GlcNAcβ-. Repeating Gal(β1→4)GlcNAc and GlcNAc(β1→3)Gal are necessary for i antigenicity, and the linear ceramide hexasaccharide chain Gal(β1→4) GlcNAc(β1→3)Gal(β1→4)GlcNAc(β1→3)Gal(β1→4)Glc-Cer, a common component of umbilical cord, but not of adult, erythrocytes, is highly i antigenic.

10.7.2 Synthesis of the I antigen

The i antigen is converted to I antigen on adult erythrocytes and mucins produced by some other cell types. This conversion is catalysed by the action of a branching enzyme β1,6 N-acetylglucosaminyltransferase (I-β1,6 GlcNAc-T), of which there are at least two, which adds a branching (β1→6)GlcNAc residue to an interchain Gal to form a branching lactosamine structure, the I antigen. A gene for the branching enzyme has been mapped to chromosome 9q21. The I antigen is a precursor for the synthesis of the ABH determinants on erythrocytes.

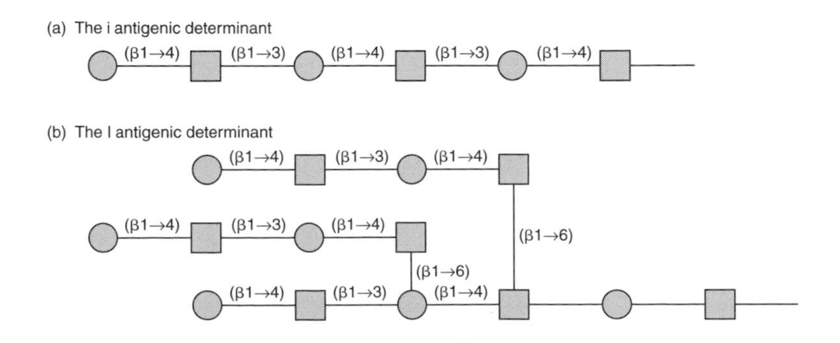

Figure 10.6

The i and I determinants. (a) The i antigenic determinant is repeating GlcNAc(β1→3)Gal(β1→4) GlcNAcβ-units. (b) The I antigenic determinant is a branching structure, derived from the i antigen, and containing the same repeating GlcNAc(β1→3)Gal(β1→4)GlcNAcβ- units.

10.7.3 Developmental regulation of synthesis of the i and I antigens

There is a reciprocal relationship between the amount of I and i antigen found on erythrocytes, with fetal erythrocytes having proportionately larger amounts of i antigen and less I antigen. The amount of i antigen present declines over the first 18 months of life, and virtually disappears, whereas the proportion of I antigen increases. This is assumed to be a result of developmental regulation of the I antigen I-β1,6 GlcNAc-T locus. Erythrocyte maturation is thus accompanied by sequential appearance of the i, I, H the A/B sugars on erythrocyte glycoproteins, glycolipids and gangliosides.

10.7.4 I/i antigens and autoimmunity

I and i antigens are recognized by auto-antibodies that occur transiently in high titres following *Mycoplasma pneumoniae* (anti-I) and Epstein–Barr virus (anti-i) infection, and chronically in a chronic haemolytic disorder called cold haemagglutinin disease. The auto-antibodies are interesting in that they are of low affinity and agglutinate red cells and fix complement only at temperatures below 37°C; they are thus termed 'cold agglutinins'.

10.8 P^k, P and P_1 antigens

All human P-related antigens are associated with membrane glycosphingolipids (*see Chapter 9*). Their structures are given in *Figure 10.7*. They are found on erythrocyte membranes, and in other tissues, notably urothelium (urothelium has a barrier function in segregating hypertonic urine within the confines of the bladder; as such it is built upon a lipid, water-impermeable, barrier).

The apparent structural similarity of this collection of oligosaccharides is misleading because the pathways by which they are synthesized are different, although they are derived from a common lactosylceramide precursor (Gal(β1→4)Glc-Cer).

10.8.1 The Pk antigen

The UDP-Gal:lactosylceramide α1,4 galactosyltransferase enzyme, the P^k transferase, catalyses the synthesis of the P^k antigen Gal(α1→4)Gal(β1→4)Glc-Cer from the precursor Gal(β1→4)Glc-Cer.

10.8.2 The P antigen

The P^k antigen acts as a precursor for synthesis of the P antigen, GalNAc (β1→3) Gal(α1→4)Gal(β1→4)Glc-Cer; by the action of a β1,3 GalNAc-T (the P transferase).

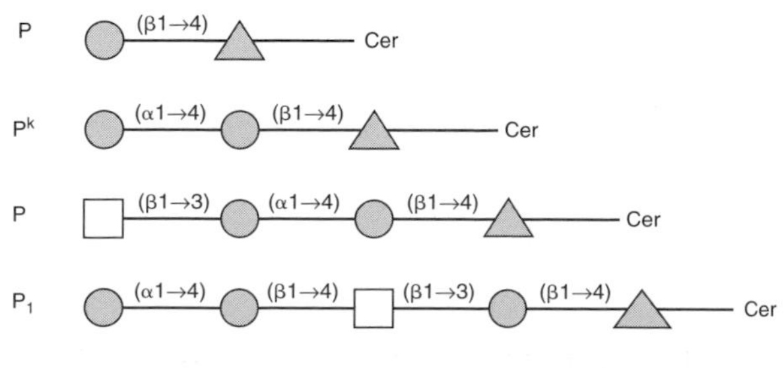

Figure 10.7

P-related determinants.

Table 10.2 P blood group types

Blood group	Distribution	Structure	Genetic basis
P_1	Commonest	P and P_1, plus small amounts of unconverted P^k	All enzymes active
P_2	Common	P and P^k antigens, but no P_1 determinants	Inactive P_1 transferase
P_1^k	Rare	High levels of P^k with normal P_1, no P determinants	Inactive P transferase
P_2^k	Rare	High levels of P^k, no P or P_1	Inactive P and P_1 transferase
p	Rare	No P, P_1 or P^k	Inactive P^k and P_1 transferases

10.8.3 The P_1 antigen

The P_1 antigen Gal(α1→4)Gal(β1→4)GalNAc(β1→3)Gal(β1→4)Glc-Cer is produced by building upon the same lactosylceramide precursor, Gal(β1→4)Glc-Cer, but this is converted first to paragloboside, Gal(β1→4)GalNAc(β1→3)Gal(β1→4)Glc-Cer. This structure is then acted upon by the P_1 transferase to yield the P_1 antigen Gal(α1→4)Gal(β1→4)GalNAc(β1→3)Gal(β1→4)Glc-Cer. Individuals nullizygous at the P_1 transferase locus synthesize normal amounts of P and P^k antigens, but are deficient in P_1 determinants, and are said to be blood type P_2.

10.8.4 Synthesis of P blood group antigens

The different P blood group types are summarized in *Table 10.2*. The most common is P_1 in which individuals synthesize both P and P_1 sugars, plus small amounts of unconverted P^k. P_2 individuals do not produce the P_1 transferase, and therefore produce normal levels of P and P^k antigens, but no P_1 determinants. Rare P_1^k individuals cannot convert P^k into P and thus synthesize unusually high levels of P^k with normal P_1. Rare P_2^k individuals lack both active P transferase and P_1 transferase and lack both P and P_1 antigen synthesis, with increased P^k. Deficiency in P, P_1 and P^k synthesis results from inactive P^k and P_1 transferase, and the P phenotype.

10.8.5 The genetics underlying the P blood group system

The genetics underlying synthesis of the P blood group sugars is poorly understood and the glycosyltransferases responsible for their synthesis have not been identified.

10.9 'Abnormal' blood group sugars: T, Tn, Cad and Tk

T, Tn, Cad and Tk are not normally exposed on the erythrocytes of healthy individuals. They may be 'unmasked' as a result of infection, as in T and Tk, defects in the mechanisms of glycosylation, as in Tn, or as an inherited defect, as in Cad. They are often associated with diseases such as cancer in which their presence may be of prognostic significance (*see 16.6.2*).

10.9.1 The Thomsen-Friedenreich (T or TF) antigen

Desialylation of erythrocyte glycoconjugates by the sialidase activity of organisms such as *Clostridium perfringens*, *Vibrio cholerae* and influenza virus, results in the exposure of the disaccharide Gal(β1→3)GalNAc-O-Ser/Thr. This structure is known as the Thomsen-Friedenreich, TF, or more usually T antigen (*see 5.6.3*). If donor blood samples are contaminated by bacteria that produce sialidases (also known, but not rigorously, as neuraminidases), this causes desialylation of the erythrocytes and consequent exposure of the T antigen. This results in a transfusion reaction in the recipient. The T antigen is also often found on cancer cells (*see 16.6.2*). It is recognized by the lectin from peanut, *Arachis hypogaea* (PNA; *see Table 13.1*) which makes PNA a convenient tool for its detection.

10.9.2 Tn antigen

The Tn antigen GalNAc-O-Ser/Thr, results from the addition of the first GalNAc residue to the polypeptide backbone in O-linked (mucin type) glycoprotein synthesis (*see 5.6.1*). Under normal circumstances this structure is elongated and the Tn antigen is therefore referred to as cryptic. Rare Tn individuals, who lack β1,3 galactosyltransferase activity, have Tn on the surface of their erythrocytes. It is also associated with many cancers, in which its presence can be associated with a poor prognosis (*see 16.6*). *Salvia sclarea* lectin (*see Table 13.1*) is a convenient tool for the detection of Tn, but it is also recognized, by several other GalNAc-binding lectins, although less selectively.

10.9.3 Cad and Sd^a antigens

A family of β1,4 GalNAc-T enzymes exist, including the one responsible for blood group A determinant synthesis, and are responsible for synthesis of the Cad and Sd^a antigens.

Cad and Sd^a antigens share the trisaccharide determinant GalNAc(β1→4)[NeuAc(α2→3)]Galβ-, and appear to be closely related to each other. They have different sub-terminal core sequences: the Sd^a determinant is GalNAc(β1→4)[NeuAc(α2→3) Gal(β1→4)GlcNAc(β1→3) Gal-R, whereas the Cad structure is GalNAc(β1→4)[NeuAc(α2→3)]Gal (β1→3)[NeuAc (α2→6)]GalNAc-R. Both therefore share a common tetrasaccharide, GalNAc(β1→4) [NeuAc(α2→3)]Galβ-.➲

> ➲ The Sd^a determinant is GalNAc(β1→4) [NeuAc(α2→3)]Gal(β1→4)GlcNAc(β1→3)Gal-R. The Cad determinant is GalNAc(β1→4)[NeuAc (α2→3)]Gal(β1→3)[NeuAc(α2→6)]GalNAc-R. Both share the tetrasaccharide GalNAc(β1→4)[NeuAc (α2→3)]Galβ-

Sd^a is a common erythrocyte antigen, although individuals who do not synthesize this determinant appear to suffer no detrimental effect. Cad is a rare blood group carbohydrate antigen carried on sialoglycoproteins and gangliosides of human erythrocytes. Synthesis of Cad is regulated during differentiation and development. Both may be detected by binding of *Dolichos biflorus* agglutinin (DBA; *see Table 13.1*).

10.9.4 Tk antigen

The presence of the Tk antigen results from exposure of erythrocytes to the endo-β-glycosidases of organisms such as *Bacteroides fragilis, Aspergillus niger* and *Candida albicans* as a result of infection.➲ The Tk determinant is GlcNAc(β1→3)[GlcNAc(β1→6)]Gal and may be conveniently detected by *Griffonia simplicifolia* isolectin II (GSA II; *see Table 13.1*).

> ➲ The Tk determinant is GlcNAc(β1→3)[GlcNAc (β1→6)]Gal-

10.10 Fucosylation

L-fucose is a common terminal monosaccharide at the non-reducing end of glycoprotein oligosaccharide chains; it may also be attached to sub-terminal residues; and also occurs in glycolipids. It is, for example, part of the ABO blood group (*see 10.5*) and Lewis sugars (*see 10.6*). Fucosylation is usually the penultimate or ultimate step in the synthesis of oligosaccharide chains, and Fuc is most commonly attached by an (α1→3) or (α1→4) glycosidic bond, and less commonly by an (α1→2) glycosidic bond.

10.11 Sialylation

Sialylation is one of the most common and biologically important modifications that oligosaccharides undergo and, as such, it is the subject of the next chapter (*see Chapter 11*).

10.12 Further modification

When glycoproteins eventually leave the Golgi apparatus, they may be further modified by glycosidases (*see 2.2*) found in tissues and in serum, including α- and β-galactosidases, α-fucosidase, α-mannosidase, β-glucuronidase, N-acetylgalactosaminidase and N-acetylglucosaminidase.

Further reading

Henry, S., Oriol, R., Samuelsson, B. (1995) Review: Lewis histo-blood group system and associated secretory phenotypes. *Vox Sang.* **69:** 166–182.

King, M.-J. (1994) Blood group antigens on human erythrocytes – distribution, structure and possible functions. *Biochim. Biophys. Acta* **1197:** 15–44.

Sialic acids

11

11.1 Occurrence

Sialic acids are charged monosaccharides which are common constituents of the oligosaccharides of vertebrates and some invertebrate species. The majority of sialic acid in higher animals is bound up in glycoconjugates. Sialylation of oligosaccharide chains is a common and physiologically important event, and sialic acids are possibly the most biologically important monosaccharide units of glycoconjugates. Sialic acid often occurs as the terminal monosaccharide of N-linked (*see Chapter 4*) and O-linked (*see Chapter 5*) oligosaccharide chains of glycoproteins, glycosphingolipids (*see Chapter 9*) and also of GPI anchors (*see Chapter 12*). Both its negative charge and its terminal position involve it in numerous biological processes.

Sialylation appears to be an early modification in evolutionary terms, occurring at the beginning of the development of multicellular animals.➲ It is common in vertebrates and complex invertebrates, such as *Echinoderms* and *Priapulida*, but not commonly present in plants, prokaryotes or most simple invertebrates (*see Chapter 7*). Exceptions include the documented presence of sialic acids in developing *Drosophila melanogaster* embryos, occasionally in other insects, and in the polysaccharides of the capsules of some bacteria. There has been speculation that the genes coding for bacterial sialic acid transferases may have occurred through gene transfer from eukaryote hosts; another explanation might be independent evolution from genes involved in the synthesis of 3-deoxy-D-manno-octulosonic acid (KDO), an acidic bacterial monosaccharide with structural similarities to sialic acids (*see 3.2.5.5*).

➲ Sialylation is a common and functionally important modification of the glycoconjugates of higher animals which evolved with the development of multicellular animals.

11.2 Types, modifications and linkages of sialic acids

The term sialic acid, derived from the fact that it was first discovered in saliva, embraces all derivatives of neuraminic acid which share a basic 9-carbon carboxylated skeleton, illustrated in *Figure 11.1*. The chemical backbone can be modified by many substitutions, making sialic acids the most structurally varied of any naturally occurring monosaccharides.

More than 40 different forms of sialic acids have been described in living organisms.➲ The 'primary' sialic acids are Neu5Ac, which occurs most commonly in mammalian tissues, and is illustrated in *Figure 11.1*, and 2-keto-3-deoxy-D-glycero-D-galactononononic acid (KDN), which is illustrated in *Figure 11.2*. The only difference in the structure of Neu5Ac and KDN is a substitution at the 5 carbon position.

➲ The basic structure of sialic acid can be modified in ≈40 different ways.

It is believed that all other types of sialic acids are derived metabolically from these two forms.➲ Neu5Ac, or N-acetylneuraminic acid, is properly called 2-keto-5-acetamido-3,5-dideoxy-D-glycero-D-galactononulosonic acid, and replacement of the 5-amino group by

➲ All types of sialic acid are modifications of Neu5Ac and KDN.

Functional and Molecular Glycobiology, Susan A. Brooks, Miriam V. Dwek and Udo Schumacher

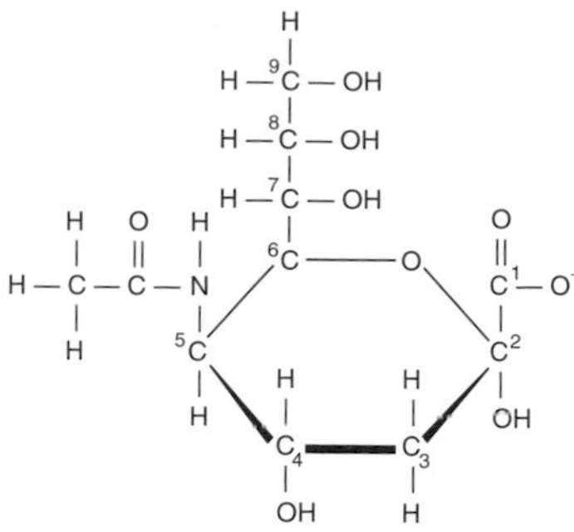

Figure 11.1

The structure of α-D-NeuAc. Sialic acids are based on a 9 carbon structure; the nine carbon atoms are labelled (1) to (9).

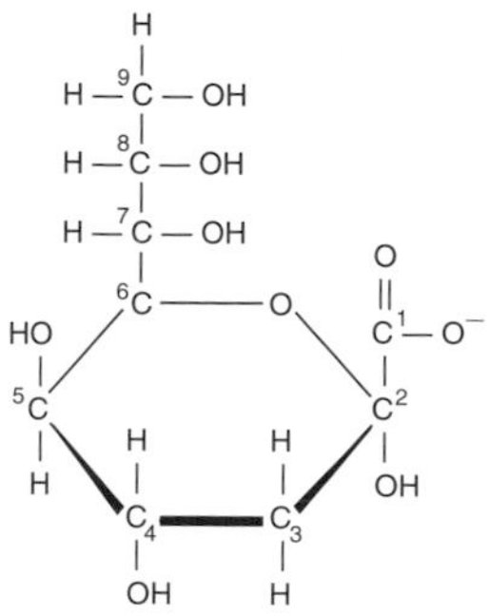

Figure 11.2

The structure of KDN.

a hydroxyl group results in KDN, properly called 2-keto-3-deoxy-D-glycero-D-galactonononic acid, or 2-keto-3-deoxy-nonolusonic acid. The 5-carbon position carries an N-acetyl group in Neu5Ac, or a hydroxyl group, in KDN. Because of these excessively long chemical names, the abbreviated terms are normally used.

Sialic acids, based on these primary structures, can be free, acetylated, lactylated, sulfated, phosphorylated or methylated. The carboxylate group at the 1-carbon position is usually ionized at physiological pH, but is also sometimes found in lactone ester with hydroxyl groups of adjacent monosaccharides. This 5-N-acetyl group of Neu5Ac can be either hydroxylated to give N-glycolylneuraminic acid (Neu5Gc), or can be de-N-acetylated to give neuraminic acid (Neu). Neu, Neu5Ac, Neu5Gc and KDN may be additionally modified by substitutions at the hydroxyl groups on the 4-, 7-, 8- and 9-carbons, and O-acetylation may occur at the 9-carbon instead of the 7-carbon position. Unsaturated and dehydro- forms of sialic acids have also been described. As so many complex substitutions and conformations are possible, a simple nomenclature is used to describe them, as outlined in *Table 11.1*.

Modification of sialic acids occur in defined compartments within the cell. Most Neu5Gc, for example, is synthesized from CMP-Neu5Ac in the cytosol by the action of an iron-dependent hydroxylase which utilizes the common electron transport chain of cytochrome b_5 and b_5 reductase. Modifications involving O-acetyl esters and other hydroxyl groups, which are mediated through the action of, as yet, poorly characterized enzymes, occur typically in the lumen of the Golgi apparatus and associated organelles after transfer of sialic acids onto the oligosaccharide chain of glycoconjugates.

Table 11.1 Summary of the nomenclature of sialic acids

Abbreviation	Denotes
Neu	Core neuraminic acid
KDN	Core 2-keto-3-deoxy-nonulosonic acid
Ac	Acetyl
Gc	Glycolyl
Me	Methyl
Lt	Lactyl
S	Sulfate
Numbers	Indicate location relative to 9-carbon position, e.g. Neu5Ac
X	Used in place of number when substitution is present, but type unknown, e.g. SiaX
Sia	Generic abbreviation for sialic acid when exact type is not known

Free sialic acids occur mostly in the β-anomeric ring structure, nucleotide-bound sialic acid, for example, CMP-Neu5Ac, also occurs in a β-configuration, whereas glycoconjugate-bound sialic acid occurs in mostly α-anomeric form. Sialic acid residues are found most commonly in the terminal positions of the oligosaccharide chains of glycoconjugates, and their presence usually prevents further chain extension. Sialic acids are most commonly (α2→3) linked or (α2→6) linked to D-Gal or (α2→6) linked to GlcNAc, (α2→6) linked to GalNAc, or (α2→8) linked to other sialic acids. ➲ The different linkages and substitutions, result in hundreds of possible conformations of oligosaccharides containing sialic acids.

➲ Sialic acid is attached to the terminal position of the oligosaccharide chain in (α2→3), (α2→6) or (α2→8) linkages. Sialylation usually prevents further glycan chain extension.

11.3 Biosynthesis and metabolism of sialic acids

Microorganisms and the cells of higher organisms synthesize sialic acids using a complex biochemical pathway. Central to this pathway is the metabolism of Neu5Ac. Neu5Ac is synthesized *de novo* within the cell, it is then activated to form the corresponding nucleotide-sugar, the activated nucleotide-sugar is subsequently appended to the oligosaccharide chain of the glycoconjugate, and finally it may sometimes be removed again and degraded.

An overview of the biosynthesis of Neu5Ac and CMP-Neu5Ac is given in *Figure 11.3*. This complex pathway usually begins with D-Glc and proceeds as follows: the numbers in brackets refer to the stages labelled accordingly in *Figure 11.3*.

- Glc is activated to D-glucose-6-phosphate (Glc-6-P).
- Glc-6-P is isomerized to D-fructose-6-phosphate (Frc-6-P) [1].
- Frc-6-P is transaminated with the amino donor L-glutamine to form D-glucosamine-6-phosphate (GlcN-6-P) [2].
- GlcN enters the pathway by one of three possible routes [3], all of which result in the production of ManNAc [4]:
  a. most commonly [3a], GlcN is phosphorylated to produce D-glucosamine-6-phosphate (GlcN-6-P); then acetylated by the action of acetyl-CoA to N-acetyl-D-glucosamine-6-phosphate (GlcNAc-6-P); then isomerized to N-acetyl-D-glucosamine-1-phosphate (GlcNAc-1-P). This is activated to UDP-GlcNAc. UDP is then released in a two-stage process and GlcNAc is epimerized into ManNAc;
  b. GlcN is acetylated to GlcNAc [3b], then phosphorylated to GlcNAc-6-P and converted to ManNAc as described above;
  c. GlcN is acetylated to GlcNAc then directly epimerized to ManNAc [3c].

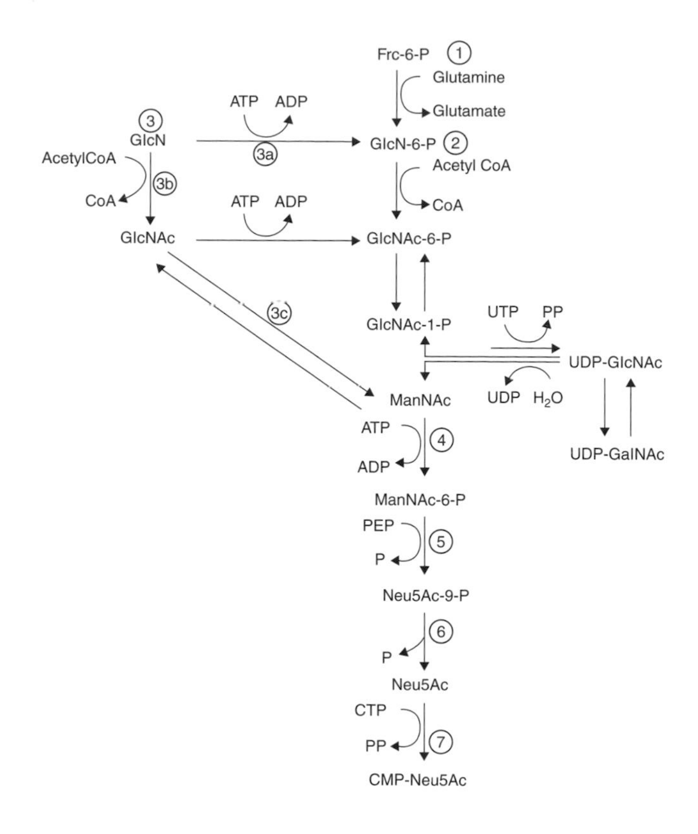

Figure 11.3

The biosynthesis of Neu5Ac and CMP-Neu5Ac (*see 11.3*).

- ManNAc is phosphorylated to N-acetyl-D-mannosamine-6-phosphate (ManNAc-6-P) [4].
- ManNAc-6-P reacts at position C1 with D-phosphoenolpyruvate (PEP) to form N-acetyl-D-neuraminic acid-9-phosphate (Neu5Ac-9-P) [5].
- Neu5Ac-9-P is dephosphorylated to Neu5Ac [6].

KDN is derived from an analogous condensation of Man-6-P, instead of ManNAc-6-P.

The free sialic acids produced by this pathway are then activated by reaction with CTP to form the nucleotide donor CMP-Neu5NAc [7], illustrated in *Figure 11.4*, and two molecules of inorganic phosphate. This occurs in the nucleus where there is a high concentration of CTP.➲ It is mediated through the action of CMP-Neu5Ac synthase (also called Neu5Ac cytidyltransferase), which is expressed ubiquitously in all prokaryotic and eukaryotic cells, and located in the nucleus in mammalian cells (all other enzymes involved in sialic acid synthesis are located in the cytoplasm; the reason for this nuclear location is unclear). CMP-Sia is then transported to the Golgi apparatus and pumped across its membranes by the action of a specific antiporter.

> ➲ CMP-NeuAc is synthesized in the nucleus.

11.4 Sialylation of oligosaccharides: the sialyltransferases and regulation of their activity

Transfer of sialic acids onto oligosaccharide chains is achieved by a large family of sialyltransferase enzymes, members of which are listed in *Table 11.2*. Little is understood regarding

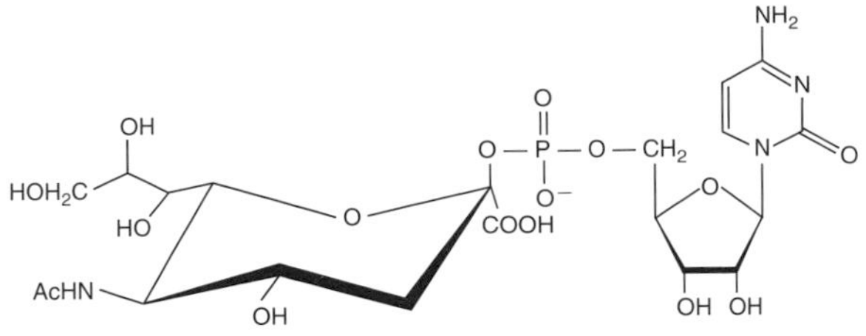

Figure 11.4

The structure of the nucleotide donor CMP-Neu5NAc.

Table 11.2 Some examples of sialytransferases and the structures whose synthesis they catalyse

Sialyltransferase	Structures synthesized
Mammalian	
ST3Gal-I	Sia(α2→3)Gal(β1→3)GalNAc-R
ST3Gal-II	Sia(α2→3)Gal(β1→3)GalNAc-R
ST3Gal-III	Sia(α2→3)Gal(β1→3)GlcNAc-R > Sia(α2→3)Gal(β1→4)GlcNAc-R
ST3Gal-IV	Sia(α2→3)Gal(β1→3)GalNAc-R & Sia(α2→3)Gal(β1→4)GlcNAc-R
ST6Gal-I	Sia(α2→6)Gal(β1→4)GlcNAc-R & Sia(α2→6)Gal(β1→3)GlcNAc-R
ST6Gal-I	Sia(α1→4)GlcNAc-R
ST6GalNAc-I	Sia(α2→6)GalNAc-R
ST6GalNAc-II	Sia(α2→6)[Gal(β1→3)]GalNAc-R
STGalNAc-III	Sia(α2→6)GalNAc-R
ST8Sia-I	Sia(α2→8)Sia-R
ST8Sia-II	Sia(α2→8)Sia-R
ST8Sia-III	Sia(α2→8)Sia-R
ST8Sia-IV	Sia(α2→8)Sia-R
ST8Sia-V	Sia(α2→8)Sia-R
Bacterial	
PST	Sia(α2→8)Sia-R and Sia(α2→9)Sia-R

the regulation of their expression and activity.➲ The sialyltransferase family is unusual in that its members share little structural homology except for two highly conserved sialyl motifs, the 'L-sialyl motif' and the 'S-sialyl motif', both in the catalytic domain, which may be implicated in the sugar-nucleotide recognition site.

➲ Sialyltransferases share two common 'sialyl motifs', but are otherwise structurally dissimilar.

The many modifications and linkages of sialic acids (*see 11.2*) are tissue specific and developmentally regulated, and some are found only in specific types of glycoconjugates. Synthesis must be supported by precisely controlled and complex enzymatic mechanisms, but these are not understood.

11.5 Desialylation

In vertebrate cells, desialylation of glycoconjugates occurs in the acidic compartments of the endosomal–lysosomal pathway by the action of specific sialidases (also called neuraminidases, but this is not a specific term).➲ These are a family of related glycosyl hydrolase enzymes. O-Acetyl residues are removed by sialate-O-acetylesterases, sulfate groups by glycosulfatases,

➲ Sialidases are also referred to as neuraminidases.

$$\text{Neu5Ac} \xrightarrow[\text{lysase}]{\text{Sialate-pyruvate}} \text{ManNAc} + \text{pyruvate}$$

$$\text{Neu5Gc} \xrightarrow[\text{lysase}]{\text{Sialate-pyruvate}} \text{ManNGc} + \text{pyruvate}$$

Figure 11.5

Desialylation in the lysosome.

and phosphate groups by special phosphatases. N-acetylneuraminate lysase catalyses the cleavage of Neu5Ac to ManNAc and pyruvate, as illustrated in *Figure 11.5*. These may either be further degraded or re-metabolized to build new sialic acids. The desialylated glycoconjugates may be destined for degradation or may, under certain circumstances, be returned to the Golgi apparatus for resialylation and recycling.

There is evidence for the expression of both cytoplasmic and cell-surface sialidases. Little is known of the function of cytoplasmic sialidases, as sialylated glycoconjugates have not been described in this cellular compartment. The cell-surface sialidases are thought to be responsible for abrupt shedding of surface sialic acid in response to specific cellular signals, for example, in the sudden desialylation accompanying activation of some leukocyte populations.

Many viruses, bacteria, protozoa, fungi and invertebrates produce sialidases. The sialidases of bacteria, fungi, invertebrates and vertebrates belong to the same 'family' and share the feature of an 'Asp box' with the sequence Ser-Xaa-Asp-Xaa-Gly-Xaa-Thr-Thr, the function of which is unknown. It seems likely that bacterial sialidases have arisen through gene transfer from the vertebrate host genome. Viral and protozoan sialidases are believed to be genetically unrelated to each other and to the sialidases of higher organisms, and to have arisen through parallel evolution. Protozoan *trans*-sialidases are particularly interesting in that they function in transferring sialic acid residues from host cell-surface glycoconjugates to the surface of the protozoan itself, where it functions to camouflage the microorganism from host immune attack and the complement system. *Trypanosoma cruzi* is one of the best studied examples of this phenomenon. It binds to host cells and transfers host sialic acids to its own surface by the action of a sialidase and a sialyltransferase (*see 11.5, 11.8.5, 16.2.3.1*). Organisms potentially infected by these pathogens have attempted to develop defence strategies against them, for example, O-acetyl-, N-glycolyl- and 2,3-didehydro-2,6-anhydro-modified sialic acid residues which seem to be resistant to many microorganism-derived sialidases.

11.6 Methods for analysing sialic acids

As sialylation so dramatically alters the charge, size, orientation, shape, hydrophobicity, and as a result, the biological properties of oligosaccharides and the glycoconjugates carrying them, their characterization and analysis is of special interest in glycobiology. ➲ Unfortunately, the immense complexity of the various modifications and substitutions that sialic acids are subject to means that this is a difficult task, and better methods for the release, purification and analysis of sialic acid and its derivatives are still required. Sialic acids pose special difficulties as many substitutions are labile and can alter or be lost during standard glycan release, purification and analysis procedures, especially if performed under acidic conditions (*see 14.6.1*). Sialidases are commercially available, and may be specific in their mode of action with regard to the particular type of linkage they cleave, but may not act on all substitutions (*see 14.6.2*).

➲ Sialic acids are difficult to analyse.

It is also of note that many immortalized cultured cell lines are deficient in sialylation, rendering them unsuitable for study of the biosynthesis and function of sialylated glycans.

11.7 Sialic acid-binding lectins

A number of histochemical methods have been described for the detection of sialic acids in tissues. They tend to be rather crude, and not to discriminate between different types of sialic acid and have been largely superseded by lectin-binding techniques (*see 14.2.1*).

The terminal position of sialic acids at the end of oligosaccharide chains means that they are appropriately presented as convenient binding partners for lectins. Most lectins are selective as to the linkage of the sialic acid they recognize, and many also recognize sub-terminal monosaccharides; the negatively charged carboxylate group at the C1 position is usually required for binding, and different lectins have different degrees of selectivity regarding the requirement for divalent cations.

Many sialic acid-binding lectins from diverse sources have been described, and some of them are listed in *Table 11.3*. Most sialic acid-binding lectins are derived from plants and invertebrates and have been widely used to study sialylated glycans. *Maackia amurensis* agglutinin (MAA), which recognizes Neu5Ac(α2→3)Gal, and *Sambucus nigra* (elderberry) agglutinin-I (SNA-I), which recognizes Neu5Ac(α2→6)Gal/GalNAc, the two most abundant linkages of sialic acid found in the oligosaccharides, have proved particularly useful in studies exploring sialylation. There is less published literature on the use of some of the more recently described or less readily available lectins, such as those from *Crassostrea gigas* (Pacific oyster), *Masticoproctus giganteus* (whip scorpion) and *Allomyrina dichotoma* (a type of beetle).

Some organisms that do not themselves synthesize sialylated glycans, such as lower invertebrates, insects and plants, do synthesize lectins that recognize these structures.➲ This would imply that they are involved in interaction with the glycans of other species, and roles in defence have been postulated in some cases. This is supported, for example, by the action of the lectin from the haemolymph of *Limulus polyphemus*, (horseshoe crab), which mediates the haemolysis of 'foreign' cells. Alternatively, although it appears less likely, the natural binding partners of these lectins may be as yet undetermined anionic carbohydrate structures, or even non-carbohydrate molecules, and sialic acid recognition may simply be coincidental.

> ➲ Sialic acid-binding lectins are often produced in organisms that do not themselves synthesize sialic acids. They may be involved in defence mechanisms against other sialic acid-bearing organisms.

The functions of the interactions between sialic acid-binding lectins and their naturally occurring ligands are largely unknown, but some specific examples have been studied in detail (*see 11.8.4*).

11.8 General functions of sialic acids

The presence of sialylated oligosaccharides at the cell surface is necessary for mammalian cell viability, and if cells are enzymatically desialylated, sialylation will be restored rapidly. Quantitative and qualitative differences in sialic acid is seen in health and disease, and at different stages of cell growth, differentiation, ageing and malignant transformation. For example, differences in levels of sialylation are seen in the maturation of human myeloid cells and erythrocytes, and in the cells of many organs. As the structures and linkages of the sialic acids are so variable, and their presence so widespread in higher organisms, it seems that they are likely to play multiple and crucial roles in many aspects of cell biology.

Many functions of sialic acids result from their simple physiochemical properties such as hydrophilicity, electronegativity, relatively large physical size, conformation and exposed position at the terminal end of oligosaccharide chains, but they also function as specific binding partners for lectins in some instances – their extraordinary structural diversity makes

Table 11.3 Examples of sialic acid binding lectins from diverse sources

Lectin/source	Binding partners
Vertebrate lectins	
Complement factor H	Sialylated glycoconjugates
Calcyclin (bovine heart)	Neu5Ac and Neu5Gc
Selectins	
E-selectin (cytokine activated endothelia)	sLe^x, sLe^a
L-selectin (leukocytes and other white blood cells)	Sialylated, sulfated and fucosylated glycans
P-selectin (cytokine activated platelets and endothelia)	sLe^x, sLe^a
Siglecs (immunoglobulin-like adhesion molecules binding to sialylated glycans)	Sia(α2→3)Gal(β1→3)GalNAc; Sia(α2→3)Gal(β1→3/4)GlcNAc
Sialoadhesin (macrophage subpopulations)	
Myelin associated glycoprotein MAG (neuronal tissues, e.g. Schwann cells oligodendrocytes)	Sia(α2→3)Gal(β1→3)GalNAc on neurons and oligodendrocytes
CD22 (B lymphocyte)	Neu5Ac(α2→6)Gal(β1→4)GlcNAc on B lymphocytes
CD33 (myelomonocytic cells)	Sia(α2→3)Gal(β1→3)GalNAc; Sia(α2→3)Gal(β1→3/4)GlcNAc on myelomonocytic cells
Schwann cell myelin protein SMP siglec-5	
Invertebrate lectins	
Limax flavus (cellar slug)	Neu5Ac > Neu5Gc
Limulus polyphemus (horseshoe crab)	Neu5Ac
Tachypleus tridentatus (horseshoe crab)	Neu5Ac and Neu5Gc
Carcinoscorpius rotunda (horseshoe crab)	Neu5Ac(α2→6)Gal > Neu5Ac(α2→3)Gal
Homarus americanus (lobster)	Neu5Ac and Neu5Gc
Crassostrea gigas (Pacific oyster)	Neu5Ac
Masticoproctus giganteus (whip scorpion)	Neu5Ac
Allomyrina dichotoma (beetle)	Neu5Ac
Cepaea hortensis (snail)	Neu5Ac > Neu5Gc
Achatina fulica (snail)	Neu5Ac(α2→3)Gal > Neu5Ac(α2→6)Gal
Pila globosa (snail)	Neu5Gc
Androctonus australis (scorpion)	sialyllactose
Centruroides sculpturatus (scorpion)	Neu5Ac and Neu5Gc
Macrobrachium rosenbergii (prawn)	Neu5Ac
Aphonopelma cepaehortensis (spider)	Sialoglycoproteins
Heterometrus granulomanus (scorpion)	Neu5Ac(α2→3)lactose
Penaeus monodon (prawn)	Neu5Ac
Paruroctonus mesaenis (scorpion)	Sialoglycoproteins
Protozoan lectins	
Plasmodium falciparum (merozoite binding lectin of malarial parasite)	Neu5Ac
Trypanosom cruzi (Chagas disease)	Neu5Ac
Tritrichomonas species	(α2→3) and (α2→6) linked Neu5Ac
Plant lectins	
Maackia amurensis isolectins I and II	Neu5Ac(α2→3)Gal(β1→4)GlcNAc
Sambucus nigra isolectin I (elderberry)	Neu5Ac(α2→6)Gal/GalNAc
Trichosanthes japonicum	
Triticum vulgaris (wheatgerm)	GlcNAc > Neu5Ac
Mycoplasma	
Mycoplasma pneumoniae	Neu5Ac(α2→3)Gal of polylactosamine chains
Bacterial adhesins and toxins	
Escherichia coli S-adhesin	Neu5Ac(α2→3)Gal(β1→3)GalNAc

(*continued*)

Table 11.3 *(continued)*

Lectin/source	Binding partners
Helicobacter pylori adhesins I and II	Neu5Ac(α2→3)lactose > Neu5Ac(α2→6)lactose
Mycoplasma pneumoniae haemagglutinin	O-linked sialylated tetrasaccharides
Streptococcus sanguis	
Streptococcus suis haemagglutinin	Neu5Ac(α2→3)Gal(β1→4)GlcNAc(β1→3)Gal
Bordatella bronchiseptica haemagglutinin	Neu5Ac
Pseudomonas aeruginosa haemagglutinin	Neu5Ac
Vibrio cholerae (cholera) toxin	GM1
Clostridium tetani (tetanus) toxin	Sialoglycolipids
Clostridium botulinium (botulism) toxin	
Bordetella pertussis whooping cough toxin	Neu5Ac
Viral haemagglutinins	
Influenza A and B haemagglutinin	Neu5Ac(α2→3)Gal or Neu5Ac(α2→6)Gal
Influenza C haemagglutinin-esterase	N-acetyl-9-O-acetyl sialic acid
Polyomaviruses	Neu5Ac(α2→3)Gal(β1→3)GalNAc or sialyl(α2→3)Gal(β1→3)[sialyl(α2→6)]GalNAc
Rotaviruses	Neu5Ac
Newcastle disease virus haemagglutinin-neuraminidase	
Sendai virus haemagglutinin-neuraminidase	Neu5Ac(α2→3)Gal
Fowl plague virus haemagglutinin-neuraminidase	
Coronavirus haemagglutinin-esterase	(α2→3) linked $Neu5,9Ac_2$

them especially well suited as molecular determinants of biological functions.➲ In other cases they act to mask underlying glycan structures from exposure.

Briefly, their main biological functions might be summarized as:

- influencing conformation, stability, charge and viscosity, particularly of soluble glycoproteins;
- protection of cells and molecules from attack and degradation;
- modulation of cell adhesion, aggregation and agglutination;
- as ligands for receptors on other cell types.

> ➲ Physical characteristics of sialic acids on oligosaccharide chains
> - hydrophilic
> - negatively charged
> - structurally diverse
> - physically large
> - in exposed positions.

11.8.1 Functions associated with the negative charge of sialic acids

Sialic acids, as their name suggests, are acidic, with a p*K* value ≈pH 2.➲ Under physiological conditions, they dissociate fully into their component ions, and act as anions for binding for organic cations. This is relevant in, for example, the binding and subsequent uptake of cations at the cell surface. Furthermore, the juxtaposition of negatively charged sialic acid residues on cell surfaces results in electric repulsion and may prevent cell–cell or membrane–membrane contact and/or adhesion.

> ➲ The negative charge on sialic acids results in:
> - cation binding
> - electrostatic repulsion
> - formation of calcium bridges
> - high viscosity of mucins.

This is one proposed role of the MUC1 mucin, which is an epithelial sialomucin present on many secretory epithelial cell types, and also in cancers arising from them (*see 5.10, 16.6.10*). It has been proposed that the anti-adhesion function of the repulsing negatively charged sialic acid residues may play a role in cancer cells disassociating from the primary tumour during the metastatic process.

The glomerular filtration system of the kidney is less permeable to charged molecules than to neutral molecules of a similar size. This serves to retain many anionic proteins, such as albumin, in circulation. This selectivity results from the presence of negatively charged sites on the capillary endothelium, on the coat of the podocyte foot processes, and within the basal lamina of the glomerulus. The molecule responsible for the negative charge of the podocyte foot processes has been identified, and is a sialylated glycoprotein of molecular mass 140 kDa called podocalyxin. Similarly, the negative charge of the basal lamina also results, in part, from the presence of a sialylated molecule, in this instance, collagen type IV (*see 8.2.1*), plus the highly anionic heparan sulfate (*see 8.3.3*).

Negatively charged sialic acids may also be involved in charge-mediated cell adhesion processes, through the formation of Ca^{2+} bridges. Ca^{2+} ions also bind to the sialic acids of gangliosides, which are located in clusters on neuronal and, in particular, in synaptic membranes close to calcium pumps. The sialic acid moieties may function in supplying Ca^{2+} ions to the neuronal cells, for both immediate relay of nerve signals and also neural cell adaptation associated with long term information storage.

Mucins are heavily glycosylated and, often, heavily sialylated, molecules (*see Chapter 5*). They are secreted abundantly at epithelial surfaces such as those lining the respiratory, gastrointestinal and genitourinary tract where they play multiple roles in lubrication and protection. The viscosity of mucins, necessary for their role in lubrication, is dependent to a large extent upon their charge, determined by sialylation, sulfation and in lower organisms, also by the presence of other anionic components such as uronic acids and phosphate, which attract water molecules. Sialylation of mucins also contributes to other aspects of their protective function as described below.

11.8.2 The function of sialic acids in maintaining three-dimensional conformational stability

The presence of sialic acid residues alters the conformation of glycoproteins and gangliosides, influencing their physical properties both in biological membranes and in soluble form, for example, in sialylated soluble enzymes and hormones. ➲ Sialic acid residues have a stabilizing effect on the three-dimensional structure of glycoconjugates and enzymatic desialylation often results in significant conformational changes. For example, the sialylation of fibrinogen influences its conversion to fibrin and plays a role in stabilization of the fibrin clot.

> ➲ Sialylation contributes to conformational stability of glycoconjugates.

Somatostatin is a hormone involved in the regulation of secretion of human growth hormone (hGH) from the anterior pituitary. hGH has multiple effects on cell metabolism through stimulation of protein synthesis and inhibition of protein breakdown, stimulation of lipolysis and retardation of glucose use for ATP production, which have the overall effect of stimulating the growth of skeleton and skeletal muscle during childhood and adolescence and promoting tissue repair in adults. Many of these effects are mediated through stimulation of synthesis and secretion of insulin-like growth factors (IGFs) by the liver. Desialylation of the somatostatin receptor leads to a conformational change resulting in weaker ligand binding and thus in a reduced biological effect.

11.8.3 Functions of sialic acids in protecting cells and molecules from attack and degradation

Sialic acids are abundant at the cell surface and on secreted molecules, where they are ideally situated to a role in protection of the cell against damaging environmental influences such as digestion by proteases, infection by microorganisms and phagocytosis by other cells.

In the highly acidic environment of the stomach, the negatively charged sialic acids of gastric mucins (produced by the lining epithelium) bind bicarbonate ions, maintaining a pH gradient from pH 1–2 in the stomach lumen to a protective pH 6–7 at the epithelial cell surface and thus protecting it from damage (*see 5.14.1*).

The glycosylation of protein molecules is known to protect them against proteolysis, and the presence of sialic acid residues is particularly effective in this respect. One example of this is the extensive sialylation of glycoproteins of the plasma membrane surrounding the lysosome. This is believed to protect against the degradative effects of the lysosomal hydrolases housed within. A similar role is played by the heavily glycosylated, and sialylated, mucins secreted by the epithelial cells lining the digestive tract which, in addition to protecting the mucosa from extremes of pH in the stomach, also protect against proteolytic attack by digestive enzymes.➲

➲ Sialic acids protect cells and membranes from acidic pH and enzymatic proteolysis.

Another example is sialylation of the influenza virus haemagglutinin which is associated with viral virulence through a regulatory role in hydrolysis of peptide bonds by host proteases.

In different circumstances, the protective effect of glycosylation may result from the three-dimensional conformational changes induced by the presence of the glycans, by the electrical charge associated with sialic acids, or by a physical shielding of the polypeptide backbone from the action of proteolytic enzymes. This is achieved by the oligosaccharide chain preventing enzyme access to the protein backbone.

11.8.4 Sialic acids as binding partners for lectins

A large number of sialic acid binding lectins have been described in vertebrates, invertebrates, plants, protozoa, bacteria and viruses. Some of the specific functions of these molecules in binding to sialic acids are summarized below.

1. Infection may be mediated through the recognition of host sialic acid residues by pathogen lectins. The best studied of these include infection by the bacteria *Helicobacter pylori* (*see 16.2.1*), influenza viruses (*see 16.2.2*) and malarial parasites (*see 16.2.3.3*). Bacterial toxins, such as those from *Vibrio cholerae* (cholera), *Clostridium tetani* (tetanus), *Clostridium botulinium* (botulism) and *Bordetella pertussis* (whooping cough) toxin also act through lectin recognition of sialic acid residues of gangliosides (sialic acid-containing glycolipids). The implication that recognition of sialic acid residues by lectins, and sialidases, on pathogenic organisms may play an important role in infection has led to speculation that the structural diversity of sialic acids seen in higher organisms may be the result of continuing evolutionary pressures on behalf of both pathogen and host to outwit each other. It has been proposed, for example, that the O-acetylation of sialic acids present on erythrocytes may have been a factor that gave a form of defence against the malarial parasite and was therefore advantageous and thus evolutionarily favourable. Similarly, increased sialic acid on cells lining the gastrointestinal tract during development appears also to be a response to microbial colonization.
2. The recognition of sLe^a and sLe^x by selectins is critical to leukocyte homing in the inflammatory response, and possibly in cancer metastasis (*see 13.12.3.2, 16.6.7*).
3. The complement-regulating factor H, a soluble serum lectin, recognizes sialic acid and upon binding restricts, and therefore regulates, the alternative pathway activation.
4. Recognition of the sialylated glycans of CD45 by the B lymphocyte I-type lectin, CD22, mediates homo- and heterotypic cell–cell adhesion and signalling integral to B-cell activation (*see 11.8.6.2*).
5. Sialoadhesin, expressed by macrophages, binds sialylated N- and O-linked glycoproteins and glycolipids on cells of the granulocyte lineage. This mediates the interaction between macrophages and developing myeloid precursor cells and may thus have a role in haematopoiesis (*see 13.12.4.1*).
6. Muscle laminin may have sialic acid binding activity and interact with a novel sialylated O-Man linked glycan on β-dystroglycan (*see 7.8.1*).
7. Myelin-associated glycoprotein (MAG) binds sialylated, probably O-linked, neuronal gangliosides and is involved in ensuring the proper myelination of axons (*see 13.12.4.2*). On binding to its sialylated ligand, MAG induces signal transduction pathways

associated with myelination. Furthermore, myelin-associated sialidase adhering to ganglioside GMI may be important for the formation and stabilization of the multilamellar structure of the myelin sheath.

8. Glycophorin A, a major glycoprotein of the erythrocyte membrane is heavily sialylated, and thus gives the erythrocyte surface a strong negative charge. Sialylation decreases as erythrocytes age and as they become desialylated they are cleared from circulation. Enzymatic desialylation of human erythrocytes results in their disappearance from circulation in a few hours, in comparison with their normal circulatory half-life of ≈120 days. The mechanism for the *in vivo* clearance of asialo-erythrocytes is recognition of normally sub-terminal Gal by a lectin, the mammalian liver asialoglycoprotein receptor, on Kupffer cells (*see 11.8.5.2., 13.12.3.1*).

11.8.4.1 An example of an important sialylated glycan structure: sialyl Lewis x

One of the most intensively studied sialylated glycan structures is sLe^x, one of the Lewis family of blood group sugars (*see 10.6*). ➲ sLe^x is of special interest because it acts as a binding partner for selectins, as does the closely related glycan sLe^a (*see 13.12.3.2*).

> ➲ sLe^x is a binding ligand for selectins and is involved in leukocyte homing in inflammation. sLe^x has the structure [NeuAc(α1→3)] Gal (β1→4)[Fuc(α1→3)]GlcNAc. sLe^a has the structure [NeuAc(α1→3)] Gal (β1→3)[Fuc(α1→4)]GlcNAc.

Lymphocyte adhesion to the activated endothelium lining peripheral lymph nodes is mediated through the interaction of L-selectin, a C-type lectin (*see 13.12.3.2*) expressed by lymphocytes, and its saccharide-binding partner Le^x, present on the endothelial cells. This process has been studied extensively (*see 16.3*). Briefly, endothelial cells activated by cytokines synthesize sialylated glycans including sLe^x and sLe^a (*Figure 11.6*). L-Selectin, expressed by neutrophils, monocytes and a subset of T lymphocytes, recognizes and binds, with fairly low affinity, to these sialylated structures. Low-affinity binding, coupled with the shear forces of blood flow, result in leukocyte 'rolling' along the endothelium. At this stage, other molecules, particularly members of the integrin superfamily, become involved in stronger adhesion of leukocyte to endothelium, preparatory to leukocyte extravasation.

The cells of some cancers frequently synthesize these sialylated Lewis antigens and may interact with selectins in the same way as leukocytes do, thus 'hijacking' the elegant mechanism for the arrest of leukocytes in microvascular endothelium and thus facilitating subsequent invasion into that organ during the metastatic cascade.

11.8.5 Sialic acids as masking agents

11.8.5.1 Roles in protecting microorganisms from host immune attack

Sialic acids are implicated in shielding microorganisms, especially parasitic trypanosomes (*see 11.5, 11.8.5.3*), bacteria and pathogenic fungi, from phagocytosis by the cells of the

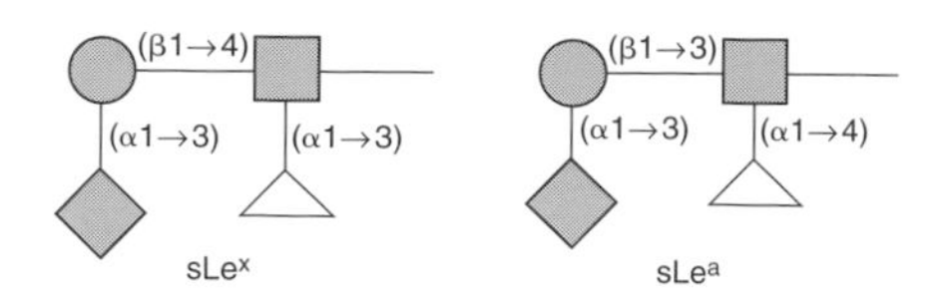

Figure 11.6

Structure of sLe^x and sLe^a.

organisms they infect. Desialylation of the pathogenic fungus *Sporothrix schenkii*, for example, increases the efficiency of its phagocytosis by human macrophages.

11.8.5.2 Roles in clearing aged and damaged cells and glycoconjugates

Sialic acids are implicated in phagocytic mechanisms for destroying ageing and damaged cells.➲ One of the best documented examples of this is the action of the mammalian liver asialylglycoprotein receptor which recognizes oligosaccharides bearing terminal GalNAc and Gal residues, monosaccharides frequently masked by terminal sialic acid (*see 13.12.3.1*). When serum glycoproteins become old and damaged, they lose their sialic acid residues, exposing normally sub-terminal GalNAc/Gal residues. The GalNAc/Gal residue is then recognized and bound by the mammalian liver asialoglycoprotein receptor, and the glycoprotein is removed from circulation. Similarly, red blood cells desialylated by viral sialidases or simply as a result of ageing, are removed from the circulation by liver and spleen macrophage lectin which recognizes exposed Gal and GlcNAc residues. Apoptosis of other cell types leads to desialylation and exposure of normally sub-terminal Gal, resulting in recognition by phagocytotic cells.

> ➲ Mammalian asialoglycoprotein receptors remove damaged and ageing desialylated glycoproteins and cells from the circulation.

11.8.5.3 Roles in modulating immune response

Desialylation is also a modulator of the immune system and complement reactivity. Animal cells, and the microorganisms that infect them, are shielded from immune recognition, and thus immune attack, by sialylation. Sialylation may therefore be a powerful protective mechanism employed by viruses and bacteria to shield themselves from the host immune system. An example of this is that the virulence of the bacteria *Neisseria gonnorrhoeae* is significantly enhanced by its sialylation, as sialic acid residues hide immunogenic bacterial porin epitopes and protect the organism against immune attack. Here, the terminal Gal(β1→4)GlcNAc of a bacterial lipopolysaccharide becomes sialylated once the bacteria is within the host. This is achieved through the action of a bacterial sialyltransferase that utilizes the substrate CMP-Neu5Ac supplied by the host.

Sialic acid is also used as a shield by the parasites trypanosomes (*see 16.2.3.1*). They become sialylated once they are within the host organism, but the mechanism is quite different to that employed by *N. gonnorrhoeae*; in trypanosomes, a bacterial *trans*-sialidase is employed to transfer terminal sialic acid from host oligosaccharides into position on the trypanosomal glycoconjugates.

Sialylation may protect animal cells from autoimmune attack, and conversely, desialylation, as a result of the action of viral and bacterial sialidases, may result in autoimmune disease.➲ Examples of this include:

> ➲ Sialylation may protect viruses and bacteria from attack by host immune defences, but conversely desialylation of cells may result in autoimmune attack.

- streptococcal infection may result in glomerulonephritis – an autoimmune damage that occurs in the kidney
- asthma and bronchopulmonary disease can result from desialylation of lung epithelia after influenza virus infection
- bacterial sialidases in the bloodstream can expose normally sialylated, and therefore cryptic, Thomsen Friedenreich (TF) antigen on red blood cells (*see 5.6.3*). Most individuals have naturally occurring antibodies against the TF antigen, and this results in destruction of the erythrocytes and anaemia
- the genetic blood cell disorders sickle cell anaemia and β-thalassaemia are related to reduced metabolism of sialic acid, and this can result in the shortened half-life of erythrocytes and consequent anaemia

- sialylation of cancer cell glycoconjugates can have an immunosuppressive effect, protecting the tumour cells from recognition, and attack, by the body's immune defences.

11.8.5.4 Sialylation as a protection against infection

Although sialic acids may be recognized as the specific binding ligands of bacterial adhesins and are therefore implicated in the infection process, in some instances synthesis of sialic acids by host cells can actually protect against infection. Examples of this include:

- adhesion of *Bacterioides intermedius* to human buccal cells and erythrocytes is enhanced when terminal sialic acid residues are stripped by sialidase treatment
- *Actinomyces viscosus* secretes its own sialidase to expose recognition sites hidden by sialic acid on the host cells
- *Pseudomonas aeruginosa*, a pathogen of particular danger to cystic fibrosis sufferers, binds specifically to a non-sialylated glycoconjugate, asialo-GM1.

11.8.5.5 Exposure and masking of sialic acid residues in the regulation of cell–cell interactions

The coordinated action of sialyltransferases (in appending sialic acids to oligosaccharides and thus masking sub-terminal monosaccharide residues) on the one hand, and sialidases (in cleaving terminal sialic acid and revealing sub-terminal monosaccharide residues) on the other hand, may be an effective method of regulating cell–cell interactions. For example, in the essential process of mammalian reproduction the presence/absence of sialic acids are important. Desialylation of sperm results in better attachment to the zona pellucida of the oocyte, and is believed to be necessary for effective fertilization (*see 5.14.6*).

Furthermore, this type of mechanism may also play a part in the action of some hormones. Many glycoprotein hormones, for example, prolactin, chorionic gonadotrophin (*see 7.6*) and erythropoietin (*see 18.5.1*), exhibit increased binding to their specific receptors when partially or fully desialylated. It is thought that sialylation/desialylation may function in mediating the efficiency of the signal transduction pathways that are initiated as a consequence of hormone–receptor binding.

11.8.6 Specific sialic acid linkages and their functions

In the preceding sections, some functions of sialic acids have been outlined. Some of what is known about the precise significance of the different linkages, (α2→3), (α2→6) or (α2→8), of sialic acids is given here.

11.8.6.1 (α2→3)-Linked sialic acid

Sialic acid (α2→3) linked to Gal is a common terminal monosaccharide on the glycans of glycoproteins and glycolipids of vertebrates. It is synthesized through the action of a family of at least five different α2,3 sialyltransferases (ST3Gal-I, ST3Gal-II, ST3Gal-III, ST3Gal-IV and ST3Gal-V) which are expressed in a tissue-dependent manner, as listed in *Table 11.4*.

A number of functions has been attributed to terminal (α2→3)-linked sialic acid, some of which have already been described in the general discussion previously, and are summarized below:

1. Glycans terminating in (α2→3)-linked sialic acid can act as binding ligands for members of the selectin family (*see 11.8.4.1*).
2. The 'capping' of sub-terminal Gal residues by (α2→3)-linked sialic acid protects the glycoconjugate from recognition, and removal from circulation, by the asialoglycoprotein receptor (*see 11.8.5.2, 13.12.3.1*).
3. (α2→3)-Linked sialylation of peripheral T lymphocytes is necessary for cell viability.

Table 11.4 Tissue-dependent expression of the α2,3 sialyltransferase family

Enzyme	Tissue expression
ST3Gal-I	Spleen, liver, bone marrow, thymus, salivary gland
ST3Gal-II	Brain
ST3Gal-III	Ubiquitous
ST3Gal-IV	Ubiquitous
ST3Gal-V	Brain, skeletal muscle, testes, liver

Table 11.5 Activity of the α2,6 sialyltransferase family

Enzyme	Activity
ST6Gal-I	Responsible for (α2→6) sialylation of serum, lymphocyte and hepatocyte glycoproteins
ST6GlcNAc-I	Glycosylates only O-glycans
ST6GalNAc-I	Glycosylates only O-glycans
ST6GalNAc-II	Glycosylate only O-glycans
ST6GalNAc-III	Substitutes the core GalNAc of O-glycans
ST6GalNAc-IV	Attaches terminal (α2→6) linked sialic acid to the chains of glycolipids

4. (α2→3)-Sialylated glycans act as ligands for the lectins of pathogens, for example, influenza virus (*see 16.2.2*), *Helicobacter pylori* (*see 13.9, 16.2.1*), and the toxin of *Vibrio cholerae* (*see 11.8.4*).

11.8.6.2 (α2→6)-Linked sialic acid

(α2→6)-Linked sialic acid is a common terminal monosaccharide of glycan chains of glycoproteins and glycolipids in most vertebrate species, although it is less common than the (α2→3) linkage. It is the most common sialic acid linkage found in N-linked glycans (*see Chapter 4*), although the (β1→4)-linked terminal Gal of the lactosamine repeat of O-glycans (*see 5.6.8, 5.6.9*) can also be modified in this way.

(α2→6)-Linked sialic acid is synthesized by the action of a family of at least six different α2,6 sialyltransferases, designated ST6Gal-I, ST6GlcNAc-I, ST6GalNAc-I, ST6GalNAc-II, ST6GalNAc-III and ST6GalNAc-IV, which attach (α2→6)-linked sialic acid to terminal Gal, and terminal and sub-terminal GlcNAc or internal GalNAc moieties. These enzymes are expressed in a tissue dependent manner, and are listed in *Table 11.5*.

(α2→6)-Linked sialylation normally precludes further chain modification, except sometimes by members of the (α2→8) sialyltransferase enzyme family (*see 11.8.6.3*).

The functions of (α2→6) sialylation of oligosaccharide chains is poorly understood, but:

1. It may be similar to (α2→3)-linked sialylation in that (α2→6) sialylation also masks subterminal monosaccharides, conceivably protecting the glycoconjugates from recognition and clearance by the mammalian asialoglycoprotein receptor (*see 11.8.5.2, 13.12.3.1*).
2. There is also some evidence that (α2→6)-linked sialic acid, as well as the (α2→3)-linked form, is a binding partner for the influenza virus haemagglutinin (*see 11.8.3, 11.8.4*).
3. A specific function of (α2→6)-linked sialic acid has been described in signal transduction events of the immune system: the B-lymphocyte I-type lectin CD22 specifically recognizes the product of the action of the ST6Gal-I enzyme, Sia(α2→6)Gal(β1→4)GlcNAc, as its binding partner. This oligosaccharide is synthesized by many leukocyte glycoproteins, including the N-glycans of the tyrosine phosphatase, CD45. This glycosylation has been suggested to play a role in specific recognition between B lymphocytes and CD45 positive T cells, neutrophils, monocytes and erythrocytes, leading to activation of a signal transduction pathway.

4. In addition to its presence on CD45, Sia(α2→6)Gal(β1→4)GlcNAc is also present on CD22, suggesting a role in homophilic adhesion and signalling. This type of receptor–ligand recognition is believed to be involved in early B-cell activation and may modulate signalling through surface IgM receptor complex. Lymphocyte activation causes an increase in the activity of the ST6Gal-I that in turn leads to an increase in (α2→6) sialylation, thus mediating cell adhesion and trafficking at discrete stages in B-cell differentiation. In further support of this hypothesis, mice lacking functional ST6Gal-I enzyme are immune deficient.

11.8.6.3 (α2→8)-Linked sialic acid

(α2→8)-linked sialic acid is of interest almost exclusively in the context of the developmental regulation of the nervous system. (α2→8)-linked sialylation is achieved by a family of at least five different α2,8-sialyltransferases (*see Table 11.2*).

ST8Sia-I, ST8Sia-III and STSia-V are responsible for synthesis of single or oligosidic (α2→8) sialic acid linkages to N-linked oligosaccharides (*see Chapter 4*) and glycolipid (*see Chapter 9*) glycan chains of molecules which are part of the ganglioside family. This is developmentally regulated, and the sialylated products are involved in the signal transduction cascade, although their precise function is not well understood. ➲

> ➲ (α2→8) linked sialic acid is a feature of:
> - N-linked oligosaccharides associated with developmental regulation of the nervous system
> - polysialic acid.

Two Golgi apparatus associated polysialyltransferases, ST8Sia-II (STX) and ST8Sia-IV (PST or PST-I) catalyse the synthesis of linear polymers of between 8 and 100 or more residues of, usually, (α2→8) linked sialic acid. This is known as polysialic acid (PSA), a highly negatively charged and thus highly hydrated molecule, which may be attached to an initial (α2→3)-, (α2→6)- or (α2→8)-linked sialic acid to an outer arm branch of an N-linked oligosaccharide. PSA may be formed of chains of KDN, Neu5Gc or Neu5Ac, which may be O-acetylated at the 7- or 9-position.

The enzymes, ST8Sia-II and ST8Sia-IV may themselves be subject to polysialylation. A simpler form of polysialylation, involving short 'oligosialic acid' chains of two or three sialic acid residues may be quite a common modification of many glycoconjugates, but their function is still poorly understood.

Polysialic acids are also synthesized by prokaryotes, but here a single enzyme is involved: this adds both the first and subsequent sialic acids to the extended oligosaccharide chain (*see 7.2.2*).

11.9 Polysialic acid

Polysialic acid (PSA) is a feature of neural cell adhesion molecule (N-CAM). N-CAM is a member of the immunoglobulin superfamily, and the PSA constituent makes up as much as one-third of the molecular mass of the molecule. PSA is also a feature of fish egg glycoproteins and capsular polysaccharides of some pathogenic bacteria. It is involved in homotypic cell adhesion mechanisms and also, as it is very heavily negatively charged, inhibits the interaction of other cell adhesion molecules by effectively repelling cell membranes of neighbouring cells. ➲ N-CAM mediates PSA-dependent modulation of homotypic and heterotypic cell-adhesion mechanisms important in the targeting of growing axons to sites of innervation, migration of neuronal cells during neuronal development, plasticity of the central nervous system (e.g. neurite sprouting, repair of damage and axon migration) and function (e.g. the ventricular conduction system of the heart and the sodium channels associated with conductivity in the brain). PSA has been mapped in mouse embryos before and during implantation. Polysialylation and sialic acid O-acetylation in neural gangliosides

> ➲ Biological roles of PSA include:
> - cell migration
> - axonal guidance
> - synapse formation
> - functional plasticity of the nervous system.

vary with development, with hibernation, and is different in cold-blooded and warm-blooded species. Polysialic acid is also implicated in the larval stages of *Drosophila melanogaster*, this is of particular interest as sialic acids are not usually found in insect cells.

Owing to the anomalously high p*K* value of the internal Sia residues, polysialic acid is very labile and can be easily degraded spontaneously, even under physiological conditions. Intramolecular self-cleavage may be implicated in regulatory mechanisms involved in the function of this molecule.

11.9.1 Biological roles of PSA

11.9.1.1 Neural cell migration

PSA has been implicated in some types of cell migration found in the nervous system, specifically cell migration involving:

- axophilic migration of neurons or glia along axon fibres
- cooperative streaming of neural precursors.

PSA involvement in cooperative streaming is illustrated by the PSA-sensitive axophilic migration of luteinizing hormone releasing hormone (LHRH) neurons along PSA-positive axons to the forebrain. Mouse mutants deficient in PSA/N-CAM show an accumulation of neural precursors at their site of generation because they are unable to migrate normally to the forebrain as the required guidance is lacking. Here, PSA is implicated in cycles of adhesion/de-adhesion as cells migrate by using each other as a substrate for elongation of their leading processes, in a kind of 'leapfrog' mechanism. Transient expression of PSA/N-CAM is also associated with the de-adhesion of cell clusters during the separation stage in development of muscle fibres.

11.9.1.2 Axonal guidance

PSA is involved in the behaviour of axons as they grow towards their targets in order to innervate them, where it is mostly involved in reducing the interactions between neighbouring axons so that they are able to respond effectively to extrinsic signalling events.

11.9.1.3 Synapse formation

PSA is implicated in the formation of plastic modifications of synaptic transmission, although not in synaptic transmission itself.

11.9.1.4 Plasticity of the nervous system

In contrast to many other molecules, PSA presence on N-CAM is retained in adulthood. Although synthesis is reduced, it persists, especially in the hippocampus and the olfactory bulb, reflecting the need to maintain plasticity of the nervous system throughout life. In the hippocampus PSA is associated with the ability to make morphological rearrangements of synapses. Furthermore, it is necessary for maintenance of circadian rhythms in animals exposed to changing day/night cycles.

11.9.2 Mechanisms of PSA action

PSA clearly has a number of quite dramatic functional effects, but the mechanisms involved are not well understood.➲ One theory is that the regulation of cell–cell interactions mediated by PSA is simply a result of its large size, which may interfere with the physical membrane–membrane apposition of neighbouring cells: the long PSA molecule would, in effect, push them apart. Alternatively, PSA might alter the interaction between receptors and ligands within the

➲ PSA acts via three different mechanisms:

- *size*: steric hindrance or clustering of receptors
- *charge*: repulsion or pH
- *affinity*: recognition by specific receptors.

same cell. Clearly, another major physical attribute of PSA is its negative charge. This could also result in repulsion between adjacent molecules. Furthermore, PSA/N-CAM has a pI of 4.5 which means that it would contribute to a zone of low pH which could affect receptor mediated interactions. Finally, PSA may act as a specific binding ligand for a receptor molecule. There is evidence, for example, for PSA-dependent binding of N-CAM to heparan sulfate (*see 8.3.3*) proteoglycans, mediating adhesion, and there is speculation that PSA might act as a binding partner for soluble growth factors presented by proteoglycans. In line with these observations, PSA has been proposed as a serum tumour marker in some malignancies.

11.10 Functional consequences of disrupted sialylation

11.10.1 Disruption of sialylation in cancer

There is often a striking change in both the types of sialic acids produced, and their levels of synthesis in cancers, and in some cases this has been shown to be associated with the metastatic potential of the cancer (*see 16.6.3*). In general, there is an increase in sialic acid in cancers, in particular (α2→6)-linked Sia. In contrast, in some cancers, such as colorectal cancer, a significant decrease in O-acetylation of Sia residues at the 9-carbon position has been reported. In most cases, the change associated with malignancy mirrors the glycosylation pattern typical of embryonic tissues associated with that site (it may therefore be described as an 'oncofetal' glycotope in this situation).➲ Increased sialylation may be the consequence of disruption in normal cellular mechanisms, including a decrease in sialidase activity. In many tumour types, including colorectal cancer, thyroid cancer and leukaemia, increased sialylation is indicative of poor prognosis. This may also be associated with an increase in number and degree of branching of N-glycans and polylactosamine chains.

➲ Sialylation is commonly disrupted in cancer.

An increase in sialic acid residues on membrane glycoproteins and gangliosides is associated with increased invasive potential and reduced intercellular adhesion, associated with metastatic competence in some cancer cell types. Neu5Gc has been proposed as an oncofetal glycotope in that it is rarely found in adult tissues, but is commonly synthesized in fetal tissues. In adult humans, it is antigenic and cancer patients often produce a strong antibody response against it.

11.10.2 Metabolic disorders involving sialic acid

Inborn errors in metabolism of sialic acid result in rare genetic disorders such as galacto-sialidosis (sialidosis) and Salla's disease (*see 16.9*).

11.10.3 Neurodegenerative disorders

There are also reports that a decrease in the activity of α2,3-ST and a resulting decrease in (α2→3)-linked Sia on serum glycoproteins accompanies Alzheimer's disease and is common in older Down's syndrome subjects. Desialylation of serum glycoproteins may be an indication of a similar decrease in sialyltransferase activity in neuronal tissue, associated with neurodegenerative disease and impaired mental function.

Further reading

Rosenberg, A. (1995) *Biology of the Sialic Acids*. Plenum Press, New York.

Rutishauser, U. (1996) Polysialic acid and the regulation of cell interactions. *Curr. Opin. Cell Biol.* **8:** 679–684.

Schauer, R., Kamerling, J.P. (1997) Chemistry, biochemistry and biology of sialic acids. In J. Montreuil. J.F.G. Vliegenthart & H. Schachter (eds), *Glycoproteins II*, pp. 243–402. Elsevier Science, Amsterdam.

Varki, A. (1992) Diversity of the sialic acids. *Glycobiology* **2:** 25–40.

Glycosylphosphatidylinositol (GPI)-anchored proteins

12

12.1 Prevalence

In the mid-1980s it was discovered that many cell-surface glycoproteins are attached to the top leaflet of the lipid bilayer via a glycosylphosphatidylinositol (GPI) anchor and such proteins were called 'glypiated proteins'. Many initial studies with GPI anchors were conducted using the membrane form of variant surface glycoprotein (mfVSG) of African trypanosomes (*Trypanosoma brucei*) (*see 16.2.3.1*). ➲ Since that time, hundreds of glypiated proteins have been found on cell membranes in species ranging from bacteria, yeasts and slime moulds to invertebrates and vertebrates, and it is thought that glypiated proteins are ubiquitous throughout the animal kingdom. In humans, 20 different GPI-anchored proteins occur on blood cells alone. The diversity of GPI-anchored proteins is indicated by *Table 12.1*.

> ➲ GPI-anchors are ubiquitous throughout the animal kingdom.

12.2 Structure

GPI-anchored proteins have distinct structural features and *Figure 12.1* provides a diagram of a GPI anchor, using data from the protein Thy-1. The protein part of the anchor is linked at its carboxyl end (via a phosphodiester linkage) to a phosphoethanolamine linked to an oligosaccharide. The oligosaccharide of the GPI anchor comprises three Man sugars and a non-acetylated glucosamine (GlcN). The core oligosaccharide is a tetrasaccharide comprising: Man ($\alpha 1 \rightarrow 2$) Man ($\alpha 1 \rightarrow 6$) Man ($\alpha 1 \rightarrow 4$) GlcN. The reducing end of the GlcN is linked via an inositol group to the phosphatidylinositol (PI) of the lipid; this part of the molecule 'anchors' it to the membrane, whereas the other end of the oligosaccharide is linked to a phosphoethanolamine group. The phosphoethanolamine is linked to the protein, which is then described as GPI anchored. The oligosaccharides and lipids of GPI-anchored proteins are processed further as the protein passes through the secretory pathway of the cell and this leads to cell-type-specific and protein-specific GPI-anchor expression.

The oligosaccharides of GPI-anchored proteins are distinct from N-linked (*see Chapter 4*) and O-linked (*see Chapter 5*) oligosaccharides. ➲ Apart from differences in the linkages of the monosaccharides,

> ➲ The oligosaccharides of GPI anchors differ from those of the N- and O-linked oligosaccharides because of the bonds between the monosaccharides and the non-acetylated form of GlcN.

Table 12.1 Examples of some GPI-anchored proteins

Enzymes	Renal dipeptidase, alkaline phosphatase, lipoprotein lipase
Antigens	Carcinoembryonic antigen, Thy-1, Blast-1, variant surface glycoprotein (Trypanosome)
Adhesion molecules	Neural cell adhesion molecule (N-CAM), guinea-pig sperm receptor (PH-20), glypican heparan sulfate proteoglycans

Functional and Molecular Glycobiology, Susan A. Brooks, Miriam V. Dwek and Udo Schumacher

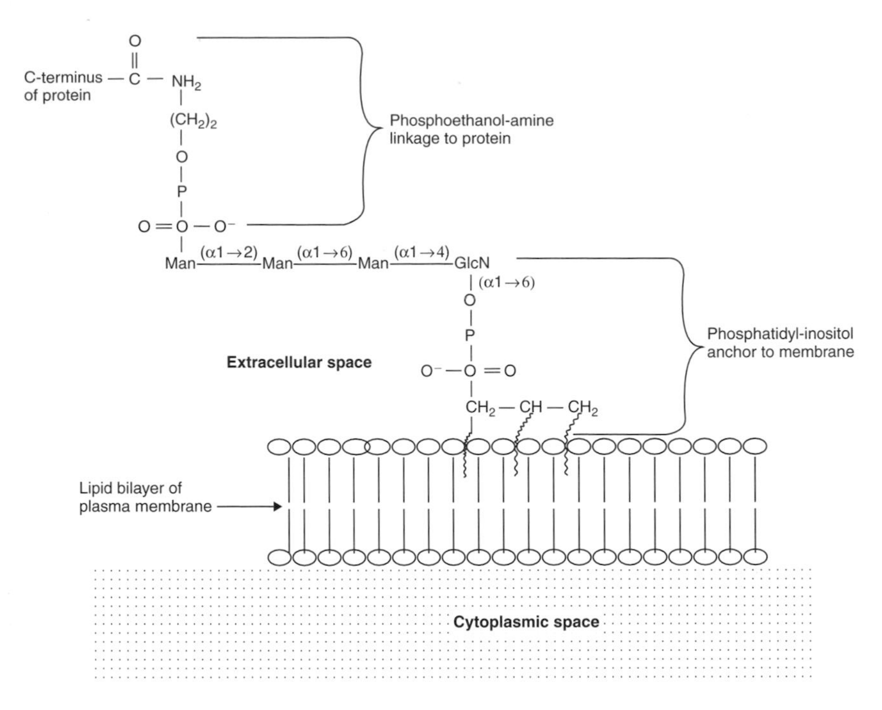

Figure 12.1

Diagrammatic representation of a GPI-anchored protein. The phosphatidylinositol group (indicated) serves to attach the protein via a glycan and a phosphoethanolamine linkage, to the membrane of the cell surface.

the most striking difference in GPI oligosaccharides is that GlcN is present in a non-acetylated form. This is a characteristic of GPI anchors and, with the exception of GlcN residues in heparin sulfate which are sulfated (*see 8.3.3*), this type of non-acetylated GlcN has not been found in other glycoconjugates.

12.2.1 Structural variation in GPI-anchored proteins

Variation in the structure of GPI anchors occurs as a result of the oligosaccharides attached to the Man_3 core as the protein passes through the secretory pathway of the cell. For example, in the mfVSG from African trypanosomes α-linked Gal is a monosaccharide constituent of the anchor oligosaccharide, whereas in Thy-1, β-linked GalNAc is added. Sialic acid may also be found as a minor glycan constituent of GPI anchors, as in prion protein. Alternatively, the fatty acid (lipid) part of the GPI anchor may vary, both in terms of chain length and the extent to which the fatty acids are saturated.

Inter- and intraspecies variation occur in the oligosaccharides and fatty acids of GPI-anchored proteins.

12.3 Models for studying the biosynthesis of GPI-anchored proteins

The biosynthesis of GPI anchors has, to a large extent, been elucidated using the mfVSG as a model (*see 16.2.3.1*). Each trypanosome contains $\approx 10^7$ VSG molecules on its surface and is therefore an abundant source of starting material for experiments. In addition to the mfVSG model, cell lines deficient in some of the enzymes involved in the biosynthesis of GPI anchors and temperature-sensitive yeast cells have also been used in *in vitro* model systems. The commercial availability of bacterial phosphatidylinositol-specific phospholipase C (PIPLC) to cleave the anchor from the protein (between the phosphate and glycerol on the anchor) has enabled demonstration of GPI-anchored proteins without the need for specialist laboratory equipment.

The study of a rare blood disorder, paroxysmal nocturnal haemoglobinuria (PNH), has led to a greater understanding of the biosynthetic steps involved in the production of GPI-anchored proteins. ➲ PNH is a genetic disorder affecting haematopoietic progenitor cells in which there is either a decrease or an absence of GPI-anchored proteins. This leads to aberrant expression of (normally GPI-anchored) proteins on haematological cells of affected individuals. One such protein is membrane inhibitor of reactive lysis (CD59). The absence of CD59 in PNH individuals leads to reduced red blood cell survival and PNH patients tend to present clinically with haemoglobin in their urine from the breakdown of red blood cells. The functional importance of GPI-anchored proteins is illustrated by the poor prognosis of patients with PNH, 85% of whom have a median survival of only 10 years after diagnosis.

➲ GPI-anchored protein biosynthesis has been studied in:
- MfVSG of trypanosomes
- mutant cell lines and yeasts
- using bacterial phosphatidylinositol-specific phospholipase C
- patients with paroxysmal nocturnal haemoglobinuria.

12.4 Biosynthetic steps

12.4.1 Synthesis of the GPI anchor

In mammals there are about twenty genes involved in the biosynthesis of GPI anchors and almost all of these have been cloned. The biosynthetic pathway for the production of GPI anchors is conserved throughout evolution with similarities and dissimilarities according to the species (*see 12.3*). Dissimilarities include differences in the substrates that the enzymes for GPI-anchor production will tolerate and differences in the way in which the fatty acid chain is reconstructed after attachment of the GPI anchor.

The main biosynthetic steps in the production of GPI-anchors are shown in *Figure 12.2*. The production of the GlcN and phosphatidylinositol (PI), *Figure 12.2* steps 1 and 2, is the first step in GPI-anchor production. Without this step, GPI-anchor production does not occur. In practice, an acetylated form of glucosamine (GlcNAc) is attached to PI. The acetyl group is then rapidly removed to leave the GlcN linked to PI. This, and the rest of the donor GPI-anchor production, takes place on the membrane leaflet of the endoplasmic reticulum (ER) in the cytoplasm of the cell. Here the Man residues are added (*Figure 12.2*, steps 3–4 and 5–6). Mannosylation occurs in a sequential manner (the Man that is utilized in these reactions is attached to dolichol; *see 4.2*). After addition of the first and the last of the Man residues, phosphoethanolamine is attached (*Figure 12.2* steps 4–5 and 6–7). It is thought that phosphatidylethanolamine is the donor molecule in this reaction. The production of the GPI donor requires energy input in the form of UDP-GlcN (*Figure 12.2* step 1).

12.4.2 Attachment of the GPI anchor to protein

Proteins to be GPI anchored contain two signal peptide sequences: an N-terminal signal peptide that directs the protein into the lumen of the ER via the signal recognition particle and a C-terminal sequence that signals attachment of the anchor. ➲

After the GPI anchor has been formed, it is transferred to the C-terminus of the protein. This step occurs ≈1–2 min after protein production has been completed and therefore GPI-anchor attachment is a posttranslational event. This step occurs in the lumen of the ER and is shown in *Figure 12.2* in steps 7–8.

➲ A protein that is intended to be GPI anchored contains a signal sequence of amino acids at both its N- and C-terminus.

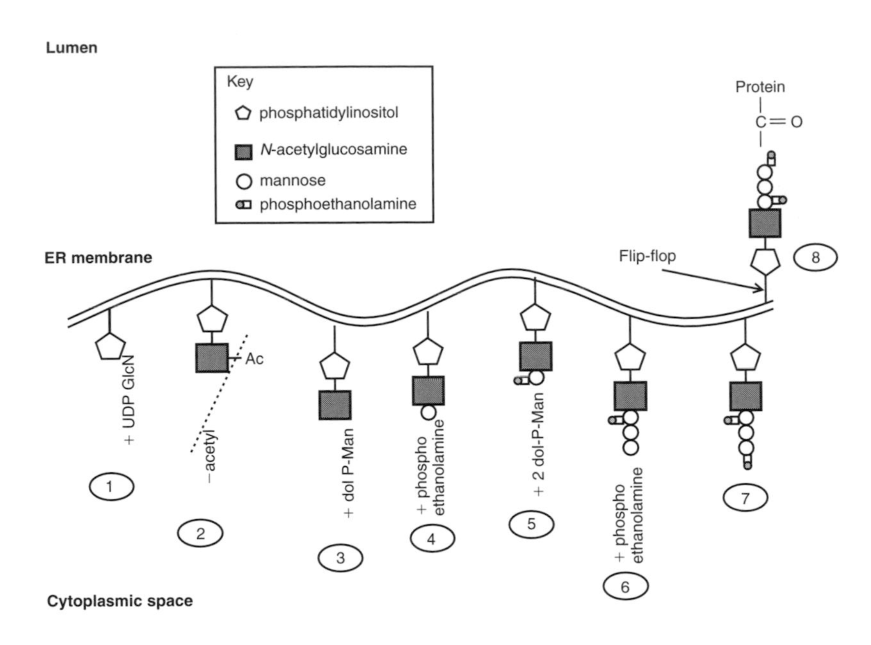

Figure 12.2

The main biosynthetic steps involved in the production of a GPI-anchored protein. A detailed explanation of the process is found in the text, the numbers refer to the steps as described in the text. Briefly, a phosphatylinositol has an N-acetylglucosamine monosaccharide attached. For this step, the activated nucleotide donor sugar (UDP-GlcNAc) is required. The GlcNAc moiety is then de-acetylated and further monosaccharides added. The glycans and phosphatidylinositol group are then attached to phosphoethanolamine and the entire structure is flipped into the lumen of the ER. The GPI anchor is subsequently linked to the C-terminal region of the protein and the protein relocated to the cell membrane in which it is anchored.

The C-terminal sequence has been shown to contain three sites: a hydrophobic region of the primary translation product, this is between 12 and 20 amino acids, a polar (spacer) region of between 8 and 12 amino acids and a cleavage site (where the anchor is attached). The cleavage site comprises six amino acids Ala, Asn, Asp, Cys, Gly or Ser and any of these can accept attachment of the anchor block. A transamidase enzyme (possibly Gpi8p) cleaves the peptide chain and attaches the GPI anchor to the protein. The fate of the original C-terminus signal peptide is unknown. The main features of the C-terminal region of GPI-anchored proteins are shown in *Figure 12.3*. The frequent expression of particular amino acids in the C-terminal sequence of GPI-anchored proteins have enabled researchers to develop algorithms to putatively identify GPI-anchored proteins.

After the GPI-anchored protein has been produced, it is transported through the ER for further posttranslational modifications and to the *cis*-Golgi where it is incorporated into lipid domains and delivered to the plasma membrane. ➲ The posttranslational modifications include modifications of the lipids and the glycans attached.

➲ A GPI anchor is attached to a protein in the rough ER as a posttranslational event.

12.5 Glycosylation of GPI-anchored proteins

Proteins that contain GPI-anchors may be N-glycosylated (*see Chapter 4*) and O-glycosylated (*see Chapter 5*). As the first step in the biosynthesis of N-linked oligosaccharides is a

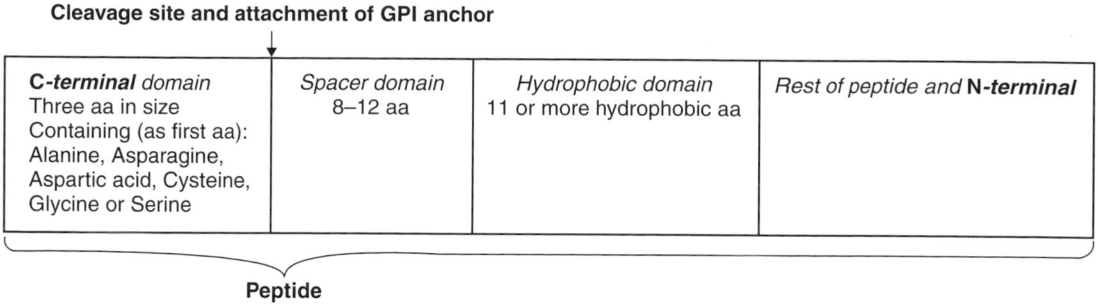

Cleavage site and attachment of GPI anchor

C-terminal *domain* Three aa in size Containing (as first aa): Alanine, Asparagine, Aspartic acid, Cysteine, Glycine or Serine	*Spacer domain* 8–12 aa	*Hydrophobic domain* 11 or more hydrophobic aa	*Rest of peptide and* ***N-terminal***

Peptide

Figure 12.3

The main features of the C-terminal region of GPI-anchored proteins include a C-terminal domain, a spacer domain and a hydrophobic domain, as indicated. These features of GPI-anchored proteins have enabled putative 'rules' to be constructed to aid the identification of proteins that have the potential to be GPI anchored.

co-translational event, it had been assumed that the addition of a GPI anchor would not affect the normal N-linked oligosaccharide repertoire of such proteins. Recent studies of recombinantly produced Thy-1 and CD59, however, suggest that the C-terminus signal is important not only for successful GPI-anchor addition but also that it influences the co-translational addition of N-linked oligosaccharides.

12.6 The function of GPI anchors

The function of GPI anchors is unclear, probably because the function of the proteins that contain the GPI anchors is often unknown. It has been suggested that GPI-anchored proteins exhibit greater extracellular mobility than their polypeptide-anchored counterparts but at present, no conclusive evidence of this exists.

Many functions have been ascribed to GPI anchors on proteins and some of these are summarized in *Table 12.2*. The ubiquitous expression of GPI-anchored proteins throughout the animal kingdom suggests an important function for these molecules and studies with knock-out mice who lack the *PIG-A* gene (thought to encode for enzyme that transfers GlcNAc to PI, step 1 in *Figure 12.2*) show that deficiencies in GPI anchor are associated with a lethal phenotype. This suggests that GPI anchors are an essential prerequisite for normal development.

Other proposed functions for these molecules include:

- protein secretion
- maintenance of cellular polarity
- control of protein turnover
- signal transduction.

Studies of the GPI-anchored proteins of parasites such as *T. brucei* have led to the suggestion that they are involved in maintaining cell shape and viability. Studies with other parasites, for example, those responsible for the human diseases malaria (*Plasmodium*) and toxoplasmosis (*Toxoplasma*) illustrate that GPI anchors are abundant constituents of the cell membrane. One proposed function of the GPI-anchored proteins of these parasites is that they may enable the parasite to evade the innate immune system of the host. This might occur by the GPI anchor sterically hindering interaction of the immunoglobulins with the cell membrane, but still allowing smaller nutrient molecules to pass through the membrane to maintain parasite viability and enable replication.

Table 12.2 Functions ascribed to GPI anchors on proteins

Presentation of proteins in lipid-bilayer
Enabling mobility of proteins in extracellular domain
Rapid release mechanism for proteins into serum
Locating proteins to apical cell surface in epithelial cells
Signal transduction

Further reading

Boccuni, P., Del Vecchio, L., Di Noto, R., Rotoli, B. (2000) Glycosyl phosphatidylinositol (GPI-) anchored molecules and the pathogenesis of paroxysmal nocturnal hemoglobinuria. *Crit. Rev. Oncol./Hematol.* **33:** 25–43.

Ferguson, M.A., Williams, A.F. (1988) Cell-surface anchoring of proteins via glycosyl-phosphatidylinositol structures. *Annu. Rev. Biochem.* **57:** 285–320.

Vidugiriene, J., Menon, A.K. (1993) Early lipid intermediates in glycosylphosphatidylinositol anchor assembly are synthesized in the ER and located in the cytoplasmic leaflet of the ER membrane bilayer. *J. Cell Biol.* **121:** 987–996.

Carbohydrate-binding proteins (lectins)

13

13.1 What are lectins?

A simple definition of lectins is that they are proteins or glycoproteins which have the ability to bind to carbohydrates with great specificity and selectivity. They are probably produced by all living organisms.

The term 'lectin' was first coined in 1954; it is derived from the Latin verb '*legere*', which means to select, pick or choose, and refers to the ability of lectins to bind with great selectively to saccharides. It has replaced earlier terms such as 'agglutinin', 'haemagglutinin' or 'phytohaemagglutinin', which were originally used to describe the ability of these substances to cross-link and agglutinate cells.

The most widely accepted definition of a 'lectin' is that adopted by the Nomenclature Committee of the International Union of Biochemistry which states that a lectin is 'a carbohydrate binding protein of non-immune origin, that agglutinates cells and/or precipitates glycoconjugates'.

The definition implies that lectins are:

- multivalent – two or more carbohydrate-binding sites are required in order for them to cross-link cells in agglutination, or glycoconjugates in precipitation
- not enzymes – their multivalency distinguishes them from enzymes that have carbohydrates as a substrate (e.g. glycosyltransferases, glycosidases, etc.), as these usually possess a single carbohydrate-binding site. Such enzymes also possess the ability to catalyse the chemical modification of carbohydrate residues under the right conditions, whereas lectins leave their carbohydrate binding partners unaltered
- not antibodies – that they are of 'non-immune origin' distinguishes them from antibodies directed against carbohydrate antigens, which can also behave in a similar way in the agglutination of cells and precipitation of glycoconjugates.

However, as knowledge about lectins, and their relationship to other related proteins, has accumulated there has been much discussion about the inadequacy of this definition and suggestions for its improvement.➲ One suggested modification is that lectins should have '*at least one binding site*' to take into account that some lectins, such as Siglecs and selectins (*see 13.4.2*) may only have a single carbohydrate-binding site. A simpler definition is that '*lectins are carbohydrate binding proteins other than enzymes or antibodies*'.

➲ Lectins are carbohydrate-binding molecules which are not enzymes or antibodies.

13.2 A brief history

13.2.1 Discovery

Lectin-induced cell agglutination was first described by Stillmark, during work for his doctoral thesis in 1888 at one of the oldest universities in Tsarist Russia, the University of Dorpat in Estonia. He was investigating the actions of toxins, and isolated a toxic extract from seeds of a

Functional and Molecular Glycobiology, Susan A. Brooks, Miriam V. Dwek and Udo Schumacher

member of the Euphorbiaceae, *Ricinus communis*, the castor oil plant.➲ He tested its effects on a variety of cells, including erythrocytes, liver cells, leukocytes and epithelial cells, and recorded an agglutination reaction 'like in clotting', as illustrated in *Figure 13.1*. Cells from different sources reacted differently, and the agglutination could be inhibited by the presence of serum. He named the substance 'ricin'. He obtained similar, but different, reactions with an extract from *Croton tiglium*, which he termed 'crotin'. A number of studies on the agglutinating properties of toxic extracts followed over the next 15 years, including those from *Abrus precatorius* seeds ('abrin') and *Robinia pseudoacacia* ('robin'). At the time, the agglutinating action of the

➲ Lectins were first discovered through their ability to bind to the glycoconjugates on the surface of cells, and cross-link them in an agglutination reaction.

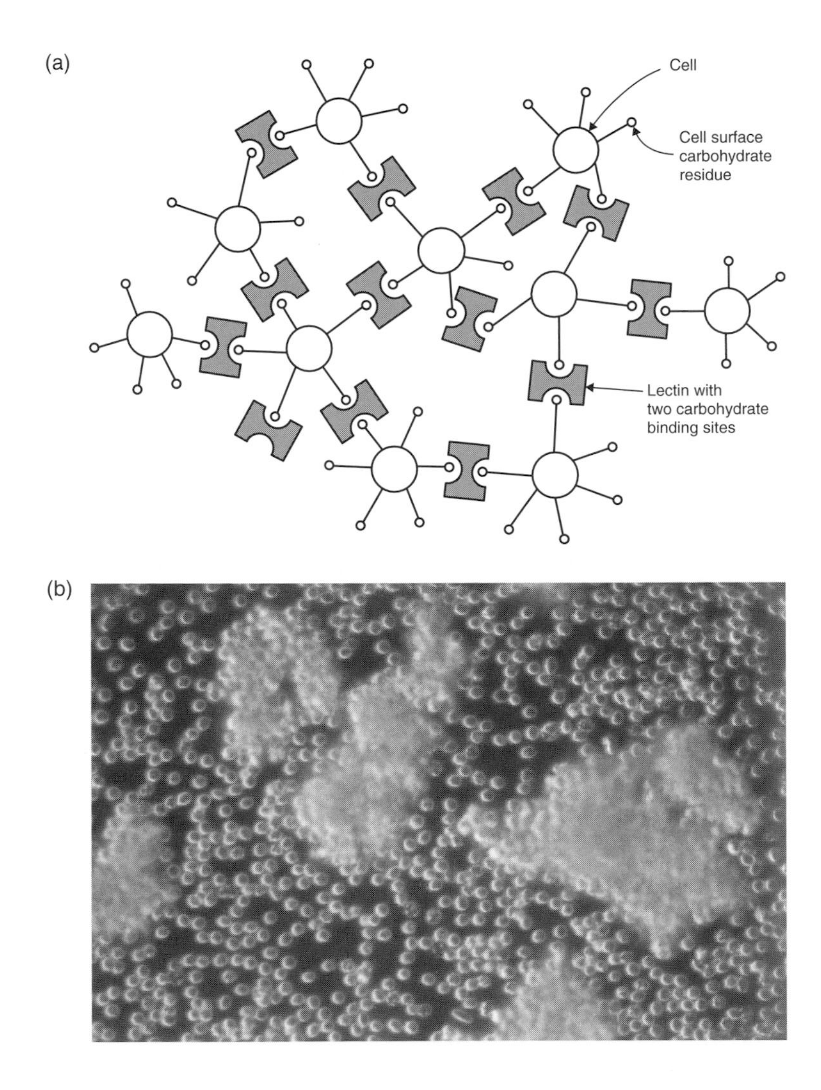

Figure 13.1

Lectin-mediated cell agglutination. (a) Cell-surface carbohydrate residues are recognized by the lectin which, because it has multiple carbohydrate-binding sites, cross links the cells, (b) photomicrograph of blood group A erythrocytes agglutinated by the GalNAc-binding lectin from *Dolichos biflorus* (DBA).

extracts was ascribed to the toxins present in them, and it was more than half a century before the toxins and the lectins responsible for cell agglutination were separated. This has not always proved to be possible as some lectins are also toxins.

13.2.2 Contribution of lectins to early immunological studies

In the 1890s, lectins were important tools in the establishment of fundamental principles of immunology. For example, rabbits fed repeated small doses of the seeds of *Abrus precatorius* developed immunity to the toxin abrin, and those fed with *Ricinus communis* seeds developed immunity to ricin. Proteins in the animals' serum (antibodies) were shown to neutralize the toxic activity of the specific extract.

13.2.3 The number of 'haemagglutinins' grows

In the first decade of the twentieth century, non-toxic agglutinins were identified in extracts of plants from the Leguminosae and Viciaceae, such as beans, peas and lentils. It was demonstrated that agglutination could be inhibited by gastric mucin, although it was not realized that this was a result of carbohydrate binding. There followed a period of intense research activity aimed at the identification of 'haemagglutinins'. Most were discovered in plants of the Leguminosae, Euphorbiaceae and Solanaceae families, but the first reports of agglutinins from invertebrates, snake venoms, fungi, bacteria and viruses also appeared at this time.

13.2.4 Realization of carbohydrate-binding properties and blood group specificity

➲ Some lectins are blood group specific and some are mitogenic.

The observation in 1936 that agglutination by an extract of *Canavalia ensiformis*, 'Con A', could be inhibited by a simple sugar led to the first suggestion that the agglutinin receptor may be carbohydrate.➲ The report in 1947 that an extract of Lima beans (*Phaseolus lunatus* syn. *limensis*) agglutinated the erythrocytes of some individuals strongly and others not at all led to the realization that the lectin was specific in agglutinating blood group A cells only. This concept of blood group-specific lectins was actively pursued by others, and resulted in renewed and intense interest in 'agglutinins' during the 1950s and 1960s. The mitogenic properties of an extract of *Phaseolus vulgaris* seeds, termed 'phytohaemagglutinin' or PHA was first made in 1960. The selectivity of wheatgerm agglutinin, WGA, in clumping malignant but not normal cells was discovered in the 1960s, and this led to an increasing interest in the use of lectins in cancer research.

13.2.5 Purification of the first lectin, and uses for many research purposes

The development of affinity isolation procedures in the 1960s led to the complete purification of a lectin, Concanavalin A, for the first time in 1965. In the following years, there was an expanding of interest in the use of lectins, mostly as mitogens following the description of the mitogenic activity of PHA-L resulting in an expansion in the knowledge about lymphocyte physiology. Lectins were also used to investigate changes in cell-surface organization during the cell cycle, transformation to malignancy and other aspects of experimental cell biology. Increasingly, lectins were used in the purification and characterization of saccharides and glycoconjugates. These applications continue to be relevant.

13.3 Nomenclature

13.3.1 The historical development of the nomenclature

The earliest lectins were given names based on their (botanical) source, such as abrin (from *Abrus precatorius*) or ricin (from *Ricinus communis*) and these terms are still in use. Once

➲ Historical names for lectins include 'agglutinin', 'haemagglutinin' and 'phytohaemagglutinin'.

their agglutinating principles were established, lectins were usually termed 'agglutinins', 'haemagglutinins', or, if derived from a plant source, 'phytohaemagglutinins', and these terms are also still sometimes used.➲ It is important to realize that lectins are a diverse and heterogeneous group of proteins which were initially rather artificially grouped together on the basis of their carbohydrate-binding properties. However, 'families' of phylogenetically related lectins have been distinguished (*see 13.4*).

Today, most lectins from plants and lower organisms are named after the organism from which they are extracted; sometimes the Latin binomial is used, for example, *Dolichos biflorus* lectin, and sometimes the common name, for example, peanut lectin. These terms are usually abbreviated, so that *Dolichos biflorus* lectin may be abbreviated to DBA, which stands for '*Dolichos biflorus* agglutinin' or PNA for 'peanut agglutinin'.

Occasionally, the nomenclature may be confusing. The lectin from *Phaseolus vulgaris*, the red kidney bean, for example, is usually known as PHA, which actually stands for 'phytohaemagglutinin'; again, a historical term for this substance. Some lectins are also mitogens and their names and the abbreviations thereof, reflect this; for example, the lectin from pokeweed, *Phytolacca americana*, is commonly referred to as 'pokeweed mitogen' or PWM.

13.3.2 Isolectins

In many cases, more than one lectin has been derived from a single source.➲ These often have different carbohydrate-binding preferences (and other properties, such as mitogenicity) to each other. They may be variously designated by Roman numerals, letters of the alphabet and/or numbers, or abbreviations. For example, the two lectins from *Ulex europaeus*, the gorse, are termed UEA-I and UEA-II. *Griffonia simplicifolia* contains five tetrameric, closely related isolectins composed of combinations of A and B subunits; these are termed GSI-B_4 or GSA-I-B_4, GSI-AB_3 or GSA-I-AB_3, GSI-A_2B_2 or GSA-I-A_2B_2, GSI-A_3B or GSA-I-A_3B, GSI-A_4 or GSA-I-A_4. A completely distinct lectin, GSA-II, which has quite different properties, is also derived from the same source. To make matters even more confusing, *Griffonia simplicifolia* is also sometimes called *Bandeiraea simplicifolia* (the same is true for *Lotus tetragonalobus* which is also sometimes called *Tetragonalobus purpurea*, and other similar examples exist). *Phaseolus vulgaris*, the red kidney bean, produces two lectins, one that is highly mitogenic and preferentially agglutinates leukocytes, termed PHA-L, and one that is poorly mitogenic and preferentially agglutinates erythrocytes, termed PHA-E.

➲ Several isolectins, with different carbohydrate-binding specificities, may be derived from a single organism.

13.3.3 Systematic classification

A new, systematic system for the nomenclature of lectins was proposed in 1984. This incorporates abbreviations for their source and also their carbohydrate-binding preference. Although a good idea, this system is not commonly used.

13.4 Lectin families

13.4.1 Plant lectin families

Recent advances in biochemistry and molecular biology have led to increased structural information about lectins which has improved our understanding of the evolutionary relationships between them. Plant lectins seem to belong to four principal families.

1. Legume lectins, which have a shared primary structure consisting of two large β-pleated sheets that form a scaffold to support the carbohydrate-binding domain with a smaller

β-pleated sheet holding the two larger ones together. They have diverse saccharide-binding partners and share highly conserved transition metal ion and Ca^{2+} ion-binding sites.
2. Chitin-binding lectins, present in the Gramineae, Urticaceae, Solanaceae, Papaveraceae, Euphorbeaceae, Phytolaccaceae and Viscaceae.
3. Monocot mannose-binding lectins of the Amaryllidaceae, Alliaceae, Araceae, Orchidaceae and Liliaceae, which, as their name suggests, bind exclusively to mannose.
4. Type 2 ribosome-inactivating proteins (RIPs), which possess specific rRNA N-glycosidase activity and varied carbohydrate-binding preferences. They have been described in taxonomically unrelated Fabaceae and Sambucaceae families. ➲

➲ The four major families of plant lectins are:
- legume lectins
- chitin-binding lectins
- monocot mannose-binding lectins
- type 2 ribosome-inactivating proteins (RIPs).

13.4.2 Vertebrate lectin families

The nomenclature for vertebrate lectins is more complex. They may be either membrane bound or soluble, and can be divided into groups according to similarities in sequence homology and activity. ➲ Currently, at least five main lectin 'families' are recognized: C-type, P-type, S-type, I-type and pentraxins.

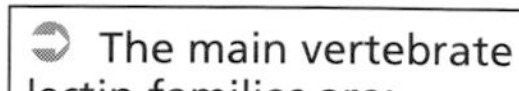

➲ The main vertebrate lectin families are:
- C-type lectins
- P-type lectins
- S-type lectins
- I-type lectins
- pentraxins.

13.4.2.1 C-Type lectins

Calcium-dependent or C-type lectins are usually expressed on the cell surface, but some (e.g. the mannose-binding lectin, MBL) are soluble, and exhibit diverse carbohydrate-binding specificities. They share the property that their binding is dependent on the presence of calcium ions. This family includes the asialoglycoprotein receptor protein that has been described in a number of species, selectins and collectins, among many others. They share a common sequence motif of 14 invariable and 18 highly conserved amino acid residues.

13.4.2.2 S-Type lectins

The S-type lectins (also called galectins or β-galactoside-binding lectins) share one or two highly conserved carbohydrate recognition domains made up of ≈135 amino acid residues. They are soluble proteins, hence S (soluble) type. The family includes at least 10 members, and the nomenclature can be confusing, for example, galectin-I may also sometimes be referred to as galaptin or L-14 and galectin-3 is also called Mac-2 or L-30.

In galectin-1 and -2, the single carbohydrate recognition domain forms 2-fold symmetric homodimers, thus producing a divalent molecule. They, and galectin-5 and -7, are the 'prototype' galectins characterized by low monomeric molecular mass (≈14 kDa) and exist as non-covalently linked homodimers. Galectin-3 is a 'chimeric' galectin because it is composed of a C-terminal lectin domain and an N-terminal accessory domain with characteristic sequences rich in proline and glycine. Galectin-4, -6, -8 and -9 are 'tandem repeat type' galectins characterized by a cross-linking peptide covalently connecting two lectin domains, and are thus constitutively dimeric.

13.4.2.3 P-Type lectins

P-Type lectins share the property of having mannose-6-phosphate as their binding ligand. They are expressed on the cell surface.

13.4.2.4 I-Type lectins

I-type lectins, which include ICAM-I, N-CAM, PECAM and the sialic acid-binding immunoglobulin-like lectins (Siglecs), are so called because they possess an immunoglobulin-like domain. Siglecs, formerly known as sialoadhesins, include sialoadhesin, CD22 and CD33 (*see 11.8.4*).

13.4.2.5 Pentraxins

Pentraxins share a cyclic pentameric structure based on identical 20–30 kDa subunits. They recognize a variety of saccharide-binding partners. There is a high degree of homology between different species, for example, human and plaice C-reactive protein share 40% homology. Lectins from invertebrates including *Limulus polyphemus* (horseshoe crab) and the tunicate *Didemnum candidum* appear, on the basis of their structure, to be related to the pentraxins.

Other animal lectin 'families' have been described, and the list is growing all the time.

13.5 Carbohydrate-binding specificity

13.5.1 Monosaccharide specificity

The carbohydrate-binding specificity of lectins is commonly described in terms of the monosaccharide, or sometimes disaccharide or oligosaccharide, that most effectively inhibits its interaction with a presumed naturally occurring complex ligand. For example, *Lotus tetragonalobus* lectin, LTA, preferentially agglutinates blood group O(H) erythrocytes (*see 10.5*), and the agglutination is inhibited by the presence of α-L-fucose. It is said to be an α-fucose-binding (or simply fucose-binding) lectin. STA the lectin from the potato, *Solanum tuberosum*, will agglutinate human erythrocytes irrespective of blood group, and the agglutination is not inhibited by any monosaccharide; it is, however, inhibited by oligosaccharides containing (β1→4)-linked GlcNAc. STA is therefore said to be specific for β-GlcNAc oligomers. Overall, plant lectins are often classified into 'families' according to their saccharide-binding preference: (i) Glc/Man specific, (ii) Gal/GalNAc specific, (iii) GlcNAc specific, (iv) Fuc specific, (v) sialic acid specific and (vi) those that bind complex-type oligosaccharides only. Many lectins do not conveniently fit into these groupings.

The binding of most lectins to, often unidentified, cell- or tissue-derived ligands can be inhibited, with varying degrees of success, by one or more monosaccharides and/or more complex oligosaccharides. These inhibitory saccharides are usually listed when the carbohydrate-binding preference of the lectin is described. Inhibitory saccharides are usually quoted in decreasing order of potency. For example, the carbohydrate-binding preference of the lectin from the jack bean *Canavalia ensiformis*, termed Con A, is often expressed as 'α-D-Man, α-D-Glc' as its binding is most strongly inhibited by α-D-Man, and inhibited to a lesser extent by α-D-Glc. Sometimes the symbol $>$ is inserted to indicate that one carbohydrate has a much greater inhibitory effect than another, or the symbol $=$ to indicate that two carbohydrates have a similar inhibitory effect. These monosaccharide inhibition experiments are important as lectins can interact with other proteins via their non-carbohydrate-binding regions through hydrophobic interactions.

13.5.2 Naturally occurring binding partners

It is important to realize that in most cases, the natural binding partners of lectins remain largely uncharacterized, and that describing their carbohydrate-binding preference in terms of monosaccharides, disaccharides or oligosaccharides, is a simplification. Lectins frequently recognize the three-dimensional structure of their ligands, and sub-terminal as well as terminal monosaccharides may be involved.

Amino acids, ions such as Ca^{2+} and Mn^{2+}, hydrophobic interactions and van der Waal's forces may also be involved in the lectin binding to a saccharide structure. This has implications when choosing lectins as tools in the laboratory, as lectins with the same nominal carbohydrate-binding specificity, for example, two 'galactose-binding lectins', may recognize subtly different binding partners.

13.6 Blood group-specific lectins

Some lectins, through their preferential recognition of human blood group sugars, are blood group specific – that is, they will preferentially bind to and cross-link in agglutination, erythrocytes of one human blood group and not others (*see 13.2.4*). For many years, blood group-specific lectins were routinely used to type human blood. In addition to the common A, B, O blood types, lectins have been described that will preferentially recognize other less common blood type classifications including the MN, Cad, T and Tn blood group sugars (*see 10.4*). Examples of blood group specific lectins are given in *Table 13.1*.

13.7 Distribution and function of lectins

It is likely that lectins have important biological functions in the organisms that produce them. This is implied by a number of observations. First, their occurrence is widespread in organisms as diverse as viruses, bacteria, fungi, plants, invertebrates, reptiles and mammals, suggesting that they have been retained through evolution. Second, a high degree of sequence homology between lectins derived from related members of the same families indicates a high level of conservation, presumably linked to function. Finally, the seeds of some plants,

Table 13.1 Some examples of blood group specific lectins

Source of lectin				
Latin name	Common name	Abbreviation	Blood group	Principle inhibitory carbohydrate(s)
Helix pomatia	Edible or Roman snail	HPA	A	α-D-GalNAc
Helix aspersa	Garden snail	HAA	A	α-D-GalNAc
Phaseolus lunatus	Lima bean	LBA	A	α-D-GalNAc
Dolichos biflorus	Horse gram	DBA	A_1	α-D-GalNAc
Vicia villosa	Hairy vetch	VVA	A	α-D-GalNAc
Vicia villosa	Hairy vetch	VVA_4	A_1 & Tn	α-D-GalNAc
Griffonia simplicifolia		GSA-I-AB_3	B	α-Gal > α-D-GalNAc
Ptilota plumosa	Red marine algae	PPA	B	α-D-Gal
Lotus tetragonalobus	Asparagus pea	LTA	O/H	α-L-Fuc
Ulex europaeus	Gorse	UEA-I	O/H	α-L-Fuc
Anguilla anguilla	Eel	AAA	O/H	α-L-Fuc
Arachis hypogaea	Peanut	PNA	T	D-Gal(β1→3)GalNAc
Griffonia simplicifolia		GSA-II	T_k	α-D-GlcNAc = β-D-GlcNAc
Salvia horminum	Salvia	SHA	T & Cad	α-D-GalNAc & β-D-GalNAc
Salvia sclarea	Clary, fetid clary sage	SSA	Tn & Cad	α-D-GalNAc
Iberis amera		IAA	M	Unknown
Vicia graminea		VGA	N & T	Clustered O-linked Gal(β1→3)GalNAc

such as the Leguminosae (peas, beans and lentils), contain considerable concentrations of lectin, as much as 0.1–5% of total protein content, again indicative of an important function. The defining property of lectins, their selective recognition of saccharide structures, fits them to, theoretically, mediate a considerable range of crucial cell–cell interactions, but in most cases their biological functions remain unknown. Below are some of the suggested roles for these ubiquitous molecules.

13.8 Occurrence and functions of plant lectins

Lectins can be detected in all parts of plants, stem, leaves, roots, flowers, bark, but the bulk of lectin expression is in the storage organs, for example, tubers or seeds. Lectins from different parts of the same plant often vary in structure and/or saccharide-binding specificity. Many commercially available lectins are derived from lectin-rich plant seeds. These seed lectins are usually present in special organelles generated from the vacuoles, called 'protein bodies' and their synthesis thus follows the general cellular pathway for storage protein synthesis.

It is surprising that little is known of the biological roles of plant lectins, despite the fact that they have been isolated for well over 100 years, are so prevalent in the organisms that produce them, and have been widely used in clinical and research applications. It seems certain that their functions are biologically important and diverse. Many putative functions are related to the defence of the plant against predators and/or infection.

13.8.1 'Immune' function

It has been proposed, owing to their similarities to antibodies, that plant lectins play a protective role in their host, acting as a primitive 'immune system'. This idea is supported by the observation that many lectins bind saccharide structures which are not found within the plant itself, implying that their natural binding partners must be produced by other organisms. Examples of this are lectins of *Sambucus nigra* (elderberry) and *Maackia amurensis* which recognize sialic acid, a monosaccharide absent in plants but a major carbohydrate component of animal tissues.

Expression of some plant lectins is under similar regulatory control as other defence proteins, and expression is upregulated in response to wounding. The rubber tree *Hevea brasiliensis* produces a rubbery latex exudate on wounding which is of commercial importance. It contains a lectin, 'hevein', which is responsible for the latex allergy suffered by people who come into frequent contact with latex, for example, hospital staff who wear latex surgical gloves. In nature, such lectins may have unpleasant or even toxic effects on predators.

13.8.2 Protection from insect attack

The role of lectins in protecting plants from insect predators has been illustrated by a number of studies. ➲ Wheatgerm and *Bauhinia purpurea* lectins are lethal to neonate corn borer (*Ostrinia nubilalis*) larvae, wheatgerm lectin has an inhibitory effect on development of weevil larvae at physiological concentrations, and pokeweed mitogen kills Southern corn rootworm (*Diabrotica undecimpunctata*) larvae. Other lectins, including wheatgerm lectin, inhibit larval growth. Lectins of the Leguminosae make their seeds unpalatable, or toxic, to insects and therefore protect them from being eaten. The lectin from the winter aconite, *Eranthis hyemalis* is toxic to larvae of *Diabrotica undecimpunctata*, a common insect pest. Lectins from the snowdrop (*Galanthus nivalis*) and the garlic plant (*Allium sativum*) are active against some chewing insects, and snowdrop lectin is toxic to aphids when introduced into transgenic plants. The development of transgenic plants expressing toxic lectins and therefore resistant to insect attack, or lectins that impair development or fecundity of insects, is a topic of some interest (*see 18.7*). There is evidence that

➲ Some lectins protect plants from insect predators.

insects exposed to food plants that produce toxic lectins develop resistance to the toxin over time.

A protein 'arcelin' in seeds of some strains of *Phaseolus vulgaris* (red kidney bean) also protects the plant against insect predators. Arcelin is a member of the legume lectin family which has lost its carbohydrate-binding activity. This raises the possibility that a function other than carbohydrate binding may be involved in lectin-induced protection against predators.

13.8.3 Protection against attack by animals

Many lectins are able to bind to glycoconjugates on cells lining the digestive tract. If these lectins are ingested they bind to the cell surface and are subsequently internalized.➲ They may be toxic to higher animals and humans. An example of this is the severe gastroenteritis in humans caused by eating undercooked red kidney beans (*Phaseolus vulgaris*), and other legume lectins. In laboratory studies, animals quickly learn to avoid experimental toxic PHA-rich diets and will sometimes choose to starve in preference to consuming the beans. Another example is elderberry (*Sambucus nigra*) bark; this is rich in a toxic lectin (which recognizes $\alpha 2 \rightarrow 6$-linked neuraminic acid) and is seldom eaten by wildlife, which instead consume the non-toxic bark of neighbouring trees.

➲ Toxic lectins protect plants from being eaten by animals.

13.8.4 Antifungal activities

In addition to their effects on predators, some lectins have anti-fungal activities. One of the first reports of this was that of wheatgerm lectin in protecting seedlings; however, more recent findings suggest that the effect is a result of contaminating chitinases. In contrast, the lectin from the rhizome of the stinging nettle, *Urtica dioica,* inhibits the growth of a number of fungal plant pathogens, although it remains unclear whether this has a protective role *in vivo*. Hevein, the lectin from the rubber tree (*see 13.8.1*) is a chitin-binding protein that can also inhibit fungal growth. Potato tuber, *S. tuberosum*, lectin inhibits the growth of several pathogenic fungi.

13.8.5 Antibacterial action

Lectins also have a role in protecting plants against bacterial infection.➲ For example, the seeds of *Datura stramonium* (thorn apple) block the movement of normally motile infective soil-borne bacteria. Many plant lectins have affinity for the unusual carbohydrates (e.g. muramic acid, N-acetylmuramic acid and muranyl dipeptides) present on bacterial pathogens (*see 7.2*). Muramic acid is similar in structure to Glc and GlcNAc, and many Glc- and GlcNAc-recognizing lectins, especially of the Viciaceae also recognize this moiety. Lectins with quite different saccharide-binding specificities, such as soyabean lectin, also, surprisingly, bind muramic acid. It has been suggested that some lectins may help protect plants against viral infection, but the evidence to support this theory remains scanty.

➲ Some lectins protect plants by acting as anti-fungal and anti-bacterial agents.

13.8.6 Symbiotic relationships

Lectins are involved in specific cell–cell recognition and adhesion mechanisms. Much research has centred on the controversial idea that Leguminosae lectins mediate the symbiotic interaction between the host plant and nitrogen fixing *Rhizobium* bacteria in their root nodules by recognizing *Rhizobium* strain-specific carbohydrate Nod factors (*see 7.4.3*).

13.8.7 Transport and storage

The carbohydrate-binding properties of plant lectins suggest that they are involved in functions such as transport or storage of sugars within the plant, maturation or germination of seeds, and cell-wall extension mechanisms. Although seed lectins are broken down during germination and thus contribute to the nutrient pool, they are degraded much later than other storage proteins, suggesting that they may have a functional role in addition to nutrient storage. Furthermore, seed lectins generally recognize oligosaccharides present on other seed storage proteins. The lectins may be involved in ordering deposition in storage bodies during seed development. This idea is consistent with the order in which different seed proteins appear: in pea *Pisium sativum* and in *Phaseolus vulgaris*, the vicilin storage proteins appear first, with lectin appearing at almost the same time, then, later, the legumins which constitute the bulk of the storage proteins. Many seed lectins also interact with saccharide structures present on the protein body membrane, which may direct storage protein deposition.

13.8.8 Cryoprotection

Lectins exert a surprising functional effect when they interact with membranes. For example, the binding of Gal-specific lectins to spinach chloroplast membranes stabilizes chloroplasts from frost damage, and it is believed that this effect is exerted through an alteration in membrane fluidity and permeability. Similarly, a cryoprotective effect has been attributed to mistletoe, *Viscum album*, lectin, the concentration of which increases during winter months.

13.8.9 Mediation of enzyme activity

Lectins from some species interact with hydrolytic enzymes in the seeds, suggesting a role in the regulation of hydrolytic enzymatic activity in storage protein degradation. The α-mannosidase of *Canavalia ensiformis* (the jack bean) is recognized by the lectin Con A present with it in the protein bodies. Similarly, the lectin from *Allium sativum* (garlic) binds to allinase, an enzyme from the same source. In rare cases, lectin binding to an enzyme results in its activation, as is the case with *Secale cereale* (rye) lectin and a rye phosphatase; it is assumed that this mechanism may be involved in the mobilization of seed reserves. *Solanum tuberosum* (potato) tuber lectin similarly activates a potato phosphatase and *Erythrina indica* lectin activates an endogenous glycosidase, and *Pleurotis ostreatus* (a type of mushroom) lectin activates a neutral phosphatase.

13.8.10 Glycoprotein trafficking

Two animal proteins, termed VIP and ERGIC, have sequence homology with legume lectins, and are involved in glycoprotein trafficking in the ER of animal cells.

13.9 Bacterial, viral and protozoan lectins

Lectins appear to be crucial to the infectivity/pathogenicity of many bacteria, viruses and protozoa (*see 16.2*). Lectin–saccharide interactions facilitate organ-specific attachment of pathogen to host cells. Members of the Enterobacteriaceae, for example, express lectins on their pili which are essential for successful adhesion to cells lining the intestine. Infection by some bacteria is prevented in the presence of an appropriate inhibitory carbohydrate; for example, infection of the lungs of mice by *Streptococcus pneumoniae* is prevented by GlcNAc. Infection of gastric mucosa by *Helicobacter pylori* is mediated by recognition of the carbohydrate blood group Lewis b (Le^b) antigen (*see 10.6*) and other saccharide structures. Of the protozoans, the importance of lectin–saccharide interactions in infectivity and organ-specific colonization is best documented in *Entamoeba histolytica* which causes amoebic dysentery in humans (*see 13.14*).

Phagocytosis of infective bacteria is mediated by bacterial lectin binding to saccharides present on phagocytic cells, or phagocytic cell lectin recognition of bacterial cell-surface saccharides, both termed lectinophagocytosis.

13.10 Slime mould lectins

Lectin expression in slime moulds is developmentally regulated and titres are low in the single-celled organism, but high in the aggregated, multicellular form. The lectins mediate the specific aggregation of single-celled slime moulds into multicellular bodies prior to production of fruiting bodies. Glycoconjugates derived from the bacteria that the slime mould, *Dictyostelium discoideum,* uses as a food source act as ligands for the slime mould lectin, discoidin I; the lectin is not involved in bacterial recognition or feeding, but instead, the organism utilizes the bacterial-derived carbohydrates to direct cellular compartmentalization of the lectin. Furthermore, discoidin I, plays a role in cell–substratum attachment during cell migration.

13.11 Invertebrate lectins

Lectins occur in many invertebrate species. These include lectins from the haemolymph of *Limulus polyphemus* (horseshoe crab), and from the haemolymph and sexual organs of *Helix pomatia* (edible snail) and from *Limax flavus* (garden snail). The function of these substances within the animal are unknown but it has been suggested that they may protect eggs or adult invertebrates from infection, or eggs from being eaten by predators. They are able to function as opsonins and cell recognition molecules in this context. Additionally, they may be functional during development.

13.12 Vertebrate lectins

In the last decade, more attention has focused on vertebrate lectins and their functions. Many molecules that have been known for some time, for example, cell adhesion molecules like N-CAM and ICAM-I (*see 13.4.2*) have carbohydrate-binding activity and are therefore classed as lectins. Furthermore, lectins often show homology to other molecules that have functions other than carbohydrate binding. Siglecs are members of the immunoglobulin superfamily, most other members of which are not involved in carbohydrate recognition. It may be relevant to think of lectins as members of other 'families', for example, as cell adhesion molecules or immunoglobulin-like molecules, which happen to function through carbohydrate binding. In addition, many 'vertebrate' lectins are also shared by invertebrates. Vertebrate lectin families are listed in *Table 13.2* and their characteristics and functions summarized.

13.12.1 S-Type lectins or galectins

Soluble, S-type (galectins or β-galactoside binding) lectins (*see 13.4.2*) have been isolated from many vertebrate tissues including those of human, rat, mouse, cow and chicken. There are at least 10 members of this family, and they are expressed in distinct patterns during development and differentiation, and in different tissues, as listed in *Table 13.3*.

Galectins do not require the presence of calcium ions, they may be extracellular or cytoplasmic, require the presence of reducing thiols for activity, and share a preference for binding β-galactosides. Their natural binding partners are probably polylactosamine-containing glycoconjugates. They show affinity for type 1 (*see 10.2.1*) or type 2 (*see 10.2.2*) Gal(β1→3/4)GlcNAc chains, fucosylation is not required (unlike the selectins which recognize the fucosylated Lewis epitopes) and substitution by Fuc of the penultimate or downstream GlcNAc abolishes binding. Terminal (α2→3) substitution by sialic acid moderately (2–4-fold) enhances binding. Substitution to form A and B blood group epitopes gives strong

Table 13.2 Families of vertebrate lectins

Family	Characteristics	Functions
C-type lectins	Ca^{2+} dependent, transmembrane or soluble	Cell–cell and cell–matrix interactions, endocytosis
S-type lectins	Soluble, recognize β-galactosides	Cell–cell and cell–matrix interactions, apoptosis, phagocytosis, immunomodulation, proliferation
I-type lectins	Mainly membrane bound, some soluble	Cell–cell and cell–matrix interactions, growth factor receptors
P-type lectins	Membrane bound, Ca^{2+} dependent or independent	Targeting of enzymes to the lysosome bind mannose-6-phosphate
Pentraxins	Ca^{2+} dependent, soluble	Neurite growth and immunomodulation

Table 13.3 Galectin expression in different vertebrate tissues

Galectin	Tissue distribution
Galectin 1	Kidney, placenta, thymus, motor and sensory neurons, skeletal and smooth muscle
Galectin 2	Gastrointestinal tract
Galectin 3	Gastrointestinal and respiratory epithelium, neurons, activated macrophages, mast cells, eosinophils, neutrophils
Galectin 4	Gastrointestinal epithelium
Galectin 5	Erythrocytes, reticulocytes
Galectin 6	Gastrointestinal epithelium
Galectin 7	Keratinocytes
Galectin 8	Brain, liver, lung, heart, kidney
Galectin 9	Liver, lung, kidney, skeletal and cardiac muscle, lymphoid tissue, thymocytes, leukocytes
Galectin 10	Eosinophils, basophils

enhancement. They may recognize isoforms of laminin (*see 8.7.2*) and fibronectin (*see 8.7.1*) which are rich in polylactosamines.

Galectins are small, 14–38 kDa, cytosolic proteins that lack a signal sequence for transport into the endoplasmic reticulum (ER) and are therefore not glycosylated, although they contain consensus sequences for N-glycosylation. Expression is developmentally regulated, and they are involved in de-adhesion events associated with embryogenesis. Some galectins are present at the cell surface, in the extracellular matrix (ECM) and in cell secretions. The secretion of galectins is an intriguing phenomenon as they do not follow the classical secretory pathway because they lack the signal sequence which would enable them to enter the rough ER (if they followed the classical secretory pathway, their active site would need to be blocked during transit through it). They have diverse biological roles including mediating cell–cell adhesion, induction of apoptosis, cellular activation or mitosis. Their role in adaptive immunity is becoming better understood.

Galectins exhibit a diversity of saccharide-binding specificities, which is consistent with adaptation to their different biological functions. Some galectins have been implicated in immune function, as listed in *Table 13.4*. Galectins may have dual function as they have been implicated in both pro- and anti-adhesion mechanisms. It has been proposed that this may occur through several possible routes.

- direct binding of galectins to their saccharide-binding partners
- bridging these glycan structures on adjacent cells or ECM proteins
- cross-linking of surface galectins by soluble saccharide ligands
- clustering of cell surface glycans on the plasma membrane of a single cell.

Table 13.4 Proposed immune functions of galectins

Galectin-1	Induction of apoptosis in T cells, modulation of complement receptor 3 (expressed by monocytes, neutrophils and macrophages)
Galectin-3	Inhibition of apoptosis, triggering mast cells and basophils in allergic inflammation, eosinophil recruitment, activation of neutrophils, regulation of monocyte differentiation
Galectin-9	Induction of apoptosis, eosinophil recruitment

They have been implicated in many processes including:

- fetal development
- migration of myoblasts on laminin
- regulation of cell growth
- tissue remodelling
- apoptosis
- tumour cell adhesion and metastasis.

The best characterized galectins are galectin-I (L-14, galaptin) and galectin-3 (L-30, Mac-2). The terms L-14 and L-30 refer to their molecular masses of 11–16 and 28–35 kDa respectively. Galectin-3 is unique in this family as it has a large, flexible N-terminal domain of 110–130 residues, containing repeat sequences rich in proline, tyrosine, glycine and glutamine. Galectin-3 is monomeric but can aggregate into oligomers at high concentrations. The galectins have similar saccharide-binding affinities for lactose and N-acetyllactosamine, but galectin-3 can accommodate larger oligosaccharides, such as polylactosamines (*see 10.3*), NeuAc (α2→3)lactose, and the blood group A determinant GalNAc(α1→3)[Fuc(α1→3)] Gal(β1→4) GlcNAc (*see 10.5*). It binds to integrins, other cell adhesion molecules and ECM components such as laminin (*see 8.7.2*) and fibronectin (*see 8.7.1*). ➲

➲ Galectin binding partners include:
- laminin
- lysosome-associated membrane proteins (LAMP-I and LAMP-II)
- lactosamine containing glycolipids of olfactory neurones
- ganglioside GM1
- $\alpha_7\beta_1$ integrin
- fibronectin
- carcinoembryonic antigen (CEA).

13.12.2 P-Type lectins

The P-type lectins, which are receptors for mannose-6-phosphate (Man-6-P), have been studied extensively. Two are found in humans, cows and rodents. They are involved in sorting lysosomal enzymes from the *trans*-Golgi network into the lyosomes. This is one of the best understood specific biological roles of saccharide–lectin interactions. The presence of Man-6-P residues on N-linked oligosaccharides (*see 4.2*) of newly synthesized lysosomal enzymes acts as a targeting signal to direct them to the lysosome. The process is mediated by two receptors in the *trans*-Golgi network, the cation-dependent Man-6-P receptor and the insulin-like growth factor II or cation-independent Man-6-P receptor.

13.12.3 C-Type lectins

The C-type lectins are the most diverse class of animal lectins. They include mammalian endocytic receptors, collectins, selectins, lymphocyte lectins, invertebrate lectins, viral lectins and snake venom lectins. At least 150 different C-type lectin genes have been identified. Some C-type lectins are described in more detail in the following sections.

The binding of C-type lectins to their saccharide-binding partners is dependent on the presence of calcium ions. The involvement of the calcium ions in lectin–saccharide binding has been studied intensively. In the case of the mannose-binding lectin (*see 13.12.3.3*), for example, there are two Ca^{2+} ions, one of which is in direct contact with the Man-binding partner. Man has equatorial hydroxyl groups at positions 3 and 4, and these coordinate with the Ca^{2+} ions at the binding site. The hydroxyl groups also act as donors and acceptors of hydrogen bonds from glutamic acid and asparagine residues that act as additional ligands for the Ca^{2+}. Other monosaccharides, such as GlcNAc, have similarly orientated 3- and 4-hydroxyl groups, and can therefore also act as binding partners for the lectin, explaining its broad binding specificity. A similar interaction between Ca^{2+} ions and hydroxyls on the monosaccharides of the glycan-binding partner have been implicated in lectin–saccharide binding of other C-type lectins, including P-selectin binding to sLe^x (*see 13.12.3.2*).

13.12.3.1 Mammalian asialoglycoprotein receptor

The mammalian asialoglycoprotein receptor is the first described and best studied C-type lectin. It was first identified in rabbit hepatocytes and is involved in the removal of desialylated plasma glycoproteins from the circulation; it recognizes GalNAc and Gal residues exposed when the normally terminal sialic acid residues are lost. Similar lectins have been found in the liver of other species including rat, mouse and human. Chicken liver contains an analogous GlcNAc receptor.

Several asialoglycoprotein receptors are found on macrophages; for example, peritoneal macrophages express a Gal- and GlcNAc-specific receptor and Kupffer cells (which are the resident macrophages of the liver) express a Fuc-specific receptor. A variety of other macrophage subtypes express a Man receptor. These molecules may play a functional role in recognition/binding of target molecules. Natural killer T cells also possess C-type lectins, but their role in target recognition is disputed.

13.12.3.2 Selectin family

The selectin family (another group of C-type lectins) are a class of cell-surface carbohydrate receptors that mediate interaction with ligands on other cell surfaces, particularly lymphocytes and circulating phagocytic cells such as neutrophils and monocytes. The genes coding for the selectins lie close together on the long arm of chromosome 1 in close proximity to a cluster of genes encoding proteins involved in the regulation of complement activity.

The selectins are involved in the complex mechanisms of adhesion between neutrophils and lymphocytes and the endothelia of blood vessels in leukocyte extravasation (*see 16.3*), and may also be implicated in cancer cell metastasis (*see 16.6*). ➲

> ➲ Selectin–ligand interactions are important in leukocyte extravasation.

The selectins share a common modular arrangement of domains, as illustrated in *Figure 13.2*: they are type I transmembrane proteins and their extracellular region is composed of an N-terminal Ca^{2+}-dependent lectin domain, an epidermal growth factor (EGF)-like domain, a variable number (2–9) of short consensus repeat (SCR) domains homologous to those found in complement binding proteins, and (except for one putative splice variant of P-selectin) a transmembrane section and short cytoplasmic tail. The molecular structure of the selectins implies that the carbohydrate-binding domain, especially in P- and E-selectin, extends well beyond the cell surface and into the bloodstream where interaction with saccharide ligands on circulating cells is maximized. L-selectin, which is shorter, is presented at the tips of microvilli on the lymphocyte surface, again resulting in its projection away from the cell surface.

As for other C-type lectins, carbohydrate binding is Ca^{2+} dependent. The location of the Ca^{2+}-binding sites are unknown, but seem to be associated with the EGF-like domain.

In contrast to the typical saccharide–lectin interaction, binding of these monomeric lectins requires a functional triad network composed of a receptor, a carbohydrate ligand and a carrier, which results in high-affinity binding between cell types. The carbohydrate and carrier protein together form a counter-receptor and the proper orientation of the oligosaccharide

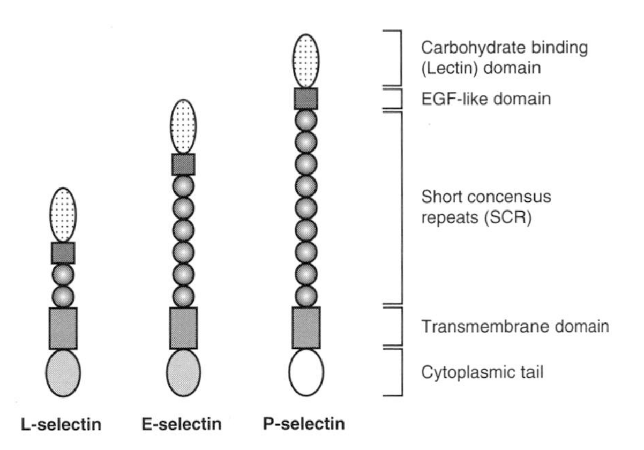

Figure 13.2

The structure of the selectins: they are type I transmembrane proteins which share a high degree of homology. They are composed of a C-type N-terminal (lectin) domain, an EGF-like motif, a variable number of short consensus repeats (SCR), a transmembrane domain and an intracellular cytoplasmic tail.

Table 13.5 Examples of selectin counter receptors

L-selectin counter-receptors
GlyCAM-I (glycosylation dependent cell adhesion molecule I)
CD34
Glycoprotein 200 (gp200)
MadCAM-I (mucosal addressin cell adhesion molecule I)
Polyanionic heparan sulphate
P-selectin counter-receptor
P-selectin glycoprotein ligand I (PSGL-I)
E-selectin counter-receptors
Cutaneous lymphocyte antigen (CLA)
E-selectin ligand I (ESL-I)

chains and the generation of 'clustered saccharide patches' (containing multiple oligosaccharides), depends upon the protein component which renders the oligosaccharides accessible to the selectin. These 'clustered saccharide patches' occur where a group of appropriate oligosaccharide binding partners are presented in close proximity on a single polypeptide backbone, for example, a mucin (*see 5.4*). Some examples of selectin counter-receptors are listed in *Table 13.5*.

The range of saccharide structures recognized by selectins is still to be fully established, but all recognize structures related to sLe^x, which contains terminal sialic acid and Fuc (*see 10.6*), as listed in *Table 13.6*. What these carbohydrate structures have in common is a N-acetyl lactosamine backbone and type 2 chain, i.e. Gal(β1$\rightarrow$4)GlcNAc (*see 10.2.2*), rather than type 1 chain, i.e. Gal(β1$\rightarrow$3)GlcNAc (*see 10.2.1*). They also carry sialic acid in a (α2$\rightarrow$3) linkage and Fuc in a (α1$\rightarrow$3) or (α1$\rightarrow$4) linkage. The importance of α1,3 fucosylation is illustrated by the human congenital disease leukocyte adhesion deficiency type II (LAD-II) in which patients synthesize proteins deficient in fucosylation and exhibit defective selectin-mediated lymphocyte trafficking (*see 16.8.1.4*). Sulfation may also be important for binding of both L- and P-selectin.

Table 13.6 Some molecules recognized as binding partners by selectins

Specific oligosaccharide structures	Common name of oligosaccharide	Type of molecule		
		E-selectin	P-selectin	L-selectin
Sia(α2→6)Gal(β1→4)GlcNacβ		−	+−	−
Sia(α2→3)Gal(β1→4)[Fuc(α1→3)] GlcNAcβ-	sLe^x	++	++	+
Sia(α2→3)Gal(β1→4)GlcNAc(β1→3) Gal(β1→4)[Fuc(α1→3)]GlcNAcβ-	CD65	+−	+	? +
Sia(α2→3)Gal(β1→4)[Fuc(α1→3)] GlcNAc(β1→3)Gal(β1→4) [Fuc(α1→3)]GlcNAcβ-	Dimeric sLe^x	++	++	? ++
Sia(α2→3)Gal(β1→3)[Fuc(α1→4)]GlcNAcβ-	sLe^a	++	++	+
SO_4^-3Gal(β1→4)[Fuc(α1→3)]GlcNAcβ-	3′ sulfo-Le^x	+	++	++
SO_4^-3Gal(β1→3)[Fuc(α1→4)]GlcNAcβ-	3′ sulfo-Le^a	+	++	++
SO_4^-3Gal(β1→′1-ceramide)	Sulfatide	−	+	++
Undefined binding to broader molecular types				
Heparin sulfate proteoglycans		−	++	++
Chondroitin sulfate		−	+−	+−
Polyphosphomannan		−	−	++
Fucoidin		−	++	++
Dextran sulfate		−	++	++
Sialylated, fucosylated mucins		++	++++	?
Sialylated, sulfated, fucosylated mucins		++	?	++++
Sialylated glycoproteins		++++	++++	?

P-Selectin

P-Selectin is a glycoprotein expressed by platelets and endothelial cells. It was originally called granule membrane protein 140 (GMP-140) or platelet activation-dependent granule to external membrane protein (PADGEM) and is now designated CD62P. It is a single chain glycoprotein of 140 kDa and is stored in α-granules of resting platelets or in Weibel-Palade bodies of resting endothelial cells. It is rapidly transferred to the cell membrane upon cell stimulation by thrombin, histamine, oxygen radicals and other stimulatory molecules. Maximum expression occurs 5–30 min after stimulation. There are two transmembrane forms of the molecule with complement-binding regions of different lengths, and a soluble form lacking the transmembrane domain.

> ➲ After stimulation, P-selectin is expressed by platelets and endothelial cells and mediates the adhesion of platelets and phagocytic cells to the endothelium.

P-Selectin mediates the adhesion of platelets to endothelial cells, and phagocytic cells (monocytes and neutrophils) to endothelium. ➲ In P-selectin knockout mice, circulating neutrophils are elevated, leukocyte rolling is ablated and recruitment of neutrophils to sites of inflammation is delayed. P-selectin has two high-affinity binding sites for Ca^{2+}. Binding of Ca^{2+} ions induces a conformational change in the shape of the molecule necessary for carbohydrate binding. The minimum requirement for P-selectin binding is a non-reducing type 1 chain Gal(β1→3GlcNAc) (*see 10.2.1*) or type 2 chain Gal(β1→4GlcNAc) (*see 10.2.2*) substituted at C4 or C3 of the GlcNAc residue with Fuc (i.e. Le^a or Le^x epitopes; *see 10.6*). Sialic acid (α2→3) linked to the terminal Gal (i.e. sLe^a or sLe^x; *see 10.6*) enhances binding 30-fold, whereas (α2→6)-linked sialic acid or (α1→2)-linked Fuc reduces binding affinity. P-selectin also binds 3-sulfated Gal-containing glycolipids (sulfatides) and 3-sulfated glucuronic acid containing glycolipids (sulfoglucuronyl lipids or SGNLs).

E-Selectin

➲E-selectin, also called ELAM-I or CD62E, is a 113 kDa glycoprotein transcriptively induced on the surface of endothelial cells in response to cytokine (e.g. tumour necrosis factor (TNF)α, interferon (IFN)γ or interleukin (IL)-1) activation. Expression is maximal 4–6 h following stimulation and returns to basal levels after 24–48 h. It mediates the adhesion of circulating neutrophils, monocytes, $CD4^+$ memory T lymphocytes, eosinophils and basophils to endothelium. Its expression is induced by inflammatory mediators. ➲ The minimum requirement for E-selectin binding is similar to that of P-selectin; sialylated and non-sialylated forms of Le^a or Le^x glycotypes (*see 10.6*) are the peripheral binding partners and it also binds 3-sulfated Gal-containing glycolipids (sulfatides) and 3-sulfated glucuronic acid-containing glycolipids (sulfoglucuronyl lipids or SGNLs).

> ➲ E-Selectin and P-selectin bind Le^a or Le^x glycotopes, sulfatides and sulfoglucuronyl lipids.

> ➲ E-Selectin is expressed by endothelial cells in response to cytokine stimulation. It mediates the adhesion of circulating neutrophils, monocytes, $CD4^+$ memory T lymphocytes, eosinophils and basophils to endothelium.

L-Selectin

L-selectin is a glycoprotein constitutively expressed by neutrophils, monocytes and a subset of T lymphocytes. It is also called LEC-CAM (lectin-endothelial cell–cell adhesion molecule), CD62L and T-cell homing receptor. It mediates the adhesion of T cells to endothelium.➲ The preferred saccharide-binding partners of L-selectin are less well established than those of P- and E-selectin, but sialylated, sulfated glycans are strong candidates and it has been shown, like E- and P-selectin, to recognize sLe^x (*see 10.6*).

> ➲ L-Selectin is constitutively expressed by T cells and mediates their adhesion to the endothelium.

L-selectin counter-receptors expressed on high endothelial venules and mucosal tissues act as 'address signals' to guide normal leukocyte trafficking. They are therefore termed 'vascular addressins'.

Neutrophil L-selectin is also implicated in intracellular signalling through the generation of cytosolic calcium transients, enhanced TNF and IL-8 gene expression, activation or potentiation of the oxidative burst and MAP kinase activation. Experimental ligation and cross-linking of L-selectin (by sulfatide and anti-L-selectin antibody binding), mimicking its interaction with its natural ligand, for example, results in tyrosine phosphorylation of several cellular proteins and generation of Ca^{2+} transients.➲

> ➲ Commonly used alternative names for L-selectin include:
> - LEC-CAM-1
> - **mouse:**
>   MEL-14
>   lymphocyte homing receptor
>   Ly-22
> - **human:**
>   Leu8
>   TQ1
>   DREG-56
>   LAM-1.

13.12.3.3 The collectin family

The collectin family (another group of C-type lectins) are soluble ≈350 kDa multimeric lectins with an N-terminal cysteine-rich domain, a collagen-like domain, a 'neck' or spacer region, and a C-terminal, Ca^{2+}-dependent, carbohydrate-binding domain. The term 'collectin' derives from the names of the two major domains that constitute these molecules: a collagen-like domain and a lectin domain. Each subunit is ≈32 kDa in size. They resemble the complement protein C1q in structure and function. They are involved in a primitive, antibody-independent type of immune response and after binding to yeast, viral or bacterial surface glycans they opsonise these pathogens prior to complement fixation. They are found in body fluids, for example, synovial fluid, saliva and pulmonary secretions.

Notable among the collectins is the mannose-binding lectin (MBL; also called the mannan- or mannose-binding protein, MBP), which was first described in rabbit liver and serum, and has been identified in a number of mammalian species, including rat, mouse, pig, rhesus monkey, chicken, cow and human. Both liver (L-MBL) and serum (S-MBL) forms are primarily synthesized in hepatocytes, although S-MBL is additionally produced by the kidney.

S-MBL exhibits carbohydrate-binding-mediated complement activation, and is implicated in bactericidal activity. It also acts as a direct opsonin and mediates the binding and phagocytosis of bacteria by monocytes and neutrophils. Its saccharide specificity is broad, encompassing Man, GlcNAc and Fuc, in keeping with its functional requirement to recognize a range of heterogeneous oligosaccharide structures present on the cell surface of pathogens. These monosaccharides have in common the presence of vicinal, equatorial hydroxyl groups at positions 3 and 4. MBL has minimal affinity for monosaccharides, such as Gal and SA, that lack this arrangement (*see 13.12.3*).

MBL is a homooligomer of $\approx$31 kDa subunits, each containing four primary structural regions: a short ($\approx$30 amino acids) N-terminal region rich in cysteine residues which form interchain disulfide bonds responsible for oligomerization, a (18–20 repeats of a three amino acid motif) collagen-like domain, a (30 amino acids) neck region and a ($\approx$115 amino acids) C-terminal carbohydrate-binding region. Oligomerization is necessary for complement activation. The collagen-like domain is thought to be involved in the interaction with serine proteases of the complement cascade and with cell-surface receptors such as C1q. Complement activation is by a route distinct from both the 'classical' and the 'alternative' pathways and it has been suggested that a third 'lectin-mediated' or 'MBL' pathway should be considered. There is increasing evidence that individuals with low levels of MBL are more susceptible to infections, including HIV and tuberculosis.

Other collectins include conglutinin and lectins of the pulmonary surfactants.

13.12.3.4 Calnexin and calreticulin

Two homologous lectins, calnexin and calreticulin, are molecular chaperones that participate in the quality-control systems for glycoproteins passing through the pathway of the ER and Golgi stacks. Calnexin is a type I membrane-bound lectin and calreticulin is a soluble luminal lectin. Both require Ca^{2+} for binding and are thus termed Ca^{2+} dependent. The molecular structure of calnexin has recently been elucidated, and it has features (a β-sandwich structure) characteristic of legume lectins (*see 13.4.1*). It has a monovalent lectin-like domain responsible for glycan recognition and a long arm which could interact with other proteins. Calnexin and calreticulin transiently bind to the early $Glc_1Man_{5-9}GlcNAc_2$ N-linked glycoproteins (*see 4.2*), probably through recognition of the Glc(α1→3)Man sequence, and this binding is required for correct protein folding during protein synthesis. In spite of their similar saccharide-binding specificities, they bind carbohydrates selectively. Several models have been proposed to describe their roles in protein folding, including the 'dynamic' model, described briefly below, which appears to be most consistent with the recently reported molecular structure of calnexin. The involvement of these molecules in N-linked glycoprotein synthesis is described in more detail in *Section 4.2.6.1*.

During N-linked glycoprotein synthesis, newly synthesized glycoproteins carrying three Glc residues on each glycan, $Glc_3Man_{5-9}GlcNAc_2$, are co-translationally trimmed by the ER glucosidases I and II to $Glc_1Man_{5-9}GlcNAc_2$. These $Glc_1Man_{5-9}GlcNAc_2$ monoglucosylated glycoproteins are recognized by calnexin/calreticulin and are presented to the thiol oxidoreductase chaperone Erp57. The combination of Erp57 with calnexin or calreticulin specifically modulates glycoprotein folding, perhaps by promoting formation of disulfide bonds. Glucosidase II removes the final Glc residue and thus abolishes calnexin/calreticulin–glycoprotein interaction. If the glycoprotein is correctly folded, it will then exit the ER. If it is not, it is recognized by the ER enzyme UDP-glucose: glycoprotein glucosyltransferase (UGGT) and re-glucosylated, thereby allowing it to re-associate with calnexin/calreticulin and be retained within the quality control pathway of the ER. UGGT recognizes only incompletely folded glycoproteins as its substrate, and thus acts as a quality control molecule within the glycoprotein folding cycle. This de-glucosylation/re-glucosylation cycle is repeated until correct folding is achieved. After exiting the calnexin/calreticulin cycle of de-glucosylation/re-glucosylation, correctly folded glycoproteins can undergo further trimming by ER mannosidases I and II leading to $Man_{7-9}GlcNAc_2$ glycans. Terminally misfolded glycoproteins, in conjunction with calnexin, are de-mannosylated by mannosidase I and thus targeted for degradation in the proteosome.

There is controversy as to whether calnexin/calreticulin function as lectins only, or both as lectins and as classic chaperone molecules. In the second case, calnexin/calreticulin would possess a second site, in addition to the carbohydrate-combining site, that recognizes polypeptide segments in the unfolded or partially folded glycoprotein.

13.12.4 I-Type lectins

➲I-Type lectins belong to a very large family of proteins, the 'immunoglobulin superfamily', which also contains the immunoglobulins. They are defined by the presence of an Ig-like domain and all contain a similar core of two β-sheets typical of immunoglobulin molecules. Members of the family recognize very diverse saccharide structures. Many function as cell adhesion molecules; others function as growth factor receptors in construction of the ECM.➲ Examples of members of the I-type lectin family include the hyaluronan (*see 8.3.1*) binding intercellular adhesion molecule-I (ICAM-I) which plays a functional role in neutrophil infiltration into inflamed tissue, the heparin-binding (*see 8.3.3*) fibroblast growth factor receptors (FGFR), neural cell adhesion molecule (N-CAM), and the myelin protein P0 which recognizes HNK-1-like carbohydrate structures.

> ➲ I-Type lectins belong to the immunoglobulin superfamily and are defined by the presence of an Ig-like domain.

> ➲ Examples of I-type lectins include:
> - intercellular adhesion molecule-I (ICAM-I)
> - fibroblast growth factor receptor (FGFR)
> - neural cell adhesion molecule (N-CAM)
> - myelin protein P0
> - Siglecs including sialoadhesin, CD22, CD33, myelin associated glycoprotein (MAG) and the avian Schwann cell myelin protein (SMP).

A major subgroup of I-type lectins are the Siglecs (also sometimes called sialoadhesins). They include: sialoadhesin, CD22, CD33, myelin-associated glycoprotein (MAG) and the avian Schwann cell myelin protein (SMP) (*see 11.8.4*). They recognize a common group of sialic acid carbohydrate ligands and are involved in a range of cellular recognition phenomena, particularly within the immune system. CD22, for example, mediates cell-adhesion and signalling during B-cell activation.

13.12.4.1 Sialoadhesin

Sialoadhesin is a macrophage receptor found on specific subsets of macrophages in bone marrow and lymphatic tissues, and is responsible for mediating sialic acid-dependent recognition of erythrocytes. It is also expressed by macrophages sited in the peripheral nervous system and in parts of the brain that normally have contact with plasma proteins; it is involved in response to central nervous system injury. It recognizes a range of different sialylated gangliosides as binding partners.

13.12.4.2 Myelin-associated glycoprotein

Myelin-associated glycoprotein (MAG) is a multifunctional glycoprotein. It is involved in myelination of axons. It is expressed by oligodendrocytes in the central nervous system and Schwann cells in the peripheral nervous system, and has been implicated in promotion and/or inhibition of cell adhesion and neurite outgrowth. It interacts with different neuronal cell types and binds to heparin. Furthermore, it interacts with different types of collagen and interferes with fibrillogenesis of collagen type I (*see 8.2.1*). The avian Schwann cell myelin protein (SMP) has 43.5% sequence homology with MAG, but is not considered to be a homologue of it, although it shares many similar functional properties.

13.12.5 Pentraxins

➲ Pentraxins are responsible for a primitive non-self recognition immune defence mechanism.

Pentraxins are responsible for the most evolutionarily ancient non-self recognition immune defence mechanism.➲ Some pentraxins are soluble plasma lectins, are upregulated in response to inflammation, and are involved in early-stage defence mechanisms against infection. C-reactive protein (CRP) is a member of this group and is able to precipitate pneumococcal C-polysaccharide. CRP binds and precipitates galactans in a Ca^{2+}-dependent manner. Serum amyloid P component (SAP) recognizes a ligand abundant on the surface of many infective microorganisms and is involved in host defence against infection by them. Both CRP and SAP can bind C1q with subsequent activation of the classical complement pathway.

13.12.6 Nuclear lectins

Carbohydrate-binding proteins, presumed to be lectins, are also present in the vertebrate cell nucleus. An example is the 35 kDa carbohydrate-binding protein CPB35, specific for Gal and GalNAc, which is a component of the heterogeneous nuclear ribonucleoprotein complexes (hnRNP). It has homology with various other mammalian β-D-galactoside specific lectins, but its function is not understood.

13.13 Lectins as tools in carbohydrate analysis

Hundreds of purified lectins are commercially available. Their carbohydrate-binding function makes them useful research tools for the investigation of oligosaccharide structures (*see 14.2*).

13.14 Lectins in the diagnosis of infections, drug targeting and toxic conjugates for directed cell killing

The selective interaction of lectins with their saccharide ligands might be of potential use in drug targeting, with the advantages of lower doses and fewer side-effects than conventional therapy (*see 17.5.5*). Evidence that this could be feasible comes from a number of experimental studies. For example, a complex carbohydrate (DL-lactide-co-glycolide) binds with a high degree of selectivity to mouse Peyer's patch M-cells and thus may have potential in specific delivery of oral vaccines. However, there is evidence that species-specific adaptation of different microbial populations has resulted in differences in glycosylation and that experimental results from animal models may not necessarily be applicable to humans. The carbohydrate-specific recognition of *Entamoeba histolytica*, the cause of human amoebic dysentery (*see 13.9*), may be open to exploitation in development of more effective anti-amoebic therapy through coupling drugs to an appropriate carbohydrate carrier. Furthermore, lectins are attractive tools for chimeric targeted toxins (i.e. toxin molecule attached to a lectin). Experiments have been carried out, for example, in conjugating toxins to lectin carriers in an attempt to target them to particular cell populations. To date, however, this approach has proved disappointing, and the immunogenicity of the lectins may prove to be a major obstacle to their clinical use.

Lectins have proved useful in typing bacteria on the basis of differences in their cell surface glycosylation. Evidence that some bacterial infections are mediated through specific interactions between lectins and their carbohydrate-binding partner (*see 13.9, 16.2*), has led to the suggestion that specific carbohydrate moieties may act as inhibitors of bacterial–mammalian cell adhesion, and thus prevent infection. Similarly, the phenomena of lectinophagocytosis (*see 13.9*) may be enhanced through stimulation of phagocytic cells to synthesize appropriate carbohydrates. As lectinophagocytosis does not involve a host antibody response, this may be of particular interest in immunocompromised individuals.

Lectins may also have potential in the diagnosis and treatment of virally transmitted disease. They may discriminate between viruses on the basis of their glycan repertoire, and there is evidence, for example, that snowdrop lectin may be useful in detection of HIV infection. Recent reports also suggest that relatively newly discovered lectins from orchids and *Allium* species inhibit human retrovirus replication, so that these lectins could have potential in the development of future therapeutic approaches in the targeting of viruses *in vivo*.

Further reading

Barondes, S.H. (1988) Bifunctional properties of lectins: lectins redefined. *Trends Biol. Sci.* **13:** 480–482.

Elgavish, S., Shaanan, B. (1997) Lectin–carbohydrate interactions: different folds, common recognition principles. *Trends Biol. Sci.* **22:** 462–467.

Etzler, M.E. (1998) Mini review: from structure to activity: new insight into the functions of legume lectins. *Trends Glycosci. Glycotechnol.* **10:** 247–255.

Goldstein, I.J., Hughes, R.C., Monsigny, M., Osawa, T., Sharon, N. (1980) What should be called a lectin? *Nature* **285:** 66.

Peumans, W.J., Van Damme, E.J.M. (1995) Lectins as plant defense proteins. *Plant Physiol.* **109:** 347–352.

Powell, L.D., Varki, A. (1995) I type lectins. *J. Biol. Chem.* **270:** 14 243–14 246.

Rudiger, H. (1998) Plant lectins – more than just tools for glycoscientists: occurrence, structure and possible functions of plant lectins. *Acta Anatomica* **161:** 130–152.

Sharon, N., Lis. H. (1995) Lectins – proteins with a sweet tooth: functions in cell recognition. *Essays Biochem.* **30:** 59–75.

Singh, R.S., Tiwary, A.K., Kennedy, J.F. (1999) Lectins: sources, activities and applications. *Crit. Rev. Biotechol.* **19:** 145–178.

Tools for carbohydrate analysis

14

14.1 Introduction – the range of tools and methods available

Lectins (*see Chapter 13*) are useful tools for detecting carbohydrates. They have biologically significant activity, including specific recognition of complex carbohydrate structures. These properties have led to their use in a variety of settings, including purification of specific glycoconjugates, blood group typing, investigation of cellular glycosylation in histopathology, isolation and stimulation of lymphocyte populations in immunology and identification and taxonomy of microorganisms.

Although lectins can provide useful preliminary data on the identity of the terminal monosaccharide(s) of oligosaccharide chains, more detailed biochemical analysis of carbohydrate structures requires a range of biochemical approaches. Molecular analysis of oligosaccharides has been hampered by the fact that many monosaccharides are identical in molecular mass and are very similar in terms of other physical characteristics (*see Chapter 1*) and are therefore difficult to separate and analyse. ➲ Furthermore, the different possible linkages between monosaccharide and the heterogeneity of the linear and branching structures they can form (*see 1.16, 1.17, 1.18*), means that their analysis is complex. It is currently not possible to sequence oligosaccharides in a manner analogous to the way in which proteins and nucleic acids are sequenced. One method rarely provides full details on oligosaccharide composition and linkage analysis of the constituent monosaccharides and most of the methods need to be used in combination to achieve full oligosaccharide sequence information.

The most frequently used techniques for oligosaccharide release and analysis are described in this chapter. Some are long-established and, because they utilize methods also used in other fields of biochemistry, are accessible to most life-science researchers. These sometimes require relatively large amounts of starting material, and often yield fairly crude data. Other, newer and more sophisticated techniques are also described. Some of these require much smaller amounts of sample and give more detailed analysis, but may require considerable expertise and are only accessible to specialized laboratories.

> ➲ The complexity and diversity of oligosaccharide structures and the physical similarity of the monosaccharides of which they are composed means that their analysis is technically difficult.

14.2 The use of lectins

Lectins (*see Chapter 13*) can be used to explore carbohydrate distribution in a manner analogous to that in which antibodies can be used to explore antigen expression. Currently ≈150–200 different lectins of diverse saccharide-binding specificity are available commercially and most of them at modest cost. The list is growing all the time and is likely to continue to do so as interest in glycobiology grows and there is an increasing demand for appropriate tools. They range from very well-characterized lectins in which the molecular structure and saccharide-binding characteristics are known in some detail, to others in which almost nothing is known of their properties.

Functional and Molecular Glycobiology, Susan A. Brooks, Miriam V. Dwek and Udo Schumacher

➲ Hundreds of lectins, derived mostly from plants and invertebrates, are available commercially and can be used as tools to investigate oligosaccharides.

For over 100 years, the majority of lectins have been derived from plants, and as this material is cheap, readily available and, often very abundant in lectins, almost all commercially available purified lectins are from these sources. Lectins are also available from a limited number of invertebrates including snails, slugs and marine crustaceans.➲ Lectins from snake venoms and *Anguilla anguilla* (eel) serum are also available, but other animal lectins are generally not, owing to their low tissue concentration, difficulties in obtaining suitable raw materials, and technical difficulties in purification. It is likely that in the next few years, as biotechnology advances, recombinant animal and plant lectins (*see 18.7*) will become available for research purposes, as this will open the frontiers to many new research areas.

Some examples of lectins commonly used in research are listed in *Table 14.1*, but many more are available commercially.

14.2.1 Exploring tissue distribution of glycoconjugates – applications in histology and histopathology

The synthesis of glycoconjugates by cells may change during embryogenesis, development and in disease, and changes in glycosylation are associated with incompletely understood cell division and differentiation processes. Aberrant glycosylation is a common feature in cancer,

Table 14.1 Some examples of commonly used lectins

Source of lectin			
Latin name	Common name	Abbreviation	Principle inhibitory carbohydrate(s)
Mannose binding			
Canavalia ensiformis	Jack bean	Con A	α-D-Man > α-D-Glc
Lens culinaris	Lentil	LCA	α-D-Man
Galactose binding			
Arachis hypogaea	Peanut	PNA	D-Gal(β1→3)GalNAc
Ricinus communis	Castor oil plant	RCA-I or RCA_{120}	β-D-Gal
Fucose binding			
Ulex europaeus	Gorse	UEA-I	L-Fuc(α1→2) Gal(β1→4)GlcNAc
Lotus tetragonalobus	Asparagus pea	LTA	α-L-Fuc
Pisum sativum	Pea	PSA	bi-/tri-antennary with Fuc
N-Acetylgalactosamine binding			
Glycine max	Soyabean	SBA	α-D-GalNAc
Dolichos biflorus	Horse gram	DBA	α-D-GalNAc
Helix pomatia	Roman snail	HPA	α-D-GalNAc > α-D-GlcNAc
Vicia villosa	Hairy vetch	VVA	α-D-GalNAc
N-Acetylglucosamine binding			
Triticum vulgaris	Wheatgerm	WGA	D-GlcNAc > NeuAc
Griffonia simplicifolia	–	GSA-II	α-D-GlcNAc = β-D-GlcNAc
Datura stramonium	–	DSA	GlcNAc(β1→4)GlcNAc
Sialic acid binding			
Limax flavus	Garden slug	LFA	Neu5Ac, NeuGc
Limulus polyphemus	Horseshoe crab	LPA	NeuAc

and as the interactions of saccharides with their receptors are of importance in many cellular processes, these changes may be of functional significance. Although well documented, the glycosylation of normal cells and the changes in glycan synthesis associated with disease are not well understood at the molecular level. Conventional histochemistry to detect carbohydrate structures involves use of the periodic acid Schiff (PAS) procedure. In this test, the tissue section is exposed to periodic acid, which oxidizes the hydroxyl groups on adjacent carbon atoms or adjacent hydroxyl and amino groups to produce aldehydes. The section is then stained with Schiff reagent (fuschin-sulfurous acid), which reacts with the aldehydes to yield a pink–red reaction product. The PAS reaction is useful, for example, to reveal the presence of heavily glycosylated components, such as basal lamina.

Highly acidic sulfate esters and less acidic carboxyls of sialic acids and uronic acids can be identified by a number of cationic dyes including Alcian blue, azure A, thionin, toluidine blue, aldehyde fuchsin and iron diamines. Lectins may be used to map other aspects of cellular glycosylation.

The principle of lectin histochemistry offers significant advantages in terms of selectivity over the classical PAS reaction, which yields little information on detailed molecular structure. The technique is almost identical to that used for standard immunohistochemistry, but here, instead of a primary antibody recognizing and binding cell- or tissue-bound antigen, a specific lectin is used to bind to cell- or tissue-bound saccharide. The binding of the lectin may be detected by a variety of methods, including the use of directly labelled lectin (labelled with an enzyme such as horseradish peroxidase or alkaline phosphatase, or a fluorescent compound such as fluorescein isothiocyanate (FITC) as reporter molecules), as illustrated in *Figure 14.1*; biotinylated lectin can be detected by avidin or streptavidin conjugated to an appropriate label, as illustrated in *Figure 14.2*; avidin–biotin complex (ABC), as illustrated in *Figure 14.3*; or, if an appropriate antibody directed against the lectin is available then a whole range of indirect, multistep detection systems such as the peroxidase–anti-peroxidase (PAP) or alkaline phosphatase–anti-alkaline phosphatase (APAAP) method, illustrated in *Figure 14.4*, are possible. An example of lectin histochemistry at the light microscope level is given in *Figure 14.5*. Lectin histochemistry may also be used to map oligosaccharide distribution in living cells at the confocal microscope level, as illustrated in *Figure 14.6*. ➲

> ➲ Lectin histochemistry facilitates the mapping of oligosaccharide structures in tissue sections.

Antibodies directed against carbohydrate antigens can be used in exactly the same types of methods as those described above; antibody directed against the carbohydrate structure of interest simply takes the place of the lectin. This immunohistochemistry approach is, however, of limited application as few good anti-carbohydrate antibodies are commercially available. This is probably because carbohydrate–anti-carbohydrate antibody binding affinities are often low.

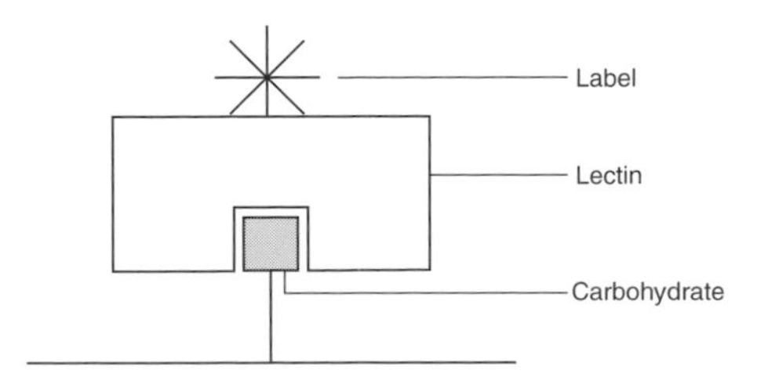

Figure 14.1

The direct method for detection of lectin binding: carbohydrate residues are detected by directly labelled lectin. The lectin may be labelled with a fluorescent or enzyme label. In the case of enzyme labels, the enzyme (e.g. horseradish peroxidase or alkaline phosphatase) converts a soluble, colourless substrate to an insoluble, coloured product.

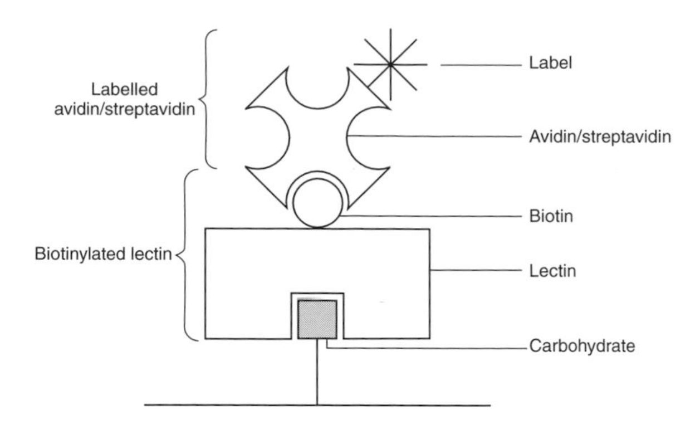

Figure 14.2

The simple avidin–biotin method for detection of lectin binding: carbohydrate residues are detected by biotinylated lectin, the binding of which is revealed by labelled avidin or streptavidin. Biotin (one of the B group of vitamins) binds to avidin (a protein derived from egg white) or streptavidin (a protein derived from *Streptomyces* species) with great avidity. The avidin/streptavidin molecule is coupled to a molecule which can act as a reporter, usually an enzyme or fluorescent label.

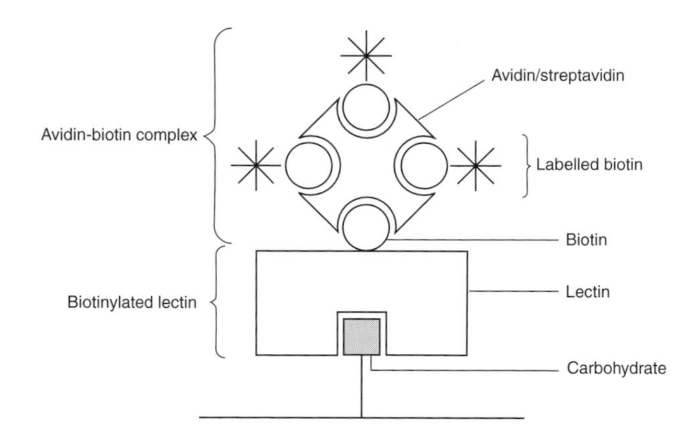

Figure 14.3

The avidin–biotin complex (ABC) method for detection of lectin binding: carbohydrate residues are detected by biotinylated lectin, the binding of which is revealed by labelled 'avidin–biotin complex' (ABC). ABC is a mixture of avidin and labelled biotin combined in such a ratio that three of avidin's four biotin binding sites are saturated, leaving the fourth free to combine with the biotin label carried by the lectin. The presence of multiple-labelled biotin molecules results in enhancement of the intensity of labelling.

As with immunohistochemistry, lectin histochemistry can also be employed at the electron microscope level to determine ultrastructural localization of glycoconjugates. Lectins directly conjugated to electron-dense labels such as ferritin or colloidal gold are commercially available for this application.

Allied to lectin histochemistry is the technique of 'reverse lectin histochemistry' in which neoglycoproteins (simple saccharide structures, often monosaccharides, chemically coupled to an inert protein carrier such as bovine serum albumin) are used to detect cell- or tissue-bound lectins. Binding of neoglycoprotein to lectin is revealed by similar methodology to that used in immunohistochemistry/lectin histochemistry. Neoglycoproteins may either be

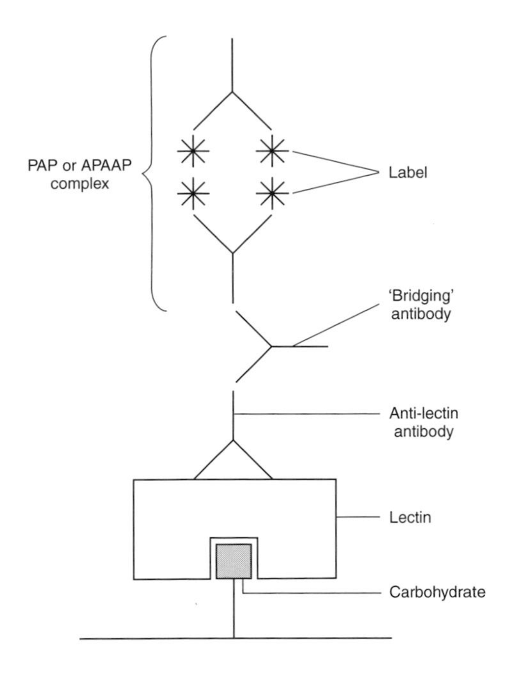

Figure 14.4

The peroxidase–anti-peroxidase (PAP) or alkaline phosphatase–anti-alkaline phosphatase (APAAP) method: carbohydrate residues are detected by unconjugated lectin, the binding of which is revealed by subsequent layering with (a) antibody directed against the lectin, (b) then a 'bridging' antibody (applied in excess so that one binding site remains unoccupied to link with the PAP or APAAP complex) which is directed against the first antibody, (c) then finally a PAP or APAAP complex. The PAP or APAAP complex is a mixture of antibody directed against the enzyme (peroxidase or alkaline phosphatase) reporter molecule, plus the reporter molecule itself. The anti-enzyme antibody is raised in the same species as the first, anti-lectin antibody, and is therefore also recognized by the 'bridging' antibody. The multiple layers and consequent 'cloud' of coloured product result in high sensitivity and strong labelling.

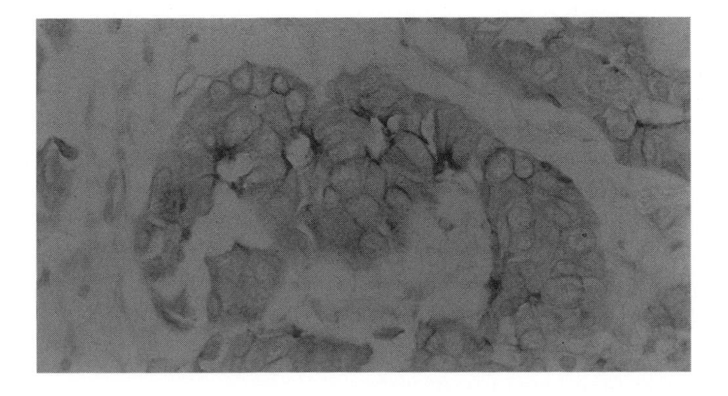

Figure 14.5

An example of lectin histochemistry at the light microscope level. A paraffin wax-embedded section of a human breast cancer labelled for binding of the lectin from *Maclura pomifera*. Labelling is localized at the cell surface, and is particularly dense at the luminal surface of duct-like structures. Some diffuse, granular cytoplasmic localisation is also seen.

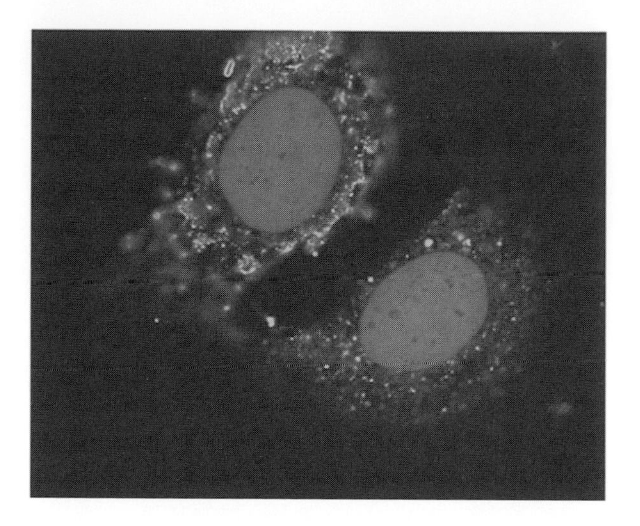

Figure 14.6

Lectin histochemistry at the confocal microscope level. Breast cancer cells in culture labelled for binding of the lectin from *Helix pomatia* (HPA). The nucleus is labelled with propidium iodide. The lectin is labelling glycoconjugates undergoing biosynthesis in the Golgi apparatus (dense perinuclear localisation) and also at the cell surface.

directly labelled (with an enzyme or fluorescent label for light microscopy or colloidal gold for electron microscopy) or biotinylated for detection with labelled avidin or streptavidin. Some authors have warned that results using this approach should be interpreted with caution.

14.2.2 Lectins for affinity purification of carbohydrates and glycoconjugates

Lectins are widely used in affinity purification of glycoproteins, glycopeptides and free glycans. ➲ Typically, the mixture to be separated is applied to a column of agarose beads on which a particular lectin has been immobilized. Lectin interaction with glycoconjugates in the mixture retards some glycoconjugates and tighter binding adsorbs others onto the column. Retarded material may be collected as a separate fraction as it emerges from the column. Tightly adsorbed material may be specifically eluted by application of an appropriate competing monosaccharide. The principles of this method are illustrated in *Figure 14.7*. Sometimes lectin binding to the oligosaccharides of naturally occurring glycoconjugates is so tight that elution of adsorbed material from the affinity column can only be achieved with difficulty, for example, by dramatically changing the pH or ion concentration of the buffer, or by the addition of detergents. All of these procedures can damage the affinity beads and/or the glycoconjugates of interest. Lectin affinity chromatography requires glycoprotein samples of ≈500 μg to 1 mg or above.

➲ Lectin affinity chromatography facilitates the purification of oligosaccharides and glycoconjugates from complex mixtures.

In 'serial lectin-affinity chromatography', a complex mixture of glycoconjugates is sequentially fractionated by separation on a series of affinity columns bearing lectins with differing saccharide-binding characteristics. This is a useful approach to separating glycoconjugates that share a similar molecular size, charge and other physical characteristics, but are distinct in their glycosylation. Many lectins are commercially available ready immobilized on a suitable

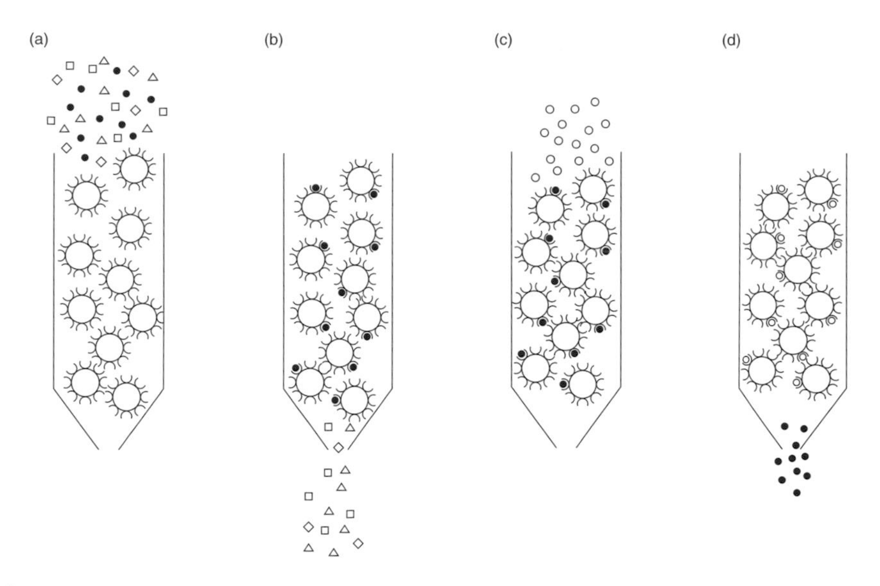

Figure 14.7

Principles of lectin affinity chromatography: (a) a mixture of glycans is applied to a column of agarose beads on which the lectin is immobilized; (b) glycoconjugates that are specifically recognized by the lectin are retarded or adsorbed, whereas others not recognized by the lectin pass through unimpeded; (c) after washing, the monosaccharide for which the lectin shows greatest affinity is applied to the column; (d) specifically adsorbed glycoconjugates are displaced by competing monosaccharide and may be collected for analysis.

affinity matrix, such as agarose or Sepharose beads. Activated beads are also available for all affinity chromatography applications, and it is a simple procedure to immobilize the lectin of choice. However, the presence of different glycoforms on a single glycoprotein makes this approach more complex than it sounds as do the possible presence of multiple glycosylation sites on a single glycoprotein.

14.2.3 Lectin-based solid-phase assays for glycosyltransferase activity

Lectins have been used in solid-phase assays for glycosyltransferase activity. A microtitre plate is coated with an appropriate acceptor for the glycosyltransferase, such as a neoglycoprotein, and the solution to be assayed for glycosyltransferase activity plus sugar nucleotide and divalent cations are then added to the wells. The glycosyltransferase transfers sugar to the immobilized acceptor thus generating a new immobilized glycoform. This can then be detected by application of an appropriate labelled (usually with alkaline phosphatase, horseradish peroxidase or a fluorescent label) second lectin. A variation on this technique is where glycosyltransferase and acceptor are allowed to react in solution and the newly glycosylated product is then captured by an immobilized antibody and detected by labelled lectin. Relatively large amounts of enzyme sample, in the region of 500 μg, are required.

Another solid-phase assay is enzyme-linked lectin assay (ELLA), which may be used for analysis and quantification of glycoproteins. This assay is based on similar principles to the standard enzyme-linked immunosorbant assay (ELISA). In its simplest form, the glycoproteins of interest are adsorbed onto microtitre plates, and their presence is detected by lectin conjugated to an appropriate label. Again, at least 500 μg to 1 mg of sample is required.

14.2.4 Blood group serology

Historically, one of the first uses of lectins was in blood group typing, and their usefulness persisted until recent years when a monoclonal antibody to type O(H) red blood cells eventually became available and superseded them. Several lectins are known which recognize blood group-specific sugars, and therefore preferentially cross-link and agglutinate blood cells of one type and not others (*see 13.2.4, 13.6*). Lectin recognition of blood group sugars such as Tn (recognized by *Salvia sclarea* and *Vicia villosa* lectin), T (recognized by peanut lectin) and Cad antigens (recognized by *Salvia horminum* and *Salvia sclarea* lectins) are of interest for research purposes (*see 10.4*).

14.2.5 Lectins in cell typing and sorting

Lectins can be used to label and separate cell populations on the basis of differences in their glycosylation (glycotype). This may be approached in a number of ways. Most simply, a lectin is mixed with a suspension of cells and agglutination of the cells is observed. Specificity is confirmed by inhibition of agglutination in the presence of an appropriate blocking sugar. Alternatively, cells may be labelled with either a fluorescent-labelled lectin or a biotinylated lectin with fluorescent-labelled avidin or streptavidin, then separated and/or analysed by flow cytometry.

14.2.6 Lectins as mitogens and toxins

Mitogenic lectins have been used extensively in immunology. For example, pokeweed mitogen (PWM) stimulates T- and B lymphocytes to divide, and PHA-L stimulates T lymphocytes to divide. *Urtica dioica* (nettle) lectin has recently been described as a useful mitogen for murine T lymphocytes. Some examples of mitogenic lectins are listed in *Table 14.2*.

At higher concentrations these, and many other lectins, are toxic and their effects may be used to select cell populations with altered glycosylation pathways. Cells are cultured in the presence of the (toxic) lectin, and those with cell-surface glycotopes recognized by the lectin are killed. Cells not recognized by the lectin, owing to their different cell-surface glycosylation,

Table 14.2 Some examples of mitogenic lectins

Source of lectin			
Latin name	Common name	Abbreviation	Principle inhibitory carbohydrate(s)
Abrus precatorius	Jequirty bean	abrin	D-Gal
Arachis hypogaea	Peanut	PNA	D-Gal(β1→3)D-GalNAc
Bauhinia purpurea	Camels foot tree	BPA	D-Gal(β1→3)D-GalNAc
Canavalia ensiformis	Jack bean	Con A	α-D-Man > α-D-Glc
Erythrina corallodendron	Coral tree	ECA	D-Gal(α1-4)D-GlcNAc
Euonymus europaeus	Spindle tree	EUA	D-Gal(α1-3)D-Gal
Glycine max	Soyabean	SBA	α-D-GalNAc
Lathyrus odoratus	Sweet pea	LOA	α-D-Man
Lens culinaris	Lentil	LCA	α-D-Man
Phaseolus limensis	Lima bean	LBA	α-D-GalNAc
Phytolacca americana	Pokeweed	PWM	(β1→4)-linked GlcNAc oligomers
Pisium sativum	Garden pea	PSA	α-D-Man
Sambucus nigra	Elderberry	SNA-I	NeuAc(α2→6) & D-Gal(β1→4)-D-Gal
Triticum vulgaris	Wheatgerm	WGA	D-GlcNAc > NeuAc
Vicia faba	Broad bean or fava bean	VFA	D-Man > D-Glc
Vicia sativa	–	VSA	D-Glc > D-Man

are tolerant to the presence of the lectin ('lectin resistant'), and survive. Selection of cell line variants with altered cell-surface glycosylation enables study of the genetic basis of glycosylation, the function of cell-surface glycoconjugates and glycoconjugate biosynthesis (*see 15.4*).

14.2.7 Gel-diffusion techniques

Gel-diffusion techniques were widely used around 20 years ago, and although now superseded by other more selective approaches are worthy of mention as they are simple to perform and can still provide useful information. Glycoproteins and polysaccharides are detectable by the technique of Ouchterlony gel diffusion using lectins. In this technique, test samples containing glycoconjugates of interest (e.g. serum samples) are placed in wells punched into an agarose gel, alongside wells containing lectin solution. Glycoconjugates and lectins diffuse into the gel, and when they meet, if they are at appropriate concentrations, they react to give an insoluble precipitate which may be viewed directly or stained (e.g. with Coomassie Brilliant Blue). An example of the results of this type of approach is given in *Figure 14.8*. This simple approach can yield preliminary information regarding concentration, molecular mass, degree of branching and heterogeneity of the sample glycoconjugates dependent upon the number, appearance and positioning of the precipitate bands.

A slightly more complex technique is an adaptation of immunoelectrophoresis called 'rocket affinoelectrophoresis'. A glycoprotein sample is separated by electrophoresis on a gel

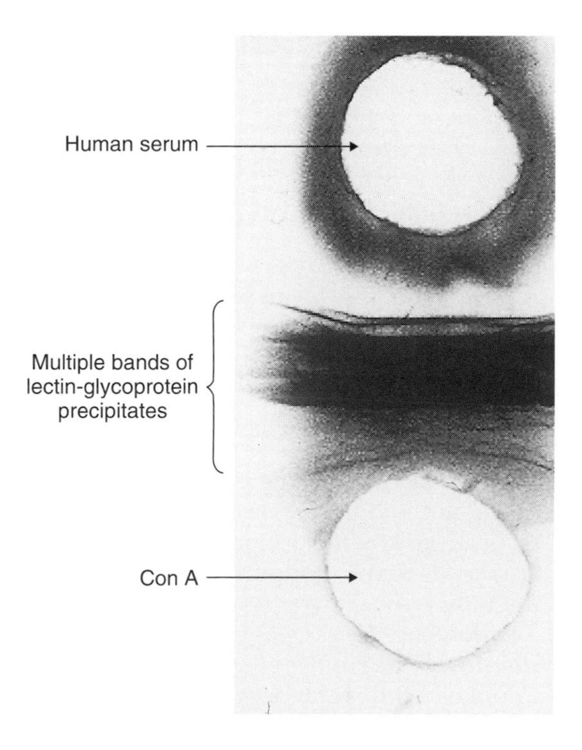

Figure 14.8

Ouchterlony gel diffusion. The upper well contained human serum and the lower well contained the Man-binding lectin from *Canavalia ensiformis* (Con A). Both solutions diffused into the gel and where the lectin recognized and bound to mannosylated serum glycoproteins, an insoluble precipitate was formed which was visualized as a dark line by staining with the total protein stain Coomassie Brilliant Blue. Multiple precipitated glycoproteins are visible.

containing lectin. As the glycoprotein is driven to move in the electric current, it encounters the lectin. If the lectin binds to the glycoprotein, an insoluble precipitate is formed in a 'rocket' shape. The height of the precipitate 'rocket' bears an approximate relationship to the concentration of the reacting glycoprotein in the original sample. In a more complex one-dimensional analysis, 'rocket affinoimmunoelectrophoresis', glycoproteins are first separated by gel electrophoresis through a strip of gel containing the lectin of interest, and then through a second strip of gel containing antibodies against the glycoproteins. Their interaction with the lectin is reflected in the degree of retardation as they pass through the lectin-containing gel. In the two-dimensional analysis, 'crossed affinoimmunoelectrophoresis', the glycoprotein sample is separated by electrophoresis in one dimension through a lectin-containing gel, then in a second dimension through an antibody-containing gel. These assays can reveal information about the heterogeneity of glycoproteins with regard to glycan content and structure.

The techniques described in this section require milligram quantities of glycoprotein sample.

14.2.8 Lectin blotting studies

Lectins are useful in blotting assays in which glycoproteins separated by sodium dodecyl sulfate–polyacrylamide gel electrophoresis (SDS–PAGE) are transferred by electroblotting (Western blotting) onto a supporting membrane which is probed for the binding of lectins with saccharide-binding properties of interest. Lectins are used in this context in much the same way as antibodies are used to probe blots for the expression of an antigen. This approach requires microgram quantities of glycoprotein sample if colormetric detection is used, and nanogram quantities if chemiluminescence detection is used. These approaches are described in more detail in *Section 14.3.4*.

14.3 Analysis of glycoproteins by SDS–PAGE and Western electroblotting

14.3.1 Separation of glycoproteins by SDS–PAGE

The separation of glycoproteins by molecular mass, driven by an electric current through a vertical cross-linked polyacrylamide gel in the presence of detergent, sodium dodecyl sulfate, called sodium dodecyl sulfate–polyacrylamide gel electrophoresis (SDS–PAGE), is the most common method used for the analysis of proteins. Different concentrations of acrylamide are incorporated into the electrophoretic gel matrix and the resulting differences in cross-linking enable gels to be constructed which separate proteins ranging in molecular mass from 10 to 200 kDa.

Some heavily glycosylated proteins do not migrate adequately in conventional SDS–PAGE owing to the large number of negatively charged species. ➲ This can be overcome by the use of a borate buffer. In this system, the boric acid binds to the monosaccharides and enables the glycoproteins to migrate in the electric field.

> ➲ SDS–PAGE can be used to separate glycosylated proteins, although heavily charged species are poorly resolved.

Following their separation, proteins are usually visualized by staining. The most commonly used stains are the (reversible) Coomassie Brilliant Blue (which detects protein concentrations in the microgram range) or silver stain (which detects protein concentrations in the nanogram range). PAS stain (*see 14.2.1*) may be used to specifically identify glycosylated proteins in polyacrylamide gels. Depending on the degree of glycosylation of the protein, 1–10 μg of protein are detectable by PAS.

Proteins separated by SDS–PAGE may be extracted from the gel for further analysis, transferred to an inert membrane (*see 14.3.2*) or, subjected to in-gel digestion, either of the protein or of the oligosaccharide from the protein (*see 14.6.4*).

14.3.2 Western blotting

➲ Western blotting is the process by which (glyco)proteins, first separated by SDS–PAGE, are transferred to an inert membrane that can then be probed with lectins or antibodies to identify oligosaccharides on the protein of interest.

The transfer of proteins from the gel to an inert membrane, usually nitrocellulose, under an electric current is termed electroblotting, also known as Western blotting. ➲ The two main systems used for electroblotting are tank (wet) blotting and semi-dry blotting. In both procedures, the (glyco)proteins in the gel are transferred to the membrane driven by an electric charge in the presence of buffer. In tank blotting, the gel and membrane are held vertically between two platinum wire electrodes in a tank of buffer; in this system the buffer needs to be cooled. Semi-dry electroblotting is, as the name implies, an almost dry process. The gel and membrane are sandwiched between layers of buffer-dampened filter paper between two (usually graphite) electrodes and in this system, separate cooling apparatus is not required.

The membrane chosen for the transfer of the (glyco)protein depends upon the techniques planned for its subsequent analysis. If the protein is to be removed from the membrane and analysed further, polyvinylidene difluoride (PVDF) or nylon membranes are preferred, but if the glycoprotein is to be analysed on the membrane itself, then the preferred membrane is usually nitrocellulose. The specification of the membrane is selected according to the molecular mass of the (glyco)protein of interest.

(Glyco)proteins may first be visualized with a reversible non-specific protein stain (e.g. Ponceau red S) and then destained prior to further analysis.

14.3.3 Detection of glycoproteins immobilized on blots

A variety of approaches may be used in the study of the glycosylation of proteins transferred to membranes. For an evaluation of the total number of glycosylated proteins, digoxigenin/anti-alkaline phosphatase labelling may be used. This system is reported to detect N- and O-linked oligosaccharides as well as glycans of GPI anchors, with a limit of sensitivity of ≈100 ng protein, depending on the degree of glycosylation. As with PAS (*see 14.2.1*), the carbohydrates are first oxidized with periodate. Instead of using Schiff's reagent, a modified digoxigenin is used which binds to periodated carbohydrate. The digoxigenin is then detected using an anti-digoxigenin antibody, the presence of which is detected using standard immunohistochemical techniques (*see 14.2.1*), using alkaline phosphatase as an enzyme reporter.

14.3.4 Analysis of the monosaccharides present on a glycoprotein

Lectins and carbohydrate-specific antibodies can be used to detect the terminal monosaccharide(s) of oligosaccharide chains of glycoproteins immobilized on blots. ➲ A limited range of anti-carbohydrate antibodies is available commercially, so lectins are more useful in this context. Lectin–saccharide binding preferences are usually quoted in terms of the monosaccharide that most effectively inhibits their binding (e.g. a Man-binding lectin or GalNAc-binding lectin) but this can be misleading as they often recognize several different monosaccharides with differing degrees of avidity, and prefer disaccharides or more complex oligosaccharide structures to simple monosaccharides (*see 13.5*). Their preferred complex binding partners are usually unknown. This may cause ambiguities in the interpretation of results, but nevertheless, lectins can be usefully employed in preliminary analysis of the glycans present on glycoproteins. The most commonly used lectins, and their nominal binding specificities, are listed in *Table 14.1*.

➲ Lectins may be used to detect glycoproteins separated by SDS–PAGE and immobilized on Western blots to provide preliminary identification of saccharide structures present.

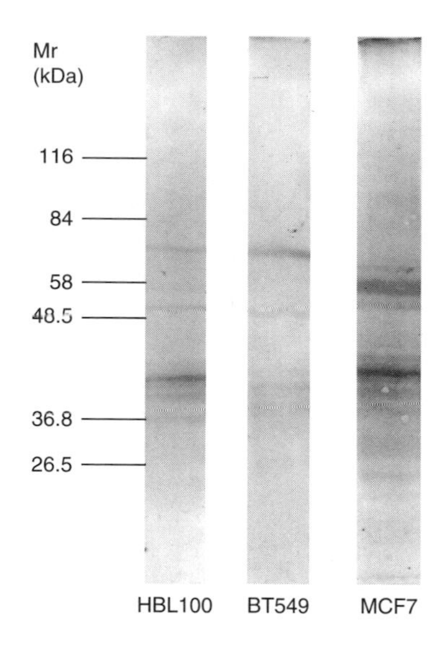

Figure 14.9

Cytosol preparations of breast cells grown *in vitro,* separated by SDS–PAGE, blotted to nitrocellulose and probed with HPA.

To visualize binding, detection systems directly analogous to those used in histochemistry (*see 14.2.1*) are used. The blot may be incubated with lectin or antibody directly conjugated to a label (direct method, *Figure 14.1*). Alternatively, binding of unlabelled lectin or antibody detected by a labelled secondary antibody (simple indirect method) or biotinylated lectin or primary antibody may be used, and binding detected by labelled avidin (simple avidin–biotin method, *Figure 14.2*). The longer and more complex methods, such as ABC and PAP/APAAP are generally not used in this context because of the potential problem of non-specific labelling from binding of reagents from the increased numbers of 'layers'. Visualization of the marker can be either colormetric (enzyme label), radioactive or fluorescent. Radioactive detection has the advantage of being very sensitive, but the most sensitive detection system available is enhanced chemiluminescence (ECL), which can be introduced at the final enzyme label development stage to produce a light reaction that is captured on X-ray film. Colormetric labelling will detect as little as a few micrograms of glycoprotein, whereas ECL is reported to be ≈ 100 times more sensitive.

An example of lectin labelling of glycoproteins transferred to a membrane after SDS–PAGE is given in *Figure 14.9.*

14.4 Proteome analysis

The human tissue/cell proteome has been estimated as $\approx 10\,000$ proteins and the characterization of tissue/cellular protein expression (proteomics) is of considerable interest. There is a need to couple glycan analysis with proteomics to understand the role of glycosylation in protein function and the significance of altered glycosylation in disease. The principle technique of proteomics is two-dimensional electrophoresis (2-DE); this approach uses isoelectric focusing and SDS–PAGE. The availability of reproducible, reliable, immobilized gradient gel strips (IPG) for isoelectric focusing enables effective separation of proteins according to their

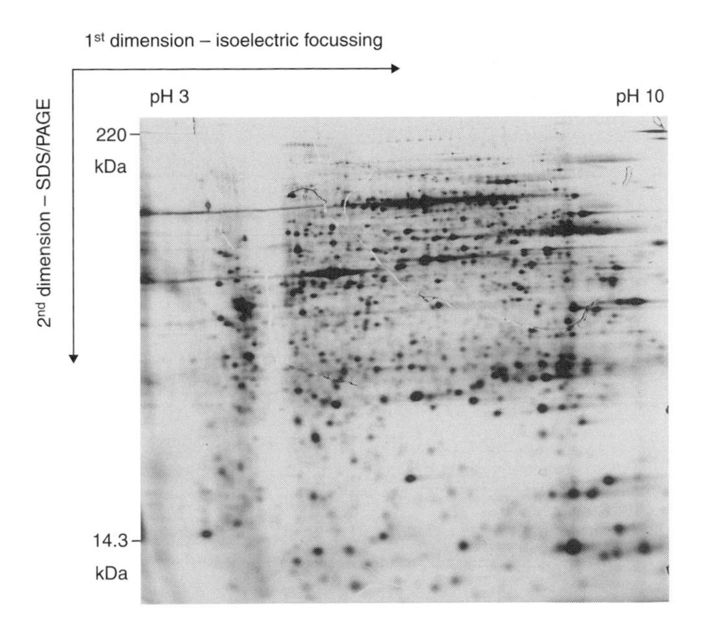

Figure 14.10

Two-dimensional electrophoresis (2-DE) separation of proteins from a breast tissue specimen by 2-DE and identification by silver staining.

isoelectric point (pI). After separation by isoelectric focusing, proteins are separated in a second dimension by slab gel SDS–PAGE. This analytical approach allows separation of between 1000 and 2000 protein spots in one experiment on an 18-cm square gel. After separation, the proteins are stained with silver stain or Coomassie Brilliant Blue. An example of a silver-stained 2-DE gel is shown in *Figure 14.10*. The use of narrow-range IPG strips further enhances the resolution of 2-DE for proteome mapping. When combined with mass spectrometry (MS) the technique provides detailed information on the (glyco)proteins present in complex mixtures, such as tissues or cellular extracts.➲ This approach is compatible with many of the methods used for glycosylation analysis described in later sections: glycoproteins can be extracted from the gel by protease digestion, and oligosaccharides released and analysed. Alternatively, the N-linked glycans can be released *in situ* using an enzyme such as PNGaseF (*see 14.6.4*).

➲ Isoelectric focusing coupled with SDS–PAGE is termed two-dimensional electrophoresis (2-DE). If combined with mass spectrometry, it can be used to identify glycans and proteins in mixtures of glycoproteins from cell or tissue extracts and body fluids.

14.5 Preparation of glycoproteins for oligosaccharide analysis

The oligosaccharides associated with glycoproteins purified by the methods described previously, for example, lectin-affinity chromatography, SDS–PAGE and Western blotting, or 2-DE can be analysed further by a range of techniques. Success requires meticulous sample preparation. In particular, it is important to ensure that the glycoprotein preparation is salt- and detergent-free and, for hydrazinolysis (*see 14.6.3*), to ensure the protein is thoroughly lyophilized by freeze-drying.

14.6 Release of glycans

14.6.1 Monosaccharides – chemical release

➲ Meticulous sample preparation is necessary for successful oligosaccharide analysis.

Monosaccharide analysis enables molar quantitation of monosaccharides in any oligosaccharide or mixture of oligosaccharides, so that the class of oligosaccharide can be determined. To undertake monosaccharide release, the oligosaccharide is cleaved, non-specifically, at the glycosidic bonds that link the monosaccharides together. Acid hydrolysis is the method of choice. Typically, methanolic hydrochloric acid is added to the glycoprotein and the sample is heated at 80°C for 18 h. ➲

Monosaccharides prepared in this way can later be analysed by high-performance liquid chromatography (HPLC; *see 14.8*) or by gas-liquid chromatography (GLC; *see 14.13.1*) with, usually, mass spectrometry (MS) detection.

If sialic acids (*see Chapter 11*) are to be analysed, they must be removed from the oligosaccharide. To do this, typically, 1–10 mg of glycoprotein is dissolved in 0.1 N HCl then heated at 80°C for 50 min to release the sialic acids. One of the disadvantages of this procedure is that it denatures the O-acyl groups of the sialic acids, so if this type of sialic acids are to be studied, different methodologies need to be adopted. For example, first use 0.01 N HCl (or formic acid at pH 2.2) for 1 hour at 70°C, then dialyse to collect the free sialic acids and repeat the process. Using this approach, it has been reported that 70–80% of O-acetylated sialic acids are released intact.

14.6.2 Monosaccharides – enzymatic release

Chemical methods provide useful information on the molar concentration of monosaccharides present in an oligosaccharide but they do not provide information regarding the linkage of monosaccharides to one another. To obtain such information, exoglycosidases are used. Exoglycosidases trim away monosaccharides from the outermost part of the oligosaccharide. In this approach, an enzyme of broad specificity is used first. For example, if the oligosaccharide is negatively charged, a sialidase may be used to identify the sialic acid linkage, such as a sialidase from *Arthrobacter ureafaciens* (which cleaves ($\alpha 2 \rightarrow 6$)- and ($\alpha 2 \rightarrow 3$)-linked sialic acids). If sialic acids are successfully removed from the oligosaccharide, then a sialidase with a different specificity can be selected, in this case, for example, sialidase from Newcastle disease virus (which cleaves ($\alpha 2 \rightarrow 8$)- and ($\alpha 2 \rightarrow 3$)-linked sialic acids).

Using the technique of exoglycosidase digestion, monosaccharides are removed sequentially from the oligosaccharide and so the anomericity of linkage (e.g. α- or β-bonds) and position (e.g. ($\alpha 2 \rightarrow 3$) sialic acid or ($\alpha 2 \rightarrow 8$) sialic acid) of one monosaccharide to the next may be deduced.

Recently, exoglycosidases have been used in combinations (arrays). This technique, also called the 'reagent array analysis method' (RAAM) is useful as a relatively fast approach for characterizing monosaccharides in N-linked oligosaccharides and it is envisaged that arrays suitable for O-linked oligosaccharides will become available in future. The system uses several enzymes in one experiment. A mixture of, for example, four enzymes may be added to the glycan of interest. Samples of glycan are also exposed to mixtures of three of the enzymes, two of the enzymes, a single enzyme, and a final sample receives no enzyme treatment. In this manner, the sample is exposed to all possible combinations of the enzyme mixture in an 'array'. ➲ After digestion (typically overnight at 37°C) the enzyme is removed and the samples analysed. The differential migration of the glycans after RAAM can be studied using electrophoresis (*see 14.9*) or the

➲ Exoglycosidases trim specific monosaccharide residues from the end of oligosaccharide chains. They can be used in arrays to determine oligosaccharide structure.

differential elution pattern can be studied using HPLC (*see 14.8*). Analysis reveals whether a particular enzyme or enzyme mixture has been successful in cleaving monosaccharide residues from the original glycan, and from this the original glycan structure can be deduced.

14.6.3 Oligosaccharides: chemical release

Anhydrous hydrazine is used to release intact oligosaccharides from glycoproteins. This chemical has been used extensively for the release of N-linked oligosaccharides from glycoproteins, typically at 95°C overnight. O-Linked oligosaccharides can also be released from some glycoproteins using gentler conditions such as 65°C for 5 h or overnight. Hydrazinolysis removes the acetyl groups from acetylated monosaccharides, so chemical procedures need to be adopted to re-N-acetylate the oligosaccharide following hydrazinolysis. One advantage in using hydrazinolysis is that oligosaccharides can be released from poorly soluble glycoproteins. For example, archival tissue specimens cross-linked by fixation in formaldehyde and stored in paraffin wax-embedded blocks may be used as the starting material for oligosaccharide release facilitating retrospective glycan analysis of surgical specimens.

The most commonly used method for the release of O-linked oligosaccharides from glycoproteins is mild alkali treatment, also called β-elimination. In this method, the O-linked oligosaccharides are released from the protein backbone with, typically, 100 mM sodium hydroxide, in 2 M sodium borohydride, overnight at 50°C.

The chemical methods described above break the protein into peptide fragments, although this occurs to a lesser extent with β-elimination than with hydrazinolysis.

14.6.4 Oligosaccharides: enzymatic release

N-Linked oligosaccharides can also be effectively released from the glycoprotein using the enzyme peptide N-glycosidase F (PNGase F). An advantage of using PNGase F is that the enzyme leaves both the protein and the released N-linked oligosaccharides intact for further analyses. ➲ A disadvantage is that the protein and enzyme need to be removed from the reaction mixture prior to oligosaccharide analysis – but this is straightforward. Another concern with the enzymatic procedure is determining whether the reaction has gone to completion; if sufficient material is available, this may be monitored using a colorimetric assay such as the phenol–sulfuric acid assay.

> ➲ The enzyme PNGase F cleaves N-linked oligosaccharides from the polypeptide backbone.

PNGase F may also be used for the in-gel release of N-linked oligosaccharide of proteins separated by SDS–PAGE (*see 14.3*) or 2-DE (*see 14.4*). There is no need to extract the glycoprotein from the gel before oligosaccharide release. If used in combination with the detection of fluorescent-labelled glycans, the starting quantity of material can be as low as 0.5 μg, depending upon the number of glycosylation sites on the protein and their occupancy.

14.6.5 Optimization of conditions and controls

In all the methods described in the preceding sections, monosaccharide and oligosaccharide release should ideally be optimized using the sample of interest. In practice, it is often the case that insufficient material is available to do this. As controls, a glycoprotein whose oligosaccharides have previously been determined should be included and, in the case of biological specimens, samples containing mixtures of glycosylated proteins with a known oligosaccharide repertoire should be included. Again, in practice, this may not always be possible.

14.7 Detecting oligosaccharides

➲ Unlike proteins and peptides, monosaccharides and oligosaccharides only weakly absorb light in the UV range and special methods for their detection are necessary.

Unlike proteins, monosaccharides and oligosaccharides only weakly absorb light in the ultraviolet (UV) range, tending to give the best, but still poor, signal at λ 214 nm. Furthermore, the immense diversity of oligosaccharide structures means that they are often present only in tiny quantities. This has prompted the development of alternative reliable, sensitive, techniques for detection of oligosaccharides. ➲ The sensitivities of the different approaches to glycan analysis are compared in *Table 14.3*.

Mechanical approaches, for example, refractive index (sensitivity for oligosaccharides ≈0.1 μmol range) and pulsed amperometric detectors (sensitivity for oligosaccharides ≈10 pmol range) have enabled improved levels of detection of native mono- and oligosaccharides compared with the use of UV detectors.

Other approaches to improve detection levels of oligosaccharides have included radioactively labelling monosaccharides and the use of radioactive isotopes to label oligosaccharides metabolically, that is during oligosaccharide production *in vitro* or *in vivo*. The oligosaccharides from protein preparations of cells labelled in this manner can be evaluated using conventional methods for monitoring isotope incorporation, for example, in the case of triturated monosaccharides, scintillation counting (sensitivity in the micromole range).

Derivitization is now the most common method for labelling mono- and oligosaccharides. Mono- and oligosaccharides occur in ring-open and ring-closed forms (*see 1.10*). This feature means a label can be incorporated at the reducing end of the oligosaccharide in a process called reductive amination. A single molecule of label is added to each mono- or oligosaccharide, this allows molar quantities to be determined.

The development and validation of fluorescent labels has enabled improved sensitivity in detection of oligosaccharides. It is now possible to detect oligosaccharides in quantities as small as 10 fmol. Many labels are available. They are usually chosen according to the techniques that will be used subsequently in their analysis. The types of commercially available fluorescent labels and their applications are listed in *Table 14.4*.

Table 14.3 The sensitivity of different methods for oligosaccharide analysis. The amount of protein required has been based on ≈20% oligosaccharide w/w protein

Analysis method	Starting amount of protein	Approximate oligosaccharide detection levels
Nuclear magnetic resonance (NMR)	100 μg	nmol
Matrix assisted laser desorption ionization/time of flight (MALDI/TOF)	10s of μg	pmol
Electrospray MS	10s of μg	10 pmol
Fast atom bombardment MS	10s of μg	10 pmol
Gel electrophoresis separation of glycans (FACE)	0.5–1 μg	100 fmol
HPLC with fluorescent label	0.5–1 μg	10 fmol
High pH anion exchange chromatography (HPAEC)		
Native glycans	100 μg	nmol
Labelled with fluorophore	0.5–1 μg	sub-pmol

Table 14.4 Techniques for glycan analysis and sensitivity of detection of oligosaccharides

Technique	Chemical name of fluorophore	Abbreviation
FACE and CE	8-aminonaphthalene-1,3,6-trisulfonic acid	ANTS
	2-aminoacridone	AMAC
	2-anthranilic acid	2-AA
	9-aminopyrene-1,4,6-trisulfonic acid	APTS
HPLC	2-aminopyridine	
	2-amino benzamide	2-AB
	pyridylamino	PA
MALDI-MS	3-(acetylamino)-6-aminoacridine	AA-Ac

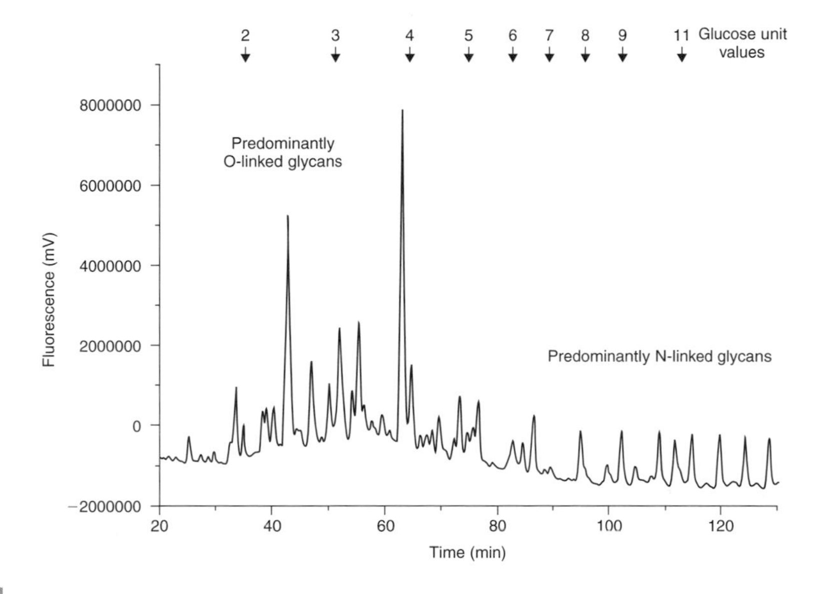

Figure 14.11

The separation of a mixture of fluorescently labelled N-linked and O-linked oligosaccharides released from a mixture of breast cancer glycoproteins with anhydrous hydrazine. In this example, the oligosaccharides were seperated by hydrophilic interaction chromatography and their elution positions compared with that of a labelled hydrolysate of dextran. The column enables the reproducible separation of glycans if used at a constant temperature (usually 30°C) and the structure of the glycans can be partially predicted with reference to the 'glucose unit value' derived from the dextran.

14.8 Oligosaccharide separation and mapping by HPLC

The use of HPLC for the separation of mixtures of N- and O-linked oligosaccharides is now well established, particularly for the mapping of oligosaccharides released from glycoprotein preparations, either chemically (*see 14.6.1*) or enzymatically (*see 14.6.2*). An example of the type of separation that can be achieved is given in *Figure 14.11*. 'Oligosaccharide mapping' is the term given to the use of different high-resolution HPLC columns, either alone or in series, for the separation of oligosaccharide mixtures. ➲ In this approach, oligosaccharides may be analysed in their native state, that is not derivitized, or after conjugation to a label.

➲ HPLC is an established method for the separation and analysis of oligosaccharides. Oligosaccharides can be separated by HPLC. This is used in combination with other techniques such as mass spectrometry to allow their structures to be 'mapped'.

High pH anion-exchange chromatography (HPAEC) has long been used for the analysis of native oligosaccharides. The oligosaccharides are maintained in a high molarity sodium hydroxide solution and detected using a pulsed amperometric detector (PAD). Columns are available for the separation of mono- and oligosaccharides. The sensitivity of the HPAEC system can be increased (to sub-picomole levels) when the mono- or oligosaccharide has been conjugated to a radioactive or fluorescent tag.

Conventional HPLC such as anion-exchange, reversed-phase and hydrophilic interaction (normal-phase) chromatography columns are also useful for the separation of N- and O-linked oligosaccharides. These columns enable charged species to be separated and the size of the structures to be estimated.

Fluorescent-labelled oligosaccharides can typically be detected in sub-picomolar concentrations and quantitation of the amount of different oligosaccharides present in a mixture of oligosaccharides can be achieved. HPLC-based methodologies for the separation of oligosaccharides may also be used in conjunction with other techniques, for example, mass spectrometry (MS; *see 14.12*) and exoglycosidase sequencing of oligosaccharides. Improved separation and sensitivity of detection is possible with microbore columns, which also enable GLC/MS (*see 14.13.1*) separation of oligosaccharides.

14.9 Separation and mapping of oligosaccharides by fluorophore-assisted carbohydrate electrophoresis (FACE)

The principle of electrophoresis for the separation of oligosaccharides has been firmly established, and a number of different techniques is available. The most commonly used system for electrophoretic separation of oligosaccharides is polyacrylamide gel electrophoresis of fluorophore-labelled saccharides (PAGEFS), also termed fluorophore-assisted carbohydrate electrophoresis (FACE).

In this system, the oligosaccharides are usually labelled with a fluorescent tag (ANTS, 2-AA, AMAC) and separated on high-density polyacrylamide gels (30–60% acrylamide gels are used).➲ The extent of cross-linking means that the running buffer must be cooled to prevent gel warping during the run. After electrophoretic separation, the oligosaccharide band patterns are visualized by illuminating the gel under UV light and capturing the fluorophore-emitted light on film/by camera with a charge-coupled device. The technique uses equipment that is available in many laboratories and lends itself to the analysis of multiple samples in one run. FACE is useful for the separation of both N- and O-linked oligosaccharides released from glycoproteins and saccharides from diverse sources, such as plants and bacteria. The sensitivity of detection of ANTS-labelled oligosaccharides is in the sub-picomolar range, but gel size limits resolution.

> ➲ Sub-picomole quantities of fluorescent-labelled oligosaccharide can be detected using ANTS.

14.10 Separation and mapping of oligosaccharides by capillary electrophoresis

Capillary electrophoresis (CE) separation is based on the migration of oligosaccharides, in fused-silica capillaries of $\approx 50\,\mu m$ internal diameter and 20–50 cm length, under an electric field of, typically, several hundred volts per cm capillary. The technique enables the separation of glycans very quickly, usually in less than 20 min. To enable separation of neutral uncharged oligosaccharides the oligosaccharides and monosaccharides are usually derivitized with a compound (ANTS or APTS) containing sulfonic groups. The sulfonic groups impart a net negative charge on the mono- or oligosaccharide and this enables the migration of the mono- or oligosaccharides in the electric field and their detection, either by UV or by in-line fluorescence detector. A relatively large amount of sample is initially required in order to obtain a concentrated solution for analysis. After separation by CE, fluorescent-labelled oligosaccharides are detectable in nanomolar quantities.

14.11 Oligosaccharide analysis by nuclear magnetic resonance

Nuclear magnetic resonance (NMR) analysis is based on the extent to which a material (in this case oligosaccharide) distorts in a magnetic field. The technique is non-destructive and so lends itself to use as a first analysis technique of a purified oligosaccharide. Full structural information can be obtained but a relatively large amount of oligosaccharide (in the order of 10–100 ng) is required and this limits the application of the technique.➲ With 500 MHz NMR spectrometers and nano-NMR probes, nanomole sensitivity can be achieved. Whereas NMR is a powerful analytical tool, the high capital cost of equipment and the requirement for specialist training in its use tends to limit its application to specialist groups.

> ➲ NMR is a powerful but specialist technique for determining structural information on oligosaccharides.

14.12 Determining the mass of an oligosaccharide using mass spectrometry

The development of mass spectrometry (MS) equipment capable of detecting small molecules in the low picomole or femtomole range, and with great accuracy in mass weight, has increased considerably the research activity in studying structural aspects of oligosaccharides using MS. According to the equipment used, varying amounts of information will be obtained regarding the oligosaccharide/glycoprotein. In practice, MS requires a substantially larger oligosaccharide sample size than HPLC. For example, to obtain the mass of fluorescent-labelled oligosaccharides 20 times the amount of oligosaccharide is required for single MS spectra compared with an HPLC chromatogram.

Matrix-assisted laser desorption/ionization mass spectrometry (MALDI/MS) is the most widely available MS equipment. It is the most simple of the MS equipment to operate, the least costly to acquire and is a very fast analytical tool. It is also relatively tolerant of salts compared with other MS detectors. The information gained on using MALDI/MS is mass weight and charge (*m*/*z*). This information usually enables assignation of the monosaccharides present in an oligosaccharide, as the mass of the most common monosaccharides in N-linked oligosaccharides is determined with a high degree of accuracy.➲ This technique requires 10–100 pmol of released glycans.

> ➲ MS provides accurate mass data on oligosaccharides from which structural information may be derived.

In this system, the mono- or oligosaccharide is mixed with a matrix that strongly absorbs UV light. Many matrices are available, for example, 2,5-dihydroxybenzoic acid which is commonly used for the analysis of N-linked oligosaccharides. The matrices suitable for charged oligosaccharide analysis give a higher signal-to-noise ratio and this results in less sensitive analysis of charged species.

The mixing process takes place on a metal plate (the target) and crystals of oligosaccharide/matrix are allowed to form. The metal target is placed into the chamber of the spectrometer and the chamber is evacuated. Pulses are fired from a laser and these hit the target and matrix/oligosaccharide ions are produced. The ions in the gas phase are directed (under an electric field) to the detector. The system can be used in a positive or negative ion mode, and this enables the analysis of both neutral and charged oligosaccharides. The time it takes for the ions to reach the detector is directly proportional to their mass weight and this aspect of the MS is termed the 'time of flight' (TOF).

Databases and the calibration of the machine with known standards enable putative compositional and thus structural assignations to be made. For linkage analysis, other types of MS, which also provide *m*/*z* information, are required. If other equipment is not available,

MALDI/MS can be used in conjunction with exoglycosidase enzymes to enable linkage analysis of oligosaccharides (*see 14.6.2*).

14.13 Methods for determining the linkage position of monosaccharides in oligosaccharides

The methods described in this section are only appropriate for relatively large amounts (>500 μg) of glycoprotein sample.

14.13.1 Gas-liquid chromatography with mass spectrometry

To obtain information on both the monosaccharides present and their linkage position to one another, the most widely used method is gas liquid chromatography with mass spectrometry (GLC/MS). In this technique, the oligosaccharide is chemically treated to incorporate a methyl group in place of all the hydroxyl groups; the monosaccharides are then released by cleavage of the glycosidic bonds. The monosaccharides contain hydroxyl groups at the points at which they were cleaved from one another. These monosaccharides are then derivitized by (i) acetylation or (ii) methylation and the resulting structures are identified by their elution time using GLC and their fragmentation pattern in MS.

The molar concentrations of monosaccharides are identified and their linkage position to one another is obtained. The limitation of this technique is that it does not provide information on the anomericity of the monosaccharide linkages. For this technique, oligosaccharide samples in the order of >10 pmol are required.

14.13.2 Fast atom bombardment mass spectrometry and electrospray ionization mass spectrometry

Fast atom bombardment (FAB) and electrospray ionization (ESI) mass spectrometry are both powerful tools for the analysis of oligosaccharides. In both systems, monosaccharide linkage and position information can be obtained from ionization fragments of the oligosaccharides. FAB is sensitive for oligosaccharides in the 10–100 pmol range, whereas ESI/MS is sensitive in the <10 pmol range. These systems can be used in a MS/MS mode, whereby ions produced in the mass spectrometer are captured in an 'ion trap' and subjected to further ionization. This can enable the structural identification of oligosaccharides.

FAB/MS, ESI/MS and GLC/MS require highly skilled personnel to operate the equipment and to interpret the spectra obtained.

14.14 Thin-layer chromatography of glycolipids

The use of thin-layer chromatography (TLC) for the separation and detection of lipids is well established. Often the lipid is crudely purified prior to the TLC, for example by phase-separation in different solvents to separate polar from non-polar lipids. To undertake TLC, silica matrices can be poured onto glass-backed slides, and although such an approach is possible, TLC is now usually undertaken using aluminium-backed, high-performance TLC plates supplied commercially. The TLC is run in either a one-dimensional or two-dimensional system. Two-dimensional separation enables greater structural information to be obtained on the (glyco)lipid of interest. After their separation, (glyco)lipids are detected by spraying the plate with a reagent that reacts with a given class of lipid. For example, resorcinol-HCl for gangliosides containing sialic acids, and orcinol-sulfuric acid for neutral glycolipids. This method is cheap in comparison with many of the techniques described previously, and as it provides good reproducibility and high sensitivity, remains widely used.

Further reading

Bigge, J.C., Patel, T.P., Bruce, J.A., Goulding, P.N., Charles, S.M., Parekh, R.B. (1995) Non-selective efficient labelling of glycans using 2-aminobenzamide and anthranilic acid. *Anal. Biochem.* **230:** 229–238.

Charlwood, J., Skehel, J.M., Camilleri, P. (2000) Analysis of N-linked oligosaccharides released from glycoproteins separated by two-dimensional gel electrophoresis. *Anal. Biochem.* **284:** 49–59.

Dwek, M.V., Brooks, S.A., Streets, A.J., Harvey, D.J., Leathem, A.J.C. (1996) Oligosaccharide release from frozen and paraffin-wax-embedded archival tissues. *Anal. Biochem.* **242:** 8–14.

Edge, C.J., Rademacher, T.W., Wormald, M.R., Parekh, R.B., Butters, T.D., Wing, D.R., Dwek, R.A. (1992) Fast sequencing of oligosaccharides: The reagent-array analysis method. *Proc. Natl Acad. Sci.* **89:** 6338–6342.

Guile, G.R., Rudd, P.M., Wing, D.R., Prime, S.B., Dwek, R.A. (1996) A rapid high-resolution high-performance liquid chromatographic method for separating glycan mixtures and analyzing oligosaccharide profiles. *Anal. Biochem.* **240:** 210–226.

Hounsell, E.F. (1998) Characterization of protein glycosylation. In E.F. Hounsell (ed.), *Glycoanalysis protocols. Methods in Molecular Biology*, Vol. 76, pp. 1–18. Humana Press, Totowa, NJ.

Rudd, P.M., Guile, G.R., Küster, B., Harvey, D.J., Opdenakker, G., Dwek, R.A. (1997) Oligosaccharide sequencing technology. *Nature* **388:** 205–207.

Modulation of glycan expression using inhibitors of N- and O-linked glycoconjugate processing and transgenic technology

15

15.1 Introduction

As described in previous chapters, many different functions have been attributed to the glycans of both N- and O-linked glycoproteins. However, one way to investigate the functional role of particular carbohydrate residues is to specifically alter the glycosylation of glycoproteins and seek a biological effect. The role of the terminal monosaccharide residues of an oligosaccharide chain are often of particular interest, as they are the ones which protrude most distantly from the protein backbone. Several experimental approaches exist by which (terminal or complete) glycosylation of cellular proteins and other glycoconjugates may be altered in order to explore the function of the glycans, and these are outlined briefly below.

15.2 Complete deglycosylation by chemical and enzymatic methods

More or less complete deglycosylation of glycoproteins can be achieved by chemical or enzymatical methods *ex vivo*; this is useful for molecular and compositional analyses of carbohydrate components of glycoproteins (*see 14.6.2*). These methods are too drastic to be applied to cells in culture or whole organisms.

15.3 Complete deglycosylation by expression in bacterial systems

Expression of the genes coding for a mammalian (glyco)protein in a bacterial system results in production of either a non-glycosylated form of the protein or a differently glycosylated form, as bacteria lack the glycosylation apparatus typical of eukaryote cells (*see 7.2, 18.4.7*).➲ As a result of this lack of glycosylation, bacterially produced glycoproteins are often synthesized as aglycon–glycoproteins, and in some instances have been shown to be functionally deficient.

➲ Bacteria do not glycosylate proteins in the same way as eukaryotes – so mammalian glycoprotein genes inserted into them will often result in the expression of aglycon–glycoprotein.

15.4 Lectin-resistant cell lines

If terminal monosaccharide residues of cells are of interest, a less labour-intensive approach to gain- or loss-of-function mutants (*see 15.9*), is the creation of lectin-resistant cell lines. These

Functional and Molecular Glycobiology, Susan A. Brooks, Miriam V. Dwek and Udo Schumacher

have been generated to enable the study of the effect of glycosylation on whole cells. Using this approach, lectins, which can sometimes be very toxic, are administered in increasing concentration in cell-culture systems over many months so that cells are selected which are devoid of the saccharide residues for which the lectin is specific (*see 14.2.6*). ➲ As lectins preferentially bind to terminal and sub-terminal monosaccharide residues this approach is particularly useful for studying the effects of altering these structures *in vitro*. Lectins are not mutagenic *per se*, so that the question of stability of glycan biosynthesis in these lectin-selected cell lines may be a factor for consideration. However, as some lectin-resistant cell lines have been shown to be stable for many years, it seems that clones with an altered glycosylation apparatus have been selected using this approach. Most of the lectin-resistant cell lines have been generated using Chinese hamster ovarian cells (CHO cells) and the Gal-binding lectin ricin (*Ricinus communis* agglutinin, RCA), which is highly toxic owing to its ribosome-inactivating activity. Alternatively, mutagens have been applied to cells and subsequently non-toxic lectins used to select the genetically altered cells. Using these approaches, more than 40 different glycosylation variants of the CHO cells have been generated.

> ➲ Lectin-resistant cell lines are useful tools to study the functional effects of terminal monosaccharide residues.

15.5 Glycosylation inhibitors

All of the methods for obtaining glycoproteins with altered glycosylation described in the previous sections are either experimentally difficult to perform, time-consuming, non-specific and/or expensive. An alternative approach for the *in vitro* analysis of the influence of glycosylation on biological function of glycoproteins is to use glycosylation inhibitors. ➲

> ➲ Inhibitors of glycosylation provide relatively fast track methods for initial functional analysis of glycans.

As the synthesis of N-linked (*Chapter 4*) and O-linked (*Chapter 5*) oligosaccharides require different metabolic pathways, and are located in different subcellular compartments, inhibitors for N- and O-linked glycosylation function in different ways. The inhibitors described in the following sections are all readily taken up by cells, and thus can be used *in vitro* and *in vivo* to alter cellular glycosylation in cell culture systems and in animal models.

15.6 Inhibitors of N-linked glycosylation

> ➲ Glycosylation inhibitors are also commonly called carbohydrate-processing inhibitors (CPIs).

The first step in the N-glycosylation of proteins is the transfer of the lipid-linked oligosaccharide precursor, illustrated in *Figure 15.1*, onto an asparagine residue immediately after the nascent protein chain has entered the lumen of the rough endoplasmic reticulum (RER). All further steps of N-glycosylation are based on modifications of this structure. N-linked oligosaccharide synthesis is described in detail in *Section 4.2*. ➲

15.6.1 Tunicamycin

Tunicamycin is an antibiotic that was first isolated from *Streptomyces lysosuperificus* by a group of Japanese scientists in 1971. This group showed that tunicamycin inhibited both the replication of a number of enveloped viruses and interfered with viral surface coat synthesis. The term tunicamycin derives from *tunica*, the Latin word for coat.

The chemical structure of tunicamycin is given in *Figure 15.2*. It acts by preventing the transfer of GlcNAc-1-P to the dolichol carrier at the initiation of N-linked glycoprotein synthesis. The position at which it acts is indicated by an asterisk on *Figure 15.1*. Thus, the

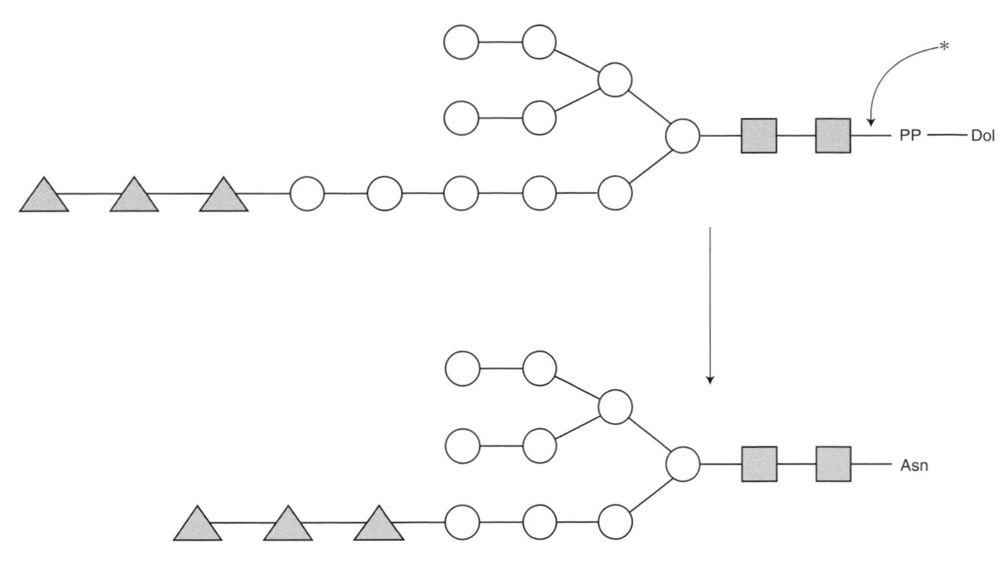

Figure 15.1

The initial step of N-linked glycosylation: a lipid (dolichol phosphate)-linked oligosaccharide is transferred to an asparagine residue of the nascent polypeptide chain. The asterisk indicates where the antibiotic tunicamycin acts.

synthesis of the lipid–oligosaccharide intermediate needed for all N-linked glycosylation steps is prevented and tunicamycin therefore blocks, at least in principle, *all* N-glycosylation.➲ It acts as a tight binding, reversible inhibitor thought to be a substrate–product transition state analogue.

➲ Tunicamycin blocks the first step of N-glycosylation.

As with any other glycosylation inhibitor, tunicamycin should be used with care. First, the dosage required to prevent glycosylation should be determined experimentally. In cell culture experiments using eukaryotic cells, glycosylation is usually inhibited in the range of 0.1–10 μg tunicamycin/ml. Tumour cells and virally transformed cells are said to be especially sensitive to tunicamycin. It is important to consider that tunicamycin can occasionally inhibit protein biosynthesis as well. In those systems, the concentration range in which inhibition of glycosylation occurs may be similar to that needed for the inhibition of glycosylation; hence it is important for each experimental system to determine the conditions under which glycosylation is abrogated while protein synthesis is not. Fortunately, the inhibition of glycosylation dominates the inhibition of protein biosynthesis in many systems, and in some systems no inhibition of protein biosynthesis is observed. The reasons for this differential action of the antibiotic are not clear. For practical purposes, each cell culture system should be tested individually for overall toxicity (resulting from inhib-ition of protein synthesis) versus specific inhibition of glycosylation.

As illustrated in *Figure 15.1*, Man residues compose the bulk of the monosaccharides of the lipid–oligosaccharide complex.➲ Hence the effectiveness of tunicamycin action in blocking N-linked glycosylation can be assessed by monitoring the incorporation of Man into glycoproteins.

➲ Man incorporation can be used to monitor tunicamycin effectiveness.

Tunicamycin has been used in many thousands of studies to determine the functional effects of N-linked glycosylation. These studies include the role of glycosylation in membrane receptor proteins, glycoprotein enzymes, viral envelope glycoproteins, tumour-associated glycoproteins, secreted glycoproteins, transport glycoproteins, and many other systems.

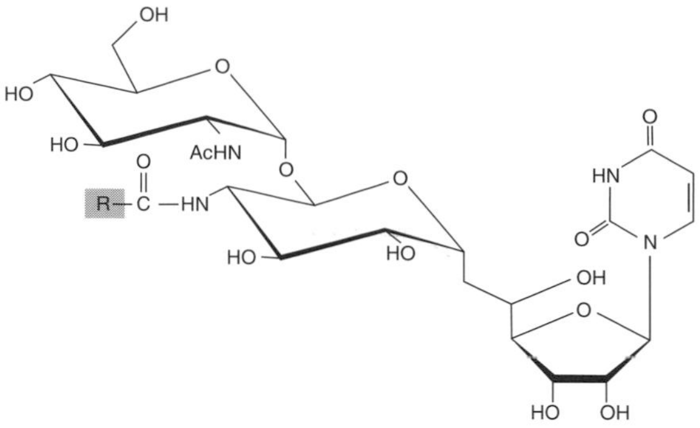

I: R=$(CH_3)_2CH(CH_2)_7CH{=}CH-$
II: $(CH_3)_2CH(CH_2)_8CH{-}CH-$
III: $CH_3(CH_2)_{10}CH{=}CH-$
IV: $CH_3(CH_2)_{11}CH{=}CH-$
V: $(CH_3)_2CH(CH_2)_9CH{=}CH-$
VI: $(CH_3)_2CH(CH_2)_{11}-$
VII: $(CH_3)_2CH(CH_2)_{10}CH{=}CH-$
VIII: $CH_3(CH_2)_{12}CH{=}CH-$
IX: $CH_3(CH_2)_{13}CH{=}CH-$
X: $(CH_3)_2CH(CH_2)_{11}CH{=}CH-$

Figure 15.2

The structural formula of tunicamycin and its analogues. This antibiotic is composed of uracil, a fatty acid, GlcNAc and aminodeoxydialdose, an unusual 11-carbon sugar called tunicamine. The analogues differ in their fatty acid composition (chain length, saturated or unsaturated, branched or not). The structures labelled I–X are different substituents covalently linked at the position marked R (shaded).

➲ Surprisingly, complete abrogation of N-linked oligosaccharide synthesis has no apparent functional consequences in many biological systems.

As the glycoproteins investigated in this way are so diverse, it is inevitable that the reported consequences of their lack of N-glycosylation are also wide ranging. ➲ However, one interesting point to emerge is that *complete* abrogation in N-linked oligosaccharide synthesis appeared to have *no* obvious functional influence on the rate of synthesis, transport and secretion of many glycoproteins. In some systems, however, this has not been the case, and examples are described briefly below.

In studies of mouse and rat plasma cells treated with tunicamycin, it has been reported that the secretion of immunoglobulins IgA, IgE and sometimes IgM is significantly, and sometimes completely, inhibited. However, in other mouse cell lines, IgM secretion has been reported to be only partially inhibited, and in some myeloma cell lines no inhibition of IgG secretion has been detected at all. These observations indicate that the inhibition of glycosylation may depend on the polypeptide chain secreted, rather than on the cell type investigated. This idea is supported by research using hepatoma cell lines, in which protein secretion can be classified into three different categories, rapid, intermediate or slow, depending on the transition time of the protein from the ER to the Golgi apparatus. The glycosylation of rapidly secreted proteins is less affected by tunicamycin than that of glycoproteins with a slower transit time.

With respect to glycosylated cell-surface transmembrane receptors, tunicamycin generally causes a decreased number of receptors to appear at the cell surface, whereas the binding affinity of the ligands to their receptor is not affected. The following receptors are examples that have been investigated: insulin receptor, gonadotrophin receptor, epidermal growth factor receptor, nerve growth factor receptor and acetylcholine receptor. The decreased number of cell-surface receptors in tunicamycin-treated cells has been attributed to a slower rate of transport of the receptors to the cell surface and an increased rate of degradation. This increased degradation may be explained by the fact that protein folding depends in part on the correct positioning and structure of the oligosaccharides, and proteins folded incorrectly because of their lack of proper oligosaccharide chains are degraded faster, probably via chaperone recognition and elimination pathways (*see 4.2.6.1, 13.12.3.4*).

One notable exception to this rule is diphtheria toxin receptor synthesis, in which the number of receptors, but not their affinity for the ligand, are affected by tunicamycin treatment. Tunicamycin-treated cells, therefore, show a decreased affinity for diphtheria toxin. Hence, these results indicate that saccharides on the cell surface play an important role in diphtheria toxin binding.

In addition to its effects on glycosylation and protein synthesis, tunicamycin can also affect transmembrane transporters. This may be an indirect effect, as these transporters are themselves glycoproteins. Tunicamycin interferes with the glucose transporter in a variety of cells and with some amino acid transporters, notably in hepatocytes.

Because of the multiple actions of tunicamycin and its profound effect on N-glycosylation biosynthesis, it is difficult to predict its impact in a particular experimental design. Careful dosage evaluation and critical data interpretation are needed when experiments with tunicamycin are performed.

15.6.2 Tunicamycin-related antibiotics

Tunicamycin is not the only antibiotic able to block the synthesis of sugar–lipid intermediates, but it is the most widely used commercially available compound. Other tunicamycin-like antibiotics include amphomycin, tsushimycin and diumycin, which all inhibit the dolichol pathway, but, as they cannot cross the cell membrane they can only be used in cell-free extracts and are hence of limited application in biological studies.

15.6.3 Castanospermine and deoxynojirimycin

During N-linked oligosaccharide synthesis, once the common N-linked oligosaccharide precursor has been covalently linked to the protein core, it is processed by the trimming of some core monosaccharides, and then extended by the addition of others. These initial steps occur within the RER. Trimming of the common precursor structure is achieved by the action of two membrane-bound glycosidases, termed glucosidase I and glucosidase II, which remove three Glc residues, illustrated in *Figure 15.3*. These glucosidases are inhibited by castanospermine and deoxynojirimycin.

15.6.3.1 Castanospermine

➲ Castanospermine blocks Golgi glucosidases thus preventing complex type chain synthesis. High mannose type glycoproteins accumulate instead.

The chemical structure of castanospermine is given in *Figure 15.4*. It is a plant alkaloid isolated from the nuts of an Australian tree, *Castonospermum australe*, and has a potent inhibitory effect on α-glucosidases. ➲ It not only blocks the ER- and Golgi apparatus-resident glucosidases I and II that act at the beginning of N-linked oligosaccharide synthesis, it also inhibits other enzymes with similar specificity, for example, intestinal sucrase and maltase. By its mode of action it has been classified as a competitive inhibitor. When used in cell culture systems, it inhibits cleavage of the three Glc residues of the common N-linked oligosaccharide precursor, and thus prevents the usual biosynthesis of complex type N-linked oligosaccharide chains. Hence, glycoproteins rich in the precursor $Glc_3Man_{7-9}(GlcNAc)_2$ structure accumulate.

15.6.3.2 Deoxynojirimycin

If the oxygen atom in the pyranose ring of Glc is replaced by a NH-group, the resulting structure is deoxynojirimycin, illustrated in *Figure 15.5*. It is synthesized by some *Bacillus* species

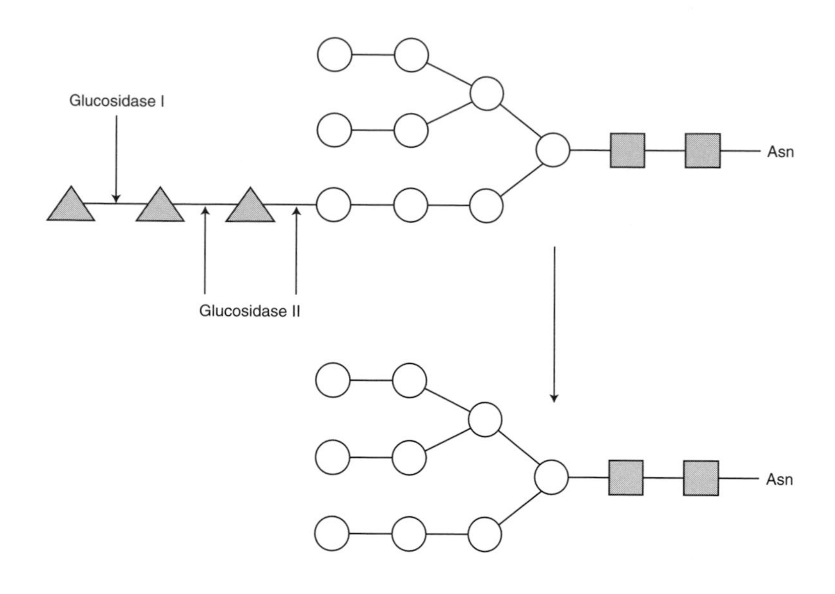

Figure 15.3

'Trimming' of the three Glc residues by glucosidase I and glucosidase II. Glucosidase I cleaves the outer (α1→2)-linked Glc, and glucosidase II removes both middle and inner (α1→3)-linked Glc.

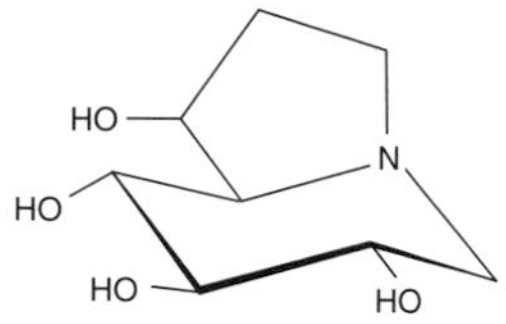

Figure 15.4

The structure of castanospermine which inhibits the action of glucosidases I and II.

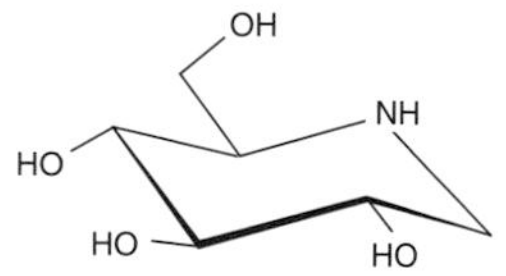

Figure 15.5

The structure of deoxynojirimycin which inhibits glucosidase I and II.

and inhibits both glucosidase I and II. The preference for glucosidase I and/or II may vary with the source from which the inhibitor was derived.

15.6.3.3 Action of deoxynojirimycin and castanospermine

Both deoxynojirimycin and castanospermine act to prevent the removal of Glc residues of the N-linked oligosaccharide precursor molecule, and thus inhibit the addition of further monosaccharide residues. This alteration of glycosylation can have considerable effects on several cellular functions including synthesis, folding and/or secretion of exocytosed proteins. Again,

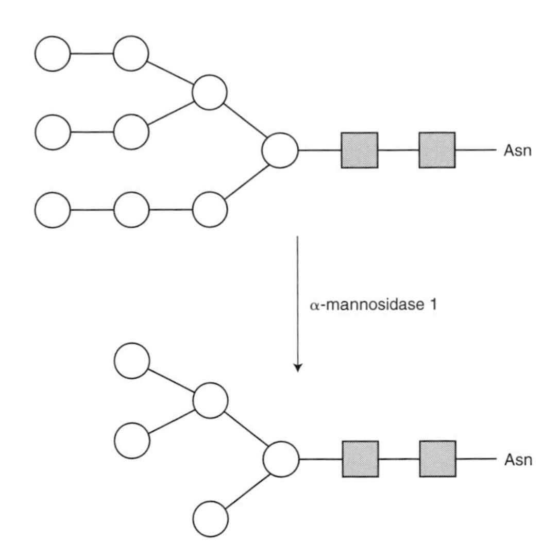

Figure 15.6

If a carbohydrate residue of a glycoprotein is processed beyond the high mannose type of side chain, the terminal Man residues are 'trimmed' by α-mannosidases resident in the ER and Golgi apparatus. The resulting oligosaccharide chain is further processed to either hybrid or complex type of side chains.

the effect may be limited to particular molecules. In the hepatoma Hep-G-2 cell line, secretion of α_1-antitrypsin is inhibited by deoxynojirimycin, but the secretion of other glycoproteins, such as coerueloplasmin and complement components (e.g. C3) are little influenced. One explanation for the reduced secretion of these proteins is that the presence of Glc residues retards their transport to the Golgi apparatus. Although this may be true for some proteins, it is not the whole story as not all glycoproteins are retarded in the Golgi apparatus when subjected to deoxynojirimycin treatment.➲

➲ Deoxynojirimycin and castanospermine act by similar mechanisms to inhibit N-linked oligosaccharide processing. They both inhibit glucosidase I and II in the RER and Golgi apparatus.

15.6.4 Deoxymannojirimycin

In N-linked oligosaccharide processing, the action of glucosidases I and II in trimming Glc residues from the common precursor molecule, is often followed by cleavage of Man residues, in order that further processing to hybrid or complex type oligosaccharide chains can proceed. Four to six Man residues are removed by specific α-mannosidases which reside in the ER and in the Golgi apparatus. This process is illustrated in *Figure 15.6*.

Deoxymannojirimycin was developed as a Man analogue to deoxynojirimycin (which inhibits the glucosidase I and II; *see 15.6.3.2, 15.6.3.3*) and indeed, as predicted, inhibits the action of α-mannosidases.➲ Thus, it prevents the formation of both hybrid and complex N-linked oligosaccharide chains. Its structure is given in *Figure 15.7*.

➲ Deoxymannojirimycin inhibits the synthesis of both complex and hybrid type N-linked oligosaccharide chains.

Inhibition of the Golgi apparatus mannosidase I/II by deoxymannojirimycin results in an accumulation of high-mannose type oligosaccharide $Man_{8-9}(GlcNAc)_2$ structures within the cells and prevents the formation of hybrid and complex chains. In several systems, its application does not lead to major changes in cellular behaviour or secretion of glycoproteins.

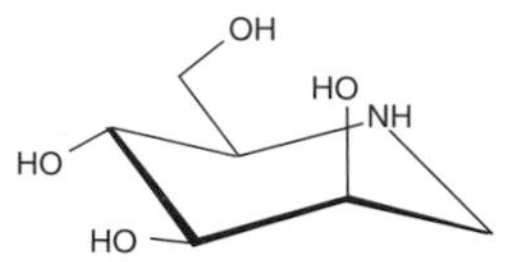

Figure 15.7

The α-mannosidase inhibitor deoxymannojirimycin. Note the structural similarity to deoxynojirimycin (Figure 15.5) which formed the parent molecule for the development of this compound.

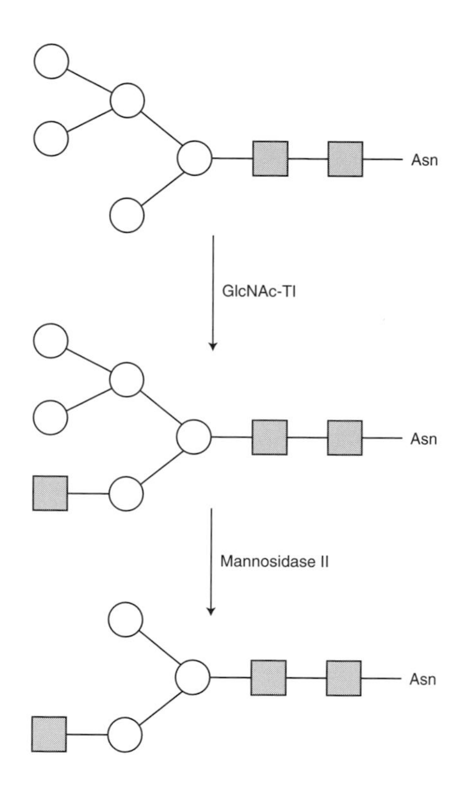

Figure 15.8

Following trimming of Man residues, the next step in N-linked oligosaccharide synthesis is the addition of one GlcNAc to a Man residue catalysed by GlcNAc-TI, then the trimming of a Man residue by mannosidase II.

In the rat, deoxymannojirimycin does not inhibit the α-mannosidase present in the RER, but does inhibit the Golgi apparatus resident α-mannosidase enzyme. Using radioactively labelled Man and deoxymannojirimycin, it is therefore possible to dissect which processing steps take place in the RER and which occur in the Golgi apparatus.

15.6.5 Swainsonine

Following trimming of Man residues in the ER and Golgi apparatus, the next step in N-linked oligosaccharide synthesis is the addition of one GlcNAc to a Man residue, as illustrated in *Figure 15.8*. Further chain extension then results in formation of complex or hybrid type N-linked oligosaccharide chains. As yet, no specific inhibitor of the GlcNAc transferase I is in widespread use.

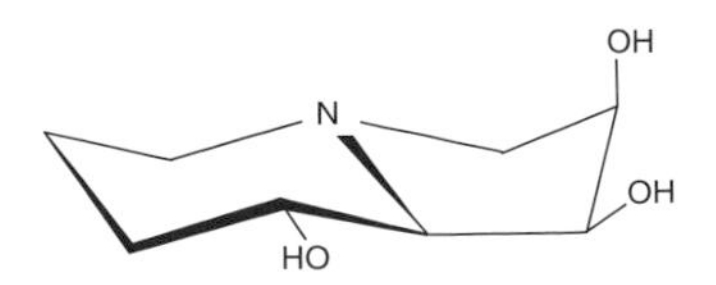

Figure 15.9

The structure of the mannosidase-II inhibitor swainsonine, an indolizine alkaloid.

Synthesis of complex type N-linked oligosaccharides requires subsequent cleavage of further core Man residues, as illustrated in *Figure 15.8*. This cleavage is catalysed by the Golgi apparatus resident mannosidase II and this enzyme is inhibited by swainsonine, which thus prevents synthesis of complex type N-linked oligosaccharides. ➲

➲ Swainsonine inhibits the formation of complex type N-linked oligosaccharide chains.

Swainsonine was the first inhibitor of glycoprotein processing described and is an indolizine alkaloid. Its chemical structure is given in *Figure 15.9*. It was first isolated from the Australian plant *Swainsona canescens* and has subsequently been detected in other plants (including some fungi and locoweed) and has been chemically synthesized. In addition to the Golgi apparatus resident mannosidase, swainsonine also inhibits the lysosomal α-mannosidase. This property of swainsonine initially led to its discovery, as animals which had ingested *Swainsona canescens* over a long period showed symptoms similar to those of human α-mannosidosis (*see 16.8*).

Swainsonine, like deoxymannojirimycin, has relatively little effect on the biological functions of glycoproteins in cell culture-based experiments but a more profound effect on the whole organisms. ➲ This observation indicates that a partially synthesized complex N-linked oligosaccharide chain (in effect, a hybrid-type chain) is usually sufficient for normal biological function. Examples in which swainsonine had no discernible effect on secretion/function of N-linked glycoproteins include thyroglobulin, surfactant glycoprotein A, H2-histocompatibility antigen and von Willebrand factor (essential for blot clotting).

➲ Modification of N-linked glycosylation by deoxymannojirimycin and swainsonine has little effect on the function of many glycoproteins studied *in vitro*.

Some biological systems have been described, however, in which swainsonine treatment has had a significant effect on function. For example, injection of the drug swainsonine into rats caused a time-dependent loss of intestinal sucrase activity. This was attributed to the altered oligosaccharide structure of the sucrase molecule, as swainsonine has no direct effect on the enzymatic activity of sucrase when applied *in vitro*. Furthermore, the interaction between *Trypanosoma cruzi* and peritoneal macrophages was also altered by swainsonine treatment. The dissemination capacity of B16-F10 melanoma cells or MDAY-D2 lymphoreticular tumour cells, both murine tumours, was also altered when swainsonine-treated tumour cells were injected into the tail vein of mice. The organ colonization by these cells was reduced, indicating that altered glycosylation had an influence on the metastatic process. In addition to its anti-metastatic effects, immunomodulatory activities (stimulation of interleukin (IL)-2 production, increase of natural killer cell activity) have been attributed to swainsonine.

15.6.6 Other inhibitors of N-linked glycosylation

So far, the 'classical' inhibitors of N-linked glycosylation have been described. The use of many other inhibitors has been reported, but these are generally not commercially available and/or experience of their usage is limited.

In addition to 'specific' inhibitors, drugs interfering with protein targeting or the secretory pathway have also been used to influence protein glycosylation. Most of them are ionophores, that is, substances that alter the ion concentration, and sometimes the pH, within an organelle. Examples of drugs of this type are monensin and brefeldin A. As their effects are not mediated by inhibition of a specific glycosylation pathway, they are not discussed here.

15.6.7 Inhibitors of N-linked oligosaccharide processing – summary and outlook

Inhibitors of N-linked glycoprotein processing enzymes have been used in numerous studies. In general, those drugs which interfere at a more elementary level with the glycosylation machinery (e.g. tunicamycin) show more dramatic effects on cellular behaviour than those that interact with more peripheral glycosylation mechanisms (e.g. swainsonine). However, the cytotoxicity of these drugs has to be taken into consideration (especially in the case of tunicamycin). When designing experiments aimed at elucidating glycosylation mechanisms, carbohydrate-processing inhibitors are often used together with radioactively labelled monosaccharides whose incorporation into glycoproteins can be monitored. Lectins (*see Chapter 13*), which could also be used to monitor the incorporation of defined sugar residues, have rarely been used for this purpose.

15.7 O-Linked glycoprotein inhibitors

In contrast to the many inhibitors of the N-linked glycoprotein-processing enzymes that are available, only one inhibitor of O-linked glycosylation has found widespread usage. ➲ *O*-benzyl-GalNAc or benzyl 2-acetamido-2-deoxy-α-D-galactopyranoside is an inhibitor of O-linked core β1,3 galactosyltransferase activity, and a substrate for the N-acetyl-β-D-glucosaminyltransferase. Its chemical structure is given in *Figure 15.10*.

O-linked glycosylation (*see 5.5*) is fundamentally different to N-linked glycosylation. The addition of O-linked sugars to the polypeptide backbone begins in the Golgi apparatus, not in the RER. In the Golgi apparatus, O-linked oligosaccharide synthesis begins with transfer of a single α-linked GalNAc to a serine or threonine residue in the polypeptide chain. This step is catalysed by one of the family of polypeptide α-GalNAc transferases (UDP-GalNAc: polypeptide α1,3-N-acetylgalactosaminyltransferase, polypeptide GalNAc-T) (*see 2.7.2, 2.11.2*). The resulting structure, GalNAcα1→Ser/Thr is called the Tn epitope.

➲ O-benzyl GalNAc inhibits the enzyme core 1 β1,3 Gal-T that catalyses synthesis of core 1 from the Tn epitope at initiation of O-glycan synthesis, as follows:

core 1 β1,3 Gal-T
(inhibited by O-benzyl GalNAc)

UDP-Gal + GalNAcα1-Ser/Thr ⟶ Gal(β1→3)GalNAcα1-Ser/Thr + UDP
Tn Core

Core 1 can be converted into 'core 2' by core 2 β1,6 GlcNAc-T which adds a (β1→6) linked GlcNAc to give a simple bi-antennary structure, GlcNAc(β1→6)[Galβ(1→3)]GalNAcα1-Ser/Thr. The reaction is as follows:

core 2 β1,6 GlcNAc-T

UDP-GlcNAc + Gal(β1→3)GalNAcα1-Ser/Thr ⟶ GlcNAc(β1→6)[Galβ(1→3)]GalNAcα1-Ser/ Thr + UDP
core 1 core 2

Thus, in the presence of O-benzyl GalNAc, no core 1 *or* core 2 structures can be synthesized. Consequently, O-linked glycoproteins rich in GalNAc residues linked to serine/threonine (the Tn antigen) (*see 5.6.1*) accumulate within the O-benzyl GalNAc-treated cells.

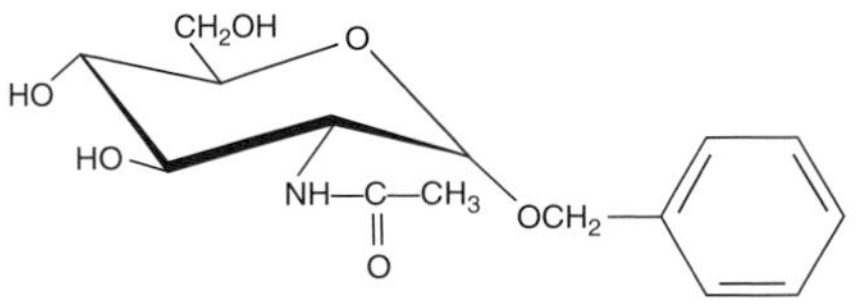

Figure 15.10

The structure of *O*-benzyl GalNAc, an inhibitor of O-linked glycoprotein processing.

Core 1 Gal(β1→3)GalNAcα1→Ser/Thr is then formed by addition of a Gal residue to the Tn epitope, catalysed by the core 1 β1,3 Gal transferase (*see 5.6.3*). This enzyme is inhibited by O-benzyl GalNAc.

As the synthesis of core 3 (*see 5.6.5*) and core 5–8 (*see 5.6.7*) structures are not synthesized upon core 1, their synthesis is not directly dependent on this pathway and is not altered in O-benzyl GalNAc-treated cells.

O-benzyl-GalNAc has been little used as an inhibitor of O-glycosylation, and most of the studies using this inhibitor have been concerned with the glycosylation of mucins – in particular, with the MUC1 mucin (*see 5.10*). ➲ The effect of O-benzyl-GalNAc on MUC1 mucin-producing cells in culture results in reduced glycosylation of MUC1 glycoprotein so that epitopes recognized by anti-MUC1 mucin monoclonal antibodies are no longer hidden by saccharide residues. In particular, sialic acid residues on mucin glycoproteins are reduced as these often form the terminal monosaccharide residue of core 1 and 2 structures. However, many more studies with this inhibitor are needed to ultimately assess its potential.

➲ O-benzyl-GalNAc is the only widely used inhibitor of O-linked glycoprotein synthesis. It inhibits core 1 and core 2 formation.

15.8 Glycosylation engineering

Although adding inhibitors of glycosylation to cell cultures and subsequently performing assays for cellular function is a relatively cheap and quick approach to studying the functional role of glycosylation, it has a number of disadvantages:

- Inhibitors such as tunicamycin can have a considerable number of side-effects not linked to changes in the glycosylation machinery (*see 15.6.1*)
- A particular inhibitor may not be specific for one 'glycozyme' only, and may actually affect the activity of a number of enzymes
- For many enzymes involved in glycosylation processing such as, for example, N-acetylglucosaminyltransferase I or GlcNAc-TI, specific inhibitors are not available.

To overcome these problems, inactivation or induction of activity of specific enzymes of glycosylation by molecular biological methods have been used, appropriately this approach has been termed 'glycosylation engineering'.

15.9 Experimental addition and deletion of glycosyltransferase genes in cell culture

If whole cells are required to study functional and dynamic aspects of glycosylation, a powerful approach is to add-in or knock-out specific glycosyltransferase genes and/or genes for other glycoprotein processing enzymes in gain-of-function or loss-of-function mutants respectively. Although experimentally this is an elegant strategy, it has not been used widely because of its complexity and costs.

15.10 Transgenic animals

Recently, several transgenic mouse models with alterations in their glycosylation machinery have been developed. Two main approaches to producing transgenic animals have been chosen:

- Gain-of-function transgenics, in which a novel glycosylation enzyme is introduced into the organism
- Loss-of-function transgenics, in which a glycosylation processing enzyme is deleted, often by homologous recombination with a functionally deficient gene.

These approaches depend on insertion or deletion, respectively, in the germline (for details of experimental techniques, the reader is referred to any of the many, standard molecular biology textbooks).

15.10.1 Gain of function transgenes

For a gain-of-function transgene, it is not enough to introduce the DNA sequence of the gene into the germline, the gene has also to be expressed by the cells of the organism. ➲ This is achieved by the insertion of a promotor sequence, which results in continuous expression of the gene and thus drives the expression of the protein. Different types of promotors have been used in different experimental systems. One promotor type is the chicken β-actin promotor, which, because actin is part of the cytoskeleton of every cell, will also induce expression of the transgene gene in all cells. Tissue-specific promoters have also been used. Examples include the murine whey acidic protein which specifically promotes expression in the mammary gland and serum amyloid P component which specifically promotes expression in the liver. Using tissue-specific promoters, the transgene will thus be expressed in a tissue- or organ-specific manner. Whilst under promoter control the transgene is expressed continuously.

> ➲ Different types of promotors allow continuous, transient or organ-specific expression of transgenic glycozymes.

Inducible promotors also exist and these allow the temporary expression of proteins. An example of such an inducible promotor is the tetracycline-sensitive promotor. When tetracycline is added to the drinking water of a transgenic animal, the transgene under the control of this promoter is switched on. This type of promotor is used when only temporary expression of the gene is desirable, or when a particular construct leads to a phenotype incompatible with normal development and therefore only transient expression is desirable.

This strategy has considerably extended our knowledge on the functional role of glycosylation. Two examples of gain-of-function mutants of particular interest are described below.

15.10.1.1 α1,3/4 Fucosyltransferase III gain-of-function transgenic mouse

The α1,3/4 fucosyltransferase III (Fuc-TIII) gain-of-function mouse, expresses the α1,3/4 Fuc-TIII gene (the 'Lewis transferase'; *see 10.6*) under control of the liver fatty acid binding protein promotor, which causes certain cells of the gastrointestinal tract to express the enzyme when they would not normally do so. ➲ Different cell types in the gastrointestinal tract of the transgenic animals synthesize different products of the Fuc-TII gene, that is, different Lewis structures. For example, undifferentiated, proliferating crypt cells and epithelial enterocytes synthesize Le^a and Le^b structures, whereas only the proliferating crypt cells synthesize Le^x. Furthermore, when cell division is induced in Le^x-negative, differentiated villus cells, they also began to synthesize Le^x. This indicates that cycling cells are, in this instance, producing glycan structures which are required for the Fuc-TIII transferase to produce Le^x structures.

> ➲ α1,3/4 fucosyltransferase III transgenic mice have been useful in elucidating the functional role of the fucosylated Lewis blood group antigens.

The gain-of-function transgenic α1,3/4 fucosyltransferase III mouse has also been used in investigations into the role of the Le^b

determinant in *Helicobacter pylori* infection. As described previously, *H. pylori* produces a lectin which is involved in attachment to the gastric mucosa (*see 13.9*). In the α1,3/4 fucosyltransferase III gain-of-function mice, the gastric mucosa cells and the pit cells synthesize Le^b determinants. Both α1,3/4 fucosyltransferase III (gastric mucosa cells with Le^b) and normal (gastric mucosa cells without Le^b) mice are able to harbour human pathogenic strains of *H. pylori* in the stomach, but adherence to the mucosa, which is a prerequisite for infection, is only observed in the Fuc-TIII transgenic mouse. Hence, this mouse serves as a useful model to study the pathogenesis of *H. pylori* infection.

15.10.1.2 α1,2 Fucosyltransferase I (α1,2 Fuc-TI), gain-of-function transgenic animals

Mice and pigs have also been used to develop transgenic animal models of potential importance in organ xenotransplantation. Lactosamine units in many species except humans, Old World monkeys and apes are commonly capped by the Gal(α1→3)Gal determinant (*see 7.13, 17.8.2*), which, because it is not present in humans, is highly immunogenic. The presence of this glycotope is responsible for the hyperacute rejection of animal xenografts in humans because of their naturally occurring anti-Gal(α1→3)Gal antibodies. Expression of α1,2 fucosyltransferase I (α1,2 Fuc-TI) in animal cells results in competition for substrates such that terminal Gal residues are replaced by Fuc. This results in an increase in synthesis of the human blood group H antigen Fuc(α1→2)Gal in these animals. Unfortunately, Fuc also displaces terminal sialic residues, and this results in exposure of normally cryptic monosaccharide residues which are themselves immunogenic in man and present further obstacles to xenograft applications.

15.10.2 Loss-of-function transgenes

15.10.2.1 *Mgat*-1 gene knockout transgenic mice

The *Mgat*-1 gene encodes the N-acetylglucosaminyltransferase (GlcNAc-TI), and a loss-of-function knockout mouse model has been developed in which this gene has been disrupted. The enzyme resides in the *medial*-Golgi compartment and its action is responsible for the formation of the basic oligosaccharide structure from which all complex- and hybrid-type N-linked oligosaccharides are synthesized (*Figure 15.8*). No specific inhibitor for this enzyme has been found (*see 15.6.5*), so ablating GlcNAc-TI has enabled progress in the study of the functional role of complex and hybrid type oligosaccharides synthesis. It is interesting to note that this enzyme is found in all metazoans and plants, but not in protozoa. Hence its evolution marks the transition between single and multicellular organisms. Plant growth is not altered when this enzyme is lacking, but knockout mice die *in utero* by embryonic day 10.5, a time point when organogenesis is initiated which includes the development of the neural tube, which fails to develop. This supports the hypothesis that complex and hybrid type N-linked oligosaccharides are indispensable for cell–cell interactions, and that the incompletely processed oligomannoside residues are insufficient to sustain normal organogenesis. ➲

> ➲ Complex and hybrid type N-linked oligosaccharides are essential to normal embryonic development in mice.

15.10.2.2 *Mgat*-2 gene knockout transgenic mice

The *Mgat*-2 gene codes for GlcNAc-TII, the enzyme that acts after mannosidase II in chain elongation of complex and hybrid type N-linked oligosaccharides (*see 4.2, Figure 15.11*). Transgenic animals deficient in this enzyme die early in the postnatal period, suggesting once again that synthesis of complex and hybrid type N-glycans is essential to normal mammalian development.

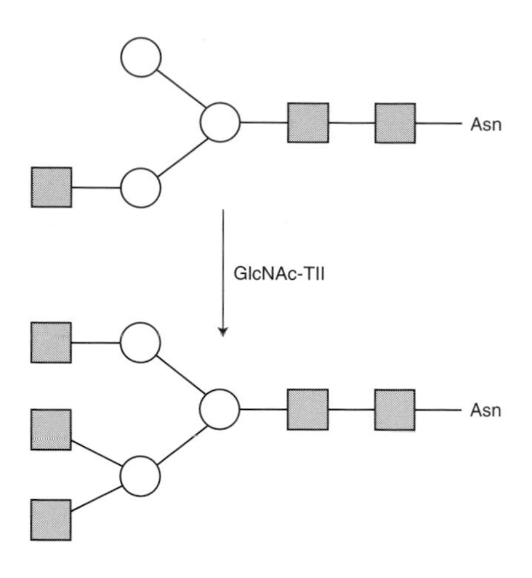

Figure 15.11

GlcNAc-TII enzyme that acts after mannosidase II in chain elongation of complex type N-linked oligosaccharides.

15.10.2.3 *Mgat*-3 gene knockout mice

➲ GlcNAc-TIII knockout mice have delayed liver tumour development but are otherwise phenotypically normal.

In contrast, *Mgat*-3 knockout mice, which lack the GlcNAc-TIII, are viable, fertile and without obvious abnormalities.➲ In these mice, hybrid-type N-linked oligosaccharides are synthesized instead of complex-type ones. However, if the mice are treated with diethylnitrosamine to induce liver tumours, tumour progression is retarded in comparison with wild-type mice.

15.10.2.4 Lessons learned from glycosyltransferase knockout models

Several other glycosyltransferase gene knockout mice have been generated, and, unexpectedly, their phenotype is often remarkably unaltered. This observation is not unique to investigations in the field of glycobiology, but is a frequent finding in other research areas as well. Much redundancy seems to exist within nature and often the lack of one specific gene has little noticeable biological effect in laboratory animals.

It is possible, however, that such alterations in gene expression may be significant when the organism is challenged, by, for example, infection or disease. Subtle analytical methods have to be devised in order to detect changes in the behaviour, immune response or tumorigenesis of cells in these knockout transgenic animals. Furthermore, glycan analysis is also required to determine whether a change in glycosylation actually results from abrogation of a specific transferase gene. This is because in all studies heterozygote mothers are used for breeding homozygous transgenic offspring. When mated with a homozygous transgenic null-GlcNAc-TI male, traces of maternal GlcNAc-TI mRNA remain present in the egg. During early stages of development, the dividing cells do not synthesize new cytoplasm (and therefore become smaller with each cell division). The small amount of residual maternally derived GlcNAc-TI may be sufficient to rescue the offspring. To detect such subtle effects, sensitive analytical tools are required.

15.10.2.5 Future prospects

Despite its limitations, transgene technology has the potential to be useful in overcoming the species barrier posed by saccharide antigens, in particular with regard to the important challenge of xenotransplantation (*see 7.13, 15.10.1.2, 17.8.2*). In combination with subtle analytical techniques for oligosaccharide structure and cellular/whole animal functional analysis, transgene technology has the potential to increase our understanding of the function of glycosylation in living systems. Some avenues, for example, the creation of double knock-outs or the combination of knocking-out an animal gene and introducing a human gene, yet to be used widely in glycobiology are likely to be areas of (fruitful) future research efforts.

Further reading

Dennis, J.W., Granovsky, M., Warren, C.E. (1999) Protein glycosylation in development and disease. *BioEssays* **21:** 412–421.

Elbein, A.D. (1987) Inhibitors of the biosynthesis and processing of N-linked oligosaccharide chains. *Annu. Rev. Biochem.* **56:** 497–534.

Elbein, A.D. (1991) Glycosidase inhibitors: inhibitors of N-linked oligosaccharide processing. *FASEB J.* **5:** 3055–3063.

Kaushal, G.P., Elbein, A.D. (1994) Glycosidase inhibitors in study of glycoconjugates. *Methods Enzymol.* **230:** 316–329.

Pan, Y.T., Elbein, A.D. (1995) How can N-linked glycosylation and processing inhibitors be used to study carbohydrate synthesis and function? In J. Montreuil, H. Schachter & J.F.G. Vliegenthart (eds) *Glycoproteins*. Elsevier, Amsterdam.

Disease processes in which carbohydrates are involved

16

16.1 Introduction

The interactions between saccharides and their receptors (lectins) are of vital importance in many normal cellular mechanisms. These interactions are being recognized as important in a range of disease processes, including, for example, infections by bacteria, viruses and parasites, in inflammatory disorders, and possibly in cancer metastasis. In addition, there are human disorders caused by disruptions in normal synthesis or catabolism of glycosylated molecules. This is an expanding subject area, but a few important examples of the involvement of disease processes in which carbohydrates play a crucial role are outlined in this chapter.

16.2 Infectious diseases

Saccharide–receptor interactions are implicated in bacterial, viral and parasitic infections, where they mediate cell attachment, colonization or invasion.➲ Many bacteria and viruses produce lectin-like molecules (*see Chapter 13*) that recognize saccharide structures of the organisms that they invade. These saccharides may be part of glycoproteins, glycolipids or proteoglycans. Understanding these interactions could lead to more effective means of preventing or treating infection, for example, through the production of synthetic binding partners to the lectin-like molecules.

> ➲ Infections are often initiated by the interaction of lectin-like molecules expressed by bacteria, viruses or parasites recognizing, and binding to, host cell-surface oligosaccharides.

16.2.1 Bacterial infections

Many bacterial infections are mediated by the interaction of adhesins, which are part of the structure of their pili or hairs, with glycoconjugates present on the target cell surface. Binding affinity between adhesin and saccharide may be quite weak, but multiple interactions through the adhesins on multiple pili result in a strong total adhesive force, perhaps analogous to the action of Velcro™. Binding often triggers signal transduction events necessary for further infection and colonization processes.

Bacteria express adhesins with different binding characteristics, associated with saccharides present on a range of possible target cells.➲ Many bacteria that colonize the large intestine produce adhesins that recognize lactoceramide Gal(β1→4)Glc-Cer, which is a common glycotope on colonic columnar epithelium, for example. One common binding partner for bacterial adhesins is GlcNAc(β1→4)Gal which, if it occurs subterminally in the oligosaccharide chain, some bacteria reveal by secreting glycosidase enzymes to cleave terminal monosaccharides. Another is Gal(α1→4)Gal, which is present on the urothelium of some individuals and is involved in the pathogenesis of bladder infections by *Escherichia coli*. Many bacterial adhesins are specific for glycan structures associated with glycolipids, although glycoprotein glycans can also be involved. The glycans within the polysaccharide

> ➲ Common oligosaccharide structures recognized by bacterial adhesins include Gal(β1→4)Glc, GlcNAc(β1→4)Gal, Gal(α1→4)Gal and sialic acids.

Functional and Molecular Glycobiology, Susan A. Brooks, Miriam V. Dwek and Udo Schumacher

chain of the proteoglycan heparan sulfate are particularly common binding partners for bacterial adhesins.

Many bacterial lectins also recognize sialic acid residues, as shown in *Table 16.1*. *Helicobacter pylori*, an infective organism implicated in the development of peptic ulcers, and *Pseudomonas aeruginosa*, responsible for respiratory tract infections in cystic fibrosis patients, both have haemagglutinins or lectins specific for sialic acids (*see Chapter 11*).

16.2.1.1 Carbohydrates in infections associated with cystic fibrosis

➲ Cystic fibrosis patients suffer infections by microorganisms including *S. aureus* and *P. aeruginosa*. Infection is mediated through their recognition of abnormally glycosylated respiratory mucins.

Altered glycosylation is an important aspect of the pathology of cystic fibrosis. ➲ Cystic fibrosis is the most common severe genetic disorder in European countries. It is caused by mutation of the cystic fibrosis transmembrane conductance regulator (CFTR) gene, a chloride transporter, located on chromosome 7. The relationship between this defect and the obstructive mucin production associated with this disorder is unclear, but the defective CFTR gene may lead to abnormal pH in the lumen of the Golgi apparatus which may, in turn, compromises glycosyltransferase and sulfotransferase activity. The disorder is characterized by excessive mucin production, particularly by epithelial cells lining the gut and lungs. Most deaths from the disorder result from colonization of lung epithelium by *Staphylococcus aureus* and, particularly, *P. aeruginosa*, which adhere to cystic fibrosis mucins. Glycosylation, in particular, sialylation and fucosylation, as well as sulfation of mucins produced by lung epithelium, is disrupted in cystic fibrosis patients, and these altered glycans may be involved in effective attachment and infection by the pathogenic microorganisms. In cystic fibrosis patients, sulfation, fucosylation and synthesis of core 1 structures with blood group H and Lewis x (Le^x) determinants, and ($\alpha2\rightarrow3$)-linked sialic acid on mucin-type glycoproteins are increased.

The Gal-binding *P. aeruginosa* agglutinin I and Fuc-binding agglutinin II, produced by the bacteria, can, in addition to facilitating bacterial adhesion, arrest ciliary beating in human airways; addition of the monosaccharides to nebulizers may therefore improve clinical management.

16.2.1.2 Bacterial capsular polysaccharides and virulence

Many bacteria produce capsular polysaccharides that help protect the bacterium from phagocytosis. In non-capsulated bacteria, cell-surface lipopolysaccharides may sometimes have a similar protective effect. An effective multivalent vaccine to *Streptococcus pneumoniae* capsular polysaccharides was developed in the 1940s, and showed 100% efficacy in preventing pneumonia in volunteer US army recruits, but interest in anti-bacterial vaccines waned with the success of antibiotics. With current concern regarding the increasing emergence of antibiotic-resistant strains of bacteria, this is an area that is once again of interest. An effective vaccine against meningococcal polysaccharides is another good example of this approach.

Table 16.1 Examples of bacterial lectins that recognize sialic acid residues

Escherichia coli S-adhesin
Helicobacter pylori adhesins I and II
Mycoplasma pneumoniae haemagglutinin
Streptococcus sanguis haemagglutinin
Streptococcus suis haemagglutinin
Bordatella bronchiseptica haemagglutinin
Pseudomonas aeruginosa haemagglutinin

16.2.1.3 Bacterial toxins

Bacteria also produce toxins which have lectin-like activity, binding to saccharide residues, usually sialic acids, at the cell surface. Some examples are given in *Table 16.2*.

16.2.2 Viral infections

Virus glycoproteins are glycosylated utilizing the host machinery of glycosylation, and the host therefore plays a major role in determining the type of structures that are produced. Viral glycoproteins may protect the virus from the host immune system, from degradation, or be involved in infectivity. ➲

➲ Many viral infections are mediated through viral lectin-like molecules recognizing host cell oligosaccharides.

Viruses rely on saccharide–receptor interactions for infectivity and pathogenicity. Some, like the herpes simplex virus, have adhesins that recognize saccharide determinants of proteoglycans, such as heparan sulfate. Tunicamycin treatment (which abrogates N-glycosylation; *see 15.6.1*) of cells prevents their infection by the virus, whereas viral binding to untreated cells is unaffected.

Other viruses, including, for example, influenza virus, rotavirus, Corona virus, polyomavirus and Sendai virus use sialic acid binding haemagglutinins in infection (*see Table 16.3*). Of these, influenza virus has been most intensively studied. Human influenza virus A and B haemagglutinins are specific for Sia(α2→6)Gal, whereas many animal influenza virus strains bind to Sia(α2→3)Gal sequences. Influenza C virus haemagglutinin is specific for 9-*O*-acetyl-N-neuraminic acid (Neu5,9Ac_2). There has been much interest in the precise binding partner of the influenza virus haemagglutinin, with a view to designing therapeutic inhibitors of viral attachment to host cells, which would prevent subsequent internalization of the virus.

Influenza virions also secrete sialidases, which are involved in preventing viral agglutination, dissociation of virions during budding from the infected cell surface, or de-sialylation of soluble mucins to improve access to membrane-bound sialic acid residues. These sialidases are of particular interest, as inhibitors of their action are inhibitors of viral infection and may have potential in therapy as anti-viral agents. The recently developed anti-viral agent Relenza™ (produced by Glaxo SmithKline), is based on this principle; it is a sialic acid analogue that is recognized by the influenza sialidase and through binding, inactivates it.

Understanding carbohydrate structures involved in infection and disease processes has implications in many other viral diseases. For example, inhibiting normal glycosylation of

Table 16.2 Some examples of bacterial carbohydrate-binding toxins

Vibrio cholerae toxin (cholera)
Clostridium tetani toxin (tetanus)
Clostridium botulinium toxin (botulism)
Bordetella pertussis toxin (whooping cough)

Table 16.3 Sialic acid binding specificity of some viral haemagglutinins

Virus	Carbohydrate binding specificity
Influenza virus A and B	Neu5Ac(α2→3)Gal or Neu5Ac(α2→6)Gal
Influenza virus C	Neu5,9Ac_2
Rotavirus group C	Neu5Ac
Corona virus	(α2→3) linked Neu5,9Ac_2
Polyoma virus	Neu5Ac(α2→3)Gal(β1→3)GalNAc or Neu5Ac(α2→3)Gal(β1→3)[Neu5Ac(α2→6)]GalNAc
Sendai virus	Neu5Ac(α2→3)Gal

the envelope N-linked glycoproteins of the human immunodeficiency virus (HIV) by treatment with the N-glycosylation inhibitor N-butyldeoxynojirimycin (NB-DNJ; *see 15.6.3*) dramatically reduces infectivity of the virus, probably by interfering with the folding of the viral coat protein. The virus can still bind to cells via interaction with CD4 receptors, but fusion of viral and cellular membranes is inhibited. A similar treatment of hepatitis B virus results in viral retention within the cell.

16.2.3 Parasitic infections

Parasites synthesize a wide array of glycosylated molecules. In many parasitic diseases, it is the host immune response, frequently raised to unusual and therefore antigenic oligosaccharide structures synthesized by the parasite, that is responsible for tissue damage. Furthermore, many parasites rely on saccharide–receptor interactions to facilitate interactions during various stages of their lifecycle, and to evade destruction by their host. This is a field of particular interest as a better understanding of the specialized glycoconjugates produced by parasites may open the way for development of diagnostic tests and in eventual development of effective vaccines.

16.2.3.1 Trypanosomiasis (Chagas disease caused by *Trypanosoma cruzi* and African sleeping sickness caused by *T. brucei gambiense* and by *T. b. rhodesiense*)

Trypanosoma cruzi possesses a specialized cell surface *trans*-sialidase which transfers ($\alpha 2 \rightarrow 3$)-linked sialic acid from host sialoglycoconjugates to terminal Gal residues on glycoconjugates on the surface of the parasite.➲ The addition of sialic acid to the parasite surface may help to protect the parasite from complement-mediated lysis and therefore enable it to survive in the host's bloodstream or decrease its antigenicity. Alternatively, sialylated parasite glycans may be recognized by host cell sialic acid-binding lectins, augmenting infectivity. Removal of host phagolysosomal proteins by the *trans*-sialidase may also be involved in stages of the parasite lifecycle where it escapes the host phagolysosome and enters the cytoplasm.

> ➲ *T. cruzi* possesses a specialized *trans*-sialidase which transfers sialic acid from host sialoglycoconjugates onto parasite cell-surface glycans.

T. brucei, lives free in host lymphatics and blood vessels. It evades attack from the host immune system by selectively expressing an ever-changing repertoire of cell-surface, glycophosphatidylinositol (GPI)-anchored (*see 12.2.1*) 55 kDa, variable surface glycoproteins (VSGs) coded for by different VSG genes, only one of which is switched on at any one time. The parasite has a dense covering of VSG, $\approx 10^7$ copies of the glycoprotein per cell. Up to ≈ 1000 VSG variants, all structurally similar, but immunologically distinct, are possible.➲ When host immunity to parasites carrying one dominant VSG variant results in their destruction, a population of parasites carrying another VSG variant expands and replicates unchecked until another specific immune response is elicited. Such antigenic variation is repeated continually allowing the parasite to persist throughout the lifetime of the host. Compounds that inhibit GPI-anchor synthesis are of interest as potential anti-trypanosomal agents.

> ➲ *T. brucei* evades the host immune response by continually changing its cell surface glycans.

16.2.3.2 Leishmanisis (caused by *Leishmania donovani*)

Leishmania donovani synthesizes a specialized, developmentally regulated lipophosphoglycan relevant to different stages in the parasite's lifecycle. Reduction of terminally exposed Gal or Man residues controls stage-specific adhesion of developing promastigotes to the sandfly midgut, and elongation of the lipophosphoglycan protects the parasite from complement-mediated damage. Intriguingly, this glycoconjugate is involved in ligand attachment to host macrophages, and also in the early survival stages within the macrophage in which it contributes to inactivation of the macrophage protein kinase C which is required for oxidative burst in these cells, and prevents phagosome–endosome fusion. Parasite strains lacking a gene involved in lipophosphoglycan synthesis lack virulence.

16.2.3.3 Malaria (caused by *Plasmodium* species)

Malaria glycoconjugate research remains in its infancy, largely owing to the difficulties in culturing the parasite at most stages of its lifecycle. Evidence suggests that the parasite does not possess mechanisms for N-glycosylation.

Free GPI lipids (*see Chapter 9*) expressed by the parasite induce expression of some host cell adhesion molecules. Much of the immunopathology of malaria is due to the action of tumour necrosis factor-α (TNFα), which is released in response to specific malarial GPI glycolipids. Research into free and protein-bound malarial GPI lipids has contributed to understanding of the pathophysiology of the disease and may eventually result in the development of new anti-malarial vaccines.

A malarial adhesin called circumsporozoite protein binds heparan sulfate on the host hepatocytes during parasite invasion of these cells. Later in the lifecycle, the parasite bears adhesins specific for Neu5Ac(α2→3)Gal sequences on the O-linked cell-surface tetrasaccharide of the transmembrane glycoprotein glycophorin on red blood cells.

Elevated titres of auto-antibodies against glycosylated structures have been detected in the sera of patients with malaria. Of special interest are IgM antibodies directed against T antigen (*see 10.9.1*), which is present on the surface of desialylated blood group O erythrocytes.

16.2.3.4 Schistosomiasis (caused by *Schistosoma* species)

Schistosomes synthesize a wide variety of glycoproteins and glycolipids, many of them heavily fucosylated, which specifically interact with host cells.➲ Different *Schistosoma* species produce a different repertoire of glycoconjugate structures, although sialylated glycans do not seem to be produced. Glycan structures are employed to evade the host's immune response at all of the various different stages in the parasite's long (up to 40 years) and complex lifecycle.

> ➲ Schistosomes synthesize many diverse glycan structures throughout their lifecycles, including fucosylated oligosaccharides like Le^x.

In infected children, 'blocking' IgM and IgG2 antibodies (anti-glycan antibodies are commonly of the IgM type) develop against the many glycan structures present on the parasite, and this may inhibit a protective immune response. These 'blocking' antibodies attach to the carbohydrate antigens on the parasite but fail to mediate an antibody-dependent cellular toxicity, whereas their presence blocks access by effector antibodies of the IgG1 and IgE classes. As infected children reach adulthood, the balance of antibody types changes to a predominantly protective type of response resulting in at least partial immunity.

The parasite excretes large amounts of a heavily glycosylated glycoproteins, carrying, in particular, poly-Le^x structures. The adult host develops autoimmunity to Le^x and this may have a potent effect in compromising host cellular immunity as part of the parasite's long-term survival strategy. These auto-antibodies induce a local anti-inflammatory and anti-thrombin effect as they are directed against host immune cells such as neutrophils.

16.3 Leukocyte recruitment in inflammation

The specific recruitment of leukocytes to sites of inflammation is mediated through saccharide–receptor interactions. This is a field that has been intensively studied and in which the molecular mechanisms are well understood. In simple terms, the sequence of events is as follows, and is illustrated in *Figure 16.1*. Under normal conditions, leukocytes

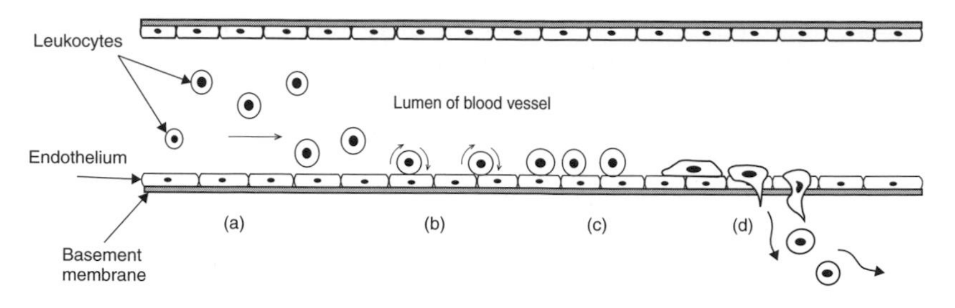

Figure 16.1

Leukocyte recruitment to sites of inflammation: (a) leukocytes travelling rapidly in the bloodstream become marginalized; (b) weak, transient adhesion-mediated by interactions between leukocyte cell-surface carbohydrates and selectins cause leukocytes to roll along the endothelium, pushed along by the shear force of the blood flow; (c) leukocyte cell-surface integrins bind firmly to ICAM-I expressed by endothelial cells mediating firm adhesion; (d) leukocytes flatten and spread, then migrate through the endothelium into local tissues.

travel rapidly in the central part of the bloodstream, carried along in the flow of the blood and without significant contact or interaction with the vascular endothelium. In conditions of inflammation, however, pro-inflammatory molecules cause dilation and leakiness of the capillary, causing the blood to thicken and its flow rate to lessen. Now leukocytes become marginalized and move more slowly; their opportunity for interaction with endothelial cells increases. The first stage involves adhesive interactions between leukocyte cell-surface sialylated and fucosylated oligosaccharides, for example sLe^x or sLe^a, and E-selectin and/or P-selectin (these are C-type lectins; *see 13.12.3*), which are expressed by the endothelial cells in response to activation by pro-inflammatory cytokines. Furthermore, adhesive interaction occurs between L-selectin expressed by the leukocyte and sialylated, sulfated and fucosylated O-glycans (such as 3′-sulfo-Le^a and 3′-sulfo-Le^x) present on endothelial cells. This selectin–saccharide interaction is necessarily quite weak and transient, so it has the effect of slowing the flow of the leukocytes, rather than stopping them completely. The leukocytes roll along the endothelium, pushed by the shear force of the faster flowing blood. The interaction may be amplified by leukocyte–leukocyte and leukocyte–platelet selectin–saccharide interactions. Once the leukocyte is in contact with the endothelium, it is stimulated by high local concentrations of substances such as interleukin (IL)-8, complement component C5a and leukotriene B4 and by platelet activating factor (PAF) to produce integrins which then bind firmly to ICAM-I on the endothelium. This mediates strong adhesion. This is followed by a phase of firm leukocyte–endothelial cell adhesion in which the white cell is stationary on the endothelium, resisting the force of the flowing blood. From here, the leukocyte cell flattens and spreads, then migrates through the endothelial cell to invade local tissues in response to chemotactic gradients.

The importance of selectin–saccharide interactions in normal recruitment of leukocytes is illustrated by the congenital disorder of fucose metabolism, leukocyte adhesion deficiency II (LAD-II; *see 16.8.1.4*). Individuals with this disorder do not produce fucosylated glycans such as those recognized by the selectins. Their leukocytes are unable to tether and roll on activated endothelium and thus their immune system is less able to combat infectious diseases. At the other extreme, some inflammatory disorders such as rheumatoid arthritis and ischaemia–reperfusion injury are characterized by excessive leukocyte recruitment; here specific inhibition of selectin–saccharide interaction may have therapeutic benefit. ➲ Furthermore, there are

> ➲ Selectin–oligosaccharide interactions play an important role in leukocyte recruitment to sites of inflammation.

distinct similarities in the way that leukocytes are recruited to sites of inflammation, and the way that cancer cells metastasize; these mechanisms are therefore of relevance to cancer research. Many leukaemia and lymphoma cells express L-selectin, and many cancers and cancer cell lines synthesize oligosaccharides, such as those of the Lewis family, recognized by selectins (*see 16.6.7*). ➲

➲ The role of selectin–receptor interactions in recruitment of lymphocytes is of interest in inflammatory disorders, cancer metastasis and LAD-II.

Intestinal inflammatory disorders such as Crohn's disease and ulcerative colitis are associated with significant changes in glycosylation of gut mucins. In particular, disturbances in sulfation and sialylation have been reported. (α2→6)-linked sialic acid increases, although (α2→3)-linked sialic acid decreases with progressive severity of inflammation. Sialyl Tn synthesis (*see 10.9.2*) is increased in ulcerative colitis and may be associated with risk of subsequent neoplasia. O-acetylation, which normally infers resistance to bacterial sialidases, is reduced.

16.4 Changes in glycosylation of leukosialin with differentiation, immune deficiency and malignancy

Leukosialin (CD43, sialophorin), a heavily glycosylated, cell membrane, O-linked, mucin-type glycoprotein (*see 5.4.2*) is expressed by various types of leukocytes. ➲ The extent and type of glycosylation, and thus the apparent molecular mass (the glycans contribute to the actual molecular mass of the molecule, plus glycosylation affects migration in an SDS–PAGE gel, thus also altering the *apparent* molecular mass), differs between cell types. For example, leukosialin from resting T lymphocytes has an apparent molecular mass of 105 kDa and leukosialin from granulocytes, monocytes and platelets has an apparent molecular mass of 130 kDa. Erythroid cells contain leukosialin with type 1 cores, but myeloid and lymphoid cells may contain both core 1 and core 2 types. In the myeloid series, acute myelogenous leukaemias make oligosaccharides with type 1 cores, whereas chronic myelogenous leukaemias and mature granulocytes make oligosaccharides with type 2 cores. Mature granulocytes synthesize complex polylactosamine structures, and a change from production of type 1 core to type 2 core is seen upon T-lymphocyte activation.

➲ Glycosylation of leukosialin changes with leukocyte cell type, maturation, activation and in leukaemia.

O-glycans present on the leukosialin of resting T lymphocytes are mainly sialylated core 1 structures with NeuAc(α2→3)Gal(β1→3)[NeuAc(α2→6)]GalNAc, but upon activation, sialylated core 2 structures NeuAc(α2→3)Gal(β1→3)[NeuAc(α2→3)Gal(β1→4)GlcNAc(β1→6)]GalNAc predominate (*see Figure 16.2*) caused by the upregulation of core 2 β1,6 GlcNAc transferase in activated T lymphocytes. The expression of this enzyme is negligible in resting T

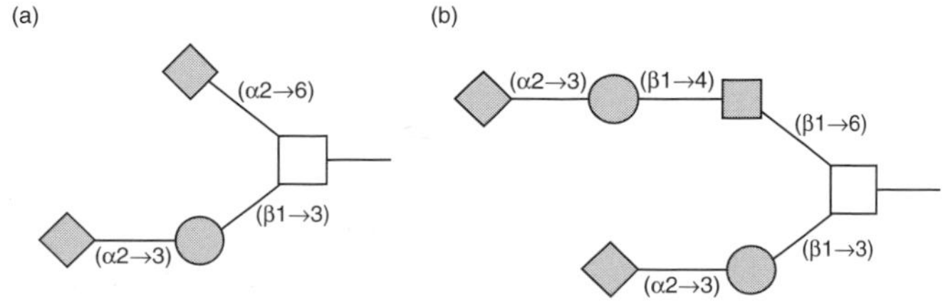

(a) (b)

Figure 16.2

O-glycans expressed by resting T lymphocytes are mainly: (a) disialylated core 1, but upon activation, (b) disialylated core 2 predominates.

lymphocytes. T-Cell activation results in a change of oligosaccharide synthesis from those based on core 1 to core 2 types, and core 2-based oligosaccharides are necessary for subsequent sLe^x formation. Thus, the synthesis of core 2 structures is significant because it results in sLe^x being found on leukosialin. sLe^x is a ligand for the E- and P-selectins expressed on activated endothelium, and thus synthesis of core 2 structures ultimately leads to interaction of activated lymphocytes with endothelial cells at inflammatory sites.

There is also a dramatic change in O-glycan synthesis on leukosialin during thymocyte development. In the thymus, immature thymocytes (immature T lymphocytes) express leukosialin with a mixture of sialylated core 1 and sialylated core 2 structures. As the T cells mature, the core 2 structures are replaced with sialylated core 1.

Resting T lymphocytes in the peripheral blood of healthy humans synthesize predominantly sialylated core 1 type O-glycans. In patients with T-cell leukaemias, sialylated core type 2 structures, similar to those produced by normal activated T lymphocytes, predominate. This, as in activated T lymphocytes, is due to upregulation of the β1,6 GlcNAc-T responsible for core 2 branching. In comparison with normal T-cell activation, the processing of this oligosaccharide is much increased in acute leukaemia cells and in some cases of chronic lymphocytic leukaemia. This change may be a reflection of the immaturity of the neoplastic cells, resembling in glycosylation the immature T lymphocytes in the thymus.

Leukosialin from various leukocyte-derived cells lines also exhibits a range of molecular masses according to its degree of glycosylation. The K562 cell line, derived from erythroleukaemia cells, synthesizes leukosialin O-glycans with almost exclusively sialylated core 1 structures while HL-60 cells, derived from malignant promyelocytes, expresses leukosialin with sialylated core 2 structures.

An increase in the synthesis of sialylated core 2 is seen in patients with some immunodeficiency disorders such as Wiskott–Aldrich syndrome (*see 16.4.1*) and HIV infection (*see 16.5*).

16.4.1 Wiskott–Aldrich syndrome

Wiskott–Aldrich syndrome is a genetic immunodeficiency syndrome caused by a defective gene at p11-q12 on the X chromosome. It affects only male children and is characterized by a reduced platelet count, problems with both cellular and humoral immunological responses, and the skin disorder eczema. Patients die at a young age from infection or leukaemia.

There is increased activity of the core 2 β1,6 GlcNAc-T in T- and B-lymphocytes and platelets resulting in increased synthesis of core 2-based O-glycans, such as sialylated core 2, predominantly on the sialoglycoprotein leukosialin.➲ This is normally associated with activated T lymphocytes (*see 16.4*). The lymphocytes of these patients do not respond normally to mitogens, which would result in their activation, and are non-functional. This phenomenon is referred to as ‘pseudoactivation’.

> ➲ In some diseases, such as Wiskott–Aldrich syndrome and HIV infection, increased synthesis of sialylated core 2 O-linked glycans are commonly found on T- and B-lymphocyte leukosialin, and this affects leukocyte function.

16.5 Glycosylation and HIV infection

Similar alterations in the O-linked glycosylation of leukosialin on T lymphocytes has been described in patients with AIDS, and the proportion of the T cells carrying sialylated core 2 O-glycans on leukosialin increases with the severity of the disease. This is because HIV infects activated T cells and immortalizes their phenotype. This phenomenon of ‘pseudoactivation’ in which T lymphocytes, although carrying oligosaccharides associated with the normal activated state, are non-functional, is similar to that seen in Wiskott–Aldrich syndrome (*see 16.4.1*).

Changes in O-glycosylation of leukosialin provoke antibody production in AIDS patients. Such antibodies may react with immature thymocytes, depleting T-cell precursors, thus contributing to the immunodeficiency associated with this disease. Production of this type of auto-antibodies may also develop in Wiskott–Aldrich syndrome (*see 16.4.1*).

One of the early steps in HIV infection is the binding of the highly glycosylated virus envelope glycoprotein gp120 to the host lymphocyte membrane. gp120 is rich in GalNAc and sialyl ($\alpha 2 \rightarrow 6$)GalNAc O-glycans. Anti-Tn, anti-sialyl-Tn, anti-Le^y and anti-blood group A antigen antibodies all neutralize HIV infectivity. Diverse N-glycans are also present, including a predominance of oligomannose and also complex oligosaccharide structures, including bisected multi-antennary N-linked oligosaccharides. Glycosylation of gp120 may alter during the course of infection, and this may in turn alter its antigenicity and, similarly, experimental alteration of the glycosylation of gp120 alters infectivity, suggesting potential therapeutic approaches.

16.6 Cancer metastasis

16.6.1 The metastatic cascade

Metastasis is the process by which cancer cells spread from their original primary site (e.g. breast) to other sites in the body (in the case of breast cancer, for example, to lymph node, liver, lung, bone or brain). Metastasis is responsible for the death of most cancer patients, and thus is of major clinical importance. The process of metastasis is often referred to as the metastatic 'cascade', as it involves a large number of different steps, all of which have to be successfully completed in an ordered sequence for metastases to become established. Several of these steps may involve interactions between saccharides and their receptors (lectins). Briefly, the steps involved in the metastatic cascade are as follows, and are illustrated in *Figure 16.3*.

1. A reduction in cell adhesive mechanisms between adjacent cancer cells, and between cancer cells and adjacent normal cells and matrix in order that cancer cells can break free from the primary tumour.
2. Migration through local extracellular membrane (ECM), involving degradation/ penetration of the basement lamina.
3. Entry into a lymphatic vessel and/or blood vessel and travel in the lymph or bloodstream to the new site, involving resistance to immune attack and survival of mechanical stresses.
4. In the case of blood-borne metastases, a complex series of events, analogous to those described in *Section 16.3* which occur in leukocyte recruitment, result in extravasation.
5. Proliferation at the new site.

There are many changes in glycosylation in malignancy, and these may have a functional role in metastasis.➲ Adhesion of leukocytes to endothelium, a process which may be mimicked by metastasizing cancer cells, is mediated through selectin–saccharide interactions and it seems likely that cancer cells may use similar molecular mechanisms (*see 16.3*). Furthermore, molecules implicated in metastatic mechanisms, such as the adhesion molecules integrins, cadherins and members of the immunoglobulin superfamily, are heavily glycosylated N-linked glycoproteins, and other adhesion molecules, for example, galectin-3, recognize oligosaccharides as binding ligands.

➲ Cancer cells may use saccharide–lectin interactions to attach to the endothelium of blood vessels in organs in which they later form metastatic tumours.

16.6.2 Altered glycosylation of cancer cells

Altered glycosylation is a universal feature of cancer cells. Changes may be multiple and complex. Glycosylation changes typically associated with cancer are summarized in *Table 16.4*.

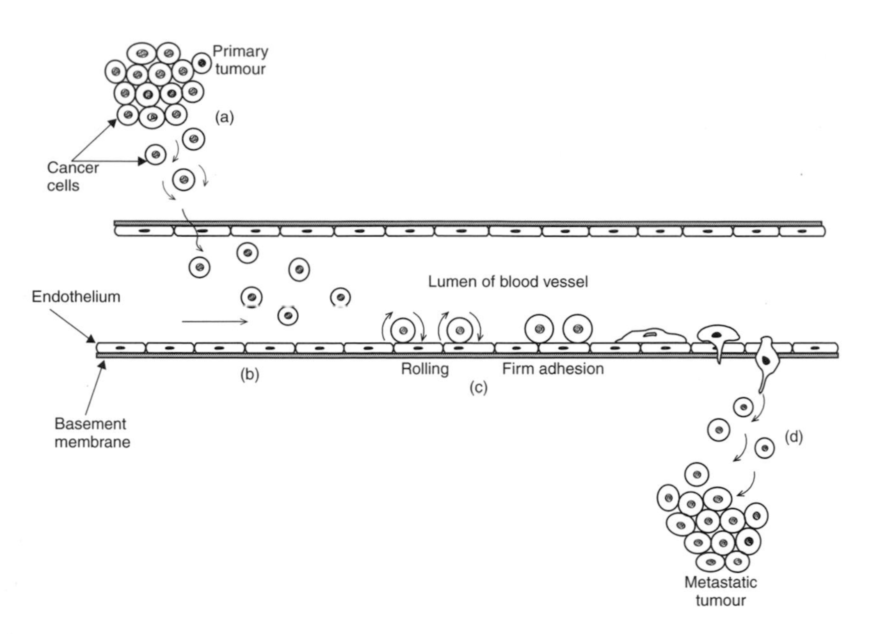

Figure 16.3

The metastatic cascade: (a) cancer cells break free of the primary tumour and migrate through local tissue; (b) cancer cells enter the bloodstream and are transported to a distant site; (c) cancer cells attach weakly to the endothelium and roll along propelled by the shear force of the blood flow, then become firmly attached, flatten and migrate through the endothelium; (d) they proliferate to form a metastatic tumour at the new site.

Table 16.4 Glycosylation changes typically associated with cancer

- The synthesis of carbohydrate structures more typically found in embryonic or foetal tissues
- Increased branching of complex carbohydrates including increased expression of complex, branching (β1$\rightarrow$6) N-linked oligosaccharides (the synthesis of larger glycans by transformed or malignant cells is sometimes referred to as Warren–Glick phenomenon, as these workers energetically studied and published extensively on this topic)
- Oligosaccharide truncation
- Disturbances in sialylation
- Reduced sulfation
- Occasionally, emergence of novel structures

Sometimes the synthesis of altered glycans is of prognostic and clinical significance. One example is that the lectin from *Helix pomatia* (HPA) recognizes (as yet incompletely characterized) glycans bearing terminal GalNAc, the presence of which is associated with the ability of the cancer cells to metastasize, and therefore with poor prognosis in a range of human cancers (*see 16.6.9*). The association between the presence of the HPA-binding glycans and metastatic competence is consistent with the glycans having a functional role in metastasis, although this has yet to be established. In general, the functional significance of altered glycosylation in cancer remains to be discovered.

New approaches to cancer therapy include attempts to neutralize circulating mucins or mucin fragments by antibodies directed against them, and also to immunize the patient with cancer-related mucin oligosaccharides to stimulate the immune response to these antigens.

Some of the better documented examples of altered glycosylation in cancer are described briefly below.

16.6.3 Changes in sialylation

An increase in sialylation is a commonly observed phenomenon in cancer cells, often manifesting itself as an increase in sialylated lactosamine Gal(β1→4)GlcNAc or sialyl-Tn GalNAc-α-O-Ser/Thr. An increase in sLe^x and sLe^a structures is also found on glycosphingolipids and on N- and O-linked glycans.

The presence of sLe^x and sLe^a is associated with tumorigenicity and metastatic potential in many tumours and in tumour model systems, possibly because these structures can act as ligands for selectins (*see 13.12.3.2*). Furthermore, increased cell-surface sialylation may result in a decrease in attachment of cancer cells to collagen type IV and fibronectin, and thus increase metastatic competence. Significantly, alterations in the type of sialic acid present, and decreases in cellular sialylation have also been reported emphasizing the complexity of sialylation and its relevance to tumour biology.

16.6.4 Increased (β1→6)GlcNAc linkages and poly-N-acetyllactosamine in cancer

In general, increases in (β1→6) GlcNAc linkages and poly-N-acetyllactosamines are associated with malignancy, although reduction in the synthesis of core 2 structures have been reported in some tumour types, notably breast cancer. Increased branching is also characteristic. Core 2 is often a preferred site of attachment for poly-N-acetyllactosamines, and these in turn provide a backbone for the synthesis of sLe^x, which is frequently increased in cancer. sLe^x is of particular relevance in that it is a ligand for E- and P-selectins, and may thus be involved in metastasis of tumour cells, through adhesion to microvascular endothelium.

An increase in (β1→6) branching of complex N-glycans, which may be detected by binding of the lectin PHA-L, has been extensively studied and reported in a number of human cancers, including those of breast, colon, oesophagus and skin (melanoma). PHA-L recognizes Gal(β1→4)GlcNAc(β1→6)[Gal(β1→4)GlcNAc(β1→2)]Man(α→ (*Figure 16.4*). The presence of this oligosaccharide, detected by PHA-L binding, is associated with metastasis and poor prognosis in cancer patients, and with the ability of cultured cell lines to metastasize in animal models. This altered glycosylation appears to be a consequence of increased expression of GlcNAc transferase V and transfection of non-tumorigenic cells with GlcNAc-TV cDNA dramatically alters their behaviour in *in vitro* assays, causing the cells to exhibit more tumorigenic features, such as reduced contact inhibition, altered substratum adhesion, increased invasiveness and, in particular, increased motility. Transfection induces non-tumorigenic immortilized cells to form tumours in nude mice, and enhances metastatic competence in cultured carcinoma cells.

16.6.5 Truncated glycans

General loss of O-linked mucin oligosaccharides, and their truncation is a common feature of many epithelial cancers, for example, those of colon and breast. This is caused by failure in one or more of the enzymes required for further oligosaccharide chain synthesis.

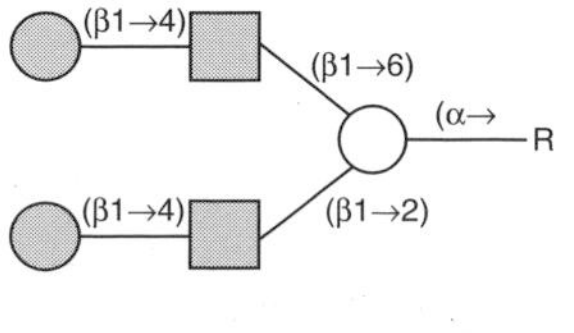

(β1→4) (β1→6) (α→ R
(β1→4) (β1→2)

Figure 16.4

The binding partner of the lectin PHA-L.

16.6.6 Tn, T, sialyl-Tn and sialyl T antigens

The first monosaccharide to be attached to the polypeptide backbone in O-linked glycan synthesis is GalNAc, and this structure, GalNAcα-Ser/Thr is the Tn epitope (*see 5.6.1*). Tn is not found on normal cells because it is acted upon by various glycosyltransferases to give a range of O-linked core structures. A common feature of tumour cells, however, is that core structure formation and subsequent elongation of O-linked glycans fails to occur normally, and aberrant, truncated O-glycans, including Tn, are detectable. The presence of Tn antigen is an indicator of aggressive biological behaviour and poor prognosis in breast cancer.

The Tn epitope may, alternatively, be sialylated by the action of the glycosyltransferase enzyme ST6GalNAc-I to give sialyl-Tn. If this occurs, no further elongation of the glycan is then possible. Sialyl-Tn has also been reported to be synthesized by a number of cancer types. This raises the possibility that immunizing cancer patients with synthetic Tn and/or sialylated Tn may stimulate an immune response directed against their tumour cells.

Another truncated O-glycan that has been identified on some cancer cells is the Thomsen–Friedenreich antigen (TF or T antigen). This is the O-linked core 1 structure, Gal(β1→3)GalNAcα-Ser/Thr (*see 5.6.3*), which is synthesized by the action of core 1 GalT on GalNAcα-Ser/Thr or Tn. In normal cells, core 1 or T antigen is usually elongated to form more complex O-glycans, but is frequently present in its simple truncated form in cancer cells. The synthesis of truncated tumour-associated O-linked glycans is summarized in *Figure 16.5*. The presence of the T antigen has been correlated with the invasive capacity of early bladder cancers. T antigen may also be sialylated by the action of various sialyltransferases to produce sialylated forms of the T antigen.

The availability of lectins and monoclonal antibodies that recognize these types of structures has made it possible to investigate their distribution in normal and diseased tissues. The lectin from peanut (*Arachis hypogaea*, PNA) recognizes Gal(β1→3)GalNAc, the T antigen, and several α-D-GalNAc binding lectins including the lectin from the hairy vetch (*Vicia villosa*, VVB_4) are convenient tools for investigating Tn. Tn and T are either not found on healthy cells, or are found at extremely low levels.

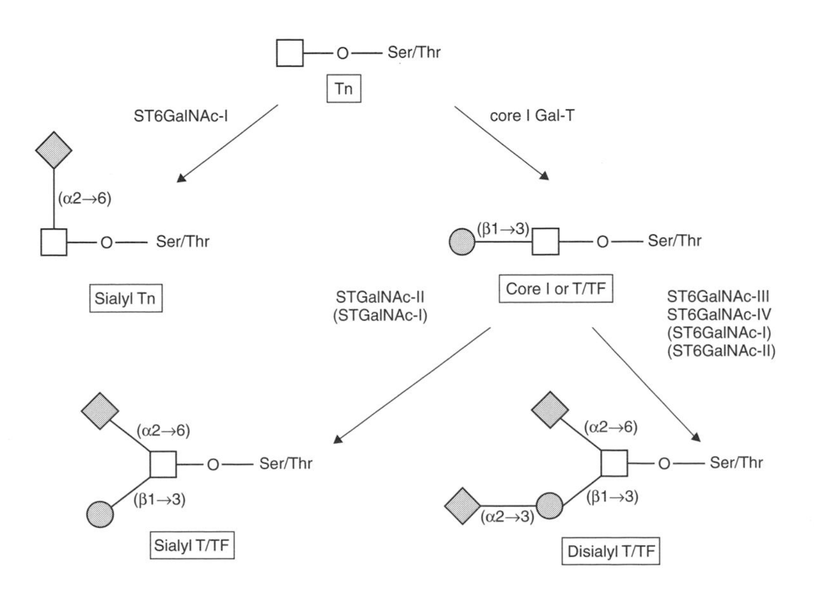

Figure 16.5

The synthesis of truncated tumour associated O-linked glycans.

16.6.7 Lewis blood group sugars

Lewis antigens (*see 10.6*) arise from fucosylation of simple type 1 Gal(β1→3)GlcNAc or type 2 Gal(β1→4)GlcNAc chains, and are frequently synthesized by cancer cells. sLe^x and sLe^a can both function as ligands for E-selectin, a selectin cell-adhesion molecule expressed on activated endothelium which has a functional role in leukocyte extravasation in inflammatory responses (*see 16.3*), and may be implicated in cancer cell arrest during metastasis. This presents the possibility of eventual development of specific anti-adhesion therapy. The prognostic significance of sLe^x and sLe^a has been demonstrated in many studies, and their use as serum markers of tumour progression and prognosis has been extensively investigated.

Sialylated forms of Le^x are selectively synthesized in some types of cancer, and as it is a potential ligand for E-selectin (a lectin-like molecule expressed on the activated endothelium of blood vessels and involved in leukocyte extravasation in inflammation; *see 16.4*), it is likely to be involved in metastasis. These glycosylation differences are due to either a difference in the expression levels of glycosyltransferases in cancer cells, or a difference in their location in the Golgi apparatus.

16.6.8 ABH blood group determinants

ABH blood group sugars are carried by glycoproteins and glycolipids on red blood cells, endothelial and epithelial cells in many tissues, and, in individuals termed 'secretors', in secretions such as mucus, semen, saliva and milk (*see 10.5.7*). Normal synthesis is often disrupted in cancer cells. Frequently, these structures are absent or truncated, as a result of deregulation of one or more of the relevant glycosyltransferases, resulting in the presence of precursors or fetal antigens not normally present in adult tissues. In cancer these are termed 'oncofetal antigens'. There is evidence that this does not result from mutation and/or inactivity of the relevant A/B glycosyltransferase genes, but instead to a downregulation of gene promoter activity. Sometimes 'incompatible' ABH sugars are present on cancer cells, for example, an individual with type A or O/H type erythrocytes may have tumour cells carrying type B sugars. This may be the result of errors in differential splicing.

16.6.9 HPA-binding oligosaccharides

Binding of the lectin from *Helix pomatia* (HPA) to cell-surface glycans is a marker of metastatic competence and poor prognosis in a number of human cancers, including those of breast, stomach, oesophagus, colorectum, prostate, lung and melanoma. HPA recognizes terminal α-GalNAc, and to a lesser extent α-GlcNAc residues. The HPA binding partners present on metastatically competent tumour cells remain incompletely characterized, but appear to be a heterogeneous range of moieties, including the Tn epitope (GalNAc-α-O-Ser/Thr) and blood group A sugar, associated with a number of cell-surface glycoproteins, and are presumed to represent truncated forms of normal structures.

16.6.10 Mucins

In normal cells, mucins are synthesized exclusively on the apical surface, and secreted into the lumen. One frequent aberration in glycosylation observed in malignancy is a loss of this ordered expression, whereby mucins may be synthesized at, and secreted from, the entire cell surface.➲ Thus, they are frequently detectable in the serum of cancer patients. A good example of this is the secretion of the mucin CA125 into the serum of ovarian cancer patients. Serum levels can be monitored to assess residual tumour burden following therapy and to warn of recurrent disease.

Mucins frequently carry sialylated and unsialylated forms of Lewis antigens, and other glycans such as sialylated and unsialylated T and

➲ In normal cells, mucins are usually synthesized only at the apical surface of cells, in cancer synthesis occurs over the entire cell surface.

Tn antigens. Mucins may also have altered structure in comparison with their normal counterparts. As mucins are sialylated and therefore heavily (negatively) charged they have the potential to radically affect the behaviour of the tumour cell. In particular, they may inhibit cell adhesion facilitating, for example, the displacement of the cancer cell from the primary tumour and preventing the destructive interaction with cells of the immune system. They may also inhibit tumour cell interactions with other molecules such as cadherins and integrins and shield the tumour cell from immune recognition.

16.6.10.1 The MUC1 gene product

Synthesis of the mucins of the *MUC* gene family (*see 5.10*) by normal cells and tissues is restricted and highly regulated. Different patterns of expression of these mucins in various human cancers reflect the loss of differentiation and gene regulation in these cells.

Of the *MUC* gene family, the MUC1 gene product, a membrane associated mucin, is synthesized by many glandular epithelial cells (except those of the small and large intestine), and expression is dramatically upregulated in lactating breast epithelium, and by breast, ovarian, some lung and colon carcinomas. The oligosaccharides attached to the MUC1 mucin protein backbone in cancer differ from those synthesized by normal cells, in that they are shorter and more frequently sialylated. Sialylated and unsialylated forms of T and Tn antigens are commonly present in cancer cell derived MUC1 mucins.

A number of monoclonal antibodies that recognize the MUC1 mucin have facilitated extensive immunohistochemical mapping of the molecule in normal and cancer tissues. ➲ In common with many other molecules, MUC1 is present at the apical or luminal surface of normal cells, but is distributed over the entire surface of cancer cells. Furthermore, the high proline content and heavy glycosylation of MUC1 means that the molecule possesses a rigid, extended extracellular portion, extending 200–500 nm away from the cell surface, that is readily accessible to encounters with neighbouring cell-surface molecules, and thus likely to play a prominent role in cell–cell interactions.➲ Expression of *MUC1* is increased by as much as 10-fold in some cancers, notably breast cancer, in comparison with adjacent normal epithelial tissue. In breast and colorectal cancer, increased *MUC1* expression is associated with tumour progression and advanced tumour stage, and is an indicator of poor patient prognosis. Radiolabelled antibodies to MUC1 have been used in radioimaging (radioimmunoscintography), and, as membrane-bound mucins have the advantage of being accessible to cytotoxic antibodies, have been proposed as targeting agents for directed therapy.

➲ Examples of antibodies recognizing MUC1

B72.3	Ca3	HMFG-2
BC1	115D8	M8
BC2	DF3	Mc5
BC3	E29	NCRC-11
BrE-2	F36/22	OM-1
C595	H23	OncM115
Ca2	HMFG-1	SM3

Of these, HMFG-1, HMFG-2 and SM-3 have been most widely reported in the literature. The peptide recognized by most of these antibodies is a three or four amino acid sequence within the peptide Ala-Pro-Asp-Thr-Arg-Pro-Ala-Pro, part of the 20 amino acid tandem repeat that is a feature of the mucin.

➲ MUC1 is highly negatively charged and protrudes 200–500 nm from the cell surface. Its presence may therefore profoundly affect cell behaviour.

In normal cells, MUC1 has been implicated in anti-adhesive mechanisms (*see 5.10.1*), and this effect may be important in cancer where MUC1 is present over the entire cell surface. Furthermore, as it is heavily sialylated, MUC1 carries a strong negative charge. This may have profound implications for the properties of the cell and destabilize normal cell–cell interactions. Indeed, electron microscopy studies have demonstrated that the cell membranes of adjacent MUC1-positive cancer cells make no direct contact with each other. A decrease in cancer cell–cancer cell or cancer cell–extracellular matrix (ECM) adhesion may contribute to cell motility, invasion and therefore metastatic competence.

MUC1 oligosaccharides act as binding partners for the I type lectins ICAM-1, expressed on endothelial cells, and sialoadhesin, expressed on macrophages (*see 13.12.4.1*). The mucin also enhances antigen presentation in T cells. Altered glycosylation attending malignancy can alter these interactions, which may in turn have an effect on tumour cell behaviour and tumour progression. Furthermore, oligosaccharide binding partners for selectins have been described on cancer-associated MUC1, which may be implicated in adhesion to endothelium during the metastatic cascade (*see 16.6*).

Cytotoxic T cells from patients with cancer can recognize the oligosaccharides of MUC1 in a non-MHC restricted fashion. It has been proposed that the repetitive epitopes of MUC1 molecule may allow cross-linking of the T-cell receptors directly, and that MUC1 carrying truncated oligosaccharide chains is recognized more readily by the humoral and cell-mediated immune response than the normally glycosylated form. This raises the possibility of developing immunogens based on MUC1 as an anti-cancer therapy and, indeed, trials of vaccination against MUC1 are underway. There may be problems as the high titres of free MUC1 shed into the serum of cancer patients may be recognized by the cytotoxic T cells, thus preventing them from binding to and destroying MUC1-positive cancer cells themselves. MUC1 may therefore actually aid the tumour in escaping immune surveillance. Circulating MUC1 also interferes with antibody-dependent cellular cytotoxic activity of eosinophils, with the function of lymphokine-activated killer (LAK) cells, and with stimulated cytotoxic T lymphocytes, perhaps by masking cell-surface antigens normally involved in immune recognition processes.

16.6.10.2 The trefoil factor family

Expression of the trefoil factor family (TFF) domain peptides (*see 5.12*), is strongly correlated with mucin expression. They are implicated in mucosal defence and healing, and expression is associated with gastric ulcers and Crohn's disease, indicating their role in repair of damaged gastric mucosa.

TFF1 and TFF2 are co-expressed with MUC1. TFF1 is expressed abundantly by ≈ 70% of breast cancers and also by many breast cancer cell lines, where its expression is oestrogen dependent. It has also been detected in cancers from a variety of organs, including stomach, pancreas, lung, uterus, mucinous carcinomas of the ovary, prostate, bladder, cervix, pancreas and medullary carcinoma of the thyroid. The role of TFF1 and TFF2 is to act as motogens and to facilitate cell flattening, migration and wound healing. This makes them of interest in malignancy where analogous processes take place. TFF1 protein expression is much lower in gastric adenocarcinomas than in adjacent apparently healthy gastric mucosa, and is absent in ≈ 50% of cancer cases. Some studies have shown a correlation between reduction in TFF1 expression and staging. Similarly, loss of expression of TFF3 is associated with advanced tumour stage. Thus, levels of TFFs may be useful prognostic indicators of disease and potential diagnostic tools.

16.6.10.3 Other membrane associated mucin-like molecules in cancer

In common with MUC1, other genetically unrelated mucin-like glycoproteins, carry long, stiff rod-like, O-glycans that protrude sometimes hundreds of nanometres away from the cell surface (*see 5.10.1*). Like MUC1, if they are overexpressed in cancer they may mask other cell surface molecules from immune recognition and interfere with cell adhesion processes. Examples include epiglycanin and ASGP-1.

Epiglycanin

Epiglycanin has a molecular mass of ≈ 500 kDa and extends up to 500 nm above the cell surface. It contains 75–85% of carbohydrate by weight and bristles with over 500 short O-linked oligosaccharides. Most are of the T antigen Gal(β1→3)GalNAcα-O-Ser/Thr, or the Tn epitope

GalNAcα-O-Ser/Thr type. It is unusually low in sialylated structures, does not contain Fuc, and may contain at least one N-linked oligosaccharide. It is produced in large amounts by cultured mouse mammary adenocarcinoma cells, where it is copiously shed into the culture fluid, or into ascites of tumour bearing animals.

ASGP-1, ASGP-2 and SMC (rat Muc4)

Ascites sialoglycoprotein-1 (ASGP-1) like epiglycanin, has been described in cultured mammary tumour cells, this time of rat origin, and is secreted into tissue culture fluid and in ascites of tumour-bearing animals. The two mucin-like molecules are, however, very different. ASGP-1 has a molecular mass of ≈570–690 kDa, depending on carbohydrate content, which varies between ≈67–73% of the mass of the molecule. ASGP-1 is glycosylated with much more complex oligosaccharides than those of epiglycanin, these are mostly tetrasaccharides or larger and commonly sulfated. ASGP-1 is not a transmembrane molecule, but is associated with the cell membrane through the unusual mechanism of complexing with a highly glycosylated (N-linked, ≈45% carbohydrate), high Man content, ≈120–140 kDa transmembrane subunit ASGP-2, which can act as a ligand for the receptor tyrosine kinase ErbB2. The heterodimeric glycoprotein complex of ASGP-1 and ASGP-2 is termed sialomucin complex (SMC or rat Muc4). The binding between the subunits is tight, but non-covalent. SMC is coded for by a single gene, transcribed into a 9 kb transcript, translated into a large (≈300 kDa) polypeptide precursor, which is proteolytically cleaved and further processed to yield the mature ASGP-1/ASGP-2 complex. The gene coding for SMC is a homologue to the human *MUC4* gene (*see 5.10.4*). Several functions have been suggested for SMC, including physical protection of exposed epithelia (e.g. in the trachea and colon), and modulation of signalling through the EGF family of receptors (through interaction with ErbB2). Furthermore, aberrant expression has been reported in tumour cells and results in reduced recognition by natural killer cells and in a reduction in cell–cell adhesion, leading to increased metastatic capacity.

16.6.11 Adhesion molecules expressed by cancer cells

The most important cell adhesion molecules with respect to metastasis are probably the cadherins, integrins, members of the immunoglobulin superfamily and galectins.➲ Expression of these molecules influences the interactions of tumour cells with other cells and with components of the ECM. There is evidence that alterations in the glycosylation of these molecules attends malignant transformation and affects their function.

> ➲ Altered glycosylation of integrins, cadherins, members of the immunoglobulin superfamily and galectins may be of functional importance in cancer metastasis.

16.6.11.1 Integrins

The glycosylation of integrins is essential to their function in binding laminin and fibronectin. Differences in the glycosylation of these molecules, in particular in respect to Le^x structures and sialylation, on cancer cell lines is associated with changes in metastatic behaviour, such as interaction with, and cell spreading on, endothelial cell basement membrane components.

16.6.11.2 Galectins

Galectins are S-type lectins (*see 13.12.1*) which recognize Gal, especially in poly-N-acetyllactosamines, and are involved in cell–cell and cell–matrix interactions. Increased expression of galectins, in particular of galectin-3 and galectin-1, has been associated with organ colonization and cancer metastasis, possibly through the interaction of these molecules expressed on endothelium with polylactosamines on cancer cells. Galectin-3 may be a 'de-adhesion molecule' in cancer. Galectin-3 also binds to the heavily N-glycosylated basement membrane component laminin. Galectin-3 expression in metastatic cells weakens the interaction between the cancer cell and the ECM by binding soluble galectin-3 to laminin polyllactosamine residues, thus stimulating secretion of matrix metalloproteinases and resulting in degradation of the basement membrane.

16.6.12 An example of altered glycosylation in malignancy: choriocarcinoma and human chorionic gonadotrophin

Human chorionic gonadotrophin (hCG) is a glycoprotein hormone produced by trophoblast cells of the placenta. It is secreted into the bloodstream and can be detected in the urine during pregnancy. Its presence in urine is the basis of the pregnancy test. It is also detectable in the serum and urine of patients with trophoblastic diseases. ➲ There is evidence that hCG is differently glycosylated in malignant conditions compared with benign trophoblastic diseases and pregnancy. Experimental studies illustrate that an affinity column of *Datura stramonium* lectin (DSA), which recognizes (β1→4)-linked GlcNAc oligomers and N-acetyllactosamine sequences, is effective in discriminating between the differently glycosylated forms of hCG in urine samples of individuals with malignant disease and pregnant controls or those with benign trophoblastic disease. This is clinically relevant as it may be a useful tumour marker to aid diagnosis and monitor treatment success and disease recurrence.

> ➲ Glycosylation of hCG is altered in choriocarcinoma, and it may therefore be a useful tumour marker.

In pregnancy, and in patients with the benign trophoblastic disease hyatidiform mole, hCG, secreted into the blood, is glycosylated in a constant and predictable manner: five acidic N-linked oligosaccharide chains are attached at defined sites on the polypeptide chains of the α- and β-subunits. The hCG oligosaccharides secreted by the malignant trophoblastic disease choriocarcinoma are different. Here, an increase in sialylated and unsialylated forms of core 2 oligosaccharides are detectable. hCG produced by healthy pregnant women and women with hydatiform mole, in comparison, contains almost exclusively sialylated core 1. Although the differences appear superficially to be complex, they can be explained by alterations in activity of just two glycosyltransferases. In choriocarcinoma, the activity of the Fuc-T responsible for formation of Fuc(α1→6)GlcNAc is increased, and the core 2 GlcNAc-TIV, responsible for the formation of GlcNAc(β1→4)Man(α1→3) synthesis is ectopically expressed and its enzyme specificity is modified in comparison with that found in normal tissues. In the invasive form of the hyatidiform mole, GlcNAc-TIV is also ectopically expressed, but its substrate specificity is similar to that seen in normal tissues.

16.7 Altered glycoforms of IgG in rheumatoid arthritis

In healthy individuals, the oligosaccharide pattern of serum IgG remains constant. In patients with rheumatoid arthritis, a proportion of circulating IgG molecules have decreased levels of galactosylation of their N-glycan chains, some, termed G_0, carrying no Gal at all. ➲ They also contain some bisected, fucosylated N-glycan chains. Disease severity is correlated with the degree of altered glycosylation, and spontaneous remission of the disease during pregnancy is characterized by an increase in galactosylation to normal levels. The level of agalactosyl IgG is a prognostic marker and a useful tool for monitoring disease severity and activity.

> ➲ Agalactosyl IgG is characteristic of rheumatoid arthritis.

Table 16.5 Alterations in glycosylation in selected autoimmune diseases other than rheumatoid arthritis

Disease	Altered glycosylation
Systemic lupus erythematosus	A shift to the G_0 glycoform of IgG
IgA nephropathy	Abnormal O-linked glycosylation of IgA1
Autoimmune haemolytic anemia	Decrease in G_0 glycoform of IgG
HIV infection	Reduced sialylation of T-cell surface glycoproteins, including CD43
Graves disease	Expression of normally absent Gal(α1→3)Gal (thyroid autoimmunity)
Insulin-dependent diabetes	Accumulation of advanced glycation products (non-enzymatic)
Myasthenia gravis	Increased expression of galectin-1 in the thymus

The significance of glycosylation of IgG to the pathophysiology of this multifactorial disease remains unclear, but the lack of Gal exposes underlying GlcNAc residues, leading to decreased binding of IgG to C1q and Fc receptors. In contrast, *in vitro*, aggregated agalactosyl IgG is recognized by the mannose-binding lectin (MBL), triggering complement activation and this may exacerbate the inflammatory process.

The mechanism underlying this glycosylation change is unclear, but it may result from decreased activity of a B-cell-specific β1,4 Gal-T. This remains controversial as different research groups have obtained conflicting results, and there is no evidence of an inverse relationship between levels of β1,4 Gal-T activity and levels of agalactosyl IgG. Furthermore, β1,4 Gal-T knockout mice do not develop rheumatoid arthritis. It is striking, however, that several spontaneous and induced animal models of rheumatoid arthritis also show alterations in IgG glycoforms.➲ Furthermore, evidence of the existence of several different β1,4 Gal-Ts makes the situation more complex.

➲ A shift to the G_0 glycoform of IgG, as well as being present in rheumatoid arthritis, is also characteristic of some inflammatory conditions, including tuberculosis, osteoarthritis and Crohn's disease.

Altered glycosylation appears to be involved in a number of other autoimmune disorders, as listed in *Table 16.5*.

16.8 Inherited disorders in glycan biosynthesis

A number of inherited disorders in glycan synthesis has been described. Some of the better understood disorders are outlined below.

16.8.1 Congenital disorders of glycosylation or carbohydrate-deficient glycoprotein syndromes (CDGs)

Congenital disorders of glycosylation or carbohydrate-deficient glycoprotein syndromes (CDGs) are a family of clinically heterogeneous autosomal recessive genetic defects in assembly, attachment or processing of N- and/or O-glycans. They result in multi-systemic diseases characterized by defective glycosylation of glycoproteins. Various subclassifications exist. The features of these CDGs are summarized in *Table 16.6*, and this section will focus on describing in more detail the features of CDG-I and -II.

Table 16.6 Congenital defects of glycosylation (CDG)

CDG	Defect in	Approx. no. of cases reported worldwide	Symptoms	Treatment
CDG-Ia	phosphomannomutase 2 or PMM2 locus	≈300	Very heterogeneous, includes brain abnormalities, hypotonia, dysmorphy, failure to thrive, delay in motor and language development Liver dysfunction, susceptibility to infection, death within first few years of life Dysmorphy, abnormal distribution of subcutaneous fat	none
CDG-Ib	phosphomannose isomerase or PMI locus	≈10	No psychomotor or developmental symptoms Liver–intestinal disease including liver fibrosis, protein-losing enteropathy, thromboses, bleeding, low blood sugar, vomiting, diarrhoea	mannose
CDG-Ic (or V)	ER Glc-TI	≈11 families	Neurological symptoms including wakening of muscles of the trunk, moderate psychomotor retardation, epilepsy, recurrent infections, decreased blood clotting factors	
CDG-Id	ER Man-TVI	1	Retarded psychomotor development, epilepsy, small skull, optic atrophy, brain and corpus callosum hypotrophy	
CDG-Ie	ER dolichol-phospho-mannose	4	Psychomotor retardation, epilepsy, muscle weakness, failure to thrive, dysmorphy	
CDG-IIa	Golgi GlcNAc-TII	4 families	Severe mental retardation, dysmorphy	
CDG-IIb	glucosidase I	1	Similar to trisomy 18 (Edward's syndrome) phenotype – low birthweight, multiple dysmorphic features including characteristic skull shape, clenched hands with overlapping index and fifth fingers, single palmar crease, rockerbottom feet, short sternum, malformations of heart, kidney and other organs	
	Import of GDP fucose into Golgi	3	Similar to CDG-Ie plus recurrent infections and increased white blood cell count Neutrophils lack sLe^x	fucose

16.8.1.1 CDG-I

CDG-I, first described in 1980, is the most common CDG, with ≈300 patients described in the medical literature, and comprises defects in synthesis/attachment of N-glycans. ➲ The N-glycans appear to have normal structures, but there are reduced numbers of chains attached to the proteins. The disorder occurs worldwide, and affects males and females equally. In the neonatal period, sufferers have slow head movements, and later, in infancy, other manifestations of the disorder appear, including abnormal eye movements, axial hypotonia and hyporeflexia, abnormal distribution of body fat, hypogonadism, and sometimes severe kidney, lung and heart problems. Feeding problems and diarrhoea result in delayed development and failure to thrive; skeletal abnormalities and skin problems also frequently occur. Severe neurological and mental development problems, including psychomotor retardation, epilepsy, stroke-like episodes, peripheral neuropathy, severe delay in language and motor development are also seen and sufferers consequently have a lowered IQ of 40–60. The symptoms of the disorder are underpinned by histological abnormalities in several different organ systems, resulting from defective N-glycosylation of glycoproteins throughout the body.

> ➲ In CDG-I N-glycans have normal structures, but there are reduced numbers of chains attached to the proteins.

Variants of type I CDG, called types Ia to Ie, are recognized on the basis of quantitative analysis of their serum transferrin glycoforms in isoelectric focusing and chromatofocusing, which reflect differences in negative charge due to abnormal sialylation. More is known about types Ia, Ib and Ic, and they are described briefly below. ➲

Type Ia CDG is autosomal recessive, and is due to a mutation in the phosphomannomutase 2 (PMM) gene, of which there are two, located on chromosomes 22q13 and 16p13. The defective enzyme results in reduced synthesis of mannose-I-phosphate, the precursor for the synthesis of GDP-Man. There is therefore a decrease in synthesis of N-glycans and also of GDP-Fuc, GPI-anchors, and O- and *C*-mannosylated proteins. Patients typically show between 5 and 30% of normal levels of activity of the enzyme. Fewer dolichol-linked oligosaccharides are synthesized, and consequently their glycoproteins carry less N-linked glycan chains.

Type Ib CDG is caused by a deficiency in the enzyme phosphomannose isomerase, the enzyme that synthesizes mannose-6-phosphate. Patients respond well to therapy with oral mannose. Patients are mentally normal, and do not show signs of neurological disease, but suffer from a range of symptoms including severe protein-losing enteropathy, coagulation problems, chronic diarrhoea, hypoglycaemia, convulsions and hepatic fibrosis.

Type Ic CDG is caused by a deficiency in the Glc-T that converts Dol-PP-$(Man)_9(GlcNAc)_2$ to Dol-PP-$(Glc)_1(Man)_9(GlcNAc)_2$. Symptoms differ between affected individuals, but may include psychomotor retardation, ataxia, hypotonia, recurrent infections, coagulopathy. Symptoms are usually milder than type Ia, presumably because in type Ia, synthesis of N-glycans GDP-Fuc, GPI-anchors, and O- and C-mannosylated proteins are all affected, whereas the Glc-T defect has more limited consequences for glycosylation.

➲ Enzyme deficiencies responsible for CDG types Ia to Ic

Disorder	Defective enzyme
Ia	phosphomannomutase 2
Ib	phosphomannose isomerase
Ic	Dol-P-Glc: Dol-PP-$(Man)_9(GlcNAc)_2$ α1,3-glucosyltransferase (Glc-T)

16.8.1.2 CDG-II

CDG-II is rarer, and only a handful of cases have been reported. CDG-II patients have defects in the processing of N- and O-glycans in the Golgi apparatus. Patients have normal phosphomannomutase levels but are defective in GlcNAc-TII activity in their leukocytes and fibroblasts.➲ The mutation in the GlcNAc-TII gene may be as simple as a single base substitution. Patients have a characteristic dysmorphic face with coarse features and low set ears, widely spaced nipples, a ventricular septal defect, odd stereotypical behaviour patterns (e.g. hand washing, sticking out the tongue), hypotonia and limb weakness, limited speech, infections and epilepsy.

> ➲ CDG-II is caused by defective GlcNAc-TII.

16.8.1.3 Galactosaemia

Inherited defects in the three enzymes responsible for galactose metabolism lead to a group of disorders termed galactosaemias. Classical galactosaemia results from deficiency in Gal-I-P-uridyl transferase. Symptoms include failure to thrive, cataracts, cognitive disability, ataxia, ovarian dysfunction in females, enlarged liver and jaundice. Some symptoms are partially relieved by a lactose-free diet.

16.8.1.4 Leukocyte adhesion deficiency II (LAD-II) syndrome

The rare leukocyte adhesion deficiency II (LAD-II) syndrome is characterized by stunted growth, recurrent, non-pyrogenic infections, leukocytosis, delayed wound healing, severe mental retardation, morphological and skeletal abnormalities and neurological symptoms. Many of these symptoms are manifestations of the underlying defects in glycosylation whereby patients are unable to synthesize blood group H or Lewis blood group antigens, suggesting a defect in fucose metabolism.➲ It is thought that these individuals are unable to synthesize GDP-Fuc. LAD-II neutrophils therefore lack ligands for E-, L- and P-selectin (*see 16.3*) and are unable to extravasate normally, resulting in a defective immune response to infection. They may also have functionally significant defects in O-linked α-fucosylation of proteins involved in developmental regulation (*see 7.8.2*).

> ➲ Inability to synthesize GDP-Fuc is characteristic of LAD-II. Individuals cannot make the fucosylated ligands for the selectins, and thus have leukocytes that are unable to extravasate normally.

16.8.1.5 Congenital dyserythropoietic anaemia (CDA) type II, or hereditary erythroblastic multinuclearity with a positive acidified serum lysis test (HEMPAS)

There are three congenital dyserythropoietic anaemia (CDA) disorders, characterized by ineffective erythropoiesis, bone marrow erythroid multinuclearity and tissue siderosis. Type II is the most common, and is also called hereditary erythroblastic multinuclearity with a positive acidified serum lysis test (HEMPAS).

HEMPAS is an autosomal recessive disease caused by ineffective erythropoiesis in the bone marrow. Males and females are equally affected. It is thought that ≈130–300 cases exist worldwide, but this may be an underestimate as there is evidence that HEMPAS alone may produce minimal symptoms and go undetected, only becoming apparent when associated with another genetic or acquired disorder. Glutathione reductase, acetylcholine esterase and adenylcyclase deficiencies, hairy cell leukaemia, acquired tocophenol deficiency and parvovirus infection have all been reported in association with HEMPAS.

Patients produce fragile erythrocytes that are susceptible to lysis and this phenomenon is associated with hyperplasia of the erythroid precursors and the occurrence of bi- or multinucleated erythroblasts in the bone marrow. This latter feature is striking and may be as a result of failure in cell division due to incomplete glycosylation of membrane glycoproteins.

There is an abnormality in the structure of the plasma membranes of cells of the erythroid lineage. Erythrocytes have a shortened half life of ≈7–34 days in HEMPAS patients, compared with ≈120 days in normal individuals.

Symptoms of this disorder are a result of ineffective erythropoiesis, including erythroid hyperplasia, enlarged spleen, gall stones, diabetes, jaundice and anaemia. Anaemia is usually mild, but severely affected patients require repeated blood transfusions. Diabetes is common and may be due to incomplete glycosylation leading to exposure of normally cryptic antigens and autoimmunity. HEMPAS patients also commonly suffer from mental and sensory abnormalities.

The syndrome results from a structural defect in the N-glycans attached to specific erythrocyte membrane glycoproteins which are necessary for normal function. The Man(α1→6) arm of N-glycans is not properly processed precluding the processing of complex type structures normally initiated by the β1,2 GlcNAc-TII. As a result, hybrid N-linked oligosaccharide structures are processed with a reduced number of antennae and poly-N-acetyllactosamine chains.➲ The glycosylation defect may be a result of low β1,2 GlcNAc-TII and/or α-mannosidase II and/or β1,4 Gal-T activity. The genetic basis of this disorder is poorly understood and different patients express different levels of activity of the relevant transferase enzymes. The defect may be genetically heterogeneous and also be associated with genes for transcription factors regulating glycosyltransferase expression levels, or with proteins affecting membrane trafficking.

➲ HEMPAS results from failure of β1,2 GlcNAcTII and/or α-mannosidase II and/or β1,4 Gal-T activity resulting in N-glycans with a reduced number of antennae and poly-N-acetyllactosamine chains.

Band 3 and band 4.5 with incomplete glycosylation of polylactosaminoglycans and an accumulation of polylactosaminylceramides in erythrocytes have been reported. The presence of the I antigen (*see 10.7*) is moderately increased, and presence of the i antigen (*see 10.7*) is increased 3–5-fold. There is an accumulation of high mannose type N-glycans and an absence of polylactosaminyl structures.

16.8.1.6 Ehlers–Danlos syndrome

Ehlers–Danlos syndrome is a heterogeneous group of inherited disorders of the connective tissue caused by defects in glycosaminoglycan chain synthesis that manifest as failure to thrive, hyperextensible skin, skeletal problems including hypermobile joints, hypotonia, bruising easily, teeth abnormalities, delayed speech and motor development, and progeroid appearance (progeroid means premature aging). The molecular basis of the disorders are incompletely understood but may result from defects in several enzymes involved in glycosaminoglycan chain biosynthesis. A partial deficiency in decorin, a dermatan sulfate proteoglycan (*see 8.4.3*) that binds to collagen fibrils, has been reported in some cases.

16.8.1.7 Glanzmann's thrombasthenia

Glanzmann's thrombasthenia is caused by a defect in GPIIb/IIIa platelet membrane glycoprotein complex. Several mutations in the genes coding for the glycoproteins involved in this complex, which result in their altered glycosylation have been described. Aberrant glycosylation results in reduced platelet binding to fibrinogen.

16.9 Inborn errors in carbohydrate catabolism: abnormal catabolism of glycoconjugates – the lysosomal storage diseases

Lysosomes contain a full range of enzymes, proteases plus exo- and endoglycosidases, necessary for the catabolic breakdown of glycoconjugates. The degradative enzyme pathways involved in N-glycoconjugate catabolism are well understood, whereas those responsible for O-glycoconjugate catabolism are not.

Table 16.7 Errors in carbohydrate catabolism – examples of lysosomal storage diseases

Disorder	Deficiency in	Glycoconjugates affected	Clinical symptoms
Sialidosis Type I: (mucolipidosis I)	α-sialidase	glycoproteins glycolipids	Juvenile or adult presentation, progressive, visual handicap and jerky movements, mental retardation
Sialidosis Type II:	α-sialidase		Coarse face, severely malformed bones, enlarged intestines
Galactosialidosis	sialidase and β-galactosidase	glycoproteins glycolipids	Late juvenile or early adult form of sialidosis II Skeletal dysplasia, macular cherry red spot, mild mental retardation, enlarged intestines. Coarse facies, early death
β-galactosidosis (GM_1 gangliosidosis and Morquio type B disease)	β-D-galactosidase	glycoproteins glycolipids	May occur in infantile, juvenile or adult forms. Infantile form: severe retardation of psychomotor development, oedema, skeletal deformities, hepatomegaly and coarse facial features. Juvenile form: similar to infantile form, but with milder symptoms. Adult form: progressive dystonia of limbs
Tay–Sachs disease (GM_2 gangliosidosis)	β-hexosaminidase A	glycoproteins glycolipids	Severe form: retinal abnormalities, psychomotor retardation and severe and rapid motor and mental retardation fatal in early childhood. Less severe form: slower onset of milder symptoms, mostly relating to central nervous system
α-mannosidosis	α-mannosidase	glycoproteins	
Type I			Severe infantile form characterized by rapid mental deterioration with enlarged spleen and liver, psychomotor retardation, facial coarsening, corneal clouding, skeletal problems, death before 12 years of age
Type II			Presents in adolescence/adulthood as mental retardation and hearing loss accompanying psychomotor retardation, facial coarsening, corneal clouding, skeletal problems. Milder than type I, slowly progressive
β-mannosidosis	β-mannosidase	glycoproteins	Rare. Ranges for mild symptoms including mild facial dysmorphism and mild bone abnormalities to severe form with quadraplegia and infant death
Fucosidosis	α-L-fucosidase	glycoproteins glycolipids	Heterogeneous symptoms
Type 1			Severe, general deterioration in neurological functions in early childhood leading to a vegetative state and death
Type 2			Chronic and less severe, usually beginning with mild neurological and skin abnormalities

(continued)

Table 16.7 (*continued*)

Disorder	Deficiency in	Glycoconjugates affected	Clinical symptoms
Aspartylglucosaminuria (AGU)	1-aspartamido-β-N-acetyl-glucosamine-amidohydrolase or aspartylglucosaminidase	glycoproteins	Initial arrested speech development and slight muscular hypotonia from age 2–4 years, followed by gradual regression of motor and intellectual skills. Mental retardation is profound by adulthood. Adults usually have a gargoyle-like appearance
Schindler disease	α-N-acetyl-galactosaminidase	glycoproteins	
Type I			Brain atrophy leading to rapid and profound regression in development and psychomotor skills towards the end of the first year of life. Blindness
Type II			Mild intellectual impairment
Sandhoff disease (GM_2 gangliosidosis variant O)	β-hexosaminidases A and B	glycolipids	Infantile form has symptoms similar to Tay–Sachs disease; juvenile form characterized by speech difficulties, cerebellar ataxia and progressive mental retardation. Adult cases without noticeable symptoms exist
Gaucher's disease	β-glucosceramidase	glycolipids	Accumulation of Glcβ1,1Cer in macrophages and other cells
Type I, adult			Lymphadenopathy, enlarged liver and spleen, erosion and fracture of long bones, anaemia, bleeding disorders
Type II, infantile/cerebral			Central nervous system disorders, severe mental dysfunction, early death
Type III			Juvenile onset with involvement of liver and spleen. Later neurological changes leading to death
Fabry's disease	α-galactosidase	glycolipids	Severe pain, corneal opacity, death from cerebrovascular or renal disease
Krabbe's disease	β-galactoceramidase	glycolipids	Early onset; progression to severe motor and mental retardation
Metachromatic leukodystrophy	arylsulfatase A	glycolipids	Mental regression, peripheral neuropathy, seizures, dementia. Infantile, juvenile or adult onset
Hurler's disease (MPS) IH[a]	α-L-iduronidase	GAGs	Onset at age 6–8 months, lethal by 10 years. Enlargement of liver and spleen, stunted growth, mucopolysaccharidosis skeletal abnormalities, visual impairment, mental retardation, vascular problems, hearing loss and brain oedema
Hunter's disease mucopolysaccharidosis (MPS) II	iduronate-2-sulfatase	GAGs	Mild form: normal intelligence, short stature, life expectancy 20–60 years. Severe form: organomegaly mental retardation, death by 15 years
Sanfilippo's disease			Mental retardation, skeletal deformities, distorted facial appearance, enlargement of spleen and liver

Table 16.7 *(continued)*

Disorder	Deficiency in	Glycoconjugates affected	Clinical symptoms
Type A/MPS IIIA	heparan-N-sulfatase	GAGs	
Type B/MPS IIIB	N-acetyl-α-glucosaminidase	GAGs	
Type C/MPS IIIC	acetyl CoA: α-glucosaminide acetyltransferase	GAGs	
Type D/MPS IIID	N-acetylglucosamine-6-sulfatase	GAGs	
Morquio type A disease/MPS IVA	N-acetylhexosamine-6-sulfatase	GAGs	Clouding of cornea, distorted facial appearance, enlargement of spleen and liver mental retardation, skeletal deformity. Milder forms exist
Morquio type B disease/MPS IVB	β-galactosidase	GAGs	Symptoms similar to Morquio type A disease
Maroteaux-Lamy disease/MPS VI	N-acetylgalactosamine 4-sulfatase	GAGs	Corneal clouding, survival to teens. Milder forms exist
Sly's disease/MPS VII	β-glucuronidase	GAGs	Wide spectrum of severity and symptoms
Scheie's disease	α-L-iduronidase		Less severe than Hurler's disease. Onset after 5 years of age. No mental retardation, nearly normal height and life expectancy
Glycogenosis II or Pompe's disease	α-glucosidase	glycogen	Heterogeneous presentation, often neurological and cardiac symptoms. Accumulation of glycogen in swollen vesicles causing mechanical damage in all tissues. Sufferers usually die in early childhood
I-Cell disease (inclusion cell disease) or mucolipidosis II	N-acetylglucosamine 1-phosphotranferase		Severe, presenting as psychomotor retardation, radiological abnormalities. Coarse facial features, gingival hyperplasia, and death within the first year of life. Non-assembly of the carbohydrate-dense, membrane-bound 'inclusion bodies' in the cytoplasm of, principally mesenchymal cells
Pseudo-Hurler polydystrophy or mucolipidosis III			Clinically less severe form of I-cell disease
Salla disease	Failure in the normal transport cellular sialic acid mechanisms		Ataxia in infancy and moderate psychomotor retardation
Infant sialic acid storage disease (ISSD)	Failure in the normal cellular sialic acid transport mechanisms		More severe form of Salla disease; newborns with ascites, enlarged liver and spleen, severe mental and motor retardation, coarse facial features

A number of human and animal lysosomal storage disorders, called glycoproteinoses, mucopolysaccharidoses and glycolipidoses result from genetic deficiency in one of the lysosomal proteins, a glycosidase, cofactor or membrane carrier molecule, necessary for catabolism of glycoconjugates. Hydrolysis of N-glycans is normally achieved through the sequential enzymatic removal of monosaccharides from the glycan chain, starting at the terminal end and moving into the glycan chain. Failure in this mechanism, at any point in the process, results in a blockage of the degradative pathway and resultant accumulation in the lysosome of partially degraded oligosaccharide fragments. Cytolysis of affected cells results in elevated urinary excretion of these products and clinical symptoms, which vary depending on the disorder concerned, and may be severe, sometimes fatal. Many of these disorders have similar symptoms, most also present clinically with a range of severities from mild to extremely severe. Commonly, infantile forms are the most severe with later onset heralding a milder clinical picture. Some examples of these disorders are listed in *Table 16.7*.

Further reading

Axford, J. (1997) Glycobiology and medicine: an introduction. *J. R. Soc. Med.* **90:** 260–264.

Becker, D.J., Lowe, J.B. (1999) Leukocyte adhesion deficiency type II. *Biochem. Biophys. Acta – Mol. Basis Dis.* **1455:** 193–204.

Brockhausen, I. (1999) Pathways of O-glycan biosynthesis in cancer cells. *Biochim. Biophys. Acta* **1473:** 67–95.

Brockhausen, I., Schutzbach, J., Kuhns, W. (1998) Glycoproteins and their relationship to human disease. *Acta Anatomica* **161:** 36–78.

Brooks, S.A. (2000) The involvement of *Helix pomatia* lectin (HPA) binding N-acetylgalactosamine glycans in cancer progression. *Histol. Histopathol.* **15:** 143–158.

Dall'Olio, F. (1996) Protein glycosylation in cancer biology: an overview. *J. Clin. Pathol.* **49:** M126–M135.

Ilver, D., Arnqvist, A., Ogren, J., Frick, I.M., Kersulyte, D., Incecik, E.T., Berg, D.E., Covacci, A., Engstrand, L., Boren, T. (1998) *Helicobacter pylori* adhesin binding fucosylated histo-blood group antigens revealed by retagging. *Science* **279:** 373–377.

McEver, R. (1997) Selectin–carbohydrate interactions during inflammation and metastasis. *Glycoconjugate J.* **14:** 585–591.

Mengeling, B.J., Turco, S.J. (1998) Microbial glycoconjugates. *Curr. Opin. Struct. Biol.* **8:** 572–577.

Montrieul, J., Vliegenthart, J.F.G., Schachter, H. (1996) Glycoproteins and disease. In A. Neuberger & L.L.M. Van Deenen (Series eds) *New Comprehensive Biochemistry*, Vol. 30. Elsevier, Amsterdam.

Parekh, R.B., Dwek, R.A., Sutton, B.J., Fernandes, D.L., Leung, A., Stanworth, D., Rademacher, T.W., Mizuochi, T., Taniguchi, T., Matsuta, K., Takeuchi, F., Nagano, Y., Miyamoto, T., Kobata, A. (1985) Association of rheumatoid arthritis and primary osteoarthritis with changes in glycosylation pattern of total serum IgG. *Nature* **316:** 452–457.

Carbohydrate-based therapeutics

17

17.1 Introduction

Carbohydrates and glycoconjugates are relevant to many aspects of human health and disease (*see Chapter 16*). Consequently, they are also implicated in many aspects of medical treatment. In this chapter, some key areas, summarized below, are explored.

- Carbohydrates or glycoconjugates may act as drugs, or be relevant to drug efficacy or delivery. A simple example is the use of glucose as a treatment for hypoglycaemia in diabetics. Other examples are glycopeptides, alkaloids and antibiotics. In some cases, glycosylation is necessary for, or enhances, the therapeutic effect. Carbohydrates may also be used as agents in drug delivery, targeting or imaging.
- Saccharide–receptor interactions are implicated in many disease processes, including infections, inflammatory responses and cancer metastasis. Approaches designed to disrupt such interactions are relevance to the treatment of these disorders.
- Rare genetic disorders of carbohydrate synthesis or metabolism (*see 16.8*) may in some cases be treated with carbohydrate moieties or by enzyme replacement therapy designed to restore the deficient pathways.
- A major barrier to organ xenotransplantation is the Gal(α1$\rightarrow$3)Gal epitope on the cells of many potential donor species, an area of intense research interest.
- Alterations in glycosylation in cancer are well documented and suggest that there is potential for the design of anti-cancer therapies and vaccines based on the abnormal glycotypes of these cells.

17.2 Carbohydrate-based drugs and therapeutic compounds

Industrial quantities of pure, complex oligosaccharides for therapeutic use can be synthesized by chemical or enzymatic techniques. As saccharide–receptor interactions are often weak, many naturally occurring saccharides are presented as polyvalent clusters, resulting in a strong overall interaction. Consequently, much effort has gone into the development of appropriately polyvalent carbohydrate arrays in drug design. Examples of carbohydrate-based drugs include heparin-based anticoagulants, polysaccharide vaccines and aminoglycoside antibiotics. Some selected examples are described below.

17.2.1 The anti-coagulant, heparin

Heparin (or heparin sulfate) (*see 8.3.3*) [L-iduronate(α1$\rightarrow$4)GlcNAc2-6disulfate(α1$\rightarrow$4)-]$_n$ is the longest established carbohydrate drug, having been first used as an anti-coagulant in 1935. The activity of the proteoglycan heparin in inhibiting blood-clot formation relies on the presence of a simple pentasaccharide in its structure. The pentasaccharide binds to a lysine-rich region on ATIII, a serine protease, which inhibits other protease clotting factors. Heparin binding accelerates thrombin and factor Xa inactivation and the clotting cascade is arrested.

Functional and Molecular Glycobiology, Susan A. Brooks, Miriam V. Dwek and Udo Schumacher

17.2.2 Azasugars for treatment of diabetics

In diabetes, it is desirable to reduce the flux of glucose absorbed through the gut wall after eating. This may be achieved by inhibitors of glucose hydrolases, such as azasugars like acarbose, which have proved successful in diabetic therapy by slowing conversion of dietary carbohydrates into glucose. Miglitol is a derivative of 1-deoxynojirimycin (*see 15.6.3*) an inhibitor of maltase and sucrase, and is also effective in diabetes treatment.

17.2.3 Mannitol as a diuretic

Mannitol is an osmotic diuretic carbohydrate that is used to reduce cerebral oedema and thus intercranial pressure after head injury, and to promote excretion of barbiturate drugs, for example, following overdose. It is filtered into the primary urine and not re-absorbed during the passage of the primary filtrate through the nephron. It therefore remains as an osmotically active substance within the primary filtrate and thus binds water.

17.2.4 The elasto-viscous properties of hyaluronan and other glycosaminoglycans

Hyaluronan, a naturally occurring polysaccharide that is a constituent of vertebrate glycosaminoglycans (*see 8.3.1*) has remarkable elasto-viscous properties. These have been successfully exploited in surgery in which the elastic, lubricating and cushioning properties of this polymer have proved useful. An example is the use of hyaluronic acid products in eye surgery in which they prevent dehydration and thus tissue damage.

Loss of hyaluronan and other extracellular matrix (ECM) components, such as chondroitin sulfate, keratan sulfate and dermatan sulfate, leads to many chronic conditions such as osteoarthritis and macular degeneration in the eye. Carbohydrate-based substances aimed at replacing these components or stimulating their regeneration are available. They include dietary supplements, such as D-glucosamine and chondroitin sulfate for arthritis relief (which are as yet of unproven value) and injections of hyaluronan directly into joints, which is therapeutically beneficial for synovial fluid replacement and in osteo-arthritis and, although not commonly used in human medicine, is a popular treatment for race horses.

17.3 Therapeutic glycoconjugates

Glycosylation of naturally occurring molecules is often critical to function and activity. Similarly, it is understood that the action of therapeutic compounds may be significantly altered by glycosylation. Properties such as serum half-life, solubility, immunogenicity and binding specificity can be altered dramatically. Enzymatic remodelling of recombinant glycoproteins promises to be a useful tool for refining pharmacological activity (*see Chapter 18*).

17.3.1 Erythropoietin for the treatment of anaemia

A classic example of a recombinant glycoprotein used in human therapy is erythropoietin which is administered to treat anaemia caused by bone marrow suppression in, for example, patients undergoing chemotherapy. It induces proliferation and differentiation in erythroid progenitor cells. Naturally occurring erythropoietin has four sialylated complex N-linked oligosaccharides. ➲ Correct glycosylation of recombinant erythropoietin is essential for its biological activity and incorrectly glycosylated molecules have only ≈10% of the full biological activity. This is because the incorrectly glycosylated molecules are rapidly cleared from circulation by Gal/GlcNAc/Man receptors in hepatocytes and macrophages.

➲ Correct glycosylation of recombinant erythropoietin is essential for activity as the incorrectly glycosylated molecule is rapidly cleared from circulation.

17.3.2 Glycosylation enhances the efficacy of morphine

Synthetically altered glycosylation can affect the selectivity, potency or action of a drug, and may have beneficial or detrimental effects. In the case of morphine, artificial glycosylation has a beneficial effect evidenced by the fact that morphine-6-glucuronide is more effective than the native molecule.

17.3.3 Glycoconjugate antibiotics

Many antibiotics are glycoconjugates. These include members of the aminoglycoside family, such as streptomycin, gentamycin, tobramycin, netilmycin and abekacin which are effective against aerobic Gram-negative bacilli, and also *Staphylococcus aureus* and *Enterococcus* species. Here, again, altering the glycosylation of the molecule may have either beneficial or detrimental effects, and alteration of the glycosylation of antibiotics has proved a useful method for developing new and active compounds in the face of emerging bacterial antibiotic resistance. Changing the glycosylation of vancomycin, for example, has yielded a product which is effective against previously resistant strains of *Enterococci*.➲

➲ Many antibiotics are glycoconjugates; altering their glycosylation may be a way of creating new antibiotics to overcome antibiotic resistance developed by pathogenic bacteria.

Everninomycins are new glycoconjugate antibiotics belonging to the orthosomycin family which are used to treat methicillin-resistant *Staphylococci* and vancomycin-resistant *Enterococci*. Saccharomycin A and B are active against multidrug resistant Gram-positive bacteria.

Some of the glycoside antibiotics are also used as anti-neoplastic agents in cancer therapy (*see 17.10*). Daunorubicin and doxorubicin contain tetracycline ring structures covalently linked to an unusual amino sugar, daunosamine, which is attached to the tetracycline ring via an O-glycosidic bond. The antibiotics are similar in composition and exert their action by interacting with the DNA. Daunosamine plays an essential role in DNA binding. Daunorubicin is mainly effective in haematological neoplasias. Doxorubicin is also effective in acute leukaemias and lymphomas, but is additionally active in a number of solid neoplasms (e.g. breast cancer, sarcomas, lung cancer, neuroblastomas).

Bleomycin and its derivatives are also antibiotics that are used as anti-cancer agents.➲ Bleomycins are water-soluble glycopeptides. Different bleomycins carry alternative terminal amine moieties. They block the cell cycle and are effective against a number of tumours which are otherwise difficult to treat, for example, squamous cell carcinoma of the lung, and head and neck cancers. They are toxic to the lungs as they cause pulmonary fibrosis as an unwanted side-effect. They are probably most effective in combination chemotherapy.

➲ Some antibiotics act as anti-proliferative agents and are thus also used an anti-cancer drugs.

Mithramycin is another antibiotic which is used in cancer therapy. However, its severe side-effects limit its use.

17.3.4 Glycoconjugates for treatment of neurological disorders

Topiramate is a derivative of D-fructose which is efficacious in the treatment of epilepsy when used in conjunction with traditional anti-convulsant drugs. It increases the inhibitory effect of gamma aminobutyrate (GABA) on neurons and inhibits the excitatory effect of the α-amino-3-hydroxy-methylisoxazole-4-propionic acid (AMPA) type of glutamate receptor.

A drug formulation called Sygen™ is based on the ganglioside GM1 and is used to treat acute trauma, such as stroke and spinal cord injury, and chronic neurological diseases, for example, Parkinson's disease, however, its efficacy is not generally accepted.

17.4 Carbohydrates in imaging studies

Radiolabelled 2-deoxy-2-fluoro-D-glucose is used as an imaging agent for positron emission tomography (PET scan), and is particularly useful in imaging the central nervous system to enable the diagnosis of sites of injury and neurological disorders, including those caused by substance abuse, and also for research into brain function. It is taken up by the neurons in place of glucose, but is not metabolized and thus accumulates in regions of the brain where there are high levels of neuronal activity.

There is interest in the concept of using monoclonal antibodies directed against oligosaccharides of cancers to image primary tumours and their metastases. Potential target glycans include the Tn epitope (*see 5.6.1, 10.9.2, 16.6.6*) and the MUC1 mucin (*see 5.10.1, 16.6.10.1*).

17.5 Carbohydrates in drug delivery

17.5.1 Polysaccharides as binders, fillers and coatings

Polysaccharides such as starch (*see 3.4.1*) and cellulose (*see 3.2.1*) and their derivatives have useful properties as binders, fillers, coatings and slow-release agents for orally administered drug preparations and have long been used by the pharmaceutical industry.

17.5.2 New generation drug delivery compounds

Other saccharides such as cyclodextrins and their derivatives also show great promise as new drug delivery compounds, particularly for drugs with poor solubility, as they have a hydrophilic outer surface, enhancing solubility, with a hydrophobic inner core appropriate for complexing with the drug itself.

17.5.3 Bioadhesive delivery agents

The polysaccharide chitosan consists of (1→4)-linked 2-acetomid-2-deoxy-β-D-glucopyranose and 2-amino-β-D-glucopyranose units. It is not only a mucoadhesive polymer, but also acts as an absorption enhancer across mucosal epithelia; its N-deacetylated derivatives have also been employed in this way.

Another polysaccharide used as a bioadhesive vesicle is hyaluronate (*see 8.3.1*). By chemical modification of hyaluronate, polymers are available in various forms (microspheres, sponges, films, etc.). At present, studies are underway to demonstrate the clinical effectiveness of these polymers in the delivery of drugs, for example, the vaginal delivery of calcitonin in postmenopausal women as a means of preventing osteoporosis. The use of hyaluronic acid as a basis for a drug delivery system in the eye is in the development stage.

17.5.4 Transport across epithelial cell membranes

Other very common problems in drug delivery include the fact that many drugs are macromolecules, that they may be poorly lipid soluble, or are easily degraded. These aspects present problems in transporting them across biological barriers such as cell membranes. Gold–carbohydrate complexes have been used in the treatment of arthritis and appear to have advantages over other gold formulations, possibly through an enhanced ability to destabilize cell membranes, as a consequence of their amphiphilic nature.

17.5.5 Use of lectins for liposome targeting and drug delivery

Several common problems are encountered in attempts to successfully administer drugs by conventional oral or nasal routes. One of the most basic is that time available for drug absorption may be too short, and the drug may thus be degraded or mechanically swept away before being absorbed.

Bioadhesive drug delivery, in which the drug is coupled to a carrier molecule that adheres to the biomembrane of the target cell may offer an approach that enables improved drug delivery. Lectins (*see Chapter 13*) may be useful in this respect in that they could be used to attach specifically to gut or nasal mucosa cells through recognition of glycoconjugates present at the cell surface. This adhesion could allow the drug time to cross the plasma membrane successfully.

➲ Lectins may be useful for drug delivery in a number of ways:
(i) as bioadhesives,
(ii) in proteolysosomes,
(iii) to target bacteria cell surface glycans.

One successful method of improving drug delivery is to incorporate the drug into a liposome. Liposomes comprise a lipid bilayer (unilamellar liposomes), or several concentric lipid bilayers (multilamellar liposomes) encapsulating an aqueous space. They can be used for the delivery of both hydrophobic and hydrophilic drugs.

Lectin-directed drug delivery may also be allied to the use of liposomes.➲ In this case, the lectin is conjugated directly to the surface of the liposome to produce a proteo-liposome. Once again, this exploits the specific binding property of lectins to saccharides of the target cells, such as gut or nasal mucosa. Lectin–saccharide interaction facilitates the attachment of the liposome to the plasma membrane. Experimentally, lectins have been used successfully to target liposomes to a number of different cell types, such as chicken erythrocytes, mouse spleen cells, HeLa cells, mouse fibroblasts and bacterial cells.

Another application of lectins is in directing drugs to the surface of bacteria responsible for human disease. The cell walls of Gram-positive bacteria are rich in peptidoglycans and those of Gram-negative bacteria are rich in lipopolysaccharides. These cell surface glycoconjugates are potential binding sites for lectins, which may therefore offer a means of targeted drug delivery.

17.6 Genetic disorders of carbohydrate synthesis or metabolism

Rare genetic mutations result in abnormalities in glycosylation or carbohydrate metabolism and are the cause of severe symptoms (*see 16.8, 16.9*).➲

➲ Genetic disorders resulting from deficiencies in carbohydrate-processing pathways may sometimes be treated by simple administration of the missing carbohydrate; in lysosomal storage disorders, enzyme replacement therapy may be successful.

17.6.1 Treatment with carbohydrates

Some of these disorders are simply and effectively treated by administration of carbohydrate. For example, oral mannose is effective in the treatment of carbohydrate-deficient glycoprotein (CDG) syndrome type 1b, in which phosphomannose isomerase is inactive, and treatment with mannose restores normal glycan biosynthesis (*see 16.8.1.1*).

Similarly, in leukocyte adhesion deficiency II (LAD-II) syndrome, in which patients are unable to synthesize blood group H or Lewis blood group antigens because of a failure to synthesize GDP-Fuc, L-fucose may have a role in treatment. LAD-II neutrophils lack ligands for E-selectin, L-selectin and P-selectin and are unable to traffic normally, and consequently patients have a defective immune response and suffer repeated infections (*see 16.9.1.4*).

17.6.2 Enzyme replacement therapy

Lysosomal storage disorders result from genetic deficiency in DNA coding for one of the lysosomal proteins, usually a hydrolase (e.g. a glycosidase), or sometimes a cofactor or membrane carrier molecule, necessary for catabolism of glycoconjugates. This results in blockage of the degradative pathway and accumulation in the lysosome of partially degraded oligosaccharide fragments, plus elevated urinary excretion and variable, sometimes very severe, clinical symptoms (*see 16.9*). Theoretically, it should be possible to provide therapy for some lysosomal storage disorders by introducing the missing lysosomal protein. Targeting of such proteins to the lysosome is achieved by incorporation of a Man-6-P group into the N-linked glycans, which is recognized by a lysosome Man-6-P receptor.

This approach has proved effective in Gaucher's disease (*see 16.9*), or β-glucosceramidase deficiency, which results in accumulation of Glc(β1→1)Cer in macrophages and other cells.

Enzyme purified from human placenta can be treated with glycosidases to reveal terminal mannose residues for which reticuloendothelial cells have receptors; this 'remodelled' enzyme is then targeted to the affected macrophages. Alternatively, recombinant enzyme-bearing N-glycans with terminal Man can be produced in a baculovirus expression system and employed in the same manner. An alternative approach to treating Gaucher's disease is the use of drugs to inhibit glycosyltransferase activity and thereby modulate the build up of glycolipids (*see 17.6.3*).

Another promising avenue of research is aimed at the treatment of mucopolysaccharidosis I (MPSI) or Hurler's disease, a disorder arising from lack of α-L-iduronidase. The disorder usually becomes apparent by age 6–8 months, and is fatal by 10 years. Polysaccharides build up in various organs resulting in an enlargement of the liver and spleen, stunted growth, skeletal abnormalities, visual impairment, brain oedema resulting in mental retardation, and hearing loss. A dog model of gene therapy for this condition showed impressive results and the first human patients have been treated. The company involved in these developments has identified 35 other enzymes, associated with other genetic carbohydrate synthesis, storage or metabolism disorders, that may be amenable to this type of intervention.

17.6.3 Substrate deprivation by glycosylation inhibitors in lysosomal storage diseases

The imino sugar N-butyldeoxynojirimycin (NB-DNJ) (*see 15.6.3*) inhibits the first step in glycolipid biosynthesis. It has been used in lysosomal storage diseases to attempt to inhibit the rate of synthesis of glycolipid to levels at which the residual activity of the mutant catabolic enzyme is sufficient to prevent pathological storage. This approach has given promising results in mouse models of Tay Sachs disease and Sandhoff disease. Clinical trials of this approach in patients with Gaucher's disease have shown beneficial effects, including significant decreases in liver and spleen volumes, gradual improvement in haematological parameters and other disease activity markers, diminished glycolipid biosynthesis and storage (*see 17.6.2*).

17.7 Anti-inflammatory therapies

Inflammatory disease results from various conditions including arthritis and tissue damage following cardiac ischaemia, stroke and other types of reperfusion injury. They can progress to self sustaining autoimmune disease.

17.7.1 Inhibition of selectin–ligand interaction

The interaction of selectins (saccharide-binding proteins expressed at the cell surface of endothelial cells), with saccharide ligands present on the surface of leukocytes is a crucial event in leukocyte adhesion to endothelial cells during their recruitment to sites of inflammation (*see 13.12.3.2*). Thus, inhibition of this interaction, possibly through introduction of competing inhibitory saccharides, has the potential to block these interactions and thus reduce leukocyte recruitment where it is undesirable.

An example of such a scenario is in reperfusion injury. sLe^x is recognized as a ligand by all three selectins, E-, P- and L-selectin, involved in leukocyte recruitment and the inflammatory response, and thus appears to be a promising inhibitor of selectin-mediated adhesion. sLe^x has been successfully produced using chemi-enzymatic synthesis (*see 18.4.10*). Derivatives of sLe^x have been tested as potential inhibitors of the inflammatory response, and have given promising results in animal models of both acute and chronic inflammation, even at surprisingly low concentrations. One of the best known is Cylexin™, but unfortunately it has given disappointing results in clinical trials and its use has therefore been discontinued.

17.7.2 Glyco-mimetics of sLe^x

Glyco-mimetics are small molecules based on the minimum functional domain of a saccharide structure. Glyco-mimetics of sLe^x have, potentially, improved stability or serum half-life over the naturally occurring glycan and are actively being sought. A glyco-mimetic of sLe^x has been developed which has higher affinity for E-selectin than the naturally occurring oligosaccharide. Glyco-mimetics of other saccharides are under development and may prove useful in the therapy of a range of different diseases.

17.7.3 Downregulation of glycosyltransferases

An alternative approach to treating chronic inflammatory disorders may be to downregulate production of the glycosyltransferase enzymes involved in synthesis of sLe^x, such as the Fuc-TVII glycosyltransferase. This, at present, is a theoretical scenario, but may become feasible in the future.

17.7.4 Mannose-6-phosphate

Mannose-6-phosphate shows promise as an anti-inflammatory agent for the treatment of arthritis, psoriasis, diabetes and multiple sclerosis. It is believed to inhibit degradative lysosomal enzymes (e.g. heparanase) which enable the passage of leukocytes through vascular basement membranes and thus are implicated in the exacerbation of inflammatory conditions. These enzymes may adhere to the basement membrane through the mannose-6-phosphate receptor and thus be displaced by soluble mannose-6-phosphate.

17.7.5 (β1→4)-D-Mannan as an anti-inflammatory agent

A (β1→4)-D-Mannan isolated from the plant *Aloe vera* has immunostimulatory properties, possibly through activation of the macrophage mannan receptor, and is used to encourage wound healing and treat inflammatory conditions such as ulcerative colitis and inflammatory bowel disease.

17.8 Transplant rejection mediated by carbohydrates

17.8.1 The ABO blood group system

Carbohydrates are important determinants in cell/tissue rejection. Organ transplantation between individuals from incompatible ABO blood groups (*see 10.5*), for example, results in a strong tissue rejection response. However, recent research shows that ABO incompatibility does not represent an insurmountable barrier to organ transplantation. In two-thirds of patients that receive organ transplants from ABO-incompatible donors, a hyperacute rejection reaction occurs within a few hours, whereas in the remaining one-third of patients, for unknown reasons, long-term graft survival is reported.➲ Furthermore, the results of both clinical and experimental studies suggest that if titres of naturally occurring anti-A or anti-B antibodies can be depleted (this can be achieved by passing plasma or blood over an affinity column bearing the appropriate blood group antigens) for a relatively short time (days rather than weeks), then ABO-incompatible organs will survive, even when antibody levels return to normal. This process is called anergy, adaptation or accommodation. The underlying mechanism is unknown, but the masking of potentially antigenic oligosaccharides has been suggested. In experimental systems, short-term (again, days rather than weeks) continuous infusion of the appropriate A or B trisaccharide binds and 'neutralizes' circulating anti-A or anti-B antibodies preventing them from attacking the graft.

➲ If hyperacute rejection in ABO-incompatible transplants can be temporarily delayed, the process of 'accommodation' takes place and the graft will survive.

17.8.2 The Gal(α1→3)Gal epitope

The epitope Gal(α1→3)Gal is of interest because it is not synthesized by humans, apes and Old World monkeys, but is a common glycotope in many other species, and therefore represents a significant barrier to xenotransplantation (*see 7.13*).

➲ Xenotransplantation is defined as the transplantation of tissues between different species. It could provide a solution to the shortage of human organs available for transplant.

➲ A concordant xenograft is one between species that is rejected in a time frame similar to that of a recipient of the same species – for example, chimpanzee to human. A discordant xenograft is one between different species that undergoes immediate (within minutes) hyperacute rejection – for example, pig to human.

Xenotransplantation, the transplantation of organs from other species, such as pigs, into human recipients, is a topic of interest.➲ There is a massive shortage of human donors, it has been estimated that the current demand for donor organs is at least 50 000–60 000 organs per year worldwide, and in many cultures, for example, Japan, there are cultural and/or religious barriers to the transplantation of organs from human cadavers. In the USA alone, around 70% of patients requiring transplanted organs never receive them because of lack of availability. The idea of transplanting animal organs, such as heart, liver and kidneys, into human recipients is therefore an attractive one; although clearly there are ethical implications.➲ Furthermore, one very significant barrier to xenotransplantation is the Gal(α1→3)Gal epitope on glycoproteins and glycolipids of potential animal donors. Xenotransplantation from Old World primate species more closely related to man and lacking the Gal(α1→3)Gal epitope has been tried, with poor results, as outlined in *Table 17.1*. This approach is probably not viable for many other reasons, including the limited number of potential animal donors. Pigs appear to have a number of attractive features as potential donor animals, as outlined in *Table 17.2*.

Humans have high titres, as much as 1% of all circulating immunoglobulins, of naturally occurring antibodies to Gal(α1→3)Gal, which lead to immediate and powerful hyperacute rejection of xenotransplanted organs. In the few cases in which xenotransplantation of pig organs into humans has been attempted, the results have been very poor, as shown in *Table 17.1*. Median survival of experimental xenotransplants into non-human primates is less than a month, and the longest surviving transplant is less than 100 days.

Table 17.1 Examples of poor survival of xenotransplants into human recipients

Year	Donor animal	Organ	No. of cases	Graft/recipient survival
(a) Primate xenotransplants				
1964	Chimpanzee	Kidney	13	1 day to <9 months
1964	Monkey	Kidney	1	10 days
1964	Baboon	Kidney	7	4/5 days to <2 months
1965	Chimpanzee	Kidney	3	2 months
1969	Chimpanzee	Kidney	2	1 month and 9 months
1977	Chimpanzee	Heart	1	4 days
1977	Chimpanzee	Heart	1	4 days
1984	Baboon	Heart	1	20 days
1993	Baboon	Liver	2	<70 days
(b) Pig xenotransplants				
1968	Pig	Heart	1	<5 minutes
1995	Pig	Liver	1	1 day
1992	Pig	Heart	1	<1 day
1996	Pig	Heart	1	<1 day

Hyperacute rejection cannot be controlled by the immunosuppressive drugs used to maintain transplanted human organs.➲

➲ Hyperacute rejection is characterized by microvascular thrombosis and interstitial haemorrhage leading to immediate graft rejection without graft function. It is associated with the presence of preformed antibodies and complement activation.

17.8.2.1 Modification of donor tissues

Eventually, development of genetically modified donor animals may circumvent this problem. The obvious approach is the development of transgenic animals lacking the crucial α1,3 galactosyltransferase (α1,3 Gal-T) activity through expression of null mutations at the α1,3 Gal-T locus. To render the α1,3 Gal-T enzyme non-functional, the gene would need to be deleted, interrupted or replaced. This could be achieved either by interfering with the coding region or with the regulatory sequences, so that active enzyme is not synthesized. Potentially, this could be achieved by homologous recombination of the gene within the embryonic stem cells to eliminate expression, or by introduction of anti-sense cDNA constructs into an embryo to inactivate the gene, or by 'knocking-out' the α1,3 Gal-T gene in a pig cell, followed by nuclear transfer into an egg and cloning the pig carrying the disrupted gene.➲ Until such an animal is produced, we will not know whether lack of function of the α1,3 Gal-T gene will have any deleterious effects. A knockout mouse model with inactive α1,3 Gal-T gene has been developed, and these animals do not synthesize the Gal(α1→3)Gal epitope.➲ The mice are essentially normal and healthy with the exception that they tend to develop cataracts in the eyes. However, pig tissues synthesize much larger amounts of the Gal(α1→3)Gal sugar than mice, and it is possible that this disaccharide sequence may have an important, though as yet unidentified, role in these animals, possibly in embryogenesis or development. 'Knocking-out' the α1,3 Gal-T gene in selected organs only might be an alternative approach.

➲ The longest surviving xenotransplants into humans have occurred when closely related species such as chimpanzees have been the donor.

➲ The Gal(α1→3)Gal epitope is often referred to in the literature as the α-Gal epitope, the α-galactosyl epitope or sometimes the 'Galili antigen' (after the author of the first report of the antigen).

There is the possibility that there might exist an as yet undiscovered, rare, naturally Gal(α1→3)Gal-negative pig, analogous to the rare human 'Bombay' ABO blood type *(see 10.5.8)*, the incidence of which in human populations is estimated to be something like 1 in 10 000 individuals. In view of the very close relationship between the α1,3 Gal-T gene and the AB blood group transferase gene, this seems to be feasible. Finding and breeding such an animal might offer an alternative to producing genetically modified Gal(α1→3)Gal-negative animals.

Table 17.2 Advantages of pigs as organ donors in comparison with non-human primates

	Pigs	Non-human primates
Period to reproductive maturity	≈6 months	years
Length of pregnancy	112–116 days	longer
Number of offspring	5–12	usually 1–2
Cost of maintenance	low	very high
Ease of maintenance	easy	difficult
Knowledge of tissue typing	great	minimal
Blood group compatibility	unimportant	problematic
Experience of genetic engineering	great	none
Availability of pathogen-free animals	yes	no
Ethical considerations	probably fewer	great
Risk of transmitting viruses and other harmful pathogens	lower	high

Another possibility is to produce transgenic animals that express high levels of activity of the blood group H α1,2 fucosyltransferase; the idea being that type 2 chain precursors would be preferentially modified by H blood group sugars and thus unavailable for modification by addition of Gal(α1→3)Gal. Transgenic mice and pigs of this type have been developed with ≈70% reduced Gal(α1→3)Gal synthesis. The pigs show increased tolerance to human serum, but it seems unlikely that such partial suppression of the Gal(α1→3)Gal epitope would be sufficient to completely abrogate rejection *in vivo*. Competition with other glycosyltransferase enzymes may also be feasible, for example, if α2,3 sialyltransferase is expressed early in the Golgi apparatus it will attach a sialic acid residue to the 3 position of the terminal Gal residues on complex N-glycans thus blocking (α1→3)-galactosylation. This approach has been shown to reduce Gal(α1→3)Gal synthesis in cultured cells.

The possibility of removing Gal(α1→3)Gal from the vasculature of the pig organ by α-galactosidase prior to xenotransplantation has also been considered. However, effective permanent removal of the antigenic glycans using this type of approach is generally thought to be impractical as the glycotope would be re-synthesized by the endothelial cells. Furthermore, following transplantation, it is likely that the cleaved α-Gal would be replaced by the pig donor organ α1,3 Gal-T. A more promising approach may be to deplete the α-Gal by transfecting a gene encoding an α-galactosidase into the vascular endothelium of the donor organ, or alternatively to transfer the H gene, as discussed above, or a competing human α1,4 Gal-T gene. Methodology for transfecting endothelial cells of the donated organ prior to transplantation are under development.

17.8.2.2 Suppression of immune reaction

Human antibodies against the Gal(α1→3)Gal structures can be neutralized by α-Gal-oligosaccharides in a similar way to that described for ABO incompatible transplants (*see 17.8.1*). In this approach, methyl α-Gal provides only weak inhibition, the disaccharide Gal(α1→3)Gal is much more effective, and the synthetic trisaccharide, Gal(α1→3)Gal(β1→4)GlcNAc is particularly effective (≈10 times more efficient than the disaccharide), even in very low (millimolar) concentrations.

Experiments have been performed using soluble Gal(α1→3)Gal disaccharide to attempt to competitively inhibit the hyperacute rejection process. When human blood is perfused through pig kidneys in the presence of the disaccharide a significant effect in improving both function of the organ and also the integrity of the tissue structure is observed. However, there remain significant pathological changes, and it is considered that this type of approach alone is insufficient to overcome the problems in patients.

Eventually α-Gal glyco-mimetics may provide specific, non-toxic, effective and immunogen-selective immunosupression for patients with xenotransplants, although current experimental evidence from pig–baboon transplants has proved disappointing. This is a promising avenue of research, however, as evidence from transplants across the ABO blood group barrier (*see 17.8.1*) suggest that short-term suppression could lead to successful long-term accommodation of the graft.

Accommodation might also theoretically be achieved by short-term depletion of serum anti-Gal(α1→3)Gal antibodies. Affinity depletion would require the availability of relatively large quantities of purified Gal(α1→3)Gal disaccharide. This can be purified from pig gastric mucin, which is readily available. Alternatively, there is progress in the field of enzymatic synthesis of oligosaccharide structures, or it may eventually be possible to synthesize this structure in transgenic organisms such as plants (*see 18.5*).

17.8.2.3 Induction of immune tolerance

The human antibody response to the Gal(α1→3)Gal epitope might be abrogated, and potential transplant recipients therefore made tolerant of their grafts, if their maturing lymphocytes were to be exposed to the antigen presented as a 'self' antigen. Maturing B lymphocytes exposed to 'self' antigens undergo programmed cell death and are eliminated. Attempts to do this might involve introducing pig bone marrow cells into graft recipients or by transducing the recipient's bone marrow cells with a viral vector that contains the α1,3 Gal-T gene.

17.8.2.4 Future perspectives in xenotransplantation

If this type of approach were to be successful, it would have many advantages over human organ transplantation: there would potentially be an almost unlimited number of organs available on demand, meaning that waiting times for replacement organs would be shorter, transplantation could be carried out as a routine, scheduled surgical procedure and not as an unplanned emergency procedure, organs would be totally viable and ischaemic time would be minimized, donor organs would hopefully be free of infective pathogens and costs would be less.➲ This is an area of intense research and clinical interest and it seems very likely that the next decade may well see the first successful xenotransplants into human recipients. However, the concern of endogenous pig retroviruses, which may infect the human recipient, is at present also limiting the prospects of this approach.

> ➲ It seems likely that the next decade may see the first successful pig to human xenotransplants.

17.9 Inhibitors of microbial, viral and parasitic infection

Many bacteria and viruses employ saccharide–receptor interactions to mediate cell attachment, colonization or invasion. They produce lectin-like molecules, termed adhesins, which recognize oligosaccharide structures present on the cell surface of the organs they invade. Infection by many parasites, too, are mediated through lectin–saccharide interactions.

The concept of blocking bacterial attachment by competitive inhibition using soluble oligosaccharides of identical or similar structure to the ligands recognized by the adhesins is an attractive one.➲ It has, however, proved to be difficult to isolate large quantities of the glycoconjugates of interest. A promising technique is the cloning of glycosyltransferases into bacterial cells to facilitate synthesis of very large quantities of relevant oligosaccharides for testing as anti-adhesive pharmaceutical agents (*see 18.5.4*).

Because increasing the valency of an oligosaccharide usually enhances its efficacy in this situation, there has been much interest in the development of oligodendromers. One, called 'starburst', for example, is based on polyamine dendrimers which are monodispersed glycomacromolecules in which oligosaccharides decorate a multi-layered, branched 'starburst'. The number of branches, which double with each successive layer, can be precisely controlled, as can the carbohydrate groups contained within the dendrimer. Potentially, such macromolecules could be custom-designed to precisely fit the topology of the adhesins on the bacterial cell surface. Other similar approaches include the generation of neoglycopolymers with defined periodic carbohydrate substitution, and multivalent galabiosides (dimers of galabiose with an aromatic aglycone linker, or trimers/tetramers coupled by tricarboxylic or tetracarboxylic acid).

> ➲ Specifically inhibiting bacterial adhesion through competitive inhibition by synthetic glycans is an attractive idea.

Some specific examples of human infections in which this and other carbohydrate-based approaches look promising are detailed below.

17.9.1 Influenza

Human influenza virus A and B produce Sia(α2→6)Gal-binding haemagglutinins which enable them to bind to epithelia of the human respiratory tract (*see 11.8.4, 16.2.2*). Multivalent sialylated glycoconjugates, including polymers, liposomes and neoglycoproteins have been developed which may be able to block viral attachment to epithelia. One approach, which may be applicable to other similar infection processes, is the production of libraries of polymers decorated with sialic acid residues in addition to other diverse, largely hydrophobic, non-carbohydrate structures. These were screened for efficacy in blocking influenza virus binding, and several were shown to be extremely effective, presumably because of the fortuitous arrangement of sialic acids and the non-carbohydrate components in relation to the binding site of the virus.

Sialidase inhibitors are therapeutic tools against influenza virus infection as they prevent viral replication and reduce the severity and duration of the infection. ➲ The sialic acid analogue Relenza™ (produced by Glaxo SmithKline), which is administered as an inhalant, is recognized by the influenza sialidase and through binding, inactivates it. This drug has recently become available for human influenza therapy (*see 16.2.2*).

➲ The drug Relenza™ blocks influenza virus sialidase.

17.9.2 *Helicobacter pylori* infection

Helicobacter pylori infection is associated with chronic gastritis and gastric ulcer. Attachment is mediated through the Lewis b (Le^b) antigen on epithelial cells of the gastrointestinal tract recognized by an *H. pylori* bacterial adhesin. Other oligosaccharides may also be recognized. Human milk, which is rich in a variety of free sugars and sialylated and fucosylated glycoconjugates, may play a role in protecting infants from infection. The idea of developing carbohydrate-based prophylactics for preventing and managing *H. pylori* infection remains a possibility. However, clinical trials have failed to produce convincing results of benefit.

In the past, it has been difficult to study *H. pylori* infection, as it is solely a human pathogen and no animal models existed. ➲ Recently, transgenic mice have been produced which express the human Lewis α1,3/4 Fuc-T on their gastric epithelium (*see 15.10*). *H. pylori* binds in a Le^b-dependent manner to the mucosa of these animals. The number of *H. pylori* organisms isolated from the stomach of normal and transgenic mice was equivalent. In the transgenic animals, *H. pylori* were found both in the mucus and in the gastric pit associated with the surface of the mucosal cells, however, in normal mice *H. pylori* were only associated with the mucus. Cellular attachment by the bacteria resulted in the activation of an immune response including the generation of antibodies directed against *H. pylori* cell-surface Le^x structures. The antibodies were also reactive with parietal cells, and caused parietal cell damage, loss of differentiation and cell death. Lymphoid aggregates, typical of long-term human infection were also apparent. It is hoped that the utilization of such an animal model will contribute to our understanding of this widespread infection, and provide a means for development of effective prophylactic and therapeutic interventions.

➲ Development of a transgenic mouse model promises to be helpful in increasing our understanding of *H. pylori* infection in humans.

17.9.3 HIV infection

The human immunodeficiency virus-1 (HIV-1) recognizes CD4 on most cells. In addition, it binds to a saccharide ligand on some CD4-negative cells, for example, neurons and colonic epithelium. In an experimental system, analogues of this saccharide structure, GalCer, are capable of blocking HIV infection. This may be a promising area for future research efforts in the development of new approaches to prevent or control HIV infection.

17.9.4 Imino sugar derivatives as antiviral agents

The imino sugars N-butyldeoxynojirimycin (NB-DNJ) and N-nonyl-deoxynojirimycin (DNJ) (*see 15.6.3*) are inhibitors of the α-glucosidases involved in N-linked oligosaccharide trimming, causing some proteins to be misfolded and retained within the endoplasmic reticulum. These agents have an antiviral effect against several viruses including bovine viral diarrhoea virus and hepatitis B virus. There is controversy as to whether the antiviral effect is due to disruption in glycosylation. For example, hepatitis B virus (HBV) contains three N-glycosylated surface proteins that are essential for the formation of infectious virus. NB-DNJ has been shown to have antiviral activity in a woodchuck model of chronic HBV infection. However, this effect appears to be attributable to a mechanism other than an impact on glycoprocessing.

17.9.5 Bacterial toxins

The attachment of bacterial toxins, through saccharide recognition, to gastrointestinal epithelium underlies the severe symptoms of diseases such as cholera. Carbohydrate compounds are under development to remove secreted toxins and are designed to be used in conjunction with antibiotic treatment. Bacterial death after treatment with antibiotics normally results in a significant release of harmful bacterial toxins, which can be made harmless by their attachment to the carbohydrate drug. The idea has been applied to the treatment of haemolytic uraemic syndrome (a condition characterized by microangiopathic haemolytic anaemia, thrombocytopenia and acute renal failure, caused by platelet thrombi which mainly form in the microcirculation of the kidney) using a product consisting of thousands of synthetic oligosaccharide structures linked to an inert carrier. Toxins bind oligosaccharides present on the carrier, and the carrier matrix ensures safe passage of the product into the gut, retention in the gastrointestinal tract, and high concentrations of oligosaccharides at the site of infection.

A similar approach has been applied to the treatment of *Clostridium dificile*-associated diarrhoea which occurs when the gut microflora is altered following antibiotic therapy, and the treatment of 'traveller's diarrhoea'. A 'carbohybrid', that is an organic molecular core bearing carbohydrate groups which mimics the three-dimensional structure of a complex oligosaccharide, has been developed. The active carbohydrate groups are chemically linked to a carrier and are thus able to bind saccharide-recognizing receptors of the bacterial toxins and neutralize them. ➲ The product resembles the natural oligosaccharide in conformation, but has improved resistance to enzymatic hydrolysis, thus achieving a longer half-life and improved effect.

> ➲ Synthetic carbohydrate compounds are recognized by bacterial toxins, and thus inactivate them.

17.9.6 Vaccines against bacterial carbohydrates

Vaccines directed against bacterial capsular oligo- and polysaccharides have great potential in stimulating immunity to disease-causing microorganisms. Some have been used successfully for decades. These include vaccines against *Neisseria meningitidis, Salmonella typhi* and *Streptococccus pneumoniae*. With the increasing and worrying emergence of more and more antibiotic-resistant strains of bacteria, this approach may prove of great clinical importance in the future.

Carbohydrate-based vaccines are generally ineffective in very young children and in immunocompromised patients. This can be overcome by conjugating the glycan to an immunogenic polypeptide carrier. Successful applications of this type of approach include development of a vaccine against *Haemophilus influenzae* type b (Hib) and *Streptococci pneumoniae*. A completely synthetic glycan–peptide vaccine has been developed to combat *Shigella* infection (which causes dysentery).

17.9.7 Parasitic infections

The involvement of saccharide–receptor recognition in parasitic infections (*see 16.2.3*) also suggests the possibility of development of specific carbohydrate-based anti-infective or treatment modalities.

17.9.8 Post-surgical infections

A complex carbohydrate product that stimulates the immune system by targeting the β-glucan receptor on neutrophils, but does not stimulate the production of inflammatory cytokines, has been tested in colorectal surgery patients as a novel approach to preventing post-surgical infection, with promising results.

17.9.9 Anti-fungal agents

An anti-fungal therapy, initially designed to target *Aspergillus* by disrupting its saccharide coat and preventing spores developing, is under development.

17.10 Anti-cancer therapies and vaccines

17.10.1 Chemotherapy drugs

Many cancer chemotherapy drugs, such as doxorubicin, etoposide and bleomycin are glycoconjugates. Some are antibiotics (*see 17.3.3*).➲ Manipulating glycosylation has the potential to develop more effective forms of these compounds, for example, this is the case with epirubicin and MEN 10755, both semi-synthetic analogues of doxorubicin, which are effective against tumours resistant to the parent compound. Altering the glycosylation of naturally occurring chemotherapeutic agents may also have effects on other properties, for example, by reducing their toxicity or side-effects. Glycosylated pro-drugs based on cytotoxic alkylating agents such as cyclophosphamide and isophosphoramide are converted into active drugs *in vivo* by either enzymatic or acid hydrolysis of labile glycosidic linkages.

➲ Altering the glycosylation of chemotherapy drugs may enhance their effect.

17.10.2 Anti-cancer vaccines

As cellular glycosylation is disrupted in cancer (*see 16.6*) and cancer cells frequently synthesize oligosaccharide structures not normally present on healthy adult cells, the idea of an anti-cancer vaccine directed against cancer-associated glycotopes is an attractive one. Glycosphingolipids (*see Chapter 9*), truncated glycans such as the Tn epitope (GalNAc-O-Ser/Thr), sialyl Tn (*see 5.6.1, 16.6*), and mucin-associated oligosaccharides, like those of the MUC1 mucin (*see 5.10.1, 16.6.9.1*) are particularly strong candidates for this type of approach. A vaccine composed of sialyl Tn, a saccharide structure absent from normal cells but produced by many cancers (e.g. breast cancers), has been covalently linked to the protein carrier keyhole limpet haemocyanin and given promising results in trials of breast, pancreas and colon cancer, under the name of 'Theratope'. A vaccine called GMK has been developed using the ganglioside GM2, a glycolipid present on the surface of ≈95% of melanoma cells, again linked to the carrier keyhole limpet haemocyanin. It has proved effective in trials of malignant melanoma.

17.10.3 Inhibition of metastasis

The clear parallels between the way in which leukocytes are recruited to sites of inflammation, which involves selectin–saccharide interactions between the leukocyte and the endothelial cells lining the blood vessels, and cancer cell migration during metastasis (*see 16.3* and *16.6*) has led to the concept that saccharides may be capable of inhibiting metastasis. In experimental models, this looks to be a promising approach, with glycosphingolipid-containing liposomes, glycosphingolipid oligosaccharide derivatives or monoclonal antibodies to carbohydrates effectively limiting metastasis in animal models. Alternatively, the concept of ortho-signalling relies on disrupting mitogenic signalling pathways in cancer cells that are regulated by glycosphingolipids and/or their derivatives, including pathways involving receptor protein kinases and protein kinase C. Pectin and hyaluronic acid may also have potential as anti-metastatic agents. Citrus pectin significantly reduces metastasis in animal models, possibly through inhibition of galectin-3, a lectin involved in cell adhesion (*see 13.12.1*). A highly sulfated oligosaccharide called PI-88 is a potent inhibitor of heparanase and appears to be effective in reducing metastasis in animal models.

17.10.4 Other anti-cancer carbohydrate-based drugs

Other potential carbohydrate-based anti-cancer agents include derivatives of swainsonine, which inhibits N-glycan biosynthesis (*see 15.6.5*) and mitolactol, a cytotoxic carbohydrate derivative and potent alkylating agent. Several naturally occurring polysaccharides exert anti-tumour properties, including the branched (β1→3)-D-glucans marketed under the name 'Schizophylan' and 'Lentinan', which are used in cancer treatment. They are believed to interact with the glucose receptors of macrophages, upregulating their phagocytic activity. However, their usage is not generally accepted.

17.10.5 Small molecule carbohydrate-processing inhibitors

The use of small molecule carbohydrate-processing inhibitors to block the synthesis of certain saccharide structures and thereby block their signalling function has promise in many diseases including cancer, inflammatory disorders and infectious diseases. This is an alternative approach to using oligosaccharides or oligosaccharide mimetics as competitive ligands. An example of this is the use of an inhibitor of α-mannosidase II which in animal studies has had impressive effects in inhibiting solid tumour growth rate, stimulating immune response and extending survival. It appears to function, at least in part, through inhibiting tumour attachment to endothelium and also invasion through the ECM.

Further reading

Hakomori, S. (1991) New directions in cancer therapy based on aberrant expression of glycosphingolipids: anti-adhesion and ortho-signaling therapy. *Cancer Cells* **3:** 461–470.

Joziasse, D.H., Oriol, R. (1999) Xenotransplantation: the importance of the Gal alpha 1,3 Gal epitope in hyperacute vascular rejection. *Biochim. Biophys. Acta – Mol. Basis Dis.* **1455:** 403–418.

Kuberan, B., Lindhardt, R.J. (2000) Carbohydrate based vaccines. *Curr. Org. Chem.* **4:** 653–677.

Carbohydrate biotechnology

18

18.1 Introduction

The functional importance of glycosylation in some manufactured glycoconjugates has led to an increased interest in carbohydrate biotechnology. Some of the recent developments in this field are outlined briefly in this chapter. They include the synthesis of oligosaccharides by enzymes from many different types of organisms in aqueous solution, in solid-phase systems and in cell-based expression systems. The directed glycosylation of recombinant glycoproteins, and the application of some of these approaches to the production of carbohydrate-based products for the food, pharmaceutical and related industries are also described.

18.2 Barriers to chemical synthesis of carbohydrate compounds – why biotechnology is so important

The diversity of synthetic and naturally occurring carbohydrate structures means that their automated chemical synthesis is often not possible or practical in the way that it is for proteins and nucleic acids.➲ Even small-scale chemical synthesis in the laboratory, if actually feasible, is frequently technically challenging, time-consuming and often prohibitively expensive. Difficulties in carbohydrate chemical synthesis spring from the potential structural diversity of these compounds (*see Chapter 1*). Furthermore, difficulties arise because carbohydrates contain several hydroxyl groups of similar chemical reactivity: in chemically synthesizing a carbohydrate structure, each hydroxyl must be distinguished from the rest in order to obtain the desired product with the correct stereochemistry and regiochemistry. Saccharide building blocks must be selectively protected, then coupled, and finally deprotected. Synthesis of carbohydrate analogues is also technically difficult as it requires specific and complex chemical modification of the synthetic route in order to introduce a specified chemical change at a particular position.

> ➲ Chemical synthesis of carbohydrates is technically complex and their isolation from natural sources difficult.

Although there have been advances in the chemical synthesis of saccharides for research and pharmaceutical applications in recent years, chemical synthesis of all but the simplest oligosaccharides remains unfeasibly complex and prohibitively expensive. Chemical synthesis of virtually any complex carbohydrate on an industrial scale is currently beyond any reasonable cost–benefit ratio. Isolation of oligosaccharides from natural sources is also technically difficult, time-consuming and expensive. For all of these reasons, biotechnological approaches to the synthesis of saccharides and glycoconjugates appear to potentially provide some of the answers to these problems.

18.3 The current state of carbohydrate biotechnology

The recognition of carbohydrates as a class of molecules of great relevance to medicine, for example, in cancer (*see 16.6, 17.4, 17.10*), inflammation (*see 16.4, 17.7.1*), autoimmune

Functional and Molecular Glycobiology, Susan A. Brooks, Miriam V. Dwek and Udo Schumacher

➲ Biotechnological approaches to the synthesis of saccharides and glycoconjugates are much more complicated than the synthesis of nucleic acids, peptides or proteins.

disease (*see 16.7*), infection (*see 16.2, 17.9*) and xenotransplantation (*see 17.8.2*), has provided impetus for research into carbohydrate-based therapeutics and research tools. One of the problems of this research area is that glycan biosynthesis in living cells is not under direct genetic control and cannot readily be altered structurally or amplified using established molecular biological techniques.➲ Carbohydrate biotechnology has, therefore, lagged behind protein biotechnology and remains, arguably, the last major area of biotechnology to be explored.

Owing to the structural complexity of glycans and their numerous functions *in vivo*, quality control of glycosylation is important during the production of glycoconjugates and is often a requirement of regulating authorities prior to their approval for use for human therapeutic applications.➲ The degree of branching, correct synthesis of glycan chains, correct sulfation and sialylation may all be critical. In glycoproteins synthesized by cultured cells, glycosylation may be subtly influenced by culture pH, the availability of nutrients and precursor molecules, the presence or absence of cytokines and hormones, and other factors. It is important to guard against the action of sialidases and other glycosidases which may, for example, be secreted in culture or released from dead cells and degrade hitherto correctly formed oligosaccharides. Lectin affinity chromatography (*see 14.2.2*) has proved to be a useful tool in separating appropriately glycosylated glycoproteins from inappropriately glycosylated glycoforms.

➲ Correct glycosylation of therapeutic glycoconjugates is often essential, and is under subtle and complex control.

Heparin (*see 8.3.3*) for human therapeutic use provides a good, illustrative example. Natural sources of heparin, such as bovine lung and liver, yield a product with unpredictable and variable degrees of efficacy. Chemical or enzymatic depolymerization of heparin from these sources produces a more reliable, but still unpredictable, product. Recently, concerns about the transmission of bovine spongiform encephalopathy (BSE) in therapeutic products to humans have become a serious consideration. Synthetic analogues of this proteoglycan produced by biotechnological means may obviate the need for use of animal-derived heparin in medicine.

18.4 Use of enzymes in aqueous solution and in solid-phase systems to produce and modify carbohydrates

Enzymes in aqueous solution and in solid-phase systems can be used to produce and modify glycans. Examples of carbohydrate-based drugs synthesized in this way are heparin and heparin-based anticoagulants (*see 18.3*), polysaccharide vaccines and aminoglycoside antibiotics (*see 17.3.3*).

18.4.1 Enzymatic synthesis of monosaccharides in aqueous solution

Aldolases have been used extensively for the synthesis of monosaccharides and their derivatives. An example of this is the use of neuraminic acid (NeuAc) aldolase to produce NeuAc derivatives. Many sugar derivatives that do not occur naturally may be synthesized using this type of approach, including mimetics of naturally occurring functional sugars, such as 3-deoxy-D-manno-octulosonic acid (KDO; *see 11.1*) and NeuAc analogues (*see 11.2*).

18.4.2 Enzymatic synthesis of oligosaccharides in aqueous solution

There is an increasing demand for pure preparations of oligosaccharides as tools for analytical purposes and therapeutic studies. Enzymatic synthesis is an attractive option for this. It can generally be carried out in aqueous solution, is relatively easy to scale-up for potential industrial application, and with the increasing availability of appropriate glycosyltransferases and glycosidases (*see Chapter 2*), is becoming an increasingly popular approach.

Sugar nucleotide regeneration and low-cost enzymatic synthesis technology has been employed in the large-scale manufacturing processes of oligosaccharides. ➲ Recombinant glycosyltransferases mediate the transfer of sugar molecules from a sugar-nucleotide donor to a growing oligosaccharide chain. Enzymatic regeneration of the sugar-nucleotide donor molecule completes the cycle. It is possible to synthesize kilogram quantities of products in this way.

➲ Soluble enzymes may be used to synthesize oligosaccharides in aqueous solution, or can be immobilized on a solid support for the same purpose.

18.4.3 Enzymatic synthesis in solid-phase systems

There is also much research into the development of solid-phase systems. A typical example of this type of approach is the enzymatic synthesis of cellulose (*see 3.2.1*).

Cellulose has a number of potential applications, including in drug design and in the development of new biomaterials. ➲ Although cellulose is abundant in nature, the extraction of pure cellulose from naturally occurring material is a long and technically complex process, and chemical synthesis of cellulose in the laboratory from basic starting materials has proved difficult. Recently, techniques have been developed to produce synthetic cellulose via a non-biosynthetic pathway using the enzyme cellulase.

➲ Although abundant in nature, isolation of purified cellulose from natural sources is difficult. Enzymatic synthesis offers a convenient alternative.

Cellulose is a linear homopolysaccharide composed of ≈2500–15 000 repeating (β1→4)-linked glucose (Glc) residues. Briefly, enzymatic synthesis uses the enzyme cellulase (derived from *Trichoderma viride*) which is added to a solution of β-cellobiosyl fluoride monomer and allowed to react under appropriate conditions.

18.4.4 Enzymatic synthesis of more complex oligosaccharides

Whereas cellulose is a simple homopolymer, the enzymatic synthesis of more complex oligosaccharide structures requires the coordinated and orchestrated action of a number of specific enzymes (*see Chapter 2*). It is difficult to reproduce these conditions in the laboratory, but much progress is being made in this direction with both sequential enzymatic synthesis reactions in aqueous and solid-phase systems, and the development of increasingly sophisticated cell-based systems.

Aqueous phase systems are simpler to develop. Gal(β1→4)GlcNAc(β1→3)Gal(β1→4)Glc (lacto-N-neotetraose or paragloboside) has, for example, been synthesized in a two-step, aqueous-phase system. Initially, the transfer of uridine diphosphate-N-acetylglucosamine (UDP-GlcNAc) to lactose [Gal(β1→4)Glc] is catalysed by recombinant β1,3 GlcNAc-T expressed by *Escherichia coli* to yield GlcNAc(β1→3)Gal(β1→4)Glc. The transfer of uridine diphosphate-galactose (UDP-)Gal by recombinant β1,4 Gal-T then results in the synthesis of Gal(β1→4)GlcNAc(β1→3)Gal(β1→4)Glc. Experimentally, 300 g quantities have been synthesized in 100 litre reactor vessels with a conversion rate from starting materials of ≈85%, and the system could theoretically be scaled up for industrial production.

The development of cell-based systems for the synthesis of complex oligosaccharides is much more technically challenging as insect (*see 7.5*), bacterial (*see 7.2*), fungi (*see 7.3*) and higher plant cells (*see 7.4*), which might potentially be harnessed for these type of applications, each synthesize fundamentally distinct repertoires of glycans, which are often fundamentally different to those produced by animal and human cells. Engineering the glycosylation pathways of these cell types to synthesize complex oligosaccharides typical of human cells is currently an area of intense research interest, but requires complex and sophisticated manipulation of the machinery of glycosylation.

Filamentous fungi cells (*see 7.3*) are, for example, easy to modify genetically and 'lack of function' mutants deficient in undesirable glycosylation steps may be readily developed; furthermore, these cells are able to synthesize complex oligosaccharides and secrete them. They offer the potential of a cheap and abundant expression system. However, it is the subsequent engineering steps, the introduction of appropriate mammalian glycosyltransferases and their orchestration, that is complex, as the enzymes must be localized in specific membrane compartments, have appropriate levels of activity and appropriate donor substrates must be available. Development of such complex cell-based systems has great potential, but it is yet to be realized. This subject is explored in more depth below (*see 18.5.1*).

18.4.5 Some examples of applications of specific glycosyltransferases

18.4.5.1 β1,4 Galactosyltransferase

The enzyme β1,4 galactosyltransferase (β1,4 Gal-T; *see 2.11.5*) has been available commercially for many years, and thus has been used extensively in enzymatic synthesis of carbohydrate structures. ➲ Examples include the synthesis of N-acetyllactosamine, poly-N-acetyllactosamine polymers and other complex oligosaccharides including sialyl Lewis x (sLe^x; *see 10.6*) and other selectin ligands (*see 13.4.2*) and alkaloids. It has been used to produce ^{13}C-enriched linear poly-N-acetyllactosamines for galectin (*see 13.4.2*) binding studies. β1,4 Gal-T functions with acceptors linked to a solid support (e.g. a Sepharose matrix) for solid-phase synthesis or on water-soluble polymers, and in combination with other enzymes.

➲ Enzymatic synthesis of oligosaccharides such as poly-N-acetyllactosamine polymers, sLe^x and other selectin ligands, and the Gal(α1→3)Gal epitope has been successfully achieved.

18.4.5.2 β1,3 Galactosyltransferase

β1,3 galactosyltransferase (β1,3 Gal-T; *see 2.11.4*) is involved in the synthesis of the Gal(α1→3)Gal epitope implicated in xenograft rejection (*see 17.8.2, 7.13*) and also the synthesis of the Lewis a (Le^a) structure (*see 10.6*). Genes and cDNAs encoding β1,3 Gal-T have been cloned from several species including cow, mouse, marmoset and pig. Two homologous, inactive human pseudogenes have been described. Viable transgenic mice expressing null versions of the β1,3 Gal-T gene have been developed which do not synthesize the Gal(β1→3)Gal glycotope. Much research is currently aimed at the development of a null β1,3 Gal-T pig, as a potential donor animal for xenotransplant organs for human use, but there is speculation that lack of the Gal(β1→3)Gal may have serious detrimental consequences in this species.

18.4.5.3 Uses of sialyltransferases

Many naturally occurring glycoconjugates are sialylated, and sialylation is especially difficult to achieve chemically. In addition to the α2,3-STs used in sLe^a and sLe^x synthesis (*see 10.6*), α2,6-STs, α2,3 transialidase and an α2,3-ST that sialylates α-linked GalNAc have been used to synthesize α-sialosides for experimental use. The α2,3-ST that specifically sialylates the Gal(β1→3/4)GlcNAc structure has been used to synthesize the complex sialoside GM3, and for the generation of bivalent sialoglycosides anchored on different linkers for ligand recognition/inhibition studies. One recent development has been the use of recombinant α2,6-ST to synthesize sialoglycopeptides as potential inhibitors of infection by the influenza virus (*see 16.2.2*).

18.4.5.4 Other glycosyltransferases

β1,4 Man-T has been used to transfer Man residues onto lipid-linked chitobiose acceptors.

18.4.6 Studies with unnatural substrates

Many glycosyltransferases will tolerate substitutions at the site of transfer. α1,3 Fuc-T (*see 2.11.3*) and the blood group A and B transferases (*see 2.11.1*), for example, will glycosylate acceptors in which the ring structure has a C—H bond replaced by a C-methyl group or even a propyl group. This promiscuity has proved useful in the synthesis of glycosides of highly hindered tertiary alcohols, for which no effective chemical methods exist for their synthesis.

Sialyltransferases are generally tolerant of substitutions at the C9 of the Sia donor, for example, with fluoresceinyl groups, facilitating cell-labelling experiments.➲ An example of this is the synthesis of fluorescent sLe^x derivatives. Similarly, mercury-labelled CMP-NeuAc as a substrate for α2,3-ST allowed the synthesis of heavy atom-labelled sLe^x (*see 10.6*).

➲ Using sialyltransferase, it is possible to synthesize oligosaccharide structures incorporating a fluorescent or heavy atom label.

18.4.7 Use of bacterial transferase enzymes

Most enzymes that have been used in these types of studies have been of mammalian origin, but there is increasing interest in the use of bacterial transferases. Bacterial glycosylation is fundamentally different from that found in mammals in that the glycans are mainly extracellular. However, bacteria are able to synthesize a wide variety of free oligosaccharides and polysaccharides in their cell wall and express a wide range of glycosyltransferases to do so (*see 7.2*). Some of these sugar structures are identical to those synthesized by humans. *Helicobacter pylori*, for example, synthesizes the Le^x, Le^a and Le^y sugars also found on human gastric mucosa (*see 16.2.1*). *Neisseria meningitidis* synthesizes polylactosamines also present in mammalian glycolipids. The ease with which bacteria can be cultured and with which mutants may be generated, linked with their glycosylation repertoire makes them an attractive source of potential glycosyltransferase enzymes for synthesis of glycans. The β1,3 GlcNAc-T from *N. meningitidis,* for example, has been used in this way to synthesize polylactosamines.

Bacterial enzymes need to be used with caution, however, as some have subtly different activities to the comparable mammalian ones. For example, *Escherichia coli* employs a single glycosyltransferase to generate the poly(α2→8)-linked NeuAc polysialic acid (PSA) capsule, whereas in mammalian cells one enzyme initiates PSA formation, and a second is responsible for chain elongation (*see 11.9*). The α2,3-ST from *N. meningitiditis* and from *N. gonorrhoreae* both use terminal α-linked, (β1→3) and (β1→4)-linked terminal Gal residues, whereas the mammalian counterpart is more restricted in its specificity. *Photobacterium damsella* expresses an α1,6-ST that catalyses sialylation of terminal Gal or GalNAc in a (2→6) linkage. In mammalian cells, separate transferases act on Gal and GlcNAc residues. Furthermore, the *P. damsella* enzyme can add (α2→6)-linked NeuAc to the Gal in the sequences Fuc(α1→2)Gal(β1→4)Glc and NeuAc(β2→3)Gal(β1→4)Glc, but no mammalian enzyme is known which is capable of doing this. As a general rule, mammalian transferases have a more defined substrate specificity than their bacterial counterparts – under different circumstances, this may be an advantage or a disadvantage.➲

Mammalian glycosyltransferases are membrane-bound Golgi apparatus-resident proteins (*see 2.1.2*), and are therefore difficult to solubilize (requiring detergents which may denature and inactivate the enzyme) and purify. Bacterial transferases, in comparison are soluble as they act on the cell wall and have to be secreted in order to do so (*see 7.2*). It is therefore possible to express bacterial transferases as functional *soluble* proteins in, for example, *E. coli*. The recombinant transferases are then easy to purify from culture supernatant in the absence of detergent.

Bacterial enzymes are also less readily inhibited by the presence of sugar nucleotide donors and their metabolites than their mammalian counterparts. This reduces the requirement for expensive sugar nucleotide cycling to keep concentrations low.

➲ Some advantages of bacterial expression systems over mammalian ones include:
- ease of culture
- easy generation of mutants
- wide glycosylation repertoire
- soluble, not membrane bound, transferases
- not easily inhibited by sugar nucleotide donor and metabolite concentration.

18.4.8 Uses of glycosidases

Glycosidases (*see 2.2*) have proved successful tools in biotechnology as they are robust, tolerate organic solvents and are relatively cheap and readily available. In these respects they have advantages over glycosyltransferases, and although their primary function is cleaving glycosidic bonds, as the reaction is a reversible one, they have also proved useful in carbohydrate synthesis. They can be used in this way if the reaction conditions are modified in the direction of synthesis rather than that of the breakdown reaction. They have the disadvantages of weak regiospecificity and competing product hydrolysis, thus they often yield a mixture of products. This is not a problem if the next stage in the synthesis is selective for only the desired substrate. Glycosidases may be used for temporary protection which, following subsequent hydrolysis, ensures that the correct linkage is made in the final product. Examples of their application include β-galactosidases for (β1→3)-, (β1→4)-, (β1→6)- and (α1→6)-linked galactoside synthesis.

Novel donor substrates have been used which favour *trans*-glycosidase reactions. For example, the combined use of β-galactosidase and galactose oxidase generates a 6-oxo-galactosyl intermediate. This compound is not hydrolysed, and therefore an enhanced yield of N-acetyllactosamine is possible from subsequent synthesis reactions.

Most glycosidases used in oligosaccharide synthesis are exoglycosidases, but endoglycosidases, such as Endo-A and Endo-M have also been used.

18.4.9 Regeneration of sugar nucleotides and prevention of product inhibition

Many enzymatic synthesis protocols incorporate a procedure for the regeneration of the sugar nucleotide.➲ This overcomes the potential problem of product inhibition of the glycosyltransferase by the resulting nucleoside di- or monophosphatases, and also limits the need for expensive sugar-nucleotide donors. Typically, such recycling systems include the coupling of glycosyltransferase-catalysed reactions with regeneration of sugar nucleotides from nucleoside phosphates. Examples of this include recycling systems for UDP-Gal incorporating UDP-Gal epimerase, Gal-1-phosphate uridyltransferase or sucrose synthetase for use with β1,4 Gal-T for the synthesis of N-acetyllactosamine.

A strategy for relieving glycosyltransferase product inhibition is to utilize phosphatase to break down the inhibitory product nucleoside-5′ diphosphate (NDP) or nucleoside-5′ monophosphate (NMP). This is less efficient than sugar-nucleotide regeneration, but is more convenient on a small scale. This strategy has also been successfully employed in the synthesis of sLe^x (*see 10.6*).

Advances in genetic engineering have led to coupled systems in bacteria for oligosaccharide synthesis and sugar donor recycling. *N. gonorrhoeae* α1,4 Gal-T has been expressed in *E. coli* to produce (α1→4)-linked globotriose, and this synthesis was coupled to a UDP-Gal-generating system in another bacterium.

➲ Product inhibition is a potential problem in the chemi-osmotic synthesis of oligosaccharides that is often overcome by sugar-nucleotide regeneration or recycling systems.

18.4.10 'One pot' strategies

In cases in which the sequential action of several enzymes is required for the production of a final oligosaccharide product, it may be possible, if the enzymes have similar requirements of pH, etc., to adopt a convenient 'one pot' strategy. An example of a successful application of this type of approach is the synthesis of Le^x using β1,4 Gal-T and α1,3 Fuc-T (*see 10.6*).➲

The chemi-enzymatic synthesis of sLe^x (*see 10.6*) is an example of this type of approach, and is of commercial interest owing to the potential for the development of anti-inflammatory drugs. Synthesis

➲ Le^x, sLe^x, core 2 and sialyl T glycotopes have been synthesized successfully using 'one pot' strategies.

begins with GalNAc linked to an aglycone carrier achieved either by chemical synthesis or by using an enzymatic transglycosidase reaction. The sLe^x tetrasaccharide is then built using β1,4 Gal-T, α2,3-ST and α1,3 Fuc-T plus the appropriate nucleotide sugar donors. Kilogram quantities of sLe^x can be produced in this way. Soluble sLe^x has been tested in clinical trials to assess its efficiency in competitively blocking the binding of circulating leukocytes to selectins expressed by activated endothelium (*see 16.3*). However, the results were disappointing. The enzymes will tolerate substrate modifications and many sLe^x analogues have been produced in an attempt to overcome the intrinsically weak natural interaction between sLe^x and its selectin receptors.

Galactosidases and glycosyltransferases acting together in a 'one pot' synthesis have been used to synthesize core 2 trisaccharide and sialyl T glycotopes (*see 10.9.1*). Here the glycosyltransferase removes the product from the glycosidase-catalysed reaction, thus driving glycosidase equilibrium in the synthetic direction.

18.4.11 Sugar donors

Microorganisms are increasingly being used for the large-scale production of sugar donors for use in chemi-enzymatic synthesis systems. UDP-Gal, for example, can be produced in kilogram quantities by bacterial fermentation of the inexpensive starting material, orotic acid.

18.4.12 Development of fusion enzymes

Genetic engineering has been used to attach a nucleotide-synthesizing enzyme to a glycosyltransferase.➲ For example, the catalytic domain of *N. meningitides* CMP-NeuAc synthetase has been linked by a nine amino acid sequence to the α2,3-ST. This fusion enzyme can then be used to synthesize (α2→3)-sialylated lactose, and will transfer N-acetyl-NeuAc, N-glycolyl-NeuAc and N-propionyl-NeuAc equally effectively. Interestingly, this hybrid enzyme will also sialylate (α1→4)-linked galactosides, which no known mammalian α2,3-ST can use as a substrate.

➲ Fusion enzymes in which a nucleotide-synthesizing enzyme is genetically engineered to be attached to a glycosyltransferase have proved successful in chemi-enzymatic synthesis of sugars.

18.4.13 Solid-phase synthesis of glycopeptides and glycolipids

Solid-phase synthesis is an established and successful technique for the production of peptides. However, adapting it for the synthesis of glycopeptides has proved challenging. This is because the conditions required for cleavage of the final product from the solid resin support may damage the acid- or base-sensitive glycan group. To this end, special cleavage protocols, for example, using protease enzymes that act under neutral pH, have been developed. There has also been the need to develop special solid supports for glycopeptide synthesis which, for example, are compatible with an aqueous environment and allow macromolecular access of the glycosylating enzymes.

In some cases, elegant combinations of chemical and enzymatic approaches to synthesis have been developed. An example of this is the synthesis of sulfated glycopeptides from the glycoprotein P-selectin glycoprotein ligand-1 (PSGL-1; *see 13.12.3.2*); here, a pre-glycosylated threonine residue is attached to a solid support, peptide assembly is carried out, the peptide is then cleaved from the support, sulfated chemically and then glycosylated enzymatically in solution.

Glycolipids have also been synthesized successfully in a solid-phase system. Here, a water-soluble polymeric solid phase is used and glycans are synthesized enzymatically. Cleavage from the support can then be achieved using ceramide glycanase catalysed transglycosylation resulting in the transfer of the oligosaccharide from the support polymer to ceramide.➲

➲ In solid-phases systems, the harsh conditions required for cleavage of the product from the support may damage the oligosaccharide.

18.4.14 Sulfation

Methods are under development for the effective sulfation of oligosaccharides and/or the peptides to which they are attached. Methods of recycling PAPs, the universal biological sulfating reagent, have been developed and a tyrosylprotein-sulfotransferase has been used to sulfate the N-terminus of the glycoprotein PSGL-1. This technology is still in its infancy, but the synthesis of sulfated oligosaccharides and glycopeptides may become feasible in the near future.

18.5 Expression cloning of glycosyltransferases

The molecular technique of expression cloning has brought about a revolution in the field of glycobiology because it facilitates the cloning of cDNA encoding a transferase without the necessity of isolating and purifying the protein.➲ This is particularly relevant to glycosyltransferases as they are present in cells in such minute amounts that isolation of pure protein for sequencing is problematic. Furthermore, glycosyltransferases are membrane-bound enzymes containing lipophilic domains (*see 2.12*), which make it difficult to purify the enzyme while preserving its activity. It is thus possible, using molecular biology, to create modified glycosyltransferases in which the lipophilic transmembrane domain (which is not necessary for catalytic activity) is deleted and which function in an aqueous environment.

➲ Expression cloning of glycosyltransferases has brought about a revolution in glycobiology.

18.5.1 Expression systems for recombinant glycoproteins

Glycans are not the primary products of gene expression but are synthesized sequentially by glycosyltransferases (*see Chapter 2*). The expression of human genes in a variety of expression systems has become a powerful technique in the production of recombinant proteins for research or therapeutic applications. Current biotechnological approaches allow proteins, such as glycosyltransferases, to be engineered efficiently.

Many current, or potential, recombinant proteins for therapeutic use are glycoproteins. In several cases, appropriate glycosylation of a protein is necessary. Sometimes, incorrect (or lack of) glycosylation may result in incorrect folding, changes in biological inactivity and/or loss of receptor function, instability, altered pharmacodynamics, rapid clearance from circulation or undesirable immunoreactivity. Therefore, a challenging aspect of recombinant glycoprotein production is the precise control of glycosylation – either to reproduce the glycosylation profile of the naturally occurring molecule, or to subtly alter its glycosylation with a view to eliminating undesirable or deleterious activities, or confer new and desirable properties.➲ This is an area of intense research activity and one which promises to be of great interest, and challenge, in the next few years. Furthermore, because glycosylation is a posttranslational event, and is subject to many varied environmental controls, it is challenging to design conditions which will yield a pure, homogeneous glycoprotein. Glycoform microheterogeneity is a common feature of products synthesized by this type of technology.

➲ Correct glycosylation of glycoproteins for potential human therapeutic use is often important to their function *in vivo*.

An example of the profound difference that glycosylation can make is illustrated by erythropoietin (*see 17.3.1*). Erythropoietin is a hormone synthesized by the kidneys and (in small amounts) by the liver. It stimulates the bone marrow to produce red blood cells. Recombinant erythropoietin produced by *E. coli* cells is active *in vitro* but very significantly less so *in vivo* owing to altered N-glycosylation that results in rapid clearance from circulation.

Alterations in glycosylation may have consequences for the immunoreactivity of a glycoprotein, which may sometimes be desirable, for example, in stimulating the immune response of cancer patients to saccharide antigens present on their tumour cells, which is discussed below, or may be undesirable, as in the case of a recombinant glycoprotein for potential therapeutic administration in which a patient immune response to the glycoprotein could be disastrous. Ninety per cent of people who are immunized with conventional recombinant hepatitis B vaccine develop a strong antibody response, whereas 10% do not. Of those failing to respond, almost all will mount a good response to a glycosylated form of the viral glycoprotein.

There are a number of ways in which attempts to change glycosylation of recombinant glycoproteins may be achieved. One consideration is the structure of the polypeptide chain itself. N-linked glycosylation can only occur where there is a tripeptide Asn-Xaa-Ser/Thr/(Cys) (*see 4.2.1*), O-glycosylation can occur where serine or threonine residues are present (*see 5.6.1*) and glycophosphatidylinositol (GPI) anchors require a C-terminal signal peptide which is subsequently cleaved as glycosylation proceeds (*see 12.4.2*). Many glycosyltransferases recognize and operate only on peptide chains containing specific appropriate signal sequences. The glycosylation status of recombinant glycoproteins can therefore be affected significantly by small changes in amino acid sequence, and this can be utilized in attempts to alter their glycosylation deliberately.

Because it is difficult to purify mammalian glycosyltransferases from natural sources, expression systems using soluble forms of recombinant enzymes are increasingly being developed. There are two features of these systems which are particularly significant: (i) there is considerable homology among transferases from species as different as, for example, bacteria or yeasts and humans, thus facilitating production of 'humanized' glycosylation in these cell types; (ii) heterologous expression allows the fine structural specificity towards the acceptor substrate to be determined and the provision of excess recombinant enzyme allows an override of normally restricted specificity, facilitating the production of non-natural glycans. In some cases, it is essential to produce correctly glycosylated glycoproteins, for other applications it is essential that certain amino acids should carry sugar chains, but the precise structures within the chains are unimportant. The choice of the expression system to be used is therefore dependent on a number of different factors. ➲

➲ Expression systems have many potential advantages, including:
- subtle alterations in glycosylation are possible, resulting in improved biological activity
- potential flexibility and adaptability of the cellular machinery of glycosylation can be exploited in the production of different glycoforms of recombinant glycoproteins
- they can be used on non-physiological, chemically synthesized, nucleotide-sugar analogues
- only those configurations are synthesized which occur in nature.

18.5.2 Mammalian cell expression systems

The glycosylation repertoire of bacterial (*see 7.2*) and insect cells (*see 7.5*) is different to that of mammalian cells, and so the use of these types of cells for glycoprotein biosynthesis suffers from inherent problems. To try and overcome these limitations, the use of mammalian host cells appears an attractive and obvious option for the production of glycoproteins for potential use as human therapeutics. This, however, is a complex field as glycosyltransferase expression and activity are highly regulated between cell and tissue types and with differentiation and disease.

Mammalian cell lines have been successfully used for the expression of human glycoproteins, and may be modified by the addition of genes encoding for human glycosyltransferases. ➲ Modification of mammalian and human cell lines by the introduction of retrovirus vectors and *ex vivo* culture of cells *per se* may have significant and unpredictable effects on glycosylation, however, and this remains a science still in its infancy.

➲ Mammalian cell lines may offer advantages over bacterial cells or insect cell lines, as the glycosylation machinery of the latter may produce different glycoforms to those synthesized by human cells.

Recombinant mammalian cell lines are already used extensively for the large-scale production of glycoproteins for human therapeutic use. Examples include the synthesis of erythropoietin, interleukin (IL)-2, interferon (IFN)-β and IFNγ. Chinese hamster ovary (CHO) cells and baby hamster kidney (BHK-21) cells are the most common mammalian cell lines used for these purposes. Usually, a heterogeneous range of glycoforms is synthesized in this way, and if destined for human therapeutic use further purification steps are necessary.

18.5.3 Microorganisms

There is progress in modifying lower organisms, such as bacteria, yeasts or fungi, to produce 'humanized' oligosaccharides on the recombinant glycoproteins they express. For cell-based systems, these cell types are convenient to grow in culture and it is possible to generate mutant and transfected strains. Microorganisms are increasingly employed for the large-scale production of oligosaccharides, and the science of microorganism metabolic engineering is progressing fast. Other advantages of this approach are the potential for large-scale production and the use of inexpensive precursors.

18.5.4 Bacterial expression systems

In spite of the fact that bacterial saccharides are located in the cell wall and are thus not directly comparable with human plasma membrane glycoproteins, bacteria have great potential for use in expression systems. They have a naturally extensive repertoire of glycosyltransferase enzymes and are able, in some cases, to synthesize oligosaccharides indistinguishable from those produced by human cells, or cells of other sorts of organisms, such as plants (*see Chapter 7*). Furthermore, bacterial expression systems offer considerable cost savings over equivalent mammalian cell-based (e.g. CHO cells) systems.

An elegant example of the potential of this approach has been applied to the synthesis of Le^x (*see 10.6*) for research and potential human therapeutic use. Sequences for human, mouse and chicken α1,3 Fuc-Ts, which have a high degree of sequence homology, were used to design degenerate oligonucleotide primers for use in the polymerase chain reaction (PCR) to probe *Helicobacter pylori* genomic DNA. A putative *H. pylori* α1,3 Fuc-T was identified. *E. coli* transfected with this gene expressed recombinant *H. pylori* transferase and was capable of synthesizing Le^x using GDP-Fuc and Gal(β1→4)GlcNAc.

Mammalian glycosyltransferases are glycoproteins. Bacterial glycosyltransferases differ significantly from their mammalian counterparts in that they are not glycosylated and mammalian transferases are thus usually inactive in bacterial expression systems. It is therefore usually necessary to employ bacterial glycosyltransferases in bacterial cell-based expression systems. An exception to this is the bovine α1,3 Gal-T which has been successfully expressed in *E. coli.* ➲ Furthermore, plant genes have been successfully expressed in bacterial systems. The *nodBC* genes of chitooligosaccharide synthesis from the plants *Azorhizobium caulinodans* and *Rhizobium meliloti* have been cloned into the bacterium *E. coli* inducing it to synthesize chitooligosaccharides successfully.

➲ Disadvantages of *E. coli* expression systems include:
- inability to perform many of the posttranslational modifications found in eukaryotic proteins
- sometimes, lack of secretion of synthesized proteins into culture medium
- limited ability to create disulfide bonds
- sequestration of the protein in inclusion bodies.

For glycoproteins that require correct glycosylation for biological activity, or for those in which tertiary structure requires disulfide bond formation, yeast or mammalian expression systems are preferable.

18.5.5 Yeast expression systems

Yeast cultivation has a long history in human nutrition and thus it is well understood, inexpensive and considered safe biologically. Fermentation systems for synthesis of glycoconjugates and oligosaccharides using yeasts such as *Saccharomyces cerevisiae* and *Pichia pastoris* have been very successful. *P. pastoris* has the advantage that glycoconjugates are secreted into the culture supernatant in which they can be conveniently harvested. Glycosyltransferases that have been successfully expressed in yeasts include β1,4 Gal-T, α2,6-ST, α2,3-ST, α1,3 Fuc-TIII, α1,3 Fuc-TVI and α1,2 Man-T. Yeast cultures have a much faster doubling time, in the order of 1.5 hours, in comparison with mammalian cell cultures, which have a doubling time of, on average, 24 hours, which means that potentially this system could be very cost-efficient. ➲

➲ Yeast cell culture is rapid, well understood and straightforward, and has potential for the expression of human glycoproteins. Yeast glycosylation pathways are too different to those of human cells for this to be realized without significant modification.

The early stages of N-linked glycosylation pathways in the endoplasmic reticulum (ER) are highly conserved between mammalian and yeast cells, and thus yeasts share the same initial N-linked glycosylation steps as mammalian cells (*see 7.3*). Furthermore, secretion of the glycosylated protein into the culture medium facilitates ease of harvesting.

However, there remain very significant problems associated with yeast expression systems. First, yeast O-linked oligosaccharide chains are entirely different to those produced by mammalian cells. Furthermore, although the initial steps of N-linked glycosylation are similar to those of mammalian cells, yeast oligosaccharides are mostly of the high mannose type, lacking the chain extension and termination steps characteristic of the glycans of mammalian cells. Divergence in N-linked glycosylation pathways in yeasts occurs in the Golgi apparatus where α-mannoside I trims the oligomannose structure $Man_5GlcNAc_2$ followed by elongation. Initial addition of GlcNAc by GlcNAc-T is absent in yeasts as well as the enzymes associated with further chain elongation. Thus, glycoconjugates for human therapeutic application produced in this way would be cleared rapidly from the circulation through recognition by the mannose-binding lectin (MBL) (*see 13.12.3.3*). *S. cerevisiae*, for example, synthesizes large polymannan chains of 50–100 Man residues which, in addition to being cleared rapidly from the circulation, also have the disadvantage of being very bulky, which may impair any biological activity. However, other yeast strains, such as *P. pastoris,* do not hypermannosylate proteins in this way and may be better suited for biosynthetic purposes. Many yeast oligosaccharides are also immunogenic to humans.

In attempts to overcome these hurdles, yeast mutants lacking polymannosylation have been developed and ambitious attempts have been made to introduce a series of human elongation enzymes into them. An α-mannosidase I-like enzyme from *A. saitoi* has been expressed in yeast cells. It is worth noting, however, that additional transferases and nucleotide-sugar transporters are also required for elongation. Encouragingly, human α1,4 Gal-T has been expressed successfully in yeasts and has shown functional activity, suggesting the presence and activity of metabolic pathways able to synthesize and transport UDP-Gal. Some examples of glycosyltransferases expressed in yeast systems are listed in *Table 18.1*. The potential for yeast cells to successfully synthesize human oligosaccharides is still remote, but remains a tantalizing possibility.

18.5.6 Filamentous fungi as production organisms

Filamentous fungi naturally synthesize and secrete copious quantities of glycoproteins, most commonly enzymes, and are easy to grow on inexpensive media. As such, they appear

Table 18.1 Examples of human-like glycosyltransferases expressed in yeast systems

α2,3-ST	*Saccharomyces cerevisiae*
α2,6-ST	*Saccharomyces cerevisiae* and *Pichia pastoris*
β1,4 Gal-T (Gal-T1)	*Saccharomyces cerevisiae* and *Pichia pastoris*
α1,3/4Fuc-T (Fuc-TIII or the Lewis enzyme)	*Pichia pastoris*

attractive candidates for expression systems. They have great potential for the production of recombinant glycoproteins, but as yet, little progress has been made in their development, possibly owing to the relatively little that is understood of fungal glycosylation pathways. Potentially useful species include *Aspergillus niger*, *A. niger* var. *awamori* and *A. oryzae*.

➲ Filamentous fungi have not been greatly exploited for the production of recombinant human glycoproteins as their glycosylation pathways are not well understood and differ from those of human cells.

Filamentous fungi synthesize both N- and O-linked complex glycans (*see 7.3*). They most commonly produce small oligomannose N-linked glycoproteins and O-linked glycoproteins. They are able to synthesize a range of complex glycans, suggesting a comprehensive array of glycosyltransferases. Furthermore, they also secrete glycosidases which may trim oligosaccharide chains on secreted glycoproteins. However, the oligosaccharide chains produced by these organisms differ significantly from the complex type high mannose chains produced by mammalian cells.➲

It appears likely that fungi express glycosyltransferases and glycosidases not present in mammalian cells. For example, a single non-substituted GlcNAc has been reported in some fungal species, which may occur as the result of trimming of more complex oligosaccharides. Fungal O-glycans differ significantly from those of mammals, varying from a single Man residue to linear chains of up to five Man residues sometimes substituted with Glc and/or Gal. Hyperglycosylation, typical of some yeast species or strains, is uncommon, but can also occur, in fungi.

Owing to the significant differences in glycosylation patterns between filamentous fungi and human cells, filamentous fungi are currently not useful for producing glycoproteins for human therapeutic use. However, as they are relatively easy to modify genetically, there is potential for the development of recombinant strains. In contrast to yeasts, filamentous fungi are capable of trimming N-glycans, and this may be a useful feature. Subsequent engineering steps including selection of strains that do not hyperglycosylate, that produce and trim the $Man_5GlcNAc_2$ core glycan, and subsequently elongate it to produce mammalian type complex glycans is feasible, but is still in the preliminary stages of development.

Modification of glycosylation pathways by the introduction of glycosyltransferases into this and other systems is a very complex undertaking, as the enzymes would need to be localized in specific membrane compartments, their activity levels would have to be high, and also the appropriate donor substrates must be available. There are definitely obstacles. For example, there is no evidence at present that fungi synthesize cytidine monophosphate (CMP)-neuraminic acid required for the sialylation of oligosaccharides. However, some progress is being made. Fuc-TV has been expressed in the filamentous fungi *A. niger* var. *awamori*, and more significantly, GlcNAc-T1 has been introduced into filamentous fungi with the idea of developing pathways to synthesize $GlcNac_1Man_5GlcNAc_2$. Eventually, engineering lack-of-function mutants to remove unwanted pathways and linking this with gain-of-function engineering to introduce new glycosylation steps may provide an effective approach.

18.5.7 Engineering glycosylation pathways in a baculovirus expression system

The baculovirus expression system has proved to be a powerful tool in molecular biology to engineer recombinant proteins in insect cells which act as a host for the recombinant viral DNA. Put simply, the virus is used as a vector to insert a gene into the DNA of the insect cell it infects. The insect cell will then produce the product coded for by the newly introduced gene.

One potential problem is that, like bacterial and fungal pathways, insect glycosylation pathways are fundamentally different to those of human cells (*see 7.5*). Insect cells produce N-linked glycoproteins with high mannose glycans and trim them to trimannosyl core structures, but generally do not convert these core structures to complex structures. N-linked glycoproteins produced in the baculovirus expression system tend, therefore, to be of the paucimannosidic type. However, there is experimental evidence that some insect cells possess the enzymes required for the synthesis of complex glycans, but seldom undertake this type of glycosylation because it relies on the presence of appropriate substrates which are not often available within the insect cell. This is probably due to competition with other processing activities and the structure of the glycoprotein being processed. It is possible that insect cells rely on this type of glycosylation only at specific times during development and/or in certain tissues or under certain stress conditions, but it suggests that under appropriate circumstances these pathways could be activated.

The ability of insect cells and insect cell lines to produce complex N-glycosylated glycoproteins has led to efforts to modify their glycoprotein-processing capability by metabolic engineering. The fact that the cells do possess the basic requirements for complex glycan processing, for example, proper spatial and temporal subcellular distributions of transferases, the ability to synthesize appropriate nucleotide sugar pools, and the ability to synthesize appropriate receptors, means that this approach has great potential. An example of this approach is the introduction of human β1,2 GlcNAc-TI into *Spodoptera frugiperda* (Sf9) cells (these cells are very commonly used for baculovirus expression), using a conventional baculovirus vector. When the cells were also co-infected with a vector encoding a gene for a glycoprotein, the glycoprotein was synthesized with an extended N-linked trimannosyl core-bearing terminal GlcNAc. Therefore, using this type of approach, it is theoretically possible to create the modifications necessary to synthesize carbohydrate residues of glycoproteins which are identical, or very similar, to those expressed in human tissues.➲ One significant advance has been the development of a special 'immediate early' vector to facilitate the expression of glycoprotein processing enzymes very soon after infection, so that they are more able to participate in glycosylation of expressed recombinant glycoproteins synthesized later after infection. An alternative approach has been to permanently alter the insect cell genome so that it constitutively expresses a human glycosyltransferase enzyme. Stably transformed insect cell lines have been developed, for example, that constitutively express mammalian β1,4 Gal-T and will produce glycoproteins with extended galactosylated N-glycan chains.

➲ In spite of differences between insect and human cell glycosylation pathways, the well-established baculovirus expression system is potentially useful for producing recombinant human glycoproteins. Some successful modifications to the systsem illustrate its potential.

18.5.8 Higher plants

Plant glycoproteins are frequently glycosylated with (α1→3)-linked Fuc and (β1→2)-linked Xyl, which are absent from mammalian glycans and are very immunogenic to humans (*see 7.4.7*). There are a number of potential genetic modifications that may alleviate this problem. One is to utilize mutant plants lacking functional GlcNAc-TI which are therefore unable to synthesize complex N-glycans from the simple $Man_5GlcNAc_2$ core. A related idea is to produce glycoproteins carrying an ER retention sequence (KDEL or HDEL) as proteins retained in the ER typically carry unmodified high mannose glycans. These mannosylated glycans would not be immunogenic to humans, but would have the disadvantage of being rapidly cleared from circulation because they would act as ligands for the mammalian macrophage Man receptor and may induce an immune response through endocytosis and class II presentation. Another approach might be to develop knockout mutants lacking the vacuolar N-acetylglucosaminidase and/or the (α1→3) Fuc-T and (β1→2) Xyl-T and possibly add in appropriate human glycosyltransferases. One major challenge would be to develop transgenic plants able to synthesize 'human' sialylated glycoproteins as plants do not synthesize sialic acids, and in addition to the necessary transferases, appropriate enzymes for synthesis and transport of CMP-Sia would be required.

18.6 Glyco-mimetic peptides

The phage display library is a useful tool in biotechnology. Random peptides are expressed on the pili of bacteria. These displayed peptide sequences are screened against other molecules, for example, lectins (*see Chapter 13*). One way of doing this is by using immobilized lectins which will recognize and bind to some displayed peptides on the bacterial pili, thus isolating them from a mixture; bacteria expressing peptides of interest may then be expanded and cloned. If a peptide is recognized by the lectin saccharide-binding site, and if binding is inhibited by the appropriate sugar structure, the peptide is said to be glyco-mimetic. This aspect of biotechnology is in its infancy, but is an exciting area of research exploration.

18.7 Potential uses of lectins in biotechnology

There is interest in the expression of lectin genes by transgenic plants for a number of potential applications, some of which are listed in *Table 18.2*, and described below.

One of the main problems in human and animal nutrition is not the production of foodstuffs, but that these resources are infected, damaged or consumed by viruses, bacteria, insects or grazing animals. The concept of reducing the vulnerability of foodstuffs by making them toxic to such pests is an attractive one. One proposal is to create transgenic crop plants expressing lectins (*see Chapter 13*) that are selectively toxic or unpalatable to potential infective organisms or pests. As insects may develop resistance to toxic lectins in their food plants with repeated exposure, it is also necessary to focus on producing transgenic plants that express other toxic principles, such as chitinases, proteases or amylase inhibitors, in addition to the lectin.

Another application of this type of technology is the expression of lectin genes to facilitate the bulk production of lectins that are usually present in only small quantities in their natural source organism, or lectins from naturally rare or species that are difficult to access.

The suggestion that some dietary lectins have beneficial effects on animals or humans raises the possibility of transgenic food plants enriched with these substances. Much research has gone into investigating the potential of snowdrop lectin in this context. Snowdrop lectin was first identified as a potential anti-pest substance as it confers resistance to insect attack. Later research using isolated snowdrop lectin in short-term feeding experiments in rats indicated that it had a positive effect in inhibiting the overgrowth of *E. coli,* suggesting a beneficial probiotic effect. As a result, the lectin gene was cloned into potatoes and the lectin protein shown to be present in the tubers. However, when the transgenic potatoes were administered in longer term feeding experiments on rats, side-effects were noted. These included defects in the immune system and reduction in brain growth in the animals. It is not clear whether these detrimental effects were due to the presence of the lectin *per se* or resulted from the genetic manipulation procedure (a plant retrovirus was used to make the transgenic plants and this may have unknown effects).

The primary, secondary and tertiary structure of legume lectins is highly conserved and there is much sequence homology among legume lectin genes (*see 13.4.1*). Differences in saccharide-binding specificities are primarily due to differences in amino acid residues residing in loops adjacent to the carbohydrate-binding site. This raises the possibility of site-specific mutagenesis to create new lectins with subtly altered specificities. These 'designer

Table 18.2 Potential applications of transgenic plants expressing lectin genes

- Protection of crop plants against insect attack/infection.
- Bulk production of lectins for research applications.
- Probiotic effect in animal/human food plants.
- Creation of 'designer' lectins with different carbohydrate binding properties.

lectins' may prove useful in numerous applications related to carbohydrate analysis (*see 14.2*), for lectin–liposomal drug delivery systems (*see 17.5.5*) and may further our understanding of the molecular basis of protein–saccharide interactions.

18.8 Some applications of carbohydrate biotechnology in the food, pharmaceutical and related industries

Oligosaccharides and polysaccharides are of interest to the brewing, food, pharmaceutical and other industries. ➲ They have countless applications in multiple roles, but perhaps their most common uses are as thickening, supporting, stabilizing and gelling agents, and also as sweeteners. Biotechnology has opened the way to produce these substances on a huge scale and also to modify their properties. Some examples of applications are described briefly below.

➲ Carbohydrates have many uses in the food and pharmaceutical industries, including as sweeteners, thickeners, gelling, supporting and stabilizing agents.

18.8.1 Gums and alginates

18.8.1.1 Xanthan gum

Xanthan gum is a high molecular mass (several million daltons) heteropolysaccharide (*see 3.2.2.1*) synthesized by *Xanthomonas* species. It has a heterogeneous structure, depending upon the *Xanthomonas* species and also upon the environmental conditions of temperature, pH and dissolved oxygen, but it is composed of Glc, glucuronic acid, Man, and sometimes Gal, plus pyruvate and acetate. It has been of commercial importance since the 1960s owing to its properties as a thickener for use in the food, pharmaceutical and cosmetic industries. It is produced by industrial-scale fermentation, and different culture conditions, including pH, temperature, oxygen concentration, as well as different *Xanthomonas* species, yield gums with differing monosaccharide composition, molecular mass, and hence different properties.

18.8.1.2 Alginates

Alginates (*see 3.2.6.4*) are naturally occurring polysaccharides produced as structural components of brown seaweeds and also as capsular components of some soil bacteria. They are linear chains of (α1$\rightarrow$4)-linked guluronic acid and (β1$\rightarrow$4)-linked mannuronic acid. They are used extensively as thickening, stabilizing and gelling agents in the food, pharmaceutical, textile and paper industries. Both seaweeds and bacteria are commercially harnessed for industrial-scale production of these substances.

18.8.1.3 Schizophyllan gum

The fungus *Schizophyllum commune* secretes a mucilage gum homoglucan consisting of a backbone of (β1$\rightarrow$3)-linked Glc units linked with single (β1$\rightarrow$6)-linked Glc side chains. This glucan, termed 'Schizophyllan', is produced commercially for production of air-tight films for use in the food industry and in medicine as a stimulator of an anti-tumour immune response.

18.8.1.4 Guar gum

Guar gum (*see 3.3.1*), a naturally occurring polysaccharide derived from the seeds of some leguminous plants, is one of the most effective viscosifying polysaccharides known and is used as a thickener in many applications. Guar gum is a galactomannan comprising linear chains of (β1$\rightarrow$4)-linked Man residues to which side chains of a single (α1$\rightarrow$6)-linked Gal residue are attached. Guar gums with different viscosities are produced by chemical modification of the natural product, but there can be problems – for example, chemical modification using transition metals is not appropriate for guar gum destined for use in food. As an alternative, it is possible to modify the gum using enzymes, such as glucose oxidase.

18.8.2 Starch products

The hydrolysis of starch (*see 3.4.1*) to low molecular mass products is catalysed by α-amylase. It is an important commercial enzyme process, as the products are widely used in the food, paper and textile industries.

18.8.3 Dextrans

Dextrans are polysaccharides of different chain lengths composed entirely of Glc residues. They are synthesized by *Leuconostoc mesenteroides* using sucrose as a substrate. They have a number of applications, including as a chromatographic support and as plasma expanders in medicine. One type of dextran, which includes a high proportion of (α1→2) linkages, withstands attack by digestive enzymes of humans and animals but is degraded by 'friendly' intestinal bacterial microflora such as *Bifidobacterium* sp. *Lactobacillus*, thus encouraging their growth to the detriment of pathogenic intestinal microorganisms.

18.8.4 Sweeteners

Several oligosaccharides are also of interest to the food industry as potential sweeteners.➲ They have advantages over sucrose in that they are, for example, low in calories, safe for diabetics to use, may encourage the growth of 'friendly' gut bacteria and be detrimental to harmful ones, may lower cholesterol and blood pressure and do not promote tooth decay. Many of these oligosaccharides occur naturally in plants, but in quantities too low for their extraction to be commercially viable. Fructo-oligosaccharides can be produced conveniently and economically from sucrose by the action of microbial fructosyltransferases. The enzymes may be dissolved in solution, immobilized on a solid support matrix, or present in a cell-culture system supported in an alginate gel. A related product, inulo-oligosaccharides, can be produced from inulin (which is a polymer of (β2→1)-linked fructose found in plants like chicory and Jerusalem artichoke; *see 3.4.2*) by enzyme digestion. Inulino-oligosaccharides share the beneficial properties of fructo-oligosaccharides, and are a soluble form of dietary fibre.

Another carbohydrate of interest to the food industry is isomaltulose (6-O-α-D-glucopyranosyl-D-fructose). This substance can also be employed as a sweetener, but is much less sweet than sucrose and results in a lower rate of monosaccharide release into the blood, making it appropriate for diabetic use and for sports food/drink products. It also provides a similar 'bulking' property as sucrose, which is important for the texture of cakes and confectionery. Chemical synthesis of this product is very difficult, but it can be readily produced by bacterial fermentation of sucrose.

Sorbitol (*see 1.15.6*) and mannitol are sugar alcohols used as sweeteners and texturing agent used widely in foods, including chewing gum and 'sugar-free mints'. Mannitol is present in many plants such as olives, mistletoe, mushrooms and lichens. It may also be produced commercially by bacterial fermentation. These compounds are useful to disguise unpleasant tastes in drug preparations. Mannitol when injected intravenously is a powerful osmotic diuretic in the treatment of intoxication and barbiturate drug overdose, and can be used to reduce brain oedema, for example, after a stroke or after brain surgery. Taken orally in large quantities, it has laxative properties.

➲ Sweet-tasting oligosaccharides such as inulino-oligosaccharides and mannitol have advantages over sucrose: they are low in calories, safe for diabetics, do not promote tooth decay and may have a probiotic effect, lower serum cholesterol levels and reduce blood pressure.

Further reading

Altmann, F., Staudacher, E., Wilson, I.B.H., Marz, L. (1999) Insect cells as hosts for the expression of recombinant glycoproteins. *Glycoconj. J.* **16:** 109–123.

Fukuda, M., Bierhuizen, M.F.A., Nakayama, J. (1996) Mini review: expression cloning of glycosyltransferases. *Glycobiology* **6:** 683–689.

Jarvis, D.L., Kawar, Z.S., Hollister, J.R. (1998) Engineering N-glycosylation pathways in the baculovirus–insect cell system. *Curr. Opin. Biotechol.* **9:** 528–533.

Johnson, K.F. (1999) Synthesis of oligosaccharides by bacterial enzymes. *Glycoconj. J.* **16:** 141–146.

Koeller, K.M., Wong, C.-H. (2000) Complex carbohydrate synthesis tools for glycobiologists: enzyme based approach and programmable one-pot strategies. *Glycobiology* **10:** 1157–1169.

Palcic, M.M. (1999) Biocatalytic synthesis of oligosaccharides. *Curr. Opin. Biotechol.* **10:** 616–624.

Yarema, K.J., Bertozzi, C.R. (1998) Chemical approaches to glycobiology and emerging carbohydrate based therapeutic agents. *Curr. Opin. Chem. Biol.* **2:** 49–61.

Index